THE SATELLITE EXPERIMENTER'S HANDBOOK

by Martin R. Davidoff, K2UBC

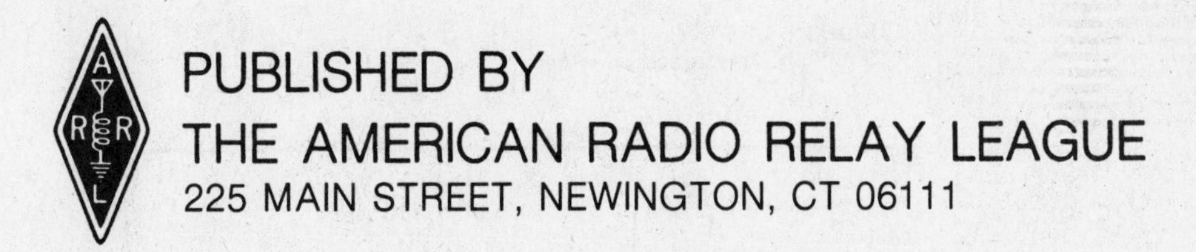

PUBLISHED BY
THE AMERICAN RADIO RELAY LEAGUE
225 MAIN STREET, NEWINGTON, CT 06111

Printed in USA

Library of Congress Catalog Card Number: 83-071699

ISBN: 0-87259-004-6

$10.00 in USA
$11.00 in Canada and elsewhere

FOREWORD

OSCAR 1, Amateur Radio's first satellite, was launched into orbit in December 1961. A small, battery-powered box, OSCAR 1 continually transmitted the Morse code identifier HI to eager ears on earth. A tremendous achievement for Amateur Radio in the early days of the Space Age, the successful mission was to be but the first of many.

The resourcefulness, ingenuity and skill of the Amateur Radio satellite community in the years since have made a fascinating story. From the California garage and basement workshops of the '60s, to the cooperative international projects of the '80s, amateurs have pursued the dream of reliable, predictable, long-distance and long-duration radio communication on vhf and higher frequencies. Each successive OSCAR has been one more step toward the realization of that dream. With the successful launch of AMSAT-OSCAR 10, the first of the "Phase III" satellites, the Amateur Radio Service entered that new era of communication. Yesterday's dreams have become today's reality.

You are a part of that reality! From setting up a modest ground station and communicating through the "birds," to understanding some of the more advanced concepts of satellite orbits and tracking, *The Satellite Experimenter's Handbook* provides all you need to know. Whether you're a beginner, an old hand at satellite work or a student of space science, this book is your launch vehicle into the fascinating journey of Amateur Radio in space.

David Sumner, K1ZZ
General Manager
Newington, Connecticut

Dedication

As the first edition of this book was going to press, the Amateur Radio community received the tragic news that ARRL President Victor C. Clark, W4KFC, had passed away suddenly on November 25, 1983. Vic was one of the world's best known and most greatly admired radio amateurs. He was a strong supporter of amateur satellites and Life Member #25 of AMSAT. Bringing people together in cooperative efforts to protect and advance Amateur Radio was one of Vic's special talents. In the hope that we may always remember to follow his example, this book is dedicated to his memory.

ACKNOWLEDGMENTS

Several members of AMSAT have made major contributions to this handbook. While it's impossible to mention everyone personally, I would like to give special thanks to Perry Klein (W3PK), Jan King (W3GEY) and Gordon Hardman (ZS1FE) for providing obscure engineering data, catching those distortions of history or technical facts that are so adept at creeping in, and suggesting numerous improvements to several drafts of the manuscript. Their assistance has been invaluable.

As this book nears completion my overwhelming feeling is one of gratitude to the entire AMSAT design team for offering moral support, and even more so for giving me the chance to join in transforming the amateur satellite dream into the amateur satellite reality. Few scientists ever get the chance to participate in a project as exciting and satisfying, or to work with individuals as dedicated, persevering or bright. To all, my sincerest thanks.

As a radio amateur and as the author, I'd like to express my appreciation to the ARRL for their continuing support of the amateur satellite program, this book being but one evidence. Steve Place (WB1EYI) acted as the editor and coordinator for all production activities. While the book clearly attests to his skill, I'd like to attest to a hidden attribute of equal importance — his continuing good humor — which kept this project enjoyable throughout.

Finally, I'd like to thank Linda, my best friend and spouse, for her support and good-natured tolerance during the many years that this project centimetered forward.

Martin Davidoff, K2UBC
December 1983

About the Author

Martin R. Davidoff, K2UBC, has been a licensed radio amateur for over 27 years. An Amateur Extra Class licensee, he was first licensed in 1956 and is a life member of both AMSAT and ARRL. Having earned his Ph.D. in Physics (Experimental, Solid State) at Syracuse University in 1971, Davidoff was employed at the Illinois Institute of Technology Research Institute where he was involved with military satellite communications systems. He is now an Associate Professor of Mathematics and Engineering at Catonsville Community College in Catonsville, Maryland, where he has taught since 1972. In addition, he was Director of the National Science Foundation Education Project relating to the use of satellites by college-level science and engineering educators in 1975, and is the author of *Using Satellites in the Classroom: A Guide for Science Educators*, published in 1978.

To the Reader:

Challenging or frustrating, fascinating or confusing — no matter how they're described, satellites certainly have added a new dimension to Amateur Radio. This text focuses on spacecraft built by, and for, radio amateurs. In addition, it contains information on weather, TV-broadcast and other satellites of interest to amateurs. Part I (Chapters 1-3) tells a story, the story of hams in space. In Part II (Chapters 4-7) we cover the information needed by those starting out in satellite communications. Finally, Part III (Chapters 8-13) presents reference material on special topics for serious experimenters, including those who want to build spacecraft.

If you're anxious to get started you may be tempted to skip the history and jump right into Chapter 4. Avoid this temptation at all costs for the "story" contains important basic technical information. By the time you reach Chapter 4, you'll find you've acquired a considerable background in satellite system fundamentals, and that the acquisition was painless. Maybe even enjoyable.

This book is actually two books in one. The first seven chapters are a beginner's manual for satellite communicators. They're meant to be read sequentially, though there are sections (so marked) that may be skimmed or skipped. The next six chapters are a reference manual that is organized so that specific topics can be read as the interest grabs you.

The beginner needs little more than a basic knowledge of communications systems to start communicating via satellite. But the advanced experimenter, seriously interested in spacecraft design, encounters a wide range of disciplines: physics (basic physics, geophysics, astrophysics), mechanical engineering (materials, heat flow) and electrical engineering (communications systems, propagation, control systems, digital electronics).

In the past, becoming knowledgeable about satellite systems has been difficult because much of the information was buried in the advanced scientific texts and journals of the various fields involved. One had to first find the information and then try to digest it. This book attempts to eliminate, or at least reduce, both of these barriers by providing the reader with (1) a wide-ranging introduction to satellite systems and (2) detailed references to further information. If I've been successful, you should have the background needed to understand the cited references by the time you turn to them.

In preparing the *Satellite Experimenter's Handbook*, I've devoted a great deal of effort to casting material in a format that readers with a background in radio communications would find comfortable. Often it's been possible to boil complicated topics down to relatively simple terms, but, at times, simplification conflicted with the clarity of fundamental ideas. Whenever such a conflict arose simplification was abandoned. As a result, you may encounter some relatively advanced mathematics in certain sections, but this material can be skipped over if you're mainly interested in the "how" and are willing to forgo the "why."

All of the references cited throughout this book were selected carefully for their clarity and significance. Although your local library may not stock all the items listed, you'll find that most libraries are willing to arrange "interlibrary loans" and obtain photocopies of articles for you at a small charge. To obtain this service, however, your references must be complete. Those provided in this text will satisfy the most stringent requirements. Another source of materials is worth considering. Many colleges and universities will grant library privileges to members of the community for a modest fee. If an institution near you has an engineering department, be sure to check out this possibility.

At first glance, you may think we've adopted a "mixed bag" or haphazard approach to the problem of units. To the mixed-bag accusation, the plea is "guilty" with compelling reasons. But our approach is certainly *not* haphazard. A clear strategy underlies all. Since the majority of beginners are probably most comfortable with English system units they're often used in the "beginners manual" (Chapters 1-7) when the values quoted are meant to convey a rough feeling for size. When detailed computations are illustated, the MKS system is used because it's easier to work with. The advanced reference material contained in the later chapters is handled almost exclusively in MKS units, but here too we have sometimes compromised. When the professional literature in an area is cast in terms of special, non-MKS, units we've gone along with general usage, since a major aim of this book is to provide readers with the background needed to master advanced material.

Now let's see if this book flies on its own merit — no launch vehicle to worry about here.

73,

Martin Davidoff

Martin Davidoff, K2UBC

TABLE OF CONTENTS

Part I

"I invite all nations to participate in a communications satellite system, in the interest of world peace and closer brotherhood among peoples of the world." — ***John F. Kennedy, 24 July 1961, Report to Congress***

Chapter 1

Enter the Space Age

The Space Age is barely a quarter century old. During this time a considerable amount of money has been spent on space exploration and the development of related technologies. The investment has clearly begun to pay off — earth satellites are the prime example. For nearly a decade most international telecommunications have been handled by satellite.[1,2,3] Daily data provided by spacecraft have revolutionized our ability to forecast weather, predict crop yields and monitor the environment. Satellites also contribute significantly to terrestrial navigation, scientific exploration, TV broadcasting and natural-resource management. And, their unique ability to monitor compliance with arms limitation treaties has contributed to international political stability. Parking spots in certain desirable regions of space have already become scarce.[4] Perhaps in the 21st century United Nations financial support will come from parking meters — leases on communications rights to orbital slots.

Satellites have already modified the way we think about ourselves, the earth, and the civilizations that inhabit it. Pictures of the earth, taken from space, have probably had a greater impact on both our awareness of this planet's limited resources and our need to pull together if we want our spaceship to continue sustaining life than all the words ever written on the subject.[5] Isaac Asimov sees satellites as causing profound, positive changes in personalized communications between earth's inhabitants — changes comparable in magnitude to the creation of speech, writing and printing. One result he expects is that "The earth for the first time will be knit together on a personal and not a governmental level."[6]

Radio Links

The radio signals linking satellites and *ground stations* (stations on, or near, the surface of the earth) are central to most satellite systems. These radio links can provide information about the spacecraft's operation and environment and form the basis of satellite communications systems. Therefore, it's important that we be aware of some basic properties of radio waves right from the beginning of our work.

The common expression, "line-of-sight," is probably most familiar in the context of light waves. In essence, it means that a person (A) who is looking at an object (B) can see the object only if nothing is in the way (i.e., if the straight line joining A and B is unobstructed). Radio signals generally adhere to the line-of-sight principle.

Taking this analogy between light and radio waves to its logical end suggests that it's impossible for two ground stations, more than a few hundred miles apart, to communicate using radio (or light) waves since the earth blocks the direct line joining them. Such a conclusion, however, is *false* for *two* important reasons. First, the line-of-sight principle is based on the assumption that events are taking place in a vacuum; realistically, it must be modified for events in the earth's atmosphere.[7] Second, we can use our ingenuity to get around (no pun intended) the principle. Just as we use mirrors to see around corners, we can use anything that reflects radio waves to bend them around obstacles. Most non-satellite, long-distance radio communication between two terrestrial stations does, in fact, involve the reflection of radio waves off the radio mirror known as the ionosphere (Fig. 1.1).

Radio reflectors are important both to those interested in long-distance communications and to scientists probing the structure of the earth's near environment. What acts as a radio mirror? Various candidates have been investigated in great detail; a partial list is given in Table 1.1. Much of the important experimental work with the reflectors listed in Table 1.1 was conducted by radio amateurs (more details and references will be found in

[1]Notes appear at the end of each chapter.

Table 1.1
Some Passive Reflectors of Radio Waves

Reflector (description)	*height above earth*	*maximum communication distance (single-hop)*	*frequencies of interest*
1) Layers of the earth's ionosphere			
a) F_2 layer (ultraviolet radiation from sun causes ionization of air molecules)	200-300 miles	2500 miles	below 100 MHz
b) E layer (ultraviolet radiation from sun)	35-70 miles	1400 miles	below 150 MHz
c) E layer (ionized trails left by *meteors* as they burn up)	35-70 miles	1400 miles	below 450 MHz
2) Aurora (ionized particles emitted by sun trapped by earth's magnetic field near north and south poles)	50-60 miles	1200 miles	below 500 MHz
3) Moon (has been used successfully by amateurs at frequencies indicated)	240,00 miles	12,000 miles	50-2300 MHz
4) Large balloon-like artificial satellites with metallic coating (see discussion of Echo I and Echo II in Chapter 2)	100-200 miles	2000 miles	above 30 MHz

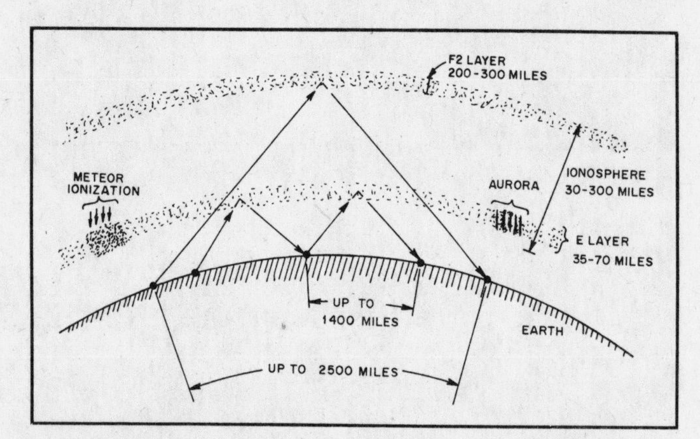

Fig. 1.1 — When various layers of the ionosphere are activated by solar radiation, solar particles, meteors or other means, they can reflect radio waves of certain wavelengths. Communication paths often involve more than one hop.

Chapters 2 and 3). To evaluate the communications effectiveness of each reflector it must be compared to alternative modes of communicating: other reflectors, telephone cables, mail service or terrestrial microwave links. Each of the reflectors mentioned exhibits one or more serious drawbacks including erratic, unpredictable behavior, the need for extremely high transmitter power and large, complex antennas for both transmission and reception, and excessive system (all transmitters plus receivers) cost.

The Satellite Relay

In the mid 1940s, long before the first satellite was placed in orbit, Arthur C. Clarke[8] published an article detailing how a satellite relay station could enable terrestrial stations to communicate over large distances. In an independent analysis, John Pierce, a physicist at Bell Telephone Laboratories, came to a similar conclusion: Active satellite relays could have a significant positive impact on long-distance communication.[9] Such a system would be most reliable if the radio frequencies used were not affected by the ionosphere.

Although a relay station can take many forms (see Chapter 12) we'll only mention one here — the transponder. A *transponder* is an electronic device that receives a small slice of the radio frequency spectrum, greatly amplifies the strength of signals within the entire slice, and then retransmits the signals in another portion of the spectrum. The radio amateur transponders used aboard satellites can handle a large number of signals of various types simultaneously, with the power of each received signal being multiplied roughly 10^{13} times (130 dB) before being retransmitted back to earth.

This brief look at the basic properties of radio waves and the active-relay concept provides the background needed to consider satellite radio links.

Satellite Radio Links

The radio signals between satellites and ground stations are often categorized as *downlinks* (signals from a spacecraft to a ground station — Fig. 1.2a), *uplinks* (signals originating at the ground and directed to a satellite — Fig. 1.2b), and *broadcast* or *communication links* (which involve both uplinks and downlinks — Fig. 1.2c). The simplest downlink, a continuous tone, can be useful to ground stations tracking the satellite and to experimenters studying radio propagation or investigating the ionosphere. A more complex downlink beacon can be used to convey *telemetry* information (measurements made by scientific and engineering instruments aboard the spacecraft) to interested ground stations. A number of different techniques are used for *modulating* telemetry beacons (superimposing the telemetry data on the radio signal). We'll look at some of these techniques in Chapter 12.

Uplinks can be used to control the operation of a satellite. For example, if a particular satellite's design permits, we could reprogram an onboard computer from earth to adjust the spacecraft's attitude (orientation in space), or to turn a beacon off temporarily to conserve energy. Ground stations equipped to control a spacecraft are called *command stations.*

Uplinks and downlinks are used together in many applications. For example, ground station A (Fig. 1.2c) may send a message to ground station B (over a non-line-of-sight path) by way of a satellite relay. The number of ground stations equipped to use a particular satellite relay may range from only a few to tens-of-thousands or more. By the late 1980s, for example, direct satellite-to-home TV systems should be operating in Western Europe, Japan and the U.S. Station A (Fig. 1.2c) would in this case be a central TV transmitter, and station B represents the millions of homes capable of receiving the TV signals. This example illustrates a broadcast link. If both stations in Fig. 1.2c are equipped to transmit and receive, we have a communications link. A multiple-access relay satellite of the type built by radio amateurs can be used by a large number of ground stations simultaneously.

In order for two ground stations to communicate through a single satellite relay, *both* must be *in range* of the spacecraft (i.e., the line-of-sight path between each station and the satellite must be unobstructed). In addition, the satellite's antenna pattern must be broad enough to include both ground stations. Fig. 1.3 shows that all stations within the range of a satellite at a given instant lie inside the circle formed by an imaginary cone whose curved surface just grazes the earth. The axis of the cone goes

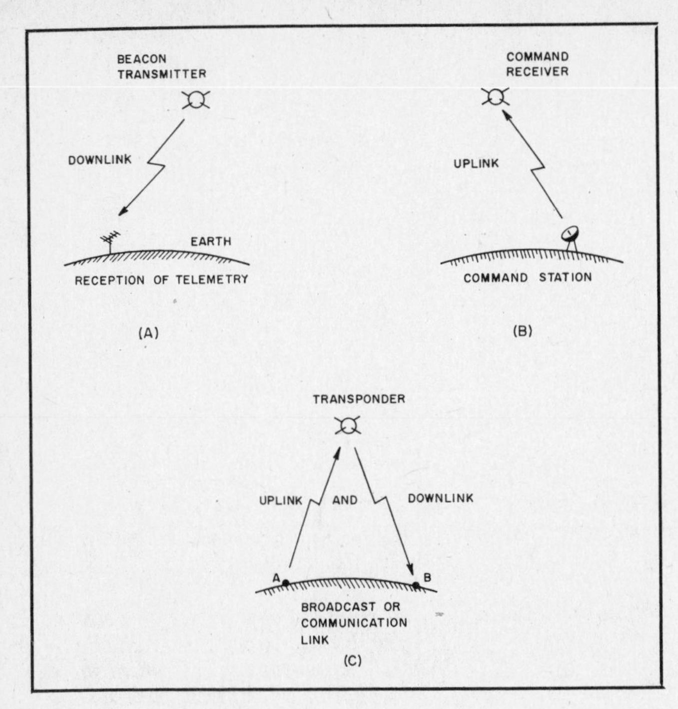

Fig. 1.2 — Satellite radio links.

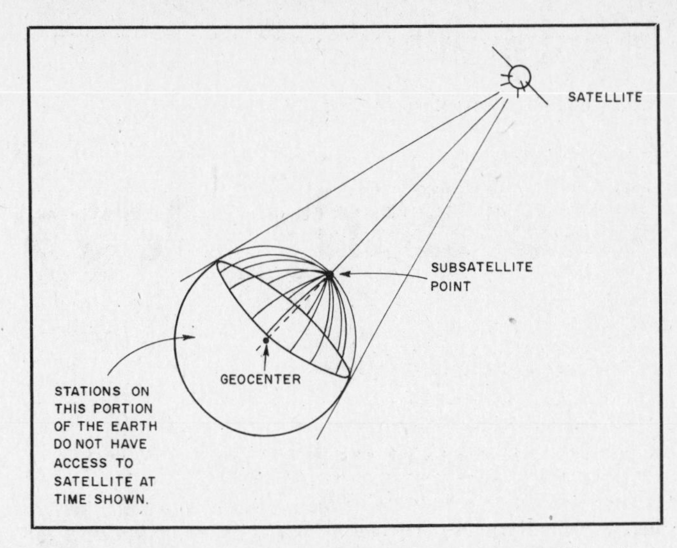

Fig. 1.3 — Only those terrestrial stations with unobstructed line-of-sight paths to a satellite are in range.

through the satellite, the *subsatellite point* (the point on the surface of the earth directly below the satellite), and the *geocenter* (the center of the earth). A satellite located 240,000 miles over Catonsville, Maryland, USA, would be in range of everything in Fig. 1.4 — essentially half the planet. The two inner circles show the portions of the earth that would be visible at lower altitudes: 22,000 miles and 900 miles. For most satellite orbits, the position of the subsatellite point constantly moves as the satellite moves along its orbit. For certain special orbits, however, the subsatellite point remains fixed at a specific longitude on the equator. We'll discuss these *geostationary* orbits further in Chapter 8.

From Figs. 1.3 and 1.4 it's clear that *maximum communications distance* (the largest terrestrial distance over which it's possible to relay a radio message) increases as satellite height increases, but it can never exceed half the circumference of the earth. (Graphs and mathematical expressions relating satellite height to maximum communications distance are presented in Chapter 9.)

These basic concepts underlie the development of artificial earth satellites over the two and a half decades since the first man-made satellite completed its first orbit. From the beginning, Amateur Radio has played a significant role.

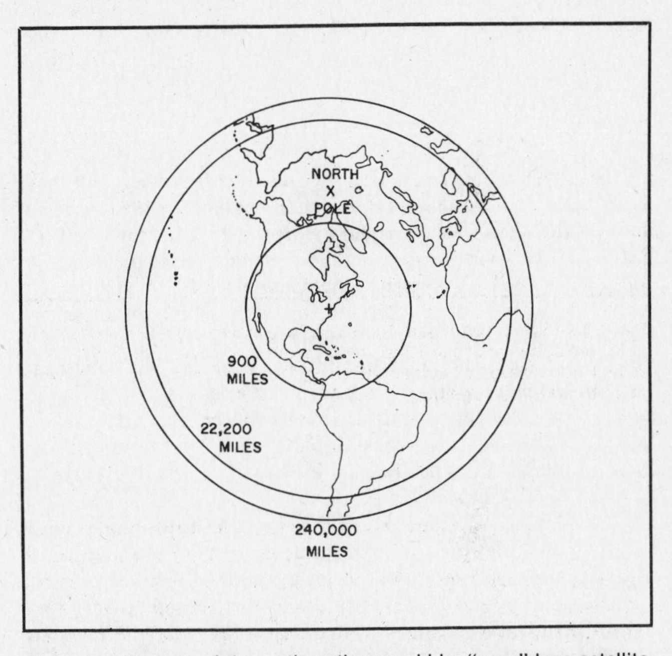

Fig. 1.4 — Regions of the earth as they would be "seen" by a satellite located directly over Catonsville, Maryland, USA (39° N, 76° W). The three distances chosen correspond to the moon (240,000 miles), an AMSAT Phase III satellite at its high point (22,200 miles) and AMSAT-OSCAR 7 (900 miles).

Notes

[1]B. I. Edelson, "Global Satellite Communications," *Scientific American,* Vol. 236, no. 2, Feb. 1977, pp. 58-68, 73.
[2]B. I. Edelson and L. Pollack, "Satellite Communications," *Science,* Vol. 195, no. 4283, 18 March 1977, pp. 1125-1133.
[3]H. L. Van Trees, E. V. Hoversten and T. P. McGarty, "Communications Satellites: Looking to the 1980s," *IEEE Spectrum,* Dec. 1977, pp. 43-51.
[4]W. L. Morgan, "Satellite Utilization of the Geosynchronous Orbit," *COMSAT Technical Review,* Vol. 6, no. 1, 1976, pp. 195-205.
[5]A. C. Clarke, "Beyond Babel," *UNESCO Courier,* March 1970, pp. 32, 34-37.
[6]I. Asimov, "The Fourth Revolution," *Saturday Review,* Oct. 24, 1970, pp. 17-20.
[7]H. S. Brier and W. I. Orr, *VHF Handbook* (Wilton, Conn: Radio Publications) 1974, Chapter 3.
[8]A.C. Clarke, "Extra-Terrestrial Relays," *Wireless World,* Vol. 51, no. 303, Oct. 1945, Chapter 3.
[9]J. R. Pierce, "Orbital Radio Relays," *Jet Propulsion,* Vol. 25, no. 4, April 1955, p. 153.

Chapter 2

The Early Days

The Dawn of The Space Age

In early October the warmth of the morning sun is very welcome in Baikonur, a village in the USSR 150 miles northeast of the Sea of Aral. But this particular morning, as the countdown for Sputnik I nears its climax, the sun is hardly noticed. One-by-one the inevitable technical hitches are dealt with and finally, at 0600 Moscow time on 4 October 1957: Zazhiganiye *(blastoff). The Space Age begins.*

Shortly after midnight (GMT), a BBC radio operator at a monitoring station near London notes the appearance of a strange "beep-beep-beep. . . " Something about the unfamiliar signal attracts his attention — though the average strength is gradually increasing, rapid fading is superimposed. The signal's frequency is drifting slowly downward, and direction-finding equipment shows the azimuth of the source to be changing rapidly. Only one conclusion is possible — the signal is coming from an artificial space satellite.

These events may not sound very spectacular today, but they generated almost unimaginable excitement back in 1957. Within minutes the news flashed round the world. In Washington, DC, it was still early in the evening. Following a week of meetings focusing on the IGY (International Geophysical Year) the Soviet embassy was holding a reception for many of the senior scientists involved. Lloyd Berkner, president of the U.S. IGY coordinating group, was paged and informed of the BBC observation and given a report of the launch just released by TASS (the Soviet news service) that identified the spacecraft as Sputnik I. When he returned to the cocktail party, Berkner announced the event to the scientists present and his Russian hosts.[1-3] It certainly must have been a lively party.

Newspaper accounts report that the world responded with "surprise and elation." In retrospect, the elation is understandable, but the surprise element seems a little misplaced. The June 1957 issue of *Radio* (Радио), a widely distributed Soviet journal on practical electronics, stated that a Sputnik (the Russian word for satellite) would soon be launched. The column provided information on the projected launch date (late September), the transmitter frequencies (20.005 and 40.010 MHz) and the type of modulation. The Russians again announced their plans at international scientific meetings in Barcelona and Washington later that summer. The sense of surprise certainly didn't arise from Soviet secrecy.

One of the transmitters on Sputnik I operated just above the 20-MHz frequency used by the United States and other countries for a worldwide network of high-power radio transmitters sending standard time signals. Therefore, the hundreds of thousands of radio amateurs and shortwave listeners owning radio receivers capable of picking up the time broadcasts were able to listen for the spacecraft. Signals from the satellite were generally so loud that they could be tuned in on even the simplest of these receiving sets. The Soviets' choice of frequencies was obviously no accident. Their interest in amateur reports was clear: "Since radio amateur observations will be of a mass character they can secure extremely important data on the satellite's flight and the state of the ionosphere."[4] It's probable that more people listened directly to Sputnik I than to any other single spacecraft launched since. (Direct satellite-to-home TV will, no doubt, invalidate this statement in the mid 1980s.)

While the 20-MHz signal of Sputnik I showed the world that very simple ground stations could be used to monitor satellite signals, this frequency was not suitable for reliably sending large amounts of information on a satellite's performance, or its environment, back to earth. Future satellites would use higher frequencies and more sophisticated techniques for forwarding telemetry data to improve reliability. As a result, it would become increasingly difficult, even for the amateur scientist with extensive Amateur Radio experience, to monitor government satellite programs directly.

The U.S. Enters Space

Barely four months on the heels of Sputnik I, on 31 January 1958, the United States launched Explorer I, its first successful satellite. Explorer I contained a scientific instrument package designed by Dr. James Van Allen of Iowa State University to measure radiation levels in space. Initially, the scientists working on the project thought their instruments had spun completely off scale. After a painstaking but unsuccessful search to pinpoint where the equipment had failed, their gloom gradually turned to elation. The scientists were forced to conclude that the instruments were actually operating properly; radiation of such unprecedented levels had been encountered that the instruments had been driven to saturation. With soaring spirits, the team of scientists began the task of mapping what was later to be known as the Van Allen Belts.[5]

Radio Amateurs and Space

Over the years, radio amateurs have taken an active and important part in space-related investigations. Many of the pioneers in radio astronomy were also hams. During the late '30s and early '40s the exploratory studies and comprehensive radio sky maps prepared by Dr. Grote Reber (W9GFZ) — using a homebuilt 32-foot diameter parabolic antenna in his Wheaton, Illinois, backyard — were significant contributions to science.[6] On 27 January 1953 Ross Bateman (W4AO) and William L. Smith (W3GKP) beamed radio signals at the moon and succeeded in hearing echos.[7] And in the late '50s, as mentioned, thousands of radio amateurs monitored signals from early Soviet and American satellites. What would radio amateurs do next?

In April 1959 Don Stoner (W6TNS), a widely known and

well-respected electronics experimenter, writing in *CQ*, suggested that amateurs undertake the construction of a relay satellite.[8] Stoner was looking far beyond placing a simple beacon in orbit; he was proposing that hams build a spacecraft containing a transponder capable of supporting two-way communications. Many said Stoner was fantasizing: After all, construction of the first government-supported satellite to use the proposed techniques (Telstar I, launched July 1962) hadn't even begun, fm repeaters were virtually unknown in the radio amateur community, and most experimenters had little or no experience with the newfangled devices called transistors. But these fantasies were the disciplined dreams of an intelligent and farsighted thinker. Although Stoner's comments were couched in humorous terms, he was serious. His note provided the spark that would lead, not many years later, to radio amateurs' placing an operational, active relay satellite in orbit — but that's getting ahead of our story.

In 1960, imaginations fired by the Stoner article, a group of radio amateurs in Sunnyvale, California, organized the OSCAR Association (*O*rbiting *S*atellite *C*arrying *A*mateur *R*adio). The aims of this pioneering club included both building amateur satellites and obtaining launches. It wasn't clear which goal would be more difficult. To arrange a launch, the U.S. government would have to be convinced that amateur satellites could serve a useful function in one or more of the following areas: scientific exploration, technical development, disaster communications, and scientific or technical education. One important factor was to help both radio amateurs and the scientific community. Most large satellites are mated to rockets having excess lift capacity; it's simpler and cheaper to ballast a rocket with dead weight than to reduce the thrust. As a result, it was possible to add secondary payloads to many missions at very little cost. Over the years, many scientific and amateur satellites have hitchhiked into space piggybacking on primary payload missions.

OSCAR I

After two years of effort by members of the OSCAR Association, the first radio amateur satellite — OSCAR I — was ready and scheduled for launch. Weighing in at 10 pounds, the spacecraft contained a 140-milliwatt beacon at 145 MHz transmitting a simple, repetitive message at a speed controlled by a sensor responding to the internal satellite temperature. Fig. 2-1 compares the OSCAR I beacon to the 10-milliwatt beacons flown on Explorer I (the first U.S. satellite) and an early U.S. Vanguard mission. Although OSCAR I did not contain a transponder, it was a significant first step toward that goal. The events and emotions surrounding the beginning of OSCAR I's 22-day sojourn in space were beautifully captured in a classic *QST* article by Bill Orr (W6SAI).[9] We let Bill tell the story.

The spirit of adventure lies buried in every man's soul. Strike the spark and ignite the soul and the impossible is accomplished. So it was on December 12, 1901 on a chill, Newfoundland morning. The first self-proclaimed radio amateur, Guglielmo Marconi, bent intently over his crude receiving instruments and heard the letter "S" transmitted across the stormy Atlantic Ocean, from a station in Cornwall.

The spirit of adventure again made its mark sixty years later on December 12, 1961. The locale this time was an experimental aerospace base on the border of the Pacific Ocean. A group of radio amateurs saw launched into orbit the first amateur radio space satellite. Born in a burst of flame, the 10-pound, home-made beacon satellite transmitted to the world that the spirit of adventure and quest that drove Marconi down the road of history was still goading the radio amateur in his eternal search after the mysteries of nature. This is the story of a small portion of that quest.

Sixty Years of Radio Amateur Communication

Marconi to the OSCAR Satellite

BY WILLIAM I. ORR, W6SAI

February, 1959: The radio amateur gazed thoughtfully for a moment at the white paper in his typewriter. Suddenly his fingers sprang into action and the keys flashed the fateful words, "Currently being tested is a solar powered six- to two-meter transistor repeater which could be ballooned over the Southwest. Can anyone come up with a spare rocket for orbiting purposes? . . . 73, Don, W6TNS."[†] *He slapped the page from the typewriter, setting in motion a chain of events that conclusively proved that truth is indeed stranger than fiction.*

The local time is 0200 on a cold, starless 1961 December morning. The location is Vandenberg Air Force Base, California. It is a cheerless, predawn moment. Inside the reinforced block house, the combined USAF and contractor crews are busy at work. The block house walls are lined with TV monitoring screens. Along one side is the launch control console. Communications, radar and propellant monitors are on; talkers and other intercommunications people are at their stations. The key personnel are locked in unison by a single communications net. All wear headsets and microphones so that they can use their hands freely. A complex network permits several simultaneous conversations. The outpouring of this network culminates in a teletype transmission to the Program Director located 170 miles away in Inglewood, California. The RTTY channel springs to life and begins to clatter: . . . FM 6565TH TEST WING VAFB CALIF TO SSD LOSA CALIF THIS IS A CONTINUOUS MESSAGE. . . . R MINUS 500 AND COUNTING. . . .

[†]Semiconductors," CQ, April 1959, p. 84.

. . . T MINUS TEN AND STILL COUNTING. . . . Tension builds up as moment of launch nears. (Left to right): Capt. Turner (USAF); Bill, W6SAI; Ray, W6MLZ; Dos, WØTSN; and Chuck, K6LFH. Chuck talks to OSCAR Control Center, WA6GFY, to make sure that traffic net to South Pole is ready for acquisition of OSCAR as it passes on initial revolution. *(Photo: USAF)*

In the cold night illuminated by a thousand lamps, the Agena-Thor aerospace vehicle sits on the reinforced launching pad. Known as Discoverer XXXVI, this intricate, calm, sophisticated spire of brute power awaits the command to hurl itself into space. From it will eject man-made satellites, orbiting the earth hundreds of miles above. One of these will be of great interest to the radio amateur. It is OSCAR.

Of the thousands of readers of Don Stoner's article, none was struck more forcibly than Fred Hicks, W6EJU, of Campbell, California. An old-timer in the communications game, Fred was now employed by a large missile contractor in the San Francisco bay area. Fred had been present in the blockhouse at Vandenberg for the first six Discoverer launches. To Fred goes all credit for grasping the true nature of Don's message, and interpreting it in terms of the full spirit of amateur radio.

Fred dropped the magazine on his desk, pushed aside a cup of coffee and reached for the telephone. He dialed a number and listened to the automatic stepping switches go through their complicated dance in the earpiece of the instrument. "Hello, Chuck? . . . Hey, buddy, did you read Don Stoner's article this month? . . . Well, he said in effect that the radio hams could build a satellite if they could only find somebody to launch it for them. . . ." The voice on the phone crackled. "Right! That's what I was thinking. Why don't you drop Don a line and get this thing organized? . . . If old K6LFH and W6EJU and their buddies can't do the job, why, nobody can!" Fred chuckled to himself as he hung up the phone. Chuck was right. Why not build a ham satellite? The idea wasn't so crazy after all. A lot could be learned from such a device. The satellite would . . . it would . . . well . . . Fred suddenly realized that such a simple, beguiling idea could not be defined and would entail a lot of work and planning to even begin to be coherent. Obviously it was a fine project for a club, or group of hams. One ham couldn't handle this "brainbuster." As H.P.M., The Old Man, might have said, "It was an idea without a handle to grab it." . . . Truly, W6EJU was blessed with the spark of adventure.

The count down begins at R minus 500 minutes and is divided into more than twenty tasks. More than 1500 separate instructions must be given from the launch console before the vehicle is ready for the great voyage into space. Guidance checks, polarity and phasing checks, vehicle erection, re-check of destruct systems, orbital electronics and control checks, propellant tank checks, telemetry operational checks, and satellite operational checks must go on in infinite, precise detail. The voice of the teletype chatters endlessly. . . .
. . . R MINUS 350 AND STILL COUNTING. . . .

13 October, 1959

"Dear Don:

"I remember you wrote an article for CQ some time ago that described a small transistorized two-meter station, and appealed for 'anyone with a space vehicle, please?' . . . Though I do not hold out any too much hope for this, I will do my best to interest certain parties . . . please send me the exact weight of the installation and space it occupies. . . . Actually, the 'Discoverer' is ideally suited to such a ham project . I will sound out the local hams . . . look for me on 14,285 kc . . . 73 and I certainly hope we can pull this off! . . . Fred, W6EJU."

The die was cast. The spark of adventure had found fuel and was burning brightly. The fateful letter was on the way; was in the mail. It would start a thousand minds dreaming and planning, and the concept would eventually involve high level decisions in the U. S. Government. Now, at this moment in time it was a gossamer; a fancy that might be lightly discarded as a mere exercise of the imagination. (After all, why not? Would not a homemade satellite be yet another convincing proof that amateur radio was indeed in the public interest, convenience, and necessity? At the very least it would be a self-educational program, introducing the great body of amateurs to space communications. Of course.)

Bob Herrin, K4RFP/6 (Launch Operations Manager), was listening on the countdown net in the communications and control laboratory, at the launch site. He joshed a few words with other technicians and engineers, intent upon their tasks. The package had been carefully placed into its egg-crate shaped compartment in the Agena second stage of the immense vehicle a few days earlier. Soon the package would fall into line in the check-off procedure that was now running at a rapid pace. Would the antenna erect itself? Would the squib fire the spring that would place the 10-pound satellite into a free orbit of its own? Would the compact, transistorized beacon spring into life, as it had done thousands of times in the shacks of the builders? Or would Oscar I merely become a footnote in the history pages of amateur radio? . . . R MINUS 180 AND HOLDING FOR FIFTEEN MINUTES chattered the teletype.

Bob looked up and his heart jumped. Even though he was an old hand at the launching game, the sound of the "hold" announcement never failed to affect him. "I hope it's only a technical hold," he wished to himself as he continued with his duties. He noticed that the black sky was breaking in the East. Daylight was near. It was always easier in daylight, for some reason. . . . R MINUS 180 AND RESUMING COUNT. . . .

15 October, 1959

"Dear Fred:

"To say I was elated to receive your letter would be the understatement of the year. However, before I allow myself to get too excited, I am going to submit a proposal to you and see what happens. . . . As you say, I hope we can pull (or is it push) this thing off. Best regards, Don."

The radio amateurs seated around the conference table grinned as Fred, W6EJU, Chairman, read the message. The first meeting of the OSCAR Committee was about to be called to order. There were: Chuck Towns, K6LFH; Bernie Barrick, W6OON; Stan Benson, K6CBK and Nick Marshall, W6OLO. These amateurs are the trail-blazers into space in the year 1959!

In Los Angeles, Don Stoner had many conversations with Ray Meyers, W6MLZ, and Henry Richter, W6VZT. Gradually a concept of a suitable radio satellite package was being pounded out. The phone bill between W6TNS and W6EJU began to grow to alarming proportions, supplemented by sideband schedules on 7 Mc. Don suggested that the rapidly growing group of hams be called the OSCAR Association: Orbital Satellite Carrying Amateur Radio! A natural name. So was OSCAR born in spirit.

At 7 A.M. Bill, W6SAI, rolled over in bed in the BOQ at Vandenberg Air Force Base, California. He reached across and shook Chuck, K6LFH, awake. "0700 local time," he said as Chuck turned his face to the wall and tried to go back to sleep. "We meet the press at 0800, and go to the pad at 1000. Today we'll either be heroes or tramps!" Chuck sat up in bed and looked at his watch. "The count down started at about two A.M.," he said. "They must be down to about R minus 180 by now."
. . . R MINUS 180 AND COUNTING. . . . CLEAR AREA TO LOAD FUEL . . . CHECK LOG TO DETERMINE FINAL ULLAGE REQUIREMENTS. . . .

The tension in the block house was quietly growing. A charged atmosphere punctuated by short commands and remarks served only to emphasize the quick passage of time. The sun would rise in a few moments and the air was growing warmer. A cool, mild breeze was coming in from the Pacific and the sky, which was not yet red, was a flat steel color. An Air Police helicopter hovered briefly by the launching site then slanted away on some mysterious mission, its huge rotor chopping the air. The Discoverer stood waiting, a white tall spire, gleaming dully in the giant light of dawn, yet bathed on all sides by spotlights. Soon it would burst into space.

21 October, 1960

"Federal Communications Commission:

We thank you for your comments regarding our proposed OSCAR

Vandenberg control and tracking station pinpoints the Discoverer as it races in orbit around the earth at 18,000 miles per hour. OSCAR satellite follows its own orbit at approximately same speed as the parent satellite. Orbital data is plotted on boards at the rear of the room from the acquisition and control consoles in the foreground. *(Photo: Lockheed)*

Ready to go! OSCAR completes its qualification tests with flying colors! At final check-out are (left to right): Gail Gangwish; Nick Marshall, W6OLO; Don Stoner, W6TNS; Chuck Towns, K6LFH; and Fred Hicks, W6EJU.

program and will attempt herein to clarify our objectives. . . . The former OSCAR Committee has been reorganized as the Project OSCAR Association . . . the Board of Directors have approved the project plans . . . the proposed satellite will be transmitting in the 2-meter amateur band, and will be electronically keyed . . . it will have a restricted life of perhaps 20 days. . . .
Fred H. Hicks, W6EJU, for Project OSCAR."

26 September, 1960

"Dear Mr. Hicks:

This will acknowledge receipt of your letter regarding Project OSCAR. . . . It appears that, with the exception of the requirement for positive control of the transmitter by the station licensee, you may be able to meet the other rule requirements in question . . . you realize that this project must receive the sanction of the other government agencies before final approval could be granted. . . . Ben F. Waple, Acting Secretary, FCC."

By now the OSCAR Association had grown to the point where items of hardware could be built and tested for the proposed satellite. Project volunteers had been assigned jobs and an OSCAR mailing list was created. Because of the press of business, W6EJU turned the chairmanship of the OSCAR program over to Mirabeau ("Chuck") Towns, K6LFH, to implement and carry on the ultimate dream of having an amateur radio station in orbit about the earth. For it was only a dream. . . .

"Really, Mr. Towns. I admit the idea has some merit to it, but I do not see what earthly good it would do to have a bunch of amateurs engage in such an effort. After all, the government has spent millions of dollars in establishing exotic tracking stations . . . really, now, let's be serious for a moment. . . ."

Bill, W6SAI, looked dully at the plate of congealed eggs and the cup of cold coffee. "To heck with breakfast," he said to Chuck. "I'm too excited to eat." The other amateurs were equally elated: Don Stoner, W6TNS, who had been invited to the launch to see his dream come true; Goodwin L. Dosland, WØTSN, President of ARRL; and Ray Meyers, W6MLZ, Director of the Southwestern Division, ARRL. Absent because of illness was Harry Engwicht, W6HC, Director of the Pacific Division, ARRL. Two hundred miles to the north Fred, W6EJU, now acting as Operations Director, and the complete OSCAR Tracking network were standing by, waiting to flash word of OSCAR orbit to waiting radio amateurs. "Let's get the show on the road," said "Dos," reaching for his overcoat. "It's almost ten minutes to eight and we have to attend the pre-launch press meeting."

The radio teletype chattered its endless song. . . .

R MINUS 150 AND COUNTING. . . . CLEAR AREA TO LOAD OXIDIZER. . . . CHECK ULLAGE REQUIREMENTS BEFORE ZEROING FLOW METER. . . .

10 November, 1960

"John Huntoon, ARRL

As I have mentioned to you, a proposal has been made to place an amateur satellite in orbit, using a future space vehicle as a 'piggy-back' carrier . . . a need exists for strong, amateur leadership from a group that represents a majority of the amateurs, rather than a small, local club. I believe that the only organization that can truly represent the amateur in this matter is ARRL. Without ARRL sponsorship, the amateur satellite program will wither and die . . . 73, Bill, W6SAI."

In the meantime, OSCAR had enlisted additional support. George Jacobs, W3ASK, Propagation and Space Communications Editor of CQ, had volunteered to be the Washington, D.C., contact man for Project OSCAR. George spent many hours discussing the project with sympathetic officials of the FCC and the State Department. He tried to discover what conditions must be met by such a unique undertaking in order to receive approval from key government officials, some of whom had only a hazy concept of the ideals and dreams of the radio amateur. George worked in close collaboration with John Huntoon, General Manager of ARRL. Finally, in the early spring of 1961, after a trip to Hq. by K6LFH and W6SAI for a conference with League officials, the ARRL adopted Project OSCAR, granting its endorsement to the project and providing important, vital backing in the name of the amateurs of the United States.

The launch site was atop a scrubby sand dune in a far corner of Vandenberg AFB. A jolting Air Force bus crossed innumerable sand dunes and washes, carrying the amateurs and reporters who would soon observe the launch. Dry bush dotted the rough landscape. Suddenly, the Discoverer atop the launch pad was visible on the horizon. It stood majestically alone, surrounded by lesser objects that emphasized its size. It was a clear white, with the motto "United States" emblazoned on it. A single plume of evaporating liquid oxygen curled lazily from one side. There was no movement about the vehicle, and the area seemed deserted and asleep. The bus, loaded with newspaper, radio and TV reporters and the group of radio amateurs ground to a halt atop a small plateau about five hundred yards from the launch site. The riders dismounted and slowly walked to a clear spot from which the Discoverer rocket was in clear view. At one corner of the plateau stood a small gasoline generator, a communications truck, a table with a battery of telephones, and a portable loud speaker plugged into the base communications system.

. . . R MINUS 80 AND STILL COUNTING. . . .

The Air Force Thor booster, standing on the launching pad had completed the touchy fueling operation in which thousands of pounds of RP-1 (a souped-up version of aircraft jet fuel) and LOX (liquid oxygen) had been pumped into it. On top of the booster, the 25-foot long Agena brought the total height of the satellite-vehicle combination to 81 feet. The sun was climbing higher in the sky and the wind had died down now, and the site was clear and warm.

. . . R MINUS 50 AND STILL COUNTING. . . . TANK PRESSURES CHECKED. . . .DESTRUCT SQUIBS ARMED. . . . RECORDERS ARE ON. . . .

"Why do you employ an 'R' count instead of a 'T' count?" asked W6SAI of Captain Barbato (USAF), the Public Information Officer.

"The R-count is in minutes and is used up to about minus ten minutes. At that time we switch to the T-count, which is run in minutes or seconds," explained the Captain. The communications truck gave notice from the Missile Flight Safety Officer that the range was clear, and that it was clear to launch.

31 July, 1961

"Secretary of State, U. S. State Department:

The American Radio Relay League, the national nonprofit membership association of amateur radio operators, requests the cooperation of the Department of State concerning space communication and experimentation by radio amateurs. A group of skilled radio amateurs on the West coast, which is incorporating as the Project OSCAR Association, has designed and constructed communications equipment suitable for launch into orbit. The Association is nonprofit and is entirely noncommercial and nonmilitary. It is affiliated with and has the full support of the American Radio Relay League . . . an informal session was held in Washington recently, with the following results:

a) Air Force representatives stated that Project OSCAR has been approved by HQ AFSC for incorporation in the Discoverer series of launchings, subject to coordination with other interested government agencies . . . it is our hope that the information contained herein will be sufficient to enable the Department of State now to undertake the procedure outlined and agreed to at the meeting — i.e., to solicit the formal concurrence of the several agencies concerned in this matter so that the project may go forward . . . (signed) John Huntoon, General Manager, ARRL."

Simultaneously, the Project OSCAR Communications link was being organized under the direction of Tom Lott, VE2AGF/W6. It was desired to have early acquisition of the OSCAR satellite by a responsible party, so various amateurs were contacted at the South Pole bases by Captain David Veazey, W4ABY USN, Assistant for Communications, Special Pro-

Directors of the Project OSCAR Association. Left to right: Fred Hicks, W6EJU; Bill Orr, W6SAI; Harley Gabrielson, W6HEK; Tom Lott, VE2AGF/W6; Chuck Towns, Jr., K6LFH (Chairman); B. Barrick, W6OON; Dick Esneault, W4IJC/W6; Harry Workman, K6JTC; and Nick Marshall, W6OLO. Not present at the time the photo was taken were Stan Benson, K6CBK; Jerre Crosier, W6IGE; Harry Engwicht, W6HC; and M. K. Caston, WA6MSO.

jects Office. Dave promised to arrange a suitable amateur tracking station to be set up on the Antarctic continent by the KC4 hams to flash back word of OSCAR, once it achieved orbit.

The crowd at the Discoverer site had grown to a small army. General Francis H. Griswold, K3RBA, Director of the National War College, Washington, D. C., had arrived. In addition, a group of scientists from California Institute of Technology had heard of the launch, and had interrupted their important work to watch the world's first home-made amateur radio satellite hurled into orbit.

. . . T MINUS 30 AND COUNTING. . . . REPORTING WILL BE BY EYEBALL AND F.M. RADAR AFTER LIFT-OFF. . . . TERMINAL COUNT WILL START AT T MINUS 11 MINUTES. . . . GUIDANCE LOCK ON COMPLETE. . . . BTL READY AND STANDING BY FOR LAUNCH. . . . RANGE GREEN. . . . T MINUS 20 AND COUNTING. . . .

The sky had clouded over and a slight overcast settled down above the poised bird. "Do you require a clear sky for launch?" asked Ray, W6MLZ. "No," replied the Public Information Officer. "This overcast won't affect the launch."

Now the news service wires were open, and Chuck, K6LFH, placed a long distance call to the OSCAR control center, WA6GFY. Was everything ready in Sunnyvale? . . . Good. . . . Good. . . . South Pole link through W4ABY and KC4USB is open. . . . W6EJU at the other end of the land line queried as to the exact time of launch. . . . "Sorry, Fred, can't announce the time until after lift-off. . . . Fred laughed, "I can tell from the sound of your voice it will be within a *very* few minutes," he said. As if to verify his words, the communications speaker over Chuck's shoulder blared into the telephone, "T minus 16 and counting!!!"

15 September, 1961

"John Huntoon, ARRL.

Reference is made to your letter of July 31, 1961, requesting the cooperation of the Department of State concerning space communication and experimentation by radio amateurs, specifically with respect to 'Project OSCAR'.

"In reply I am pleased to inform you, after consultation on this subject with other interested agencies of the Government, that the Department perceives no objection to the carrying out of Project OSCAR. . . . For the Secretary of State: Edwin M. Martin, Assistant Secretary."

T MINUS 14 AND COUNTING. . . . ONE MINUTE UNTIL START OF TERMINAL COUNT. . . . TERMINAL COUNT WILL START ON MARK. . . . MARK. . . . PHASE ONE PROCEEDING NORMAL. . . . PHASE ONE COMPLETE. . . . .PHASE TWO PROCEEDING NORMAL. . . .

Don, W6TNS, plugged his tape recorder into the a.c. outlet on the portable generator. Bill, W6SAI, climbed atop a sand dune immediately behind the plateau. The Air Force men looked to their recording cameras and the babble of voices on the press telephones rose in pitch. The Air Police helicopter scooted overhead, looping about the press area, and inquisitively shot behind a sand dune. The pulsating beat of its rotor could be heard above the noise of the preparations.

. . . The teletype pounded on in a relentless beat. . . . PHASE FOUR PROCEEDING NORMAL. . . . ORBITAL STAGE TLM AND BEACON BEING VERIFIED. . . . FUELING COMPLETE. . . . MAIN SAFETY RECEIVERS INTERNAL. . . . PHASE FOUR COMPLETE. . . . PHASE FIVE PROCEEDING NORMAL. . . .

Suddenly 'Dos,' WØTSN, laughed out loud.

"What's so funny, Dos?" asked Don. "The incongruity of the situation just struck me," said Dos. "Here I am, a radio ham and an attorney, on a launching pad in California! It's 14 below zero in Minnesota and a judge and jury are in recess until I return! Who would imagine I'd be here today watching OSCAR fly?"

Who indeed? There were many doubters and some who had damned the project with faint praise. Many times the future of the OSCAR Project looked black, as some insurmountable road block loomed ahead. The support of interested amateurs was great comfort in such moments:

PAØVF: It is with much interest that amateurs in the Netherlands were reading of Project OSCAR . . . we thank you for your kind information. . . .

GM3NQB: . . . those with whom I have talked are tremendously interested. . . .

VU2NR: . . . I would be quite happy to make any kind of observations required in regard to OSCAR . . . good luck!

LU9HAT: . . . please send me information. . . . I am a member of the local amateur satellite observers' group. . . .

ZS3G: Send us full details, as we intend building equipment for OSCAR. . . .

Indeed, there were those who believed in OSCAR. Actually, many more than was known at the time. These amateurs knew the spirit of adventure, too.

. . . PHASE FIVE PROCEEDING NORMAL. . . .ORBITAL STAGE ON INTERNAL POWER. . . . BOOSTER AND BTL ON INTERNAL POWER. . . . ENGINE SLEW COMPLETE. . . .

The missile stood silent, awaiting the final seconds before the powerful motor would burst forth. The culmination of months of work of thousands of people was rapidly approaching a climax. The atmosphere was tense on the plateau. People spoke to each other now in half-whispers, as the newsmen unfolded the story into telephones. "Put me on the air now . . . launch will be in about ten seconds."

High-speed cameras near the launch site were now whirling and the telescopic cameras at the plateau were aimed at the bird. The master tape in the Communications Center was recording every action and sound. The air was literally charged with electricity. Oblivious to the tension, Discoverer XXXVI resembled a giant finger, pointed serenely at the heavens. Within its giant frame, the tiny OSCAR package waited . . . the teletype went mad with speed. . . .

. . . . LAUNCHER CLEAR TO FIRE. . . . CLEAR TO LAUNCH. . . .RANGE CLEAR TO LAUNCH. . . . ON MARK WILL BE T MINUS TWO SECONDS. . . . MARK. . . .

November 3, 1961

"John Huntoon, ARRL:

I am pleased to advise that the Air Force will undertake to place in orbit an OSCAR package in conjunction with a military space vehicle launching. Our Space Systems Division has been instructed to accomplish the OSCAR package launching at the earliest feasible date on a non-interference basis to the performance or mission of the launch carrier vehicle. . . . Please be assured of the complete cooperation by the Air Force toward successful accomplishment of this amateur experiment. . . . (Signed) Joseph V. Charyk, Under Secretary of the Air Force."

. . . LIFT-OFF . . .

A brilliant flash of red-orange flame burst from the Discoverer. An awesome outpouring of sound marks the birth of space flight. The roar splits into frightful stridencies that beat upon the men as ocean waves attack the land with hurricane force. The red-orange ball of fire grows with astounding speed as the solemn silver shape rises on a plume of flame. Slowly, but with astounding acceleration, the flame grows, with the Discoverer at its head. The shouts of the observers are lost in the forest of noise. Now Discoverer is free of the land: It glories in its upward

flight . . . faster and faster . . . the track of flame marks its progress into the heavens . . . the program control starts to tilt the vehicle in the proper direction out over the Pacific Ocean . . . the teletype could once again be heard tapping out history . . . .
. . . GOING UP. . . .LOOKS GOOD. . . . STILL CLIMBING. . . . ON COURSE. . . . ON AZIMUTH. . . . ON COURSE. . . .

And so, on December 12, 1961, at 2042 GMT, Discoverer XXXVI was launched into orbit, carrying into separate orbit OSCAR I guided in its flight into history by the thoughts and prayers of thousands of radio amateurs who stand on the threshold of tomorrow.

The OSCAR I mission was a success in every respect. More than 570 amateurs in 28 countries forwarded observations to the Project OSCAR data reduction center, providing important information on radio propagation through the ionosphere, the spacecraft's orbit and satellite thermal design. OSCAR I clearly demonstrated that amateurs are capable of (1) designing and constructing reliable spacecraft, (2) tracking satellites and (3) collecting and processing related scientific and engineering information. Because of its low altitude, OSCAR I only remained in orbit for 22 days before burning up as it reentered the earth's atmosphere. For additional details of this spacecraft see notes 10 and 11, and Table 3.3 (which summarizes the amateur satellite program).

OSCAR II

OSCAR II was successfully launched on 2 June 1962, barely six months after OSCAR I. They were very similar, both structurally and electrically. Despite severe time pressures, however, results from the OSCAR I flight led to a number of improvements

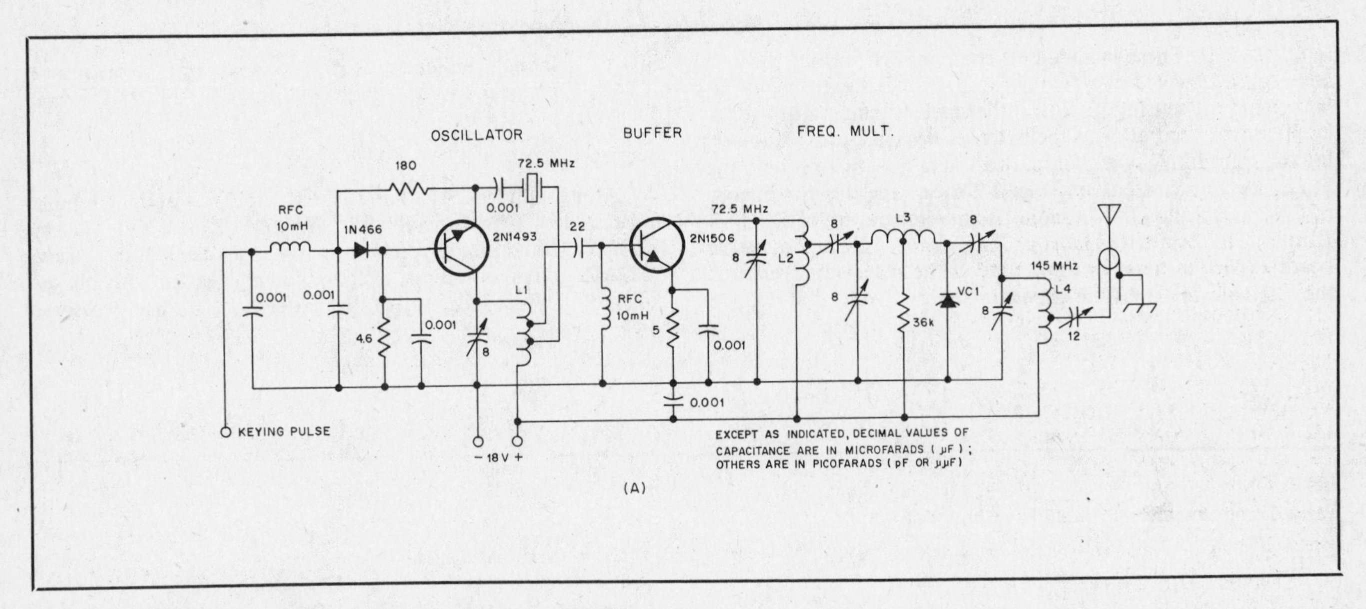

Fig. 2-1a — OSCAR I Beacon Transmitter: 140 milliwatts at 145 MHz.

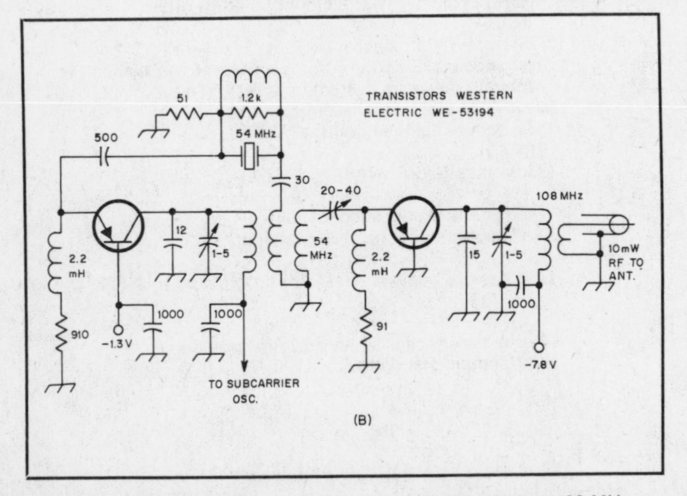

Fig. 2-1b — Explorer I Beacon Transmitter: 10 milliwatts at 108 MHz.

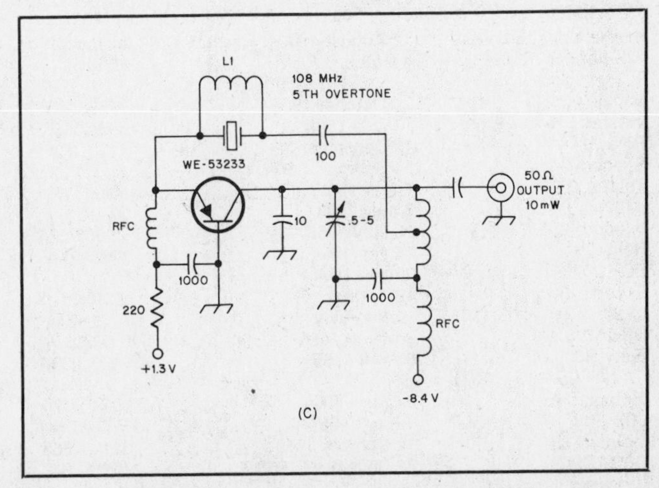

Fig. 2-1c — Vanguard Beacon Transmitter: 10 milliwatts at 108 MHz.

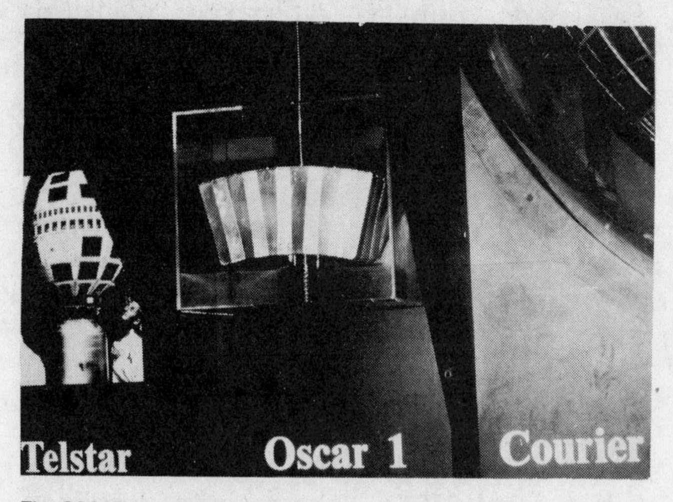

The OSCAR I that was launched in 1961 has long since burned on reentry through the atmosphere. This version, the actual backup to the first OSCAR, now resides in the Hall of Satellites at the Smithsonian's National Air and Space Museum in Washington, DC.

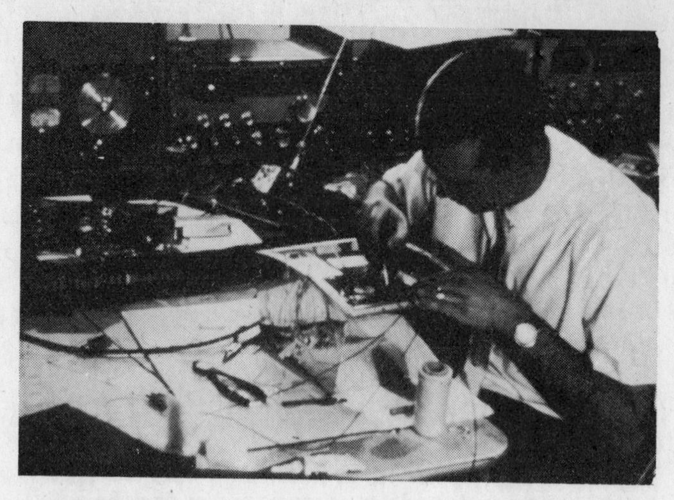

Lance Ginner, K6GSJ, prepares the first OSCAR in his garage workshop in California.

in OSCAR II. These included (1) changing the surface thermal coatings to achieve a cooler internal spacecraft environment, (2) modifying the sensing system so the satellite temperature could be measured accurately as the batteries decayed, and (3) lowering the transmitter power output to 100 milliwatts to extend the life of the onboard battery. Fig. 2-2 shows the thermal history of OSCARs I and II. The rapid rise in temperature of OSCAR II in its final orbits was most probably caused by aerodynamic heating (friction from air molecules) as the spacecraft reentered the atmosphere. The final telemetry reports from orbit 295, 18 days after launch, indicated an internal spacecraft temperature of 54° C; the outer shell was probably over 100° C by this time.[12]

OSCAR*

Along with OSCARs I and II, OSCAR* was designed, built and tested by Chuck Smallhouse (WA6MGZ) and Orv Dalton (K6VEY). Dimensionally, it was interchangeable with the earlier OSCARs, but it contained a 250-milliwatt beacon with phase-coherent keying. Because of the success of its predecessors,

Table 2.1
Early Communications Satellites (Comsats)

All carried active, real-time transponders except for SCORE, the ECHO "balloons," Courier 1B and the West Ford needles.

Satellite	*launch date*	*perigee/apogee (miles)*	*comments*
SCORE	18 December 1958	115/914	Often referred to as first comsat. However, it carried only a taped message for playback. It could *not* be used for relaying signals.
ECHO A-10	13 May 1960	—	Passive comsat (mylar balloon) failed to orbit (NASA)
ECHO 1	12 August 1960	941/1052	First successful passive comsat
Courier 1B	4 October 1960	586/767	First successful active comsat employed store-and-forward message system (non real-time)
Midas 4	21 October 1961	2058/2324	West Ford dipoles, failed to disperse
Telstar 1	10 July 1962	593/3503	First active real-time comsat (AT&T)
Relay 1	13 December 1962	819/4612	(RCA)
Syncom 1	14 February 1963	21,195/22,953	Electronics failure (NASA)
Telstar 2	7 May 1963	604/6713	
—	9 May 1963	2249/2290	West Ford dipoles, successful
Syncom 2	26 July 1963	22,062/22,750	First successful comsat in stationary orbit
Relay 2	21 Jan 1964	1298/4606	
ECHO 2	25 Jan 1964	642/816	Last passive comsat; first joint program with USSR
Syncom 3	19 August 1964	22,164/22,312	
LES 1	11 February 1965	1726/1744	
OSCAR 3	9 March 1965	565/585	First radio amateur active real-time comsat
Early Bird (INTELSAT I)	6 April 1965	21,748/22,733	First commercial comsat
Molniya 1A	23 April 1965	309/24,470	First Soviet comsat
LES 2	6 May 1965	1757/9384	
Molniya 1B	14 October 1965	311/24,855	
OSCAR 4	21 December 1965	101/20,847	First radio amateur high altitude comsat; partial launch failure

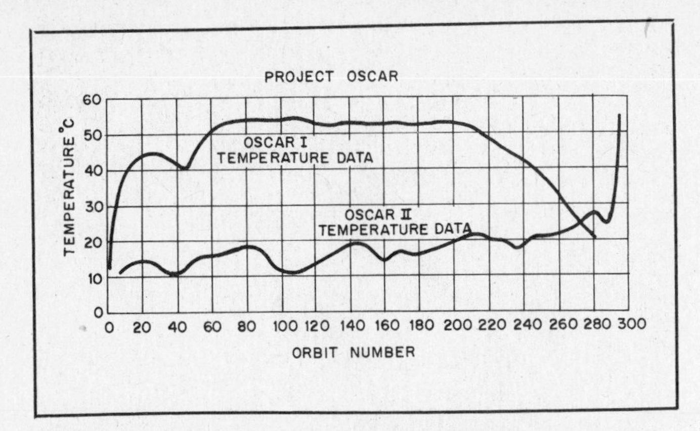

Fig. 2-2 — A comparison of the OSCAR I and OSCAR II temperature curves as derived from the telemetered data logged by nearly 1000 tracking stations. Modifications to the thermal conductivity of OSCAR II based on the flight of the first satellite provided a relatively constant package temperature until the satellite began to drop into the atmosphere of the earth at approximately orbit no. 288.

OSCAR* was never launched, as workers decided to focus their efforts on the first relay satellite — OSCAR III.[13]

What Price Success?

People often ask, "What did OSCAR I cost?" It's impossible to give a simple answer to this question. The most expensive commodity involved — the technical expertise of the radio amateurs who designed and built the spacecraft — was donated. Almost all the parts used were donated. Use of testing and machine shop facilities: donated. The main out-of-pocket expenses — long-distance phone calls, gasoline for local travel, technical books, etc., were absorbed by the volunteers. As satellites became more complex (e.g., OSCAR III) this situation had to change. In April 1962 the OSCAR Association formally incorporated as Project OSCAR, Inc. and began soliciting memberships nationally (and distributing a newsletter for interested experimenters) to help finance future satellites.

Space Communication I

During the early days of the Space Age, before the first active relay satellites were launched (see Table 2.1), the U.S. government was prudently investigating other space communication techniques. Two projects — ECHO and West Ford — were of special interest to amateurs. Project ECHO placed in orbit large (90 to 125-foot diameter) "balloons" with aluminized Mylar surfaces capable of reflecting radio signals. The West Ford project was an attempt to create an artificial reflecting band around the earth by injecting hundreds of millions of needle-like copper dipoles into orbit. Radio amateurs were quick to realize that Projects ECHO and West Ford were, by their nature, free access (i.e., anyone, anywhere could use the reflecting surface without requesting permission of the U.S. government). As always, amateurs were willing to try bouncing signals off almost anything — a large balloon, the moon, newly discovered scientific phenomena such as ionized trails left by satellites, or an Amateur Radio spacecraft.[14,15]

Project ECHO. ECHO A-10, the first in the series, never attained orbit. ECHO 1, which followed, though used successfully for communication by high power, non-amateur experimenters,

Data processing in days of yore. W6HEK, W6MKE and K6BHN coded the OSCAR II telemetry data onto punched cards for processing on the IBM computer surrounding them.

The raw data, punched cards, printouts and permanent magnetic-tape storage were state of the art in the early '60s. Today, the same analyses are performed on personal computers and programmable calculators by individual satellite users throughout the world.

Chuck Towns, K6LFH, in his own garage workshop with OSCAR II. Though the Amateur Radio Satellite Program had its roots in basements and garages, the strictest professional standards were always maintained for the final spacecraft to pass rigorous testing by the various launch agencies.

did not enable amateur communications. Interest continued because ECHO 2 looked more promising as a reflector; it was to be larger and lower than its predecessor.[16] Radio path loss calculations at 144 MHz suggested that communications might be possible by amateurs running the legal power limit (1 kW) and using large antennas. The launch of ECHO 2, originally planned for 1962, didn't occur till 1964. In the interim, both radio amateurs and the government tested their first active relay spacecraft. The overwhelming success of active relays led to the demise of Project ECHO, but not before return signals were obtained from ECHO 2 at 144 MHz by Bill Conkel (W6DNG) and Claude Maer (WØIC).[17] Rapid fading and weak signals, however, prevented two-way communications. Meanwhile, radio amateurs were refocusing their interest on other passive reflecting surfaces — the West Ford needles and the moon.

Project West Ford. Because a mechanical malfunction occurred in the dipole ejection mechanism, the first West Ford mission (October 1961) was a failure. A second test in 1963 successfully demonstrated that a belt of needles could support communication between very high power (far above amateur levels) ground stations, though the needles decayed from orbit much faster than expected. The program was discontinued because several scientific organizations seriously warned against the possible undesirable side effects of Project West Ford on future active satellite relays, the manned space program, radio astronomy and even the weather. Also, by this time, the advantages of active satellite relays had been demonstrated sufficiently.

Moonbounce. Radio amateurs have successfully communicated by using the moon, a natural satellite of earth, as a passive reflector on 50, 144, 220, 432, 1296 and 2304 MHz. Although moonbounce communication, often called EME (Earth-Moon-Earth), has always taken the highest allowable power, large antennas and super receivers, it continues to have a special attraction to radio amateurs. Today, most EME activity is concentrated on 144 MHz and 432 MHz. Signals are weak at best, but system performance seems to improve continually and ssb two-way contacts on 432 MHz are not uncommon. An early 1980 EME newsletter listed 68 stations, at least one from each continent, currently capable of two-way moonbounce communication on 432 MHz. Slight improvements in signal-to-noise ratios through small cumulative technical advances may one day change EME from a marginal mode to a highly reliable one.[18]

Satellite Scatter. Another, not very well known, space communications medium was investigated by amateurs at about this time. In 1958 Dr. John Kraus (W8JK), director of the Ohio State University Radio Observatory, noted that certain terrestrial hf beacon signals increased in strength and changed in other characteristic ways as low-altitude satellites passed nearby. He attributed the enhancements to reflection off a trail of short- lived ionized particles caused by the passing spacecraft.[19] Capitalizing on this effect, amateurs were able to locate (or confirm the position of) several silent (non-transmitting) U.S. and Soviet satellites by monitoring signals from WWV.[20]

Two electrical engineering students, Perry Klein (W3PK) and Ray Soifer (W2RS), read Kraus's work and decided to see if the effect — High-Frequency Satellite Scatter — would support communication. Calculations showed that 21 MHz was the optimal amateur frequency for tests. Their positive results received national publicity in the news media, but signals using amateur power levels proved only marginal for practical communications purposes.[21]

OSCAR III

Even as OSCAR II lifted off the launch pad, work was underway on OSCAR III, a far more complex satellite with the communications capabilities Don Stoner had dared speculate about years earlier. OSCAR III carried a 50-kHz-wide, 1-watt transponder that received radio signals near 146 MHz and retransmitted them, greatly amplified, back to earth near 144 MHz. The transponder was designed so that it would enable radio amateurs with modest equipment (normally only effective over distances under 200 miles) to communicate over paths ranging up to 3000 miles. In addition to the transponder, OSCAR III contained two beacon transmitters. One provided a continuous carrier for tracking and propagation studies, the other telemetered three critical spacecraft parameters: temperature and terminal voltage of the main battery and temperature of the transponder's final amplifier.[22-25]

OSCAR III in full-dress. Note that the two 2-m dipole antennas are constructed of flexible steel carpenter's rule material. The dark "checkerboard" areas are the solar-cell panels that are used as a battery backup and the springs shown on the top face were used to separate the spacecraft from the launch vehicle.

Because of their low initial orbits OSCARS I and II remained in space only a short time before reentering the atmosphere and burning up. A simple battery was therefore an adequate power supply to support these spacecraft for the expected mission duration. OSCAR III, however, was being placed in a higher orbit, where it would remain considerably longer. Since weight constraints severely limited the spacecraft's battery complement, consideration was given to using solar cells on this mission. Their cost and availability, and the additional complexity required of the spacecraft, precluded this approach. Nonetheless, a small bank of solar cells was used to back up the battery that powered the beacons. OSCAR III was the first amateur spacecraft to employ solar power, though only to a limited extent. To give some

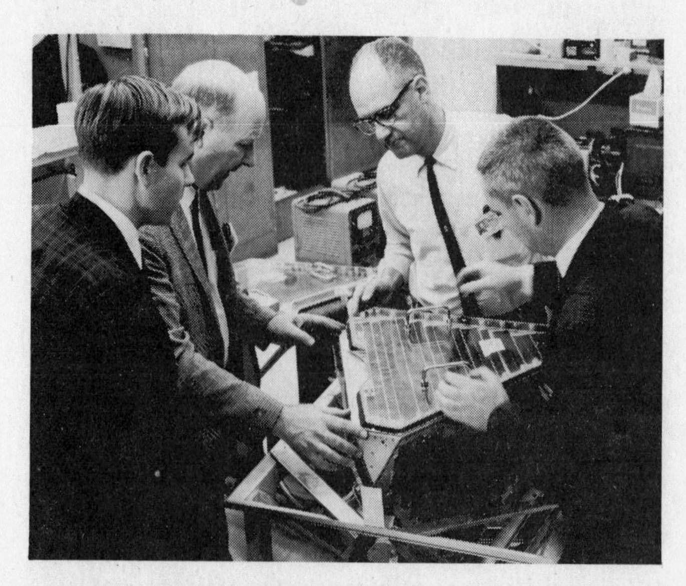

Radio Amateur Club of TRW members K6MWR, Dave Moore, W6ZPX and W6RTG make the final adjustments to the OSCAR IV satellite package, here mounted in its launch cradle.

perspective to this achievement, it should be noted that solar cells were then recent technology, having been invented only in 1954.[26]

Following the successful launch of OSCAR III (9 March 1965), the transponder operated for 18 days, during which time about 1000 amateurs in 22 countries were heard through it. A number of long-distance communications were reported, including USA (Massachusetts) to Germany, USA (New Jersey) to Spain, and New York to Alaska. The transponder clearly demonstrated that the concepts of free-access and multiple-access satellites would work. *Free-access* means that anyone licensed by his government may uplink through the spacecraft without charge and without prior notification. *Multiple-access* means that a large number of ground stations can use the spacecraft simultaneously if they cooperate in choosing frequencies and limiting power levels. The telemetry beacon, working off its own battery and the solar cells, functioned for several months.[27]

Success was to bring new challenges. It was clear that unless radio amateurs just wanted to replay yesterday's triumphs, their future satellites would need major changes. First and foremost, operating lifetimes would have to increase by 10 or 100 times to justify the major effort and expense needed to build the sophisticated spacecraft designs being considered.

OSCAR IV, the only amateur satellite to be designed around a tetrahedral frame. Intended for a 21,000-mile high circular orbit, OSCAR IV was doomed to a short operating life when the top stage of the launch rocket failed, leaving it in an elliptical orbit for which it had not been designed.

OSCAR IV

While OSCAR III was being completed, amateurs were presented with a launch opportunity aboard a Titan III-C rocket headed for a circular orbit 21,000 miles above the earth. At this height almost half the planet would be within range of the spacecraft at any time. OSCARs I, II and III had been designed to operate at lower altitudes (under 700 miles) where a transponder's output power, spacecraft antennas and attitude stabilization are much less critical. Building a spacecraft for a higher orbit, even with plenty of time and resources, is a formidable challenge. But, time wasn't available: The projected launch date was roughly a year away and the Project OSCAR crew was deeply involved in readying OSCAR III for its flight. It appeared that this once-in-a-lifetime offer might have to be passed up. Several members of the TRW (Thompson-Ramo-Woolridge) Radio Club of Redondo Beach, California, recognized the uniqueness of the opportunity and decided to undertake the project even though the constraints seemed overwhelming. (Nearly two decades have passed since the launch of OSCAR IV, and a similar launch opportunity has never again materialized.)

To meet the time schedule, the spacecraft would have to be kept as simple as possible — just a transponder and an identification beacon to satisfy Federal Communications Commission requirements. "Luxuries" such as telemetry and redundant subsystems for reliability had to be eliminated. The TRW team did, however, decide that the spacecraft would be solar powered and designed the system with a one-year-lifetime goal. The crossband transponder received (uplink) on 144 MHz and transmitted (downlink) on 432 MHz. Its design borrowed a number of ideas from a standard ranging transponder NASA used at the time. Power was set at 3 watts PEP and the bandwidth was 10 kHz.

The fear that constantly haunts satellite builders came partially true with OSCAR IV — the top stage of the launch rocket failed and the spacecraft never reached the targeted orbit. Had it achieved the intended orbit, it would have hung directly over the equator, drifting slowly eastward at just under 30° per day. Instead, OSCAR IV entered a highly elliptical orbit inclined to the equator at 26°; the high point (apogee) was 21,000 miles above earth and the low point (perigee) about 100 miles.

The rocket failure presented amateurs with a number of serious problems. Consider tracking, for example. Because of the spacecraft's height and low-power transmitter, ground stations needed high-gain (narrow-beamwidth) antennas. With the planned orbit, antenna aiming would have been simple, but with the actual orbit, it was nearly impossible. Suitable techniques for tracking satellites in highly elliptical orbits hadn't been devised. Because the downlink signal strength fluctuated rapidly, even scanning the receive antenna to peak signals didn't work well. This latter problem was closely related to the rocket malfunction. The attitude stabilization scheme chosen for OSCAR IV and the antenna configuration were based on the spacecraft's being spun off correctly from the top stage of the launch vehicle. Rocket failure meant loss of planned spin stabilization, and consequently inadequate control over the antenna's orientation. Amateurs attempting to use the transponder encountered additional difficulties that some attributed to spacecraft electronics. It's probable, however, that many of these difficulties were from the lack of attitude stabilization and could have been overcome had stabilization been achieved.

Had OSCAR IV operated a sufficient length of time, the attitude probably would have stabilized naturally (though not necessarily in a preferred orientation) and ground stations would have devised methods to overcome the tracking and radio link problems by, for example, agreeing to specified uplink power levels and using circular polarization. But, the transponder ceased operating after a few weeks. Since the spacecraft didn't have a telemetry system, we can only guess the cause: either battery failure from thermal and power supply stresses, or solar cell failure from the radiation levels encountered, both possibilities arising from the unexpected orbit.

Even with the enormous difficulties encountered, several amateurs completed two-way contacts through OSCAR IV. One contact, between a station in the United States and another in the Soviet Union, was the first direct two-way satellite communication between these two countries. The mission provided considerable information that would prove valuable in designing future spacecraft.

Although OSCAR IV was, in some respects, a major disappointment, it's important to keep in mind that the key failure occurred in the launch vehicle, a possibility that radio amateurs working with satellites must learn to accept. The amateurs who designed and built OSCAR IV did an extraordinary job, and the users whose ingenuity salvaged so much from the mission were a credit to Amateur Radio.[28-30]

Notes

[1] G. S. Sponsler, "Sputniks Over Britain," *Physics Today,* Vol. 11, no. 7, July 1958, pp. 16-21. Reprinted in *Kinematics and Dynamics of Satellite Orbits,* American Association of Physics Teachers, 335 East 45th St., NY, NY 10017. ($1)

[2] F. L. Whipple and J. A. Hynek, "Observations of Satellite I," *Scientific American,* Vol. 197, no. 6, Dec. 1957, pp. 37-43.

[3] R. Buchheim and Rand Corp. Staff, *New Space Handbook* (New York: Vintage Books, 1963), pp. 283-312.

[4] From a condensed translation of the June 1957 article in the Soviet journal *Radio:* V. Vakhnin, "Artificial Earth Satellites," *QST,* Nov. 1957, pp. 22-24, 188.

[5] From a presentation by Dr. Van Allen at a special program commemorating the 20th anniversary of the launch of Explorer I. Held at the National Academy of Sciences, Washington, DC, Feb. 1, 1978.

[6] A discussion of Dr. Reber's work and original references can be found in J. D. Kraus, *Radio Astronomy* (New York: McGraw-Hill, 1966, Chapter 1). Dr. Reber is currently living in Bothwell, Tasmania. Dr. Kraus, director of the Ohio State University Radio Observatory, is W8JK.

[7] E. P. Tilton, "Lunar DX on 144 Mc.," *QST,* March 1953, pp. 11-12, 116.

[8] D. Stoner, "Semiconductors," *CQ,* April, 1959, p. 84.

[9] W. I. Orr, "Sixty Years of Radio Amateur Communication," *QST,* Feb. 1962, pp. 11-15, 130, 132.

[10] H. Gabrielson, "The OSCAR Satellite," *QST,* Feb. 1962, pp. 21-24, 132, 134. Technical description of OSCAR I.

[11] W. I. Orr, "OSCAR I: A Summary of the World's First Radio-Amateur Satellite," *QST,* Sept. 1962, pp. 46-52, 140.

[12] W. I. Orr, "OSCAR II: A Summation," *QST,* April 1963, pp. 53-56, 148, 150.

[13] See the reference in note 12, p. 56.

[14] R. Soifer, "Space Communication and the Amateur," *QST,* Nov. 1961, pp. 47-50. The references in notes 14 and 15 treat basic concepts in a comprehensive manner, and the information contained is still of interest to experimenters involved in radio astronomy, direct reception from lunar and deep space probes, and reception of commercial satellite TV.

[15] R. Soifer, "The Mechanisms of Space Communication," *QST,* Dec. 1961, pp. 22-26, 168, 170.

[16] R. Soifer, "Amateur Participation in ECHO A-12," *QST,* April 1962, pp. 32-36. Note: ECHO 2 was known as ECHO A-12 before launch. R. Soifer, "Project ECHO A-12," *QST,* June 1962, pp. 22-24.

[17] R. Soifer, "Amateur Radio Satellite Experiments in the Pre-OSCAR Era," *Orbit,* Vol. 2, no. 1, Jan./Feb. 1981, pp. 4-7.

[18] For information on the history of EME see:
W. Orr, "Project Moon Bounce," *QST,* Sept. 1960, pp. 62-64, 158; F. S. Harris, "Project Moon Bounce," *QST,* Vol XLIV, no. 9, Sept. 1960, pp. 65-66; H. Brier and W. Orr, *VHF Handbook for Radio Amateurs* (Wilton, Conn: Radio Publications, 1974); T. Clark, "How Diana Touched the Moon," *IEEE Spectrum,* May 1980, pp. 44-48.

[19] J. D. Kraus, R. C. Higgy, and W. R. Crone, "The Satellite Ionization Phenomenon," *Proc. IRE,* Vol. 48, April 1960, pp. 672-678.

[20] C. Roberts, P. Kirchner, D. Bray, "Radio Detection of Silent Satellites," *QST,* Aug. 1959, pp. 34-35.

[21] R. Soifer, "High-Frequency Satellite Scatter," *QST,* July 1960, pp. 36-37. R. Soifer, "Satellite Supported Communication at 21 Megacycles," *Proc. IRE,* Vol. 49, no. 9, Sept. 1961.

[22] W. I. Orr, "The OSCAR III V.H.F. Translator Satellite," *QST,* Feb. 1963, pp. 42-44.

[23] A. M. Walters, "OSCAR III — Technical Description," *QST,* June 1964, pp. 16-18.

[24] A. M. Walters, "Making Use of the OSCAR III Telemetry Signals," *QST,* March 1965, pp. 16-18.

[25] W. I. Orr, "OSCAR III Orbits the Earth!," *QST,* May 1965, pp. 56-59.

[26] D. M. Chapin, C. S. Fuller, G. L. Pearson, "A New Silicon p-n Junction Photocell for Converting Solar Radiation into Electrical Power," *J. Applied Physics,* Vol. 25, May 1954, p. 676.

[27] H. C. Gabrielson, "OSCAR III Report — Communications Results," *QST,* Dec. 1965, pp. 84-89.

[28] "OSCAR IV News," *QST,* Dec. 1965, p. 41.

[29] "OSCAR IV Due Dec. 21," *QST,* Jan 1966, p. 10.

[30] E. P. Tilton and S. Harris, "The World Above 50 Mc.," *QST,* Feb. 1966, pp. 80-82.

Chapter 3

Past/Present/Future

After the disappointment over OSCAR IV, amateurs were to be treated to a string of amateur spacecraft that not only met, but exceeded, expectations. Australis-OSCAR 5, designed and built at the University of Melbourne, worked almost flawlessly, while AMSAT-OSCARs 6, 7 and 8 brought reliable two-way satellite communication to amateurs and students around the world.

OSCAR 5

The OSCAR 5 story begins in Australia. Late in 1965 several students at the University of Melbourne, mostly undergraduate members of the Astronautical Society and Radio Club, seriously began to consider building a satellite. Though none of them had any spacecraft construction experience, they were competent in electronics and mechanical design. When the California-based Project OSCAR agreed to take care of final environmental testing, locating a launch and launch operations for Australis-OSCAR 5 (A-O-5), the "down under" crew began the project in earnest. (Note: With the fifth amateur satellite being readied for flight, amateurs decided to acknowledge the advantage of Arabic numerals over their Roman counterparts — hence OSCAR 5, *not* OSCAR V.)

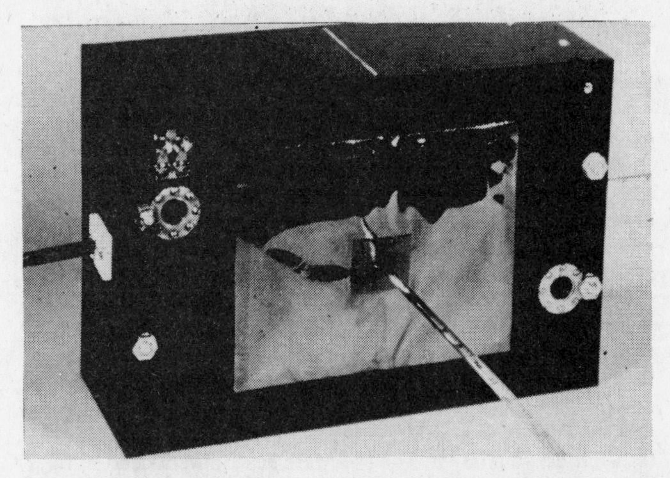

Australis-OSCAR 5 shown with its antennas deployed. Note that this satellite carried no solar cells and that actual steel carpenter rule was used for the antenna elements.

Members of the Melbourne group wanted to make a unique and significant contribution to the amateur space program but they recognized that their isolation and lack of experience dictated a relatively simple spacecraft. The design, finalized in March 1966, showed that their desire and the real constraints were compatible. A-O-5 would attempt to: (1) evaluate the suitability of the 10-m band for a downlink on future transponders; (2) test a passive magnetic attitude stabilization scheme; and (3) demonstrate the feasibility of controlling an amateur spacecraft via uplink commands. The flight hardware to accomplish these goals included telemetry beacons at 144.050 MHz (50 mW) and 29.450 MHz (250 mW at launch), a command receiver and decoder, a seven-channel analog telemetry system, and a simple manganese alkaline battery power supply. The spacecraft did *not* contain a transponder or use solar cells.

Though technical aspects of the A-O-5 project went smoothly, they turned out to be just the tip of the project's iceberg; administrative concerns were a constant frustration. Airposting a special 50-cent part from the U.S. to Australia might cost $10, and clearing the part through customs often required pages of paperwork and several trips to government offices. You probably get the picture: Technical competence isn't enough. People who build satellites also need great perseverance. Step-by-step, Australian dollar by Australian dollar, A-O-5 took shape. On June 1, 1967, 15 months after final plans were okayed, the completed spacecraft was delivered to Project OSCAR in California. A launch opportunity was targeted for early 1968. Delay followed delay, however, until the host mission was indefinitely postponed. No other suitable launch was immediately available.

So stood the situation in January 1969 when George Jacobs (W3ASK) spoke to the COMSAT Amateur Radio Club in downtown Washington, DC. Jacobs suggested that, with the space-related expertise and facilities in the area, the amateur space program might benefit from an East Coast analog of Project OSCAR. As a result, AMSAT (the Radio Amateur Satellite Corporation) was founded. Formal incorporation took place on March 3, 1969, in Washington, DC, and the first task of the new organization was arranging for an Australis-OSCAR 5 launch.

Environmental and vibration tests of A-O-5 showed that some minor changes were needed. AMSAT performed the modifications and identified a suitable host mission. Finally, on January 23, 1970, A-O-5 was launched on a National Aeronautics and Space Administration (NASA) rocket (previous OSCARs had all flown with the U.S. Air Force). Electronically the satellite performed almost flawlessly. One small glitch prevented telemetry data from being sent over the 29-MHz beacon. Since the same telemetry information was available on 144 MHz, the problem had little impact on the overall success of the mission. The magnetic attitude stabilization system worked beautifully. The spacecraft's spin rate decreased by a factor of 40 — from 4 revolutions per minute to 0.1 revolution per minute — over the first

Australis-OSCAR 5. In this view, the flexible antennas have been tied back, where they will remain during launch. At the proper time, as the satellite is separated from its launch vehicle, these elements will spring out to their full pre-cut length.

two weeks. A network of ground stations periodically transmitted commands to the satellite, turning the 29-MHz beacon on and off. Allowing the beacon to operate only on weekends helped to conserve the limited battery power. The first successful command of an amateur satellite took place on orbit 61, on January 28, 1970, when the 29-MHz beacon was turned off. The demonstration of command capabilities was to prove very important in obtaining FCC licenses for future missions.

Performance measurements of the 29-MHz beacon confirmed hopes that this band would prove suitable for transponder downlinks on future low-altitude spacecraft, and led to its use on OSCARs 6, 7 and 8. As the battery became depleted, the transmitters shut down: The 144-MHz beacon went dead 23 days into the mission, and the 29 MHz beacon, operating at greatly reduced power levels, was usable for propagation studies until day 46.[1-5]

At AMSAT, the project manager responsible for final testing, modification, and integration of A-O-5 was a young engineer named Jan King. It's hard for people not directly involved in a project of this scope to imagine the pressure on the project manager. But Jan must not have minded too much, as he went on to oversee the design and construction of AMSAT-OSCARs 6, 7 and 8, and AMSAT Phase III-A, B and C.

A-O-5 met its three primary mission objectives. In addition, careful analysis of reports submitted by ground stations that monitored the mission showed that such stations were capable of collecting reliable quantitative data from a relatively complex telemetry format. All in all, A-O-5 was a solid success. But radio amateurs wanted a transponder they could use for two-way communication and five years had passed since the last one had orbited.

Space Communication II

Deep space probes. While waiting for the next active relay satellite, radio amateurs experimented in related areas. A few constructed 2.3-GHz (S-band) microwave receiving stations to monitor the Apollo 10, 12, 14 and 15 lunar flights. During the Apollo 15 mission (August 1971) amateurs received voice transmissions from the Command Service Module as it circled the moon.[6] Although the S-band radio frequency equipment needed for monitoring manned deep-space probes is similar to that needed for listening to unmanned flights, efforts have focused on the former. Probably, this is because decoding the voice channels is much easier (and more exciting?) than extracting information from the sophisticated telemetry links. Though the late 1970s brought a long lull in the manned space program, the Space Shuttle project has generated a new spurt of activity. It's not too early to start thinking about directly monitoring the first humans landing on Mars.

ATS-1. The U.S. government launched ATS-1 (Applications Technology Satellite) into a geostationary orbit on December 7, 1966. Satellites in such an orbit appear to remain fixed above a particular spot on the equator. Of interest to radio amateurs was an experimental 100-kHz-wide, hard-limiting transponder carried by ATS-1 that received near 149.22 MHz and retransmitted at 135.6 MHz. Professor Katashi Nose (KH6IJ) of the University of Hawaii was one of the scientists working with NASA to evaluate this system. By monitoring the transponder operation, radio amateurs could learn a great deal about the performance of radio links to geostationary satellites near the 144-MHz amateur band. Amateurs were also interested in studying the performance of the hard-limiting transponder with a view toward using similar devices on future amateur missions.[7]

As of the early '80s, after more than 15 years in orbit, ATS-1 was still operational.[8] But this isn't a record. Relay I, launched in 1962 (see Table 2-1) is still often heard on 136.140 MHz and 136.620 MHz. Today's commercial satellites are being designed with projected 10-year lifetimes. Recent amateur spacecraft have shown that we should be able to obtain similar lifespans. When this occurs it may be necessary to shut off functioning older spacecraft so the uplink and downlink frequencies can be used by newer, more versatile, ones.

Modifying the ionosphere. Radio amateur interest in projects ECHO and West Ford (see Chapter 2) focused on reflecting radio signals off objects launched into space. A closely related class of experiments involves direct physical modification of the ionosphere to change its radio-reflecting characteristics. Two approaches that have received a great deal of attention involve (1) releasing chemicals, such as barium, from rockets directly into the ionosphere[9] and (2) employing very-high-power, ground-based radio transmitters, operating in the vicinity of 3-10 MHz, to produce an "artificial radio aurora" (raise the temperature of electrons in the ionosphere).[10,11] Both types of experiments are expected to continue through the '80s. The main payload on the ill-fated AMSAT Phase III-A mission, for example, was a barium-release experiment known as Firewheel.[12]

An artificial radio aurora has marked effects on propagation over the range of 20-450 MHz. The major facility for ionospheric heating, located in Plattsburg, Colorado, USA, began operating in 1970. Amateur experiments with this communications medium first took place in 1972 and are continuing. During the week of March 17, 1980, for example, the Plattsburg heater was scheduled for 20 hours of operation. Though articles suggest that one should be located within 800 miles of Plattsburg to take part in these experiments, the distance can be greatly extended by studying satellite links.

AMSAT-OSCAR 6

Amateur radio took a giant stride into the future on October 15, 1972, when AMSAT-OSCAR 6 (A-O-6) was launched successfully. Although it was more complex than all previous OSCARs combined, ground stations interested in communicating through its transponder or studying its telemetry found A-O-6 to be the easiest amateur satellite to work with. Phase II of the amateur satellite program, the age of long-lifetime satellites, was underway. While the aggregate operational time of all previous OSCARs amounted to considerably less than one year, A-O-6 was to "do its thing" for more than 4.5 years. From October 15, 1972, forward, the Amateur Radio community would have at least one transponder-equipped low-altitude satellite in operation.[13]

To understand the significance of A-O-6, we must go beyond the impressive facts and figures and look at the philosophy underlying its construction. Two ideas were central. First, the investment needed both in dollars and effort to produce spacecraft that could make significant new scientific, engineering or operational contributions to Amateur Radio was such that only long-life (at least one-year duration) satellites could be justified. Second, constructing a reliable long-life satellite required much more than replacing a battery power supply with a power system consisting of solar cells, rechargeable batteries and related control electronics. Long lifetime could be reasonably assured only if the spacecraft contained (1) a sophisticated telemetry system permitting onboard systems to be monitored, (2) a flexible command system so that various spacecraft subsystems could be activated or deactivated as conditions warranted, and (3) redundancy in critical systems. In addition, the design strategy must attempt to prevent catastrophic failure by anticipating possible failure modes and incorporating facilities for isolating defective subsystems.

AMSAT-OSCAR 6 wasn't the first amateur spacecraft to use solar power or include command and telemetry systems. But it brought each of these subsystems a quantum jump forward. For example, the command system on A-O-5 could only turn the 10-meter beacon on or off. A-O-6 recognized 35 distinct commands, 21 of which were acted on. The most sophisticated telemetry system used previously (on OSCAR 5) included seven analog channels. The A-O-6 downlink contained 24 telemetry channels. Furthermore, the spacecraft carried a newly designed processing system that greatly simplified telemetry decoding equipment requirements: Ground stations had only to copy numbers in Morse code and refer to a set of graphs.[14]

AMSAT-OSCAR 6 carried a 100-kHz-wide transponder running about 1-watt output at 29 MHz, the frequency tested as a downlink on A-O-5. The transponder was extremely sensitive. Ground stations running as little as 10 watts to a ground-plane antenna on the 146-MHz uplink would put through solid signals as long as the transponder wasn't fully loaded, or gain-compressed by users running excessive power. The ease and reliability of communicating through the transponder were enhanced greatly by the spacecraft's magnetic attitude-stabilization system, another feature pioneered on A-O-5. A-O-6 also carried a unique digital store-and-forward message system called Codestore. Suitably equipped ground stations could load messages into Codestore using Morse code (or any other digital code conforming to FCC regulations) for later playback, either continuously or on command. Codestore was often used to relay messages between Canadian and Australian command stations, and many radio amateurs outside the USA depended on Codestore for pre- and post-launch information relating to AMSAT-OSCAR 7.

Although AMSAT-OSCAR 6 turned out to be an overwhelming success, it did have problems. The 435-MHz beacon failed after about three months. It lasted long enough, however, to test 435-MHz as a downlink for future low-altitude spacecraft and to enable John Fox (WØLER) and Ron Dunbar (WØPN) to discover an interesting Doppler anomaly.[15] A second problem, one of major importance, involved mode falsing. The satellite control system turned out to be very susceptible to internally generated noise. Noise would often be interpreted as a command and turn the transponder and other subsystems on or off. Working with A-O-6 in those early months was very frustrating. It seemed as if the transponder would regularly choose the most inopportune times to shut down.

The solution to the falsing problem is an interesting story. Although the orbiting spacecraft couldn't be repaired, it was suggested that the difficulties would be minimized if a constant stream of commands directing it into the correct mode was sent to the satellite. To make this idea work, automated ground command stations had to be developed and a number of stations around the world would have to volunteer, often at significant personal cost (in time and cash), to accept responsibility for building and operating these command stations. Larry Kayser (VE3QB) was among the first to feverishly attack the falsing problem. He quickly put a command station in operation, automating it, after a fashion, with a tape-recorder control loop connected to his telephone. By ringing up his home phone he would activate the system. As Kayser tells it:

AMSAT-OSCAR 6, mounted to its launch vehicle's attach fitting, only a few hours from launch. Note the flexible "carpenter-rule" 10-m antenna elements on each side that are tied back for launch, and the tiny gold-plated piano-wire 70-cm whip antenna atop the spacecraft. After launch, upon separation from the launch vehicle, these pre-cut antennas were freed from their restraints and sprang to operating length. (The Plexiglas cover shown mounted over the front solar panel is there to protect the delicate solar cells while AMSAT and NASA personnel work around the spacecraft; it was removed prior to launch.)

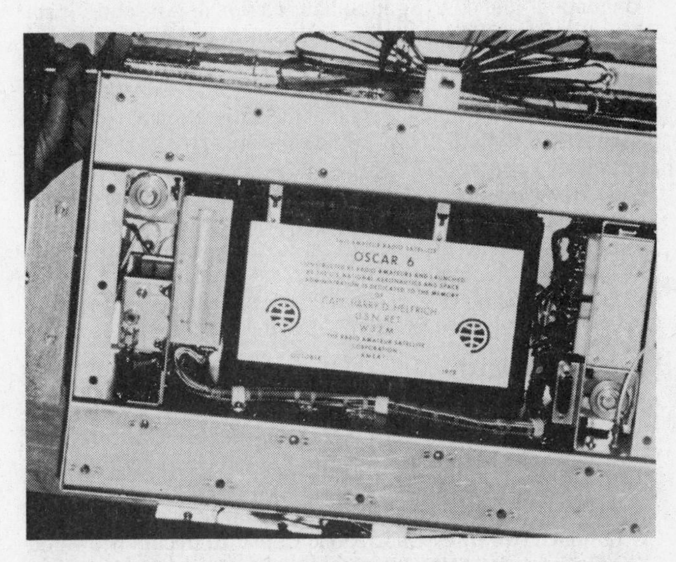

AMSAT-OSCAR 6, as shown on the plaque mounted inside the satellite structure, was dedicated to Capt. Harry D. Helfrich, W3ZM, an active AMSAT participant in the OSCAR 6 project who became a silent key shortly before the satellite was launched.

> For the next few weeks, it was not uncommon [for me] to dash for a telephone, dial a number, and hang up. This went on several times in a 10-minute period for each pass, sometimes from Montreal, Toronto, a gas station on the highway, or wherever... Full automation was certainly a more desirable way to go.

Kayser went on to design a series of systems, each one buying a little more time so that the satellite could be kept operating while more reliable techniques were developed. By August 1973, a system capable of automatically generating 80,000 commands per day was in operation. This was quite a change from the twice-weekly commands used to control the 29-MHz beacon on A-O-5.[16]

In addition to keeping A-O-6 on a reliable schedule, command stations were largely responsible for the spacecraft's 4.5-year lifespan. Without their careful management, it's doubtful that even the original one-year intended design lifetime of the spacecraft could have been reached. A-O-6 died when several battery cells failed (shorted) during its fifth year in orbit.

Subsystems for the AMSAT-OSCAR 6 spacecraft were built in the U.S., Australia and West Germany. Ground command stations were activated in Australia, Canada, Great Britain, Hungary, Morocco, New Zealand, the U.S. and West Germany. Users in well over 100 countries reported two-way communications.[17]

Though A-O-6 was awarded a free ride into space for several reasons, its potential value as an educational tool was paramount. The introduction of long-lifetime radio amateur satellites made it feasible for science instructors at all levels to incorporate class demonstrations of satellite reception into regular course work. To assist teachers pioneering this path, AMSAT and the American Radio Relay League granted funds to the Talcott Mountain Science Center in Connecticut to produce an instruction manual aimed at educators working with grades 1 through 12. The result was the well received *Space Science Involvement* manual first published in 1974. Thousands of free copies were distributed to teachers over the following six years. In 1978 a follow-up publication geared to college level instruction was published. *Using Satellites in the Classroom: A Guide for Science Educators* was produced with the financial assistance of the National Science Foundation and the Smithsonian National Air and Space Museum. The OSCAR education program includes additional activities supported by the ARRL in a continuing program of local assistance referral, personalized educational bulletins via satellite, special satellite scheduling, the publication of newsletter updates for science educators, development of a slide show library, and so on.

The outstanding OSCAR education program is probably, by itself, sufficient justification for free launches. Our emphasis on it here, however, should in no way be construed as downgrading the significance of OSCAR contributions in other areas such as emergency communications, scientific exploration and public service, which are discussed elsewhere in this book.

Jan King, W3GEY, adjusts OSCAR 7 on its perch atop a "shake-table." All OSCARs must undergo rigorous testing to prove that they will survive the rigors of launch and the hostile space environment without damaging or otherwise affecting the mission of the primary payload. The shake-table, a distant relative of your local hardware store's paint-shaker, is used for vibration tests in which the structure is subjected to the severe vibrations that will be experienced during launch. Secondary payloads, the "piggy-back riders" of the aerospace world, aren't certified for flight until they have passed such tests.

Perry Klein, W3PK, former President of AMSAT, stands in front of OSCAR 7 before closing the door to the thermal-vacuum test chamber. One of the many tests that the OSCARs must pass, the thermal-vacuum test measures a structure's cleanliness in the harsh space environment. The spacecraft, in a high vacuum, is heated to the very high temperatures it will encounter in space (in a sense boiling off impurities into their gaseous state) for several days. Then a super-cold "cold finger" (a special thermal probe) is activated and the gaseous impurities condense on its surface, where they can be measured quantitatively. Other phases of the test include several days at "room temperature" and several days at the extreme cold temperatures that the satellite will experience in space.

AMSAT-OSCAR 7

November 15, 1974 marked the beginning of another success story. AMSAT-OSCAR 7 (A-O-7) was launched and, for the first time, amateurs had two operating satellites in orbit. While A-O-6 represented a quantum leap forward technically, A-O-7 was more of an evolutionary step in technical improvement.[18] It contained two transponders, one similar to the unit flown on A-O-6 using a 146-MHz uplink and a 29-MHz downlink (known as Mode A), and the second with an uplink at 432 MHz and a downlink at 146 MHz (known as Mode B). The Mode B transponder was based on a unique design developed by Dr. Karl Meinzer (DJ4ZC). Running 8W (PEP), it featured a highly efficient method of linear frequency translation. Built in West Germany under the sponsorship of AMSAT-Deutschland, the Mode B transponder (in concert with the frequencies used, the antenna system and the magnetic attitude control) provided outstanding performance. Whereas A-O-6 demonstrated that simple grounds stations could communicate via satellite, A-O-7 showed that low-altitude satellites could, under many conditions, provide simple stations with communications capabilities over moderate distances (200-4500 miles) far exceeding any alternative mode.

The AMSAT-OSCAR 7 spacecraft carried Codestore and telemetry units nearly identical to those of A-O-6. It also contained a new high-speed, high-accuracy telemetry encoder (designed by an Australian group) that transmitted radioteletype. Beacons at 146 MHz, 435 MHz (built in Canada), and 2304 MHz were also flown. The 100-mW, 2304-MHz beacon, contributed by members of the San Bernardino Microwave Society (California, USA) was potentially one of the most interesting technical experiments aboard A-O-7. Much has been learned from this beacon, though not in the areas anticipated. Because of international treaty constraints, the FCC decided to deny amateurs permission to turn the 2304-MHz transmitter on. As a result, it was never tested. In 1979, at the World Administrative Radio Conference (WARC), the Amateur Satellite Service received several important new frequency allocations in the microwave portion of the spectrum. Although the events at the 1979 WARC and the legal constraints on the 2304-MHz beacon appear, at first glance, unrelated — are they? Might the new allocations result, in part, from the responsible, restrained manner in which radio amateurs handled the sensitive 2304-MHz beacon issue?

AMSAT-OSCAR 7 (lower left) is dwarfed by the primary payload, the ITOS-G satellite, as it sits attached to its Delta launch vehicle. The similar looking spacecraft, shown opposite to and counterbalancing the weight of OSCAR 7, is the Spanish INTASAT. Note that the flexible elements of OSCAR 7's 2-m canted-turnstile antenna protrude downward and will ride within the launch vehicle cowling.

A last look at OSCAR 7 before the cowling is secured around the trio of fellow space travellers.

The transponder frequencies chosen for the AMSAT-OSCAR 7 Mode B transponder made it theoretically possible for two ground stations to communicate by transmitting to A-O-7 on 432 MHz, having the signals relayed directly to A-O-6 on 146 MHz, and then back down to the ground on 29 MHz. Many such contacts were made when A-O-7 was in Mode B and the two satellites were physically close.[19] Never before, in any radio service, had two terrestrial stations been linked by a direct satellite-to-satellite relay.

Launched in late 1974, A-O-7 operated until mid 1981, a period covering more than six and a half years. The cessation of operation coincided with the beginning of a three-week eclipse period in which the satellite entered the earth's shadow for up to 20 minutes on each orbit. When the eclipse period began, the average spacecraft temperature dropped. It's believed that thermal stress caused a battery cell that had previously failed in the open mode to short out, placing a very large load across the solar panel output.

Flight hardware for the AMSAT-OSCAR 7 satellite was contributed by groups in Australia, Canada, USA and West Germany.

AMSAT-OSCAR 8

Each time a new amateur satellite is placed in orbit, launch-day radio networks provide information on countdown, liftoff, rocket staging and the first user reception. AMSAT-OSCAR 8 (A-O-8) was orbited successfully at 1754 UTC on March 5, 1978. Barely minutes later, those monitoring the nets heard G2BVN report reception of the 435-MHz beacon. But the suspense didn't end. The Mode A transponder on A-O-8 couldn't be turned on until the 29-MHz antenna was deployed by ground command, and the message couldn't be sent until the satellite's spin rate had decreased to an acceptable level. Pre-launch speculation was that it might take a week for the spacecraft to slow down sufficiently. Roughly seven hours after launch, however, as the satellite was passing over the East Coast of the U.S., the spin rate looked good and the decision was made to send the antenna deployment command. Hundreds of stations listening to the 80-m AMSAT net breathed a simultaneous sigh of relief as the antenna extended and the Mode A transponder responded to the "on" command.

Previous launch descriptions would have ended here as the joyous bedlam of early two-way communication began. But this time amateurs were being asked to refrain from transmitting to the spacecraft until it was fully tested in orbit. The reason was twofold: (1) An empty transponder would permit orbit determination measurements and engineering evaluation of the satellite to proceed as quickly and efficiently as possible, and (2) plans

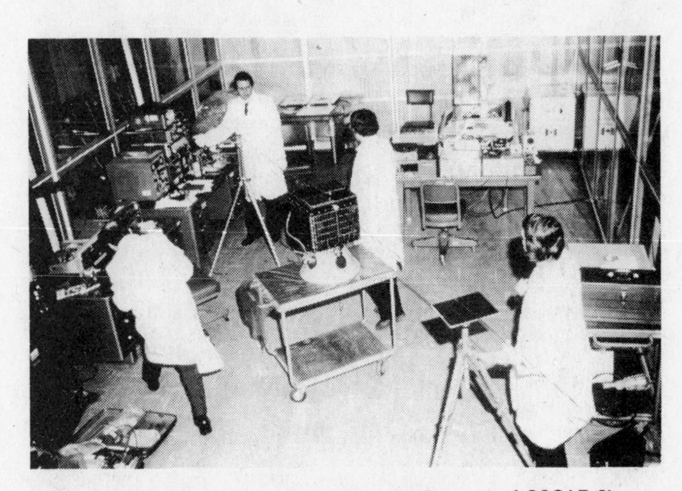

AMSAT engineers very carefully test the deployment of OSCAR 8's 10-m dipole antenna in the lab at Vandenburg AFB in the days before launch. Each antenna element, on earth command, was deployed in flight by driving pre-cut rolls of thin copper-beryllium foil through a circular opening that formed it into a long tube of precisely the proper length.

Dick Daniels, W4PUJ, applies a cleaning swab to the OSCAR 8 spacecraft on the pool table in the family room of his Arlington, Virginia home. The solar panels have not yet been mounted.

Table 3.1
Satellite Transponders

Designation	*Uplink/Downlink frequencies**	*Spacecraft*
Mode A	146 MHz/29 MHz	A-O-6, A-O-7, A-O-8, RS-1, RS-2, RS-5, RS-6, RS-7, RS-8
Mode B	435 MHz/146 MHz	A-O-7, Phase III-A, C, A-O-10
Mode J	146 MHz/435 MHz	OSCAR IV, A-O-8
Mode L	1,269 MHz/436 MHz	A-O-10, AMSAT-Phase III-C; SYNCART

Notes
*All frequencies are approximate.
1) Transponder input frequency uplink is always listed first.
2) There are no plans to use single-band transponders like the one flown on OSCAR III.

for the critical transfer-orbit stage of future missions depended on users' exhibiting such self control; AMSAT had to know if these plans were realistic. While the satellite was over North America, cooperation proved excellent. The waiting period was easier on the users than expected because monitoring the engineering tests proved intrinsically interesting. During sensitivity tests, for example, a transmitting station would announce the power levels it was using: "...10 watts...1 watt...one-tenth watt..." and the hundreds of silent, monitoring stations would witness the results first-hand in real time. The unloaded sensitivity of the transponder was remarkable. Two weeks after launch, A-O-8 was officially opened for general operation, with all systems in excellent shape.[20]

Let's backtrack a bit to look at some of the events leading to the launch of A-O-8. After A-O-7 was placed in orbit late in 1974 the AMSAT design team focused on the next major step in the radio amateur space program — building a high-altitude, long-life (Phase III) spacecraft. Early in 1977, however, when it became clear that A-O-6 was nearing the end of its lifespan, the Phase III effort was interrupted. With the fear that A-O-7 might not last until the first Phase III satellite was launched, a commitment was made to provide continuity of service to the thousands of amateurs and educators who had built Mode A ground stations and had financially supported the AMSAT satellite program. AMSAT had a serious problem; the resources (financial and volunteer) for building both spacecraft just weren't there.

To resolve this dilemma the American Radio Relay League offered to donate $50,000 to AMSAT so that an interim Phase II satellite could be built, an offer AMSAT accepted. The initial plans for A-O-8 called only for a Mode A transponder and a minimal telemetry and command system. When JAMSAT (the Japanese affiliate of AMSAT) learned of plans for A-O-8, they offered to develop a second transponder for the mission. The JAMSAT transponder (Mode J) would use an uplink at 146 MHz and a downlink at 435 MHz. AMSAT agreed to provide antennas and interface circuitry so the Mode J transponder could be included if JAMSAT could deliver the transponder in time for launch integration. The time schedule for preparing the transponder and the spacecraft was extremely tight, but both groups met their deadlines and the satellite was launched with both Mode A and Mode J transponders. AMSAT designations for various transponder frequency complements are summarized in Table 3.1. An interesting feature of the A-O-8 spacecraft is that the transponders can be operated simultaneously, as long as the batteries maintain a sufficient charge. Since a single uplink signal can then be retransmitted on both downlinks, the two modes can easily be compared. A-O-8 operated flawlessly in orbit from March 1978 through mid 1983.

Flight hardware for A-O-8 was provided by AMSAT, JAMSAT and Project OSCAR. By prior agreement, the responsibility for operating A-O-8 resided with the ARRL. A detailed technical description of A-O-8 is contained in Appendix A: Spacecraft Profiles.

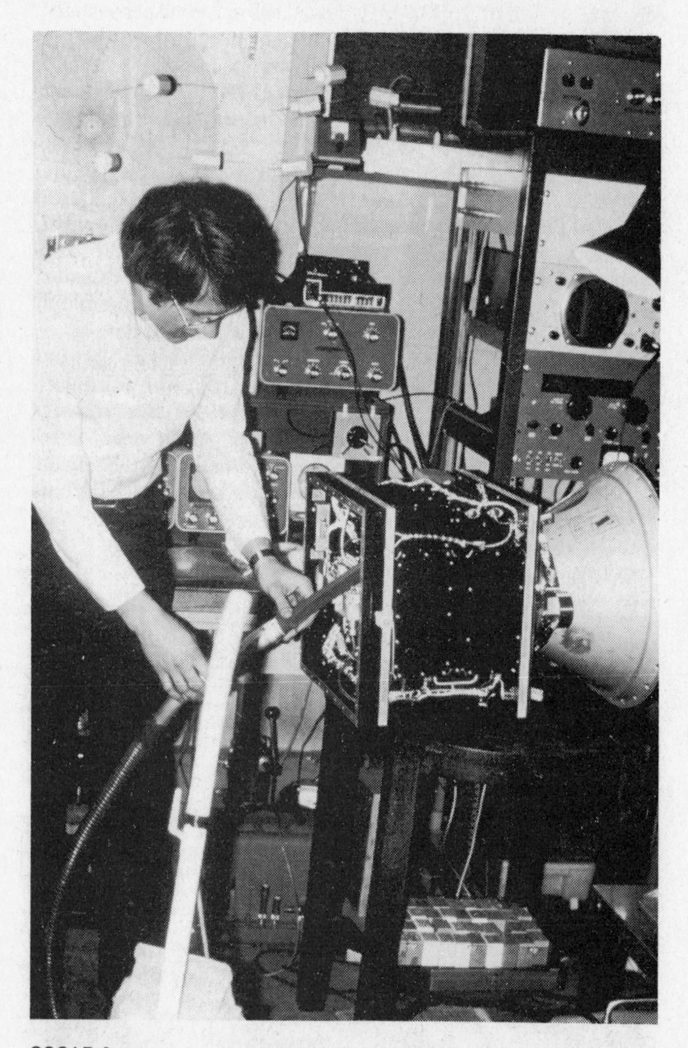

OSCAR 8 gets a thorough cleaning from W3GEY. At every step of the way, though the tools may not always be of the sophisticated laboratory type, AMSAT engineers pay meticulous attention to cleanliness. In the extreme temperatures and near-vacuum of space, minute debris from sloppy work habits could contaminate other satellites aboard the launch vehicle and even jeopardize the primary mission. AMSAT's "compulsive" care and attention have paid off well, as the record shows.

Soviet Radio Amateur Satellites: RS And ISKRA

The scene now switches to a series of radio amateur satellites built in the Soviet Union. In the mid 1970s, about the time of the joint U.S.-USSR Apollo-Soyuz earth-orbiting mission, several Soviet engineers, some of them radio amateurs, visited NASA facilities in the U.S. Their itinerary included the Goddard Space Flight Center in Greenbelt, Maryland, an area where several AMSAT designers lived. The Soviet radio amateurs and the AMSAT technical crew met, as hams are apt to do, and discussed the technical and operational aspects of the OSCAR program. During these meetings, it became clear that the Soviet amateurs were interested in producing their own radio amateur satellites. In fact, a satellite coordinating group had already been formed, and construction of prototype equipment was underway.

Though time passed — the Apollo-Soyuz program was a success in July 1975 — not much was heard of Soviet amateur satellite plans. Then, in the October 1975 issue of *RADIO,* [21] a very widely read Soviet electronics magazine, the awaited article appeared. It focused on experiments with terrestrial linear Mode A type transponders in Moscow and Kiev and discussed, for the first time in the Soviet press, the OSCAR program. Although no mention was made of Soviet radio amateur satellites, some Soviet hams were clearly laying the groundwork. Speculation as to Soviet plans wasn't officially confirmed until July 1977 when the USSR filed a notice with the International Frequency Registration Board (IFRB) of the International Telecommunication Union (ITU) announcing that a series of satellites in the amateur-satellite service would be launched.[22] This was followed by a series of *Radio* articles on the RS spacecraft.[23]

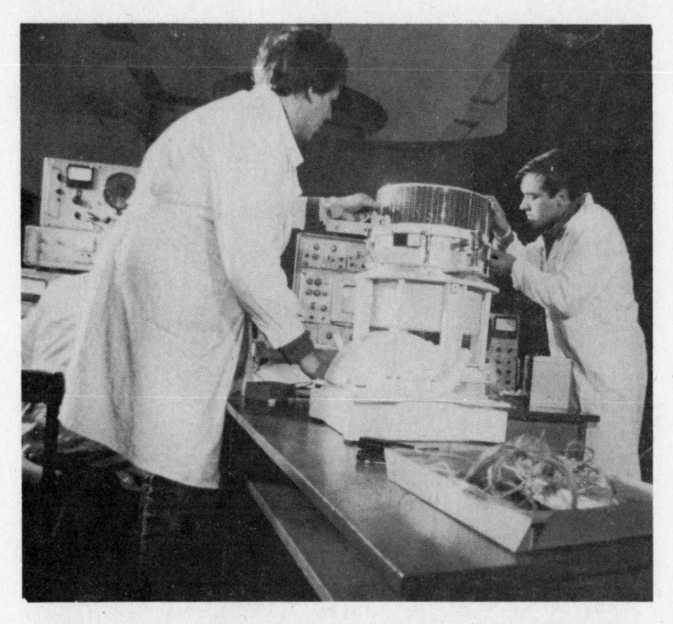

Two technicians assemble the Radio-2 artificial satellite. With Radio 1, this satellite provided communications for more than 700 Amateur Radio operators from 70 countries on all continents. Its communication range was over 8000 km (5000 miles). (Novosti photo. Provided by Embassy of the USSR.)

Finally, on October 26, 1978, a Soviet rocket lifted Cosmos 1045 and two radio amateur satellites, Radio-1 and Radio-2 (RS-1 and RS-2), into space. Each spacecraft carried a Mode A transponder. Since Soviet radio amateurs are limited to 5 watts in the 146-MHz uplink band, the transponders contained very sensitive receivers optimized for low-power terrestrial stations. Automatic shut-down circuitry protected them from excessive power drain. In operation, the protective circuits, acting like a time-delay fuse, would shut the transponder off when ground stations used too much uplink power for more than a few seconds. Unfortunately, the fuse could only be reset when the spacecraft passed near a Soviet command station. Although the Soviets appeared to make every reasonable effort to keep the satellites on and available to the rest of the world, the transponders were often off over the Western Hemisphere because of the actions of a few inconsiderate high-power users.

The Soviet approach to improving reliability through redundancy merits note: When possible, launch two spacecraft at the same time. In addition to a transponder, each satellite contained a telemetry and command system, a Codestore-like device and a power system using solar cells. The primary telemetry system used Morse code letters and numbers that identified the parameter being measured, encoded the most recent value of that parameter and indicated the status (on/off) of the transponder. Specific decoding information was provided by the Soviets a few weeks after launch. At least one of the spacecraft also contained an infrequently used high-speed digital telemetry system described as an experimental prototype being tested for future flights.

During their first few weeks in orbit, changes in the operating status of RS-1 and RS-2 (except for transponder shut down) took place only at certain times and from certain locations, strongly suggesting that command of the spacecraft was confined to Moscow and limited to normal working hours. Further observation indicated that two additional ground command stations were soon activated: one in eastern Asia and another in central Asia. The Soviets later announced that the primary command station was in Moscow, a secondary command station was in Arsen'yev (44.1° N, 133.1° E) near Vladivostok, and that a third command station, which was portable (Novosibirsk?) had been tested.

RS-1 and RS-2 were designed for passive temperature control. The techniques used were similar to those employed on OSCARs 1-8 (see Chapter 12: Environmental Control). RS-2 also included a quasi-active thermal regulating system employing a heat bridge connecting the interior of the spacecraft with a heat exchanger on the outer surface that would automatically turn on whenever the internal temperature exceeded a predetermined level. The Soviets consider the system successful and will likely use it on future flights. Technical details should be available in the near future.

The transponders aboard RS-1 and RS-2 could be kept operating for only a few months before power supply (battery) problems disabled both spacecraft. Reception of a weak telemetry beacon, believed to be RS-2, was reported, on and off, into 1981.

RS-1 and RS-2 were certainly an impressive and successful first step.[24] The Soviets have carefully analyzed both missions and used the results to plan for further RS spacecraft that should have significantly longer operational lifetimes. Insofar as is possible, the Soviets have attempted to provide information about RS spacecraft, coordinate frequencies with OSCAR satellites, and make RS spacecraft available to radio amateurs around the world. The international radio amateur community sincerely appreciates this effort.

On December 17, 1981, the Soviets simultaneously launched a set of six satellites, RS-3 through RS-8, into orbits similar to those of RS-1 and RS-2. The launch was expected, but the number of spacecraft certainly was a pleasant surprise. Beginning in March 1980 the club station at the University of Moscow, RS3A, had openly tested many of the spacecraft subsystems on the 10-m band. The tests included a Mode A transponder, a Codestore device, a Morse code telemetry system and an autotransponder (called Robot), all of which were flown. In addition, an engineering prototype of one of the spacecraft was exhibited at TELECOM-79, a large international telecommunications conference held in Geneva in 1979. Technical information on RS-3 through RS-8 will be found in Appendix A: Spacecraft Profiles.

On May 17, 1982 another Russian amateur spacecraft, ISKRA-2, was orbited, this time by hand through the airlock of

A model of the new generation Radio satellite. The 2-meter antennas on top of the spacecraft are mounted in a canted configuration, similar to OSCARs 7, 8 and 9. Ports through which the 10-meter antenna was deployed after launch are on opposite sides of the spacecraft center.

Table 3-2

Stages in the Amateur Satellite Program

Classification criteria (Note: satellites are classified according to mission objectives, not in terms of actual performance.)

1) *Function:* (experimental/developmental/operational) Distinguishes between satellites primarily designed to acquire information about spacecraft performance and those designed to satisfy needs of general users. (User needs include communications and/or long-term collection of scientific information not directly related to spacecraft performance.)

2) *Lifetime:* (long vs. short) Short missions (generally under four months) depend primarily on batteries for power. Long-lifetime amateur missions (generally at least one year) depend on solar cells for power and batteries only for short-term storage.

3) *Orbit:* (high altitude vs. low altitude) Satellites with an apogee (high point) under 1200 miles are classified as low-altitude missions. Satellites with apogees above 15,000 miles are classified as high-altitude missions. Intermediate apogees have so far been avoided to minimize radiation damage to the spacecraft from the Van Allen Belts.

Phase	*Characteristics (design goals)*	*Satellites*
Phase I	Experimental, short-lifetime, low-altitude	OSCAR I, OSCAR II, OSCAR III, Australis-OSCAR 5, ISKRA 2, ISKRA 3
Phase II	Developmental, long-lifetime, low-altitude	AMSAT-OSCAR 6, AMSAT-OSCAR 7, AMSAT-OSCAR 8, PACSAT, RS-1, RS-2, RS-3, RS-4, RS-5, RS-6, RS-7, RS-8, UoSAT-OSCAR 9
Phase III	Operational, long-lifetime high-altitude	OSCAR IV, AMSAT-OSCAR 10 AMSAT Phase III-A, C*, AMSAT SYNCART (Canada and Project OSCAR)*, Arsene (RACE, France)**

*Under construction as of mid 1983.
**In preliminary design stage as of late 1982.

the Salyut 7 space station. Real-time TV coverage of the event was provided to viewers in the USSR and adjacent countries. As a result, students at the Moscow Aviation Institute had a chance to watch the birth of the satellite they helped build. According to TASS, solar-cell-powered ISKRA-2 contained a transponder, beacon, command channel, telemetry system, and bulletin board (Codestore) facility. The novel 21 MHz up/28 MHz down transponder was intended to increase radio amateurs' store of practical information on the performance of a new combination of link frequencies. Bandwidth was 40 kHz with the input centered at 21.250 MHz and output centered at 29.600 MHz. Because of a malfunction, apparently associated with the command system receiver/decoder, the transponder was never activated in range of the U.S.

From a satellite designer's viewpoint it's simpler to construct a spacecraft for a "get-away special" type of launch than for a conventional launch since no complex satellite-rocket mechanical interface is needed. The appeal of this approach is tempered by the fact that most available orbits are at low altitudes where a satellite's lifetime and coverage are severely limited. ISKRA-2, for example, remained in space for only about seven weeks. During this period, including the hours leading up to reentry on July 9, 1982, the 29.578-MHz beacon provided telemetry. The telemetry was described in *ASR*, no. 34, May 31, 1982. Because of the limited spacecraft lifetime available it's expected that this type of launch will be used primarily for testing new hardware, links, systems, and so on.

On 18 November 1982 ISKRA-3 was placed in orbit, again from the Salyut 7 space station. Similar in design to ISKRA-2 (beacon at 29.583 MHz) it was only a partial success as it was affected by a severe overheating problem.

There's been some confusion over the names of the Soviet amateur satellites. Most Soviet literature provided in English (by TASS and in IFRB filings) refers to them as RADIO-1 through RADIO-8 of the "RS" series. They are more commonly called RS-1 through RS-8, undoubtedly because of the identifier "RS" on the downlink beacons. In the Soviet Union, where the Cyrillic alphabet is used, the symbol "P" stands for the Morse code di-dah-dit and sounds like the English "R." Similarly, the Cyrllic symbol "C" stands for the Morse code di-di-dit and sounds like the English "S." In informal Soviet literature, the satellites are often called PC-1, PC-2 and so on. The important point to remember is that the names RADIO-1, RS-1 and PC-1 all refer to the *same* spacecraft; similarly, RADIO-2, RS-2 and PC-2 all refer to the *same* satellite. RS is a particularly apt designation for the Soviet amateur satellites for several reasons. The Soviet acronym for "radio amateur satellite" is "PC" in Cyrillic (that's "RS" in English), and "RS" is an internationally recognized radio prefix assigned to the Soviet Union. The name Iskra doesn't cause confusion. It means "spark," an apt name for a spacecraft with a brief life.

Phase III

Looking back on the evolution of the radio amateur satellite program after two decades in space, we often speak of three stages: an experimental stage (Phase I), a developmental stage

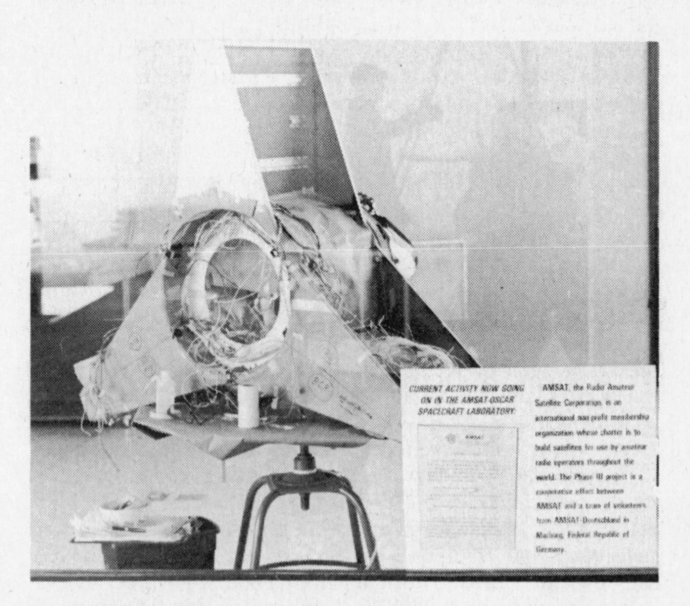

A visitor's eye-view of the Phase III-A structure through the large AMSAT Lab window. The room in which Phase III-A was built is affectionately known as the "Fishbowl" by the AMSAT personnel who assembled the satellite in full view of the public at the Goddard Space Flight Center Visitor's Center, in Greenbelt, Maryland. What appears to be the ghost of Dr. Robert Goddard watching over the satellite is the reflection of his statue: a memorial to the late rocketry pioneer for whom the Space Flight Center was named. Other rocketry and astronautics displays surround the area. *(W4PUJ photo)*

(Phase II) and an operational stage (Phase III) (see Table 3.2). The grouping is convenient even though the classification criteria aren't clearcut and some of the assignments in Table 3.2 are admittedly arbitrary. The short-lived Phase I satellites, designed to gather information on basic satellite system performance, appealed mainly to a relatively small number of hard-core radio amateur experimenters — perhaps several thousand. The communications capabilities of the long-lifetime Phase II satellites attracted a significantly larger, new group of operators to the space program — amateurs who shared a vision of the immense future possibilities for the project and who wanted to get started on the learning curve early and assist in seeing that the project succeeded. Estimates are that between 10,000 and 20,000 amateurs have communicated through a Phase II satellite. With the first Phase III spacecraft in operation, amateurs have gained access to long-distance communication capabilities of a type never before available. As a result, the number of amateurs involved in space communication should once again increase greatly.

Phase III satellites will spend most, if not all, of their time at high-altitudes. AMSAT-OSCAR 10 (dubbed AMSAT Phase III-B before launch) will, for example, be in view of about 42% of the earth when it is at apogee and will be accessible to most Northern Hemisphere ground stations more than 10 hours each day. For comparison, a Phase II satellite, such as AMSAT-OSCAR 7, is only in view of about 9% of the earth at any given time, and accessible to most ground stations for less than two hours each day.

Modern Phase III satellites are designed to provide modestly equipped ground stations with reliable, predictable, long- distance communications capabilities of a quality unmatched by any other amateur mode.[25] This objective, coupled with the large satellite-earth distances involved, leads to a chain of complex technical requirements for Phase III spacecraft: high-power transmitters, large power systems, high-gain directional antennas, attitude sensing/adjusting systems, sophisticated computer control, rocket motors, and so on.

Since the 14-MHz band is one of the most reliable and popular routes for long-distance communication currently available to amateurs, it's instructive to compare the capabilities provided by this band with the Phase III satellites. A modestly equipped 14-MHz station can communicate with any place on the earth by exploiting favorable conditions. However, if the station is interested in scheduled communications, either over specific point-to-point paths or involving a multi-point network, reliability is not very high. The unpredictable nature of the 14-MHz band isn't necessarily a negative characteristic. In fact, in many situations it's a feature that makes the band interesting and exciting.

In contrast, Phase III spacecraft will provide highly reliable, predictable and consistent communications over long paths to stations modestly equipped for the appropriate frequencies. The predictability of Phase III spacecraft will make satellite links an invaluable asset during natural disasters and in situations where getting a message through in a timely fashion is of paramount importance. Examples are general bulletins, code practice, phone patches, coordinating DXpeditions or arranging moonbounce schedules. With satellites there's no "skip zone," so multiple conversations can't take place on a single frequency. Group discussions will be greatly facilitated, however, since everyone will be able to hear everyone else.

This comparison of the 14-MHz band and Phase III satellite communication links illustrates their complementary nature. Whatever your primary interests are, don't get caught in the trap of viewing the situation as competitive with only one winner. Phase III is designed to provide a new dimension to Amateur Radio, not to replace existing options.

The launch of Phase III satellites will *not* signal the demise of Phase II. Low-altitude spacecraft have a number of unique features that radio amateurs have hardly begun to exploit. A-O-7 Mode B demonstrated that simple omni-directional receive antennas can be used at ground stations set up to listen for signals from a low-altitude spacecraft. UoSAT, discussed later in this chapter, reinforced this point even more emphatically. We've also learned that ground stations running under 1 watt to omni-antennas can access low-altitude spacecraft when they aren't fully loaded or gain-compressed by stations running too much power. Since Phase III satellites are likely to syphon off many of the higher-power stations formerly using Phase II spacecraft, it's probable that the potential of low-altitude spacecraft for supporting very-low-power communications may finally be realized. Continent-spanning contacts between stations using small hand-held units via Phase II spacecraft might become commonplace.

ESA technicians mount the AMSAT-OSCAR Phase III-A spacecraft to the CAT (Application Technology Capsule). Above Phase III-A is Firewheel, the primary payload. Its cylindrical cannisters contained lithium, barium, explosives and other compounds that when exploded would have provided a visible, "glowing," steam-like cloud enabling scientists to study the earth's magnetic-field patterning. Had the mission not crashed in the Atlantic a scant four minutes after launch (a launch vehicle problem, not an OSCAR malfunction) Phase III-A would have been separated from the CAT and been clear of the experiments long before the fireworks began. (AMSAT-DL photo).

Phase II spacecraft could also support digital store-and-forward message systems. Ground stations with microcomputers could then load messages into the spacecraft directed to a specific user or group of users, or query the satellite to see if there were any messages for them. Picture yourself coming home after a short trip and checking your computer for any automatically received Phase II "telemail." Such a system could provide

AMSAT-OSCAR Phase III-A shown mounted on the side of the ESA CAT (Application Technology Capsule) as the overall assembly is prepared for launch in Kourou, French Guiana.

worldwide coverage, on a delayed-time basis, even with satellites in very low orbits.

AMSAT Phase III-A

The launch window opened at 1130 UTC on Friday, May 23, 1980, with AMSAT Phase III-A perched atop the sleek Ariane rocket sitting on its launch pad in Kourou, French Guiana. Following nine years of planning, including four of intensive construction, AMSAT workers around the world could now only sit and listen as Dr. Tom Clark (W3IWI), president of AMSAT, relayed the countdown occurring on the northern coast of South America. Would the tropical weather clear for a liftoff? Would the rocket systems remain "go"? We all listened to the continuous reports as Phase III-A waited within the cowling of the newly developed European Space Agency (ESA) launch vehicle. The amateur spacecraft was awarded this prized position in a stiff international competition involving more than 80 applicants. Finally, at approximately 1430 UTC, the liftoff signal was given and the Ariane LO2 rose from its pad. For nearly three minutes spirits soared.

Then disaster. The words "non-nominal flying . . . problem in one engine . . . the rocket is going down . . . splashdown" dashed the hopes of thousands of amateurs monitoring the launch net. The first stage of the Ariane rocket had failed and had unceremoniously dumped Phase III-A in its final resting place several hundred feet under the Atlantic Ocean.

While many dejected AMSAT members wondered if they had just witnessed the end of the amateur space program, Clark drafted a statement objectively describing the situation to be read over the AMSAT nets scheduled later that evening.

> "What we lost on Black Friday was sheet metal, solar cells, batteries, transistors, a lot of sleep and a major portion of our lives for the last few years. What we gained over those same years was knowledge; knowledge that we could make a complicated spacecraft. Knowledge in areas of aerospace technology that none of us had before. Knowledge that we could work as a team, despite national boundaries, differences in our cultures, lifestyles and personalities. Knowledge that, from within the ranks of Amateur Radio, we could draw upon enough resources to attempt a project with a complexity rivaling commercial satellite endeavors funded at levels of tens of millions of dollars. The knowledge is still intact. We even had the forethought to purchase a duplicate set of sheet metal that constitutes the spaceframe. We have a second set of solar panels, batteries and sensors. We have on hand the documentation and artwork necessary to replicate all the printed-circuit boards. We have in place and ready to go a network of ground telecommand stations."[26]

The situation was certainly bleak but not completely hopeless.

At this point the AMSAT Board of Directors looked to the membership for guidance. Did members have the heart and confidence to continue? Being realists, the Board couldn't commit to a follow-on spacecraft without reasonable assurance of financial and moral support. Over the next several weeks the support, both financial and moral, was overwhelming. One by one, key volunteers, convinced that Phase III-B was possible, recommitted themselves to the project. Like the pieces of a giant jigsaw puzzle, the elements fell into place with a picture of Phase III-B emerging. The amateur space program would continue.

We pause now briefly to examine what was lost. Phase III-A was the first amateur satellite to carry its own propulsion system, a rocket (kick motor) accounting for roughly half the launch weight of the spacecraft. The kick motor was required for this mission because the initial (transfer) orbit had a low point (perigee) of only 125 miles. Left here, the orbit would quickly decay, causing the spacecraft to reenter the atmosphere and burn up within one year. Using the kick motor would have enabled AMSAT to (1) prolong the spacecraft's lifetime by increasing the perigee, and (2) enhance the spacecraft's utility to Northern Hemisphere ground stations by raising the inclination of the orbital plane (see Chapter 8).

It's fair to say that the Phase III-A project was more complex, required a larger financial investment, and reflected a greater total effort than all previous OSCARs combined.[27,28] It contained a 50-watt Mode B transponder and a suitably sized energy supply system; a computer for flexible control of command, telemetry, Codestore and housekeeping functions; a sophisticated attitude sensing and control system permitting the use of high-gain antennas; and two beacons sandwiching the 180-kHz-wide 146-MHz downlink. In addition to building and testing Phase III-A, AMSAT had to coordinate the development, construction and deployment of a series of ground telecommand stations capable of loading the spacecraft computer, providing real-time orbit determination and attitude control data, and reducing the relatively sophisticated telemetry to meaningful values.

Flight hardware for the project was produced in Canada, Hungary, Japan, the U.S. and West Germany. Primary responsibility for spacecraft and ground support systems resided with the U.S. and West Germany. The senior spacecraft engineer overseeing the entire construction project from the first highly speculative feasibility studies in 1971 was Jan King (W3GEY).

AMSAT-OSCAR 10 rests atop its attach fitting on a laboratory bench in preparation for packaging and shipping to the launch-vehicle integration site. Antennas occupy the top face, solar panels the side faces, sensors and antenna reflectors protrude off the ends of the arms. At the proper time in the launch-separation sequence, the satellite is literally "sprung" free of the launcher, leaving the conical attach fitting behind. (The kick motor is hidden from view within the cone).

AMSAT-DL Project Manager Karl Meinzer, DJ4ZC, and AMSAT-USA Project Manager Jan King, W3GEY pause before AMSAT-OSCAR 10. The spacecraft is shown here with solar panels and inner module covers removed to permit final assembly, wiring, alignment and testing before it is sealed for launch.

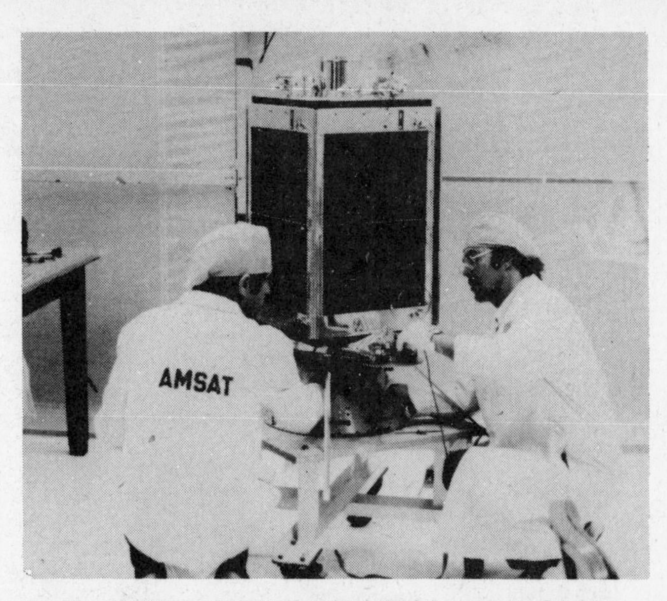

Jan King, W3GEY, and Martin Sweeting, G3YJO, apply the finishing touches to UoSAT-OSCAR 9. The spacecraft is roughly twice the size of AMSAT-OSCAR 8, having been built around the equivalent of two OSCAR 8 frames, one atop the other.

Also involved in the project from its inception was Dr. Karl Meinzer (DJ4ZC), who was responsible for the design and construction of many of the unique high-technology subsystems aboard the spacecraft. In recognition of the ESA-sponsored launch and the major technical contributions to the spacecraft by AMSAT-Deutschland members, the satellite was licensed in Germany as DLØOS. In addition to all its other outstanding features, Phase III-A is the best documented of all amateur spacecraft; AMSAT-Deutschland has produced a two-volume work containing full schematics (Vol. I) and technical descriptions (Vol. II) for all subsystems.

UoSAT-OSCAR 9

UoSAT is an amateur scientific and educational spacecraft built at the University of Surrey (England) by a group led by Dr. Martin Sweeting, G3YJO.[29] Launched into low altitude (340-mile), circular, near-polar orbit, on October 6, 1981, with the Solar Mesosphere Explorer (SME), this sophisticated Phase II satellite carries several scientific instruments and systems of interest to radio amateurs, educators and amateur scientists. The scientific payload includes a General Data Beacon at 145.825 MHz compatible with standard amateur nbfm receivers, an Engineering Data Beacon at 435.025 MHz, phase-locked hf beacons at 7, 14, 21 and 28 MHz for propagation studies, and microwave beacons at 2.4 and 10.47 GHz, also for propagation observations. In addition, the spacecraft carries a camera to send back pictures of the earth formatted to be viewed on a regular TV after minimal processing; a three-axis, wide-range, flux-gate magnetometer for measurement of the earth's magnetic field; and high-energy particle and radiation detectors. The 146-MHz telemetry beacon can be switched to Codestore or to a speech synthesizer for educational demonstrations. The synthesized voice mode will only be activated on a part-time basis because of its low data rate. All spacecraft systems are controlled by a flexible onboard computer that can be reprogrammed by ground command.

Commanding this complex spacecraft is a real challenge. Here's a typical hectic day at the Surrey command station. Last orbit of set: Collect telemetry. Between sets: Study telemetry, plan actions, write and load computer software, test software on spacecraft simulator. Next orbit (about 10 hours between sets): Collect telemetry and check status, transmit commands, verify correct receipt, instruct spacecraft to act on commands, collect telemetry. Between orbits: Study telemetry, plan actions and so on.

While the University of Surrey team perfected the commanding operation (techniques, hardware, software), checked out the spacecraft systems, and oriented the spacecraft using the new dynamic magnetic-torquing system to prepare for extending a 50-ft gravity gradient boom that would provide passive attitude control, radio amateurs around the world grew impatient at what appeared to be snails' pace progress. It was a frustrating situation for everyone, for the Surrey crew working till they dropped from exhaustion and for the amateurs who wanted to put UoSAT to practical use. Then disaster struck.

While uplinking commands on April 4, 1982 a software glitch caused *both* the 2-m and 70-cm telemetry beacons to be turned on. As a result, both satellite command receivers were desensed. When the uplink power at Surrey proved insufficient to overcome the desense, Dave Olean, K1WHS, offered the use of his powerful 2-m EME station (26 dBd antenna, nearly 0.5 megawatt EIRP) for the effort. When even this power wasn't able to solve the problem, a group of amateurs in northern California, under the leadership of Dr. Robert Leonard, KD6DG, obtained permission to activate a 150-ft.-dish antenna at Stanford University Research Institute (SRI) that wasn't currently in use. With 42 dBd gain at 70 cm (15 megawatts EIRP, 3-dB beamwidth of 0.6 degrees) the SRI dish would either get through to the command receiver or fry the spacecraft! After several months of work resuscitating drive motors, hydraulic components, and control computers — SUCCESS. On Sept. 20, 1982, after operating out of control for nearly six months, UoSAT-OSCAR-9 was salvaged and found to be in excellent health.

Many radio amateurs were disappointed to learn that UoSAT does *not* carry a transponder. A look at the reasons for this decision is in order. All amateur spacecraft to date have received free launches in recognition of their potential value for scientific work, educational applications and disaster communications. Naturally, actual amateur accomplishments will continue to have more impact than vague promises. Furthermore, a number of problems arise when one attempts to include both a transponder and scientific/educational package aboard the same spacecraft: rf compatibility, limited space for mounting antennas, system power needs and conflicting attitude-control requirements. Orbit selection also presents a conflict; a very low orbit might be excellent for certain scientific/educational purposes, but poor for communications.

Table 3-3
A Brief History of Radio Amateur Satellites

Satellite; Launch Date	*Operating Life*	*Number of Transponders*	*Transponder Bandwidth (kHz)*	*Peak Transmitter Power*	*Highest Frequency*	*Number of Beacons*	*Apogee*
OSCAR I Dec. 12, 1961	21 days	0	—	0.1 W	144 MHz	1	290 mi. 471 km
OSCAR II June 2, 1962	19 days	0	—	0.1 W	144 MHz	1	240 mi. 391 km
OSCAR III March 9, 1965	transponder, 18 days; beacon, several months	1	50	1. W	145 MHz	2	590 mi. 941 km
OSCAR IV Dec. 21, 1965	85 days	1	10	3.0 W	432 MHz	1	21,000 mi. 33,600 km
Australis-OSCAR 5 Jan. 23, 1970	52 days	0	—	0.2 W	144 MHz	2	925 mi. 1480 km
AMSAT-OSCAR 6 Oct. 15, 1972	4.5 years	1	100	1.5 W	435 MHz	2	910 mi. 1460 km
AMSAT-OSCAR 7 Nov. 15, 1974	6.5 years	2	100/50	8.0 W	2304 MHz	4	910 mi. 1460 km
AMSAT-OSCAR 8 March 5, 1978	5.3 years	2	100/100	1.5 W	435 MHz	2	570 mi. 912 km
RS-1/RS-2 Oct. 26, 1978	several months	1/1	40 kHz each	1.5 W	146 MHz	1/1	1065 mi. 1706 km
AMSAT-Phase III-A May 23, 1980	*2	2	180/180	50 W	435 MHz	2	22,400 mi 35,800 km
UoSAT-OSCAR 9 Oct. 6, 1981	*1	0	—	0.8 W	10.47 GHz	8	338 mi. 544 km
RS-3 — RS-8 Dec. 17, 1981	*1	4 + 2 Robots	40 kHz each	1.5 W	146 MHz	2 each	1050 mi. 1690 km
Iskra 2 May 17, 1982	53 days	1	40	1.0 W	29 MHz	1	210 mi. 335 km
Iskra 3 Nov. 18, 1982	37 days	1	40	1.0 W	29 MHz	1	210 mi. 335 km
AMSAT-OSCAR 10 June 16, 1983	*1	2	180/800	50 W	1269 MHz	2	22,060 mi 35,500 km

Notes:
*1 — Operational as this book went to press.
*2 — Launch vehicle malfunction, satellite did not attain orbit.
OSCAR I — First satellite built by radio amateurs.
OSCAR III — First transponder on amateur satellite.
OSCAR IV — Partial launch vehicle malfunction. Satellite did not attain desired orbit. First fully solar-powered amateur spacecraft.
OSCAR 5 — First amateur satellite which could be controlled from the ground.
RS-1/RS-2 — First Soviet radio amateur satellites.

Since several Phase II spacecraft with Mode A transponders were expected to be operating when UoSAT was launched, Dr. Sweeting thought it would be a good idea to devote a single Phase II spacecraft to radio-amateur-related scientific and educational activities. UoSAT is the result. Because of UoSAT, it will be possible to optimize other AMSAT spacecraft for communications. Funding for UoSAT was provided by grants from the British government, the University of Surrey, the British aerospace industry and British radio amateur groups. As a result, AMSAT funds can be used to support the development of communication spacecraft.

Radio amateurs interested in studying propagation will find UoSAT a fantastic tool. Amateur scientists and educators monitoring the varied instruments and anyone interested in the TV pictures will similarly benefit. Correlational studies using UoSAT may revolutionize the ability of radio amateurs to predict unusual vhf propagation. For example, if we find that the joint occurrence of certain magnetic field changes and radiation levels near the North Pole is a good predictor of 70-cm band openings in Europe and the U.S. several hours later, it will be possible to program UoSAT's computer to check automatically for the appropriate conditions and send a Codestore alert message at appropriate times. The possibilities and options are limitless.

The responsibilities and activities of both UoSAT-OSCAR 9 users and the command/operations team differ considerably from those with other OSCAR spacecraft. If a user would like to study the data provided by certain instruments, a request to the command team to turn the experiment module on at the proper times may have to be made in advance. In return for the services provided the requestor must be committed to studying the data and providing at least a synopsis to the operations group at Surrey. Technical details of UoSAT are contained in Appendix A: Spacecraft Profiles.

AMSAT-OSCAR 10 (AMSAT Phase III-B)

On 16 June 1983 AMSAT Phase III-B was successfully launched by a European Space Agency Ariane rocket along with the European Communications Satellite ECS-1. Upon separation from the launcher Phase III-B became known as AMSAT-OSCAR 10. A launch information network, probably the most extensive such network ever arranged using Amateur Radio, enabled thousands of hams on all continents to listen as the milestones unfolded during what appeared to be a perfect launch.

The OSCAR 10 beacon sprang to life on schedule about 2.5 hours after liftoff. When the first report of telemetry reception arrived from New Zealand station ZL1AOX shortly thereafter, a second sigh of relief reverberated around the world. But the feeling of exultation was quickly replaced by anxiety as analyses of the telemetry showed that the spacecraft was capturing very little solar power and running dangerously cold.

To gauge the real competence of a technical staff you have to observe how they handle the unexpected. The AMSAT crew performed flawlessly under pressure. The problems were analyzed, corrective meausures taken, and plans for orbit transfer revised before most radio amateurs realized the potentially serious nature of the situation. Weeks later, AMSAT was to learn that OSCAR 10 had been bumped twice by the Ariane rocket's third stage shortly after separation.[30] Although the impact appears to have damaged both the liquid fuel motor and the 2-m antenna, the spacecraft was successfully raised to a long lifetime orbit, and the antenna damage seems minor.

The unforeseen problems slowed the orbital-transfer and checkout phase by a scant three weeks, and on 6 August 1983, a new era in Amateur Radio began as the Mode B transponder on OSCAR 10 was opened for general use. Several more days were to pass, however, before orientation maneuvers pointed the high-gain spacecraft antennas directly at the earth so that users could observe the excellent system performance.

AMSAT Phase III-B was closely patterned after Phase III-A. Two significant changes, however, merit attention.

Instead of the solid-fuel kick motor used on Phase III-A, Phase III-B employed a high-power liquid-fuel rocket. The new motor would make it possible to (1) place the Phase III-B spacecraft in a more desirable orbit, (2) periodically adjust the orbit if desired and (3) add additional shielding to the spacecraft computer to reduce the chance of radiation damage. These changes will increase the design lifetime of the satellite to 7 to 10 years. To help achieve this goal, radiation damage studies of the computer were made a Argonne National Laboratories. The liquid-fuel motor, valued at 2 million dollars, was donated by the manufacturer, Messerschmitt-Bolkow-Blohm (MBB). It was a backup for the European Symphonie communications satellite. Because of the post launch damage the motor could only be fired once. As a result, OSCAR 10's final orbit has an inclination of 26° instead of the 57° target value. The effects are relatively minor: (1) North-south communications performance will be slightly enhanced at the expense of east-west paths; (2) it may be necessary at times to slightly mis-aim the high gain antenna in order to collect sufficient solar power; and (3) amateurs using mechanical tracking devices may find it necessary to change overlays more frequently than planned.

The second major change in the Phase III-B spacecraft is that the redundant transponder will be an 800-kHz-wide L-band unit (1269-MHz uplink, 436-MHz downlink). During Phase III-B's early years in space Mode B will be used most of the time. It's imperative, however, that we take future trends in technology and user numbers into account when planning for a spacecraft with such a long potential lifetime. The width of the Mode B transponder is constrained by its use of all available spectrum space at 146 MHz. The wide bandwidth, L-band transponder will accommodate 4.5 times as many users. Some commercial equipment for uplinking at 1269 MHz is already available at reasonable cost and more will doubtless follow now that OSCAR 10 is in orbit. Ground stations using the L-band transponder will need about 20 watts of 1269-MHz energy feeding a 4-foot diameter parabolic dish antenna, or a system producing an equivalent radiated power.

Taking these factors into account places the projected 200 to 300 thousand dollar investment in Phase III-B in an interesting perspective.

Backups for most Phase III-B systems were built to full flight specifications. These backups constitute most of a Phase III-C spacecraft. A kick motor similar to the one used on Phase III-A is already on hand for this project and efforts to identify a launch are underway. Technical details of AMSAT-OSCAR 10 are contained in Appendix A.

SYNCART

SYNCART is an acronym for *syn*chronous *a*mateur *r*adio *t*ransponder. The SYNCART project is related to the "host satellite" concept. A host spacecraft would provide power, station-keeping (maintaining position) and attitude-control functions to each passenger, and possibly a large parabolic antenna to work with passenger feeds. Because of the services provided, each passenger's package can be much simpler than a complete satellite. Several design constraints, however, demand special attention: flexibility, host protection and rf compatibility.

The SYNCART package must be easily adaptable to whatever power bus and control signals are provided by the host. It must be designed so that the host will be fully protected from any conceivable transponder failure mode. And, the compatibility design work must be thorough and conservative. Full testing in the presence of all other flight packages might not take place until after the spacecraft is in orbit or, perhaps, until just before launch when it's too late for last-minute changes.

In 1979 members of AMSAT-Canada began working on many of the major elements of a SYNCART package that was designed to be integrated into a geostationary satellite which was scheduled for a 1980 launch. Unfortunately, the host spacecraft mission was cancelled. The California-based Project OSCAR group has joined with the Canadian team to continue work on this project as the search for a new host mission goes on. Current plans are to configure the transponder to use a 1.26-GHz uplink and a 435-MHz downlink, though the design employs a modular approach that makes it relatively easy to change the frequencies of either link.

ARSENE

In 1978 a group of French radio amateurs met with the directors of CNES (French National Center for the Study of Space) to discuss the possibility of a French-built radio amateur satellite, and the educational and scientific benefits of such a project. The directors reacted very favorably and said they foresaw no difficulty in supporting the program. They would provide leftover parts and equipment not usable for future CNES programs and would offer access to pre-flight testing facilities as long as no direct financial commitment was required. This meeting led, in January 1980, to the formation of RACE (Amateur Radio Club for Space), a coordinating group linking the REF (French National Radio Amateur Association), CNES, and several technical schools and industrial enterprises.

Plans for the first French radio amateur spacecraft, ARSENE, focus on a Phase III model in a high-altitude elliptical orbit. In a tentative time schedule, they have identified a launch opportunity near 1985.[31]

Additional Spacecraft

Radio amateurs involved in the Soviet satellite program have also expressed an interest in building a high-altitude Phase III spacecraft. Since Soviet commercial satellites make considerable use of the 63° inclination elliptical orbit (the highly desirable attributes of this orbit are discussed in Chapter 8), obtaining a launch may not be a serious obstacle. In fact, a Soviet Phase III radio amateur spacecraft may not need an onboard propulsion system: It could be dropped off in a desirable orbit. Predicting a target date for this mission is pure speculation but, based on past Soviet schedules for amateur spacecraft and AMSAT experience, 1986 seems a reasonable possibility.

Several additional spacecraft and space-related activities currently being discussed are of interest to radio amateurs: PACSAT, STS and Space Mirror. PACSAT, a digital communications amateur spacecraft, will operate very much like the popular computer bulletin boards showing up everywhere. Instead of using long distance telephone calls to file your messages or pick up your "mail," however, you use a radio link when the satellite passes nearby. PACSAT will be placed in a relatively low, near-polar orbit. The Space Transportation System (STS), better known as the Space Shuttle, also presents some interesting amateur opportunities. For example, mission specialist Astronaut Dr. Owen Garriott, W5LFL, operated 2 m fm from space aboard the STS-9 Space Shuttle *Columbia* late in 1983. The Space Mirror project group is evaluating the possibility of placing a stationary rf reflector about 100 miles above the earth, held in place by radiation pressure from a ground based transmitter. The 10-meter-diameter dish-shaped reflector would be made of an ultrafine mesh of metal-coated graphite wires. Radio amateurs located within about 800 miles of the Mirror could communicate with each other using it as a passive reflector.

Mention of one other tentative mission may be premature

but I'll leave that judgment to you. First, some background. In 1973 NASA awarded a contract to the Jet Propulsion Laboratory (JPL) to study the feasibility of using a solar sail for a deep-space mission. A solar sail is, as the name implies, a large sail designed to harness photons (light) emitted by the sun for space propulsion in much the same way a sailboat uses the wind. One of the key conclusions of the JPL study was that the solar sail, by freeing us from the intrinsically inefficient rocket-expelling-propellant approach to space flight, could drastically reduce the cost of deep-space transportation. Even though the JPL study was highly optimistic about solar sail propulsion, proposals for a test flight were never funded. The World Space Foundation (WSF) was formed by a group of individuals who believed that an amateur scientific organization could oversee the construction and launch of a small-scale solar sail mission — much the same way radio amateurs have produced satellites.[32] AMSAT may be working with the WSF in engineering telemetry and command-link systems for a near-earth, solar-sail test flight. If the orbit looks appropriate it may be possible to place a transponder aboard the spacecraft.

Finances

In the good old days of OSCARs I and II when satellites were relatively simple, most flight hardware was donated. Out-of-pocket expenses incurred in building and launching a spacecraft were generally picked up by the same people doing the volunteer work. As amateur satellites grew more complex and expensive it became necessary to seek additional donations to help finance the program. In 1962 the informal OSCAR Association incorporated as Project OSCAR and encouraged hams all over the world interested in the amateur space program to support it financially by signing up as members. Insofar as possible, dues were to be used to pay for flight hardware. In addition, a commitment was made to publish an inexpensive newsletter supplying information about the satellite program for members.

In the late 1960s, the hub of amateur satellite activities shifted from southern California to Washington, DC as AMSAT began to assume the central role in satellite construction. But the goals and financial support structure for amateur activities didn't change. The financial base of the program remained dependent on a large number of modest contributions from individual donors in the amateur community. In this section, we'll take a brief look at the costs of constructing an amateur satellite, sources of support for amateur satellite activities, and the ability of the international radio amateur community to finance a large-scale, long-term satellite program.

Costs

The direct expenses involved in placing a spacecraft in orbit can be categorized as follows:

1) Launch fee
2) Technical expertise (engineering design)
3) Flight hardware (satellite parts)
4) Ground hardware (prototype subsystems, special test instruments, telecommand stations, etc.)
5) Construction (salaries or contracted costs for machining, wiring, testing, etc.)
6) Administrative (parts procurement, required technical documentation, user documentation, bookkeeping, etc.)
7) Travel, shipping, customs, communication (telephone, telex, postage)
8) Miscellaneous (launch insurance, etc.)

Note that this list contains only spacecraft-related expenses; costs of operating an organization, publishing a newsletter or providing membership services have not been included.

Launch Fees. The largest single expense associated with placing a commercial satellite in orbit is the launch. In the early 1980s a dedicated (single satellite) launch into geostationary orbit costs roughly 25 to 30 million dollars. OSCARs I-IV and 5-10 have been launched for free in recognition of their potential benefit to society in the areas of disaster communications, educational applications and scientific investigation. The message here is clear: Amateurs must actively continue to support and document such activities if they wish to receive future launches without paying commercial rates. The first eight OSCARs rode into space as secondary payloads on U.S. missions; Phase III-A and Phase III-B were mated to the Ariane rocket recently developed by the European Space Agency. NASA is now considering imposing a minimum charge for non-commercial Space Shuttle launches. Any such fee is expected to be reasonable (under $10,000). The prospects for obtaining future launches, either for free or at modest cost, look good at this time.

Technical expertise. The second largest expense in putting up a commercial satellite goes toward paying salaries to the technical staff. An estimate of the cost of having Phase III-A designed and built commercially is 2 million dollars, of which 1.5 million would be attributed to technical staff. Almost all of the high-technology engineering support going into AMSAT satellites is contributed by volunteers.

Hardware, Construction, etc. A significant percentage of the project hardware for the spacecraft and for ground testing and fabrication is donated by private sector companies, NASA and other government agencies. Such donations are often arranged by radio amateurs working in space-related industries. These amateurs, on their own initiative, volunteer to present the AMSAT case to the key people involved. Many of the tasks associated with spacecraft construction (machining, testing, administrating) are also handled by volunteers. Nevertheless, there comes a time when special items must be purchased. For example, solar cells for OSCARs 6 to 8 were donated by NASA from small stores of backups left over from other missions. The large quantity of high-efficiency cells needed for Phase III missions generally can't be obtained this way. So, solar panels in the future will have to be purchased at a cost of roughly $30,000 per spacecraft. The Phase III-A project required extensive foreign travel, as major construction was roughly split between the U.S. and West Germany, with integration and testing taking place in the U.S. Additional testing while the spacecraft was mated to the launch vehicle was done in Toulouse, France; and the launch took place from French Guiana, on the north coast of South America. All in all, AMSAT's total cash outlay directly attributable to Phase III-A amounted to roughly $210,000, certainly a large sum by amateur standards.[33]

Sources of support

We've already seen how launches, contributions of parts and equipment, free access to special test facilities, and services volunteered by dedicated individuals account for roughly 95% to 98% of the cost of placing a Phase III spacecraft in orbit. The critical remaining support must come in the form of cash: modest donations from a large number of individuals who believe in the goals of the amateur satellite program. These small donations serve several important functions in addition to providing a stable financial base. By demonstrating to government agencies being approached for launches, to companies being asked for donations, and to the volunteers working on the project that there is widespread support for the amateur space program in the radio amateur community, these small donations engender larger ones.

Financial responsibility for the amateur satellite program currently rests with AMSAT. It's important that radio amateurs understand that AMSAT and ARRL are separate organizations, each trying to serve the needs of its members as best it can. There has been, and continues to be, a great deal of cooperation between them, and ARRL has made several large donations to AMSAT. Nonetheless, the governing bodies and the financial resources of the two groups are totally distinct from one another, and at times their goals may differ. To restate an important point: Individual memberships in AMSAT are extremely important to the amateur space program — as a source of funds and because they demonstrate to ARRL, to governments and to corporations, the extent of the interest in and support for amateur satellite activities.

Financial resources

Can the international amateur community support an extensive satellite program? Consider the roughly $300,000 cost of a single Phase III spacecraft. With a 5-year design lifetime this amounts to about $60,000 per year. Since the transponder should be able to accommodate about 15,000 radio amateurs, each spending a few hours per week actively communicating, the average cost per user will be well under $5 per year.[34] This doesn't even take into account the unlimited number of scientific experimenters and educators using the beacons. To put expenses in perspective, note that $5 per year per user is close to the cost of operating the average terrestrial 2-m repeater and far less than the cost per member of a newsletter and other membership services. When looked at this way the cost of a Phase III spacecraft is a little less overwhelming.

This discussion of budget has so far glossed over one very significant consideration — this is *not* a pay-as-you-go operation. The bill for Phase III-A has to be taken into account and Phase III-B has to be paid for *before* launch. Once the Phase III program is well underway (mid 1980s) a modest membership fee should enable AMSAT to maintain a system of satellites, build new spacecraft and provide membership services (tracking and technical support information). The transition to Phase III is a difficult one, however, and every member (new or old) and every contribution (no matter how small) is critical during this period.

Notes

[1]D. T. Bellair and S. E. Howard, "Australis-Oscar," *QST,* Vol. LIII, no. 7, July 1969, pp. 58-61.

[2]D. T. Bellair and S. E. Howard, "Obtaining Data from Australis-Oscar 5," *QST,* Vol. LIII, no. 8, August 1969, pp. 70-72, 82.

[3]J. A. King, "Proposed Experiments with Australis-Oscar 5," *QST,* Vol. LIII, no. 12, Dec. 1969, pp. 54-55.

[4]R. Soifer, "Australis-Oscar 5 Ionospheric Propagation Results," *QST,* Vol. LIV, no. 10, Oct. 1970, pp. 54-57.

[5]J. A. King, "Australis-Oscar 5 Spacecraft Performance," *QST,* Vol. LIV, no. 12, Dec. 1970, pp. 64-69.

[6]P. M. Wilson and R. T. Knadle, "Houston, This is Apollo...," *QST,* Vol. LVI, no. 6, June 1972, pp. 60-65.

[7]K. Nose, "Using the ATS-1 Weather Satellite for Communications," *QST,* Vol. LV, no. 12, Dec. 1971, pp. 48-51.

[8]Science/Scope, Hughes Aircraft Co. Advertisement, *Scientific American,* Vol. 242, no. 4, Apr. 1980, p. 143.

[9]O. G. Villard, Jr. and R. S. Rich, "Operation Smoke-Puff," *QST,* Vol. XLI, no. 5, May 1957, pp. 11-15.

[10]V. R. Frank, R. B. Fenwick, O. G. Villard, Jr., "Communicating at VHF via Artificial Radio Aurora," *QST,* Vol. LVIII, no. 11, Nov. 1974, pp. 27-31, 34.

[11]V. R. Frank, "Scattering Characteristics of Artificial Radio Aurora," *Ham Radio,* Vol. 7, no. 11, Nov. 1974, pp. 18-24. (Contains an extensive bibliography).

[12]M. W. Browne, "June Space Test [Firewheel] to Outdo Moon in Brief Display," *N. Y. Times,* Vol. CXXIX, no. 44547, Tues. 8 April 1980, pp. C1, C2.

[13]J. A. King, "The Sixth Amateur Satellite," Part I, *QST,* Vol. LVII, no. 7, July 1973, pp. 66-71, 101; Part II, *QST,* Vol. LVII, no. 8, Aug. 1973, pp. 69-74, 106. This article is highly recommended for anyone interested in satellite design.

[14]P. I. Klein, J. Goode, P. Hammer and D. Bellair, "Spacecraft Telemetry Systems for the Developing Nations," 1971 IEEE *National Telemetering Conference Record,* April 1971, pages 118-129.

[15]J. C. Fox and R. R. Dunbar, "Preliminary Report on Inverted Doppler Anomaly," ARRL Technical Symposium on Space Communications, Reston, VA, Sept. 1973, pp. 1-30.

[16]L. Kayser, "SMART-System Multiplexing Amateur Radio Telecommands," ARRL Technical Symposium on Space Communications, Reston, VA, Sept. 1973, pp. 31-45.

[17]P. I. Klein and J. A. King, "Results of the AMSAT-OSCAR 6 Communications Satellite Experiment," *IEEE National Convention Record,* NYC, March 1974.

[18]J. Kasser and J. A. King, "OSCAR 7 and Its Capabilities," *QST,* Vol. LVIII, no. 2, Feb. 1974, pp. 56-60.

[19]P. Klein and R. Soifer, "Intersatellite Communication Using the AMSAT-OSCAR 6 and AMSAT-OSCAR 7 Radio Amateur Satellites," *Proceedings of the IEEE Letters,* Oct. 1975, pp. 1526-1527. M. Davidoff, "Predicting Close Encounters: OSCAR 7 and OSCAR 8," *Ham Radio,* Vol. 12, no. 7, July 1979, pp. 62-67

[20]P. Klein and J. Kasser, "The AMSAT-OSCAR D [8] Spacecraft," *AMSAT Newsletter,* Vol. IX, no. 4, Dec. 1977, pp. 4-10.

[21]S. Budin and F. Fekhel, "Amateur VHF/UHF Repeaters," *RADIO,* no. 10, Oct. 1975, pp. 14-15.

[22]Special Section No. SPA-AA/159/1273 annexed to International Frequency Registration Board Circular No. 1273 dated 12 July 1977, submitted by USSR Ministry of Posts and Telecommunications.

[23]V. Dobrozhanskiy, "Radioamateur Satellites; The Repeater: How is it Used?" *RADIO,* no. 9, Sept. 1977, pp. 23-25. Also see July, Oct., and Nov. issues of *RADIO* for additional information.

[24]R. Labutin, "The USSR 'Radio' Satellites — Preliminary Results," *RADIO,* no. 5, May 1979, pp. 7-8. For a summary of this article in English see — *Telecommunication Journal,* Vol. 46, no. X, Oct. 1979, pp. 638-639.

[25]M. Davidoff, "The Future of the Amateur Satellite Service," *Ham Radio,* Vol. 10, no. 8, Aug. 1977, pp. 32-39.

[26]T. Clark and J. Kasser, "Ariane Launch Vehicle Malfunctions, Phase III-A Spacecraft Lost!," *Orbit,* Vol. 1, no. 2, June/July 1980, pp. 5-9.

[27]J. A. King, "Phase III: Toward the Ultimate Amateur Satellite"
Part I, *QST,* Vol. LXI, no. 6, June 1977, pp. 11-14;
Part II, *QST,* Vol. LXI, no. 7, July 1977, pp. 52-55;
Part III, *QST,* Vol. LXI, no. 8, Aug. 1977, pp. 11-13.

[28]J. A. King, "The Third Generation"
Part I, *Orbit,* Vol. 1, no. 3, Sept./Oct. 1980, pp. 12-18;
Part II, *Orbit,* Vol. 1, no. 4, Nov./Dec. 1980, pp. 12-18.

[29]M. Sweeting, "The AMSAT Amateur Scientific and Educational Spacecraft — UoSAT," *Orbit,* Vol. 2, no. 2, March/April 1981, pp. 13-17. *The Radio and Electronic Engineer,* Journal of the Institute of Electronic and Radio Engineers (England), Aug./Sept. 1982, Vol. 52, no. 8/9. Special issue on: "UoSAT — The University of Surrey's Satellite." This issue is highly recommended for anyone interested in satellite design.

[30]J. Eberhart, "Satellite Hit By Its Own Rocket," *Science News,* Vo. 124, Aug. 6, 1983, p. 87.

[31]"The ARSENE Project," *Orbit,* Vol. 2, no. 1, Jan./Feb. 1981, pp. 13-15.

[32]World Space Foundation, Solar Sail Project, P.O. Box Y, South Pasadena, CA 91030. See, for example: J. Eberhart, "Riders of the Light," *Science News,* Vol. 120, no. 21, Nov. 21, 1981, pp. 328-332.

[33]T. Clark "Spacecraft Economics," *Orbit,* Vol. 1, no. 3, Sept./Oct. 1980, pp. 22-26.

[34]M. Davidoff "Cost-Performance Criteria for Evaluating Phase III Satellites," *AMSAT Newsletter,* Vol. IX, no. 1, March 1977, pp. 6-7.

Part II

Chapter 4
Getting Started

Chapter 4

Getting Started

Ground station operators often find themselves at the receiving end of the question: "How do I get started?" As with many engineering problems there is no universal answer. Radio amateurs who are considering the plunge into satellite communications should first identify the aspects that appeal to them, take stock of their equipment and experience, and then devise a flexible, progressive plan for setting up a ground station. Such plans typically involve a series of steps, each one a slightly greater commitment, taking into account such important constraints as finances and available time. As a beginner you're going to be faced with what will seem like an endless series of choices, ranging from the general, such as "which satellite and which mode?" to the specific, such as "what equipment and which antenna?" Our aim, in Chapters 4 through 7, is to provide the basic information you'll need to make these decisions.

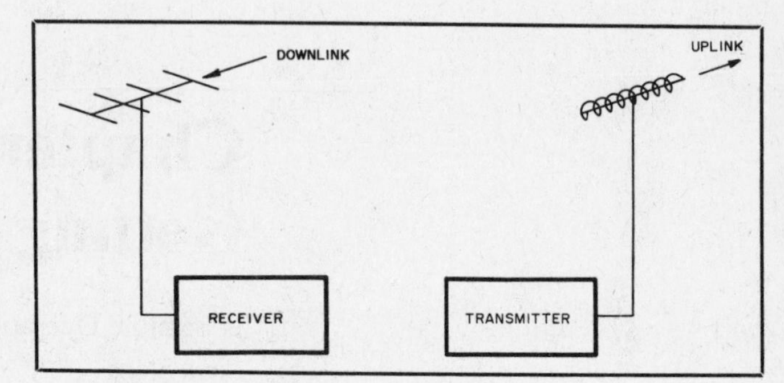

Fig. 4-1 — Simple satellite ground station.

Is Satellite Operation for You?

Before you start to invest time and money in setting up a ground station, give some serious thought to whether satellite operation is really for you. The *glamour* of space communication can be a powerful lure, but when you get down to the nitty-gritty, long-term appeal will depend on (1) your communications needs (do you really require the reliable and predictable long-distance capabilities of a Phase III spacecraft, or are you attracted to the challenge of contest-style activity on a Phase II satellite?), (2) your technical or scientific interests (do you especially enjoy the technical challenges or scientific aspects of space communications?) and (3) your time and financial resources. Locate a few local hams who have had satellite experience and ask them how they feel. Some will probably see satellites as the most exciting new dimension of Amateur Radio since the discovery, back in the early '20s, that "useless" short waves could propagate across the Atlantic. Others may firmly believe that satellite relays are as exciting as the telephone and expect them to have the same future in Amateur Radio as a-m on 20 m. After talking to advocates of both viewpoints, try to understand why they feel the way they do. It shouldn't be difficult to wangle an invitation from an active satellite user to sit in during a pass. Observing a ground station in operation first-hand is the best way to get a feel for what's involved. If the bug has you in its clutches at this stage, you may as well give in. Start making plans to set up your own ground station.

First Steps

Putting together a ground station generally involves several steps.

Step 1: Learn all you can about satellite communication.
Step 2: Choose a mode (frequency combination) and satellite.
Step 3: Set up a receive station.
Step 4: Set up for cw and/or ssb transmission.

The minimal requirement for completing Step 1 is to read Chapters 1 through 7. If you've read the first three chapters you already know a great deal about satellites, including the different capabilities of Phase II (low altitude) and Phase III (high altitude) spacecraft. Your initial ground station setup will probably be designed to give you access to a particular mode (set of uplink and downlink frequencies) or satellite. Mode and satellite selection are so closely related that it's fruitless to try to separate them. The following sections are organized by mode, but considerations relating to specific satellites and the high/low altitude choice are interwoven throughout. Be sure to complete Chapters 5 through 7 on tracking, antennas and equipment before committing yourself to a particular mode or spacecraft.

The Basic Station

A satellite ground station and an hf station are not all that different. Both require a transmitter, a transmitting antenna, a receiver and a receiving antenna, and in both cases we're usually working with cw or ssb signals. See Figure 4-1. Naturally, there are also many differences, some subtle, some obvious. As we focus on these differences, don't forget the basic, underlying similarity between the hf station and the satellite ground station.

Our analysis of the radio amateur satellite ground station will consider those uplink and downlink frequencies that will be employed in the mid 1980s. Therefore, we'll be looking at receiving equipment for 29, 146 and 435 MHz, at transmitting equipment for 146, 435 and 1269 MHz, and at antennas for all of these frequencies. As all current and planned satellite transponders are designed for cross-band operation, separate receive and transmit antennas will be required (Fig. 4-1).

Choosing the Mode

The heart of a radio amateur satellite is the transponder, a device that receives a slice of the radio frequency spectrum that

Table 4-1

Transponder Frequencies and Mode Designations Used in the Amateur Satellite Service

Mode	*Ground station transmit band*[1]	*Ground station receive band*[1]	*current satellites*[2]	*future satellites*[2]
A	145.8-146.0 MHz	29.3- 29.5 MHz		
B	435.0-438.0 MHz	145.8-146.0 MHz		
J	145.8-146.0 MHz	435.0-438.0 MHz		
L	1260-1270 MHz	435.0-438.0 MHz		

[1]Each satellite may use only a part of the band indicated.
[2]These columns should be filled in by the reader with the latest information from *QST* and *Orbit*.

is centered about a particular frequency, amplifies the entire slice, and retransmits it centered about a different frequency. For example, the incoming slice might be a 100-kHz-wide segment centered about 145.950 MHz, which is amplified a million- million times, and then retransmitted as a 100-kHz slice centered about 29.450 MHz. The transponders used on Amateur Radio satellites accept cw, ssb, fm, digital, video, or any other type of signal, and then retransmit them in the same format, but shifted in frequency. Devices of this type are called linear transponders. Today, all transponders on radio amateur spacecraft receive signals on one band and retransmit them on a different band. Each combination of frequency bands is called a *mode*. The various modes currently in use or planned for the near future are listed in Table 4-1.

A simple formula associated with each transponder enables one to predict the *approximate* downlink frequency corresponding to each uplink frequency. Actual downlink frequencies may vary by several kilohertz because of a phenomenon known as Doppler shift. Whenever the transmitter and receiver on a link are in motion relative to one another, the receive frequency is shifted from its expected value. (For a more detailed discussion of Doppler shift, see Chapter 10.) The translation formula for each satellite is given in Appendix A. Note that there are basically two types of linear transponders: the *non-inverting* type, which retransmits the entire slice as received (as in Fig. 4-2a), and the *inverting* type, which flops or reverses the slice (as in Fig. 4-2b) before retransmission. Although frequency inversion may at first appear to be an unnecessary complication, it does serve to reduce Doppler shift significantly.

We now consider some of the principal advantages and disadvantages of each mode.

Mode A — 2 m Up/10 m Down: If you have an hf receiver (or transceiver), monitoring Mode A is an easy way to start working with satellites. In the 1972-1982 time period, the majority of satellite operators began by listening to AMSAT-OSCARs 6, 7 and 8, or RS-1 through RS-8 at the upper end of the 10-m band. The popularity of this route arose because it was quick and inexpensive. Though a receive crystal for 29.0 MHz to 29.5 MHz or a 10-m preamp was often needed, many amateurs found that listening required only setting the receiver to the correct frequency at the proper time. Before you choose this path, check recent issues of *QST* or *Orbit* to determine whether the currently active satellites are carrying Mode A transponders, and update Table 4-1 for reference. As of August 1983, four operational satellites are equipped for Mode A: RS-5, RS-6, RS-7 and RS-8.

The popularity of Mode A attests to its advantages; but it also has several inherent limitations. First, 29-MHz signals are often subject to ionospheric absorption and reflection, serious disadvantages for a satellite link. The 10-m band was chosen as a downlink despite this problem to make it as easy as possible for those amateurs already owning hf equipment to become acquainted with satellite communications. The expectation was (is)

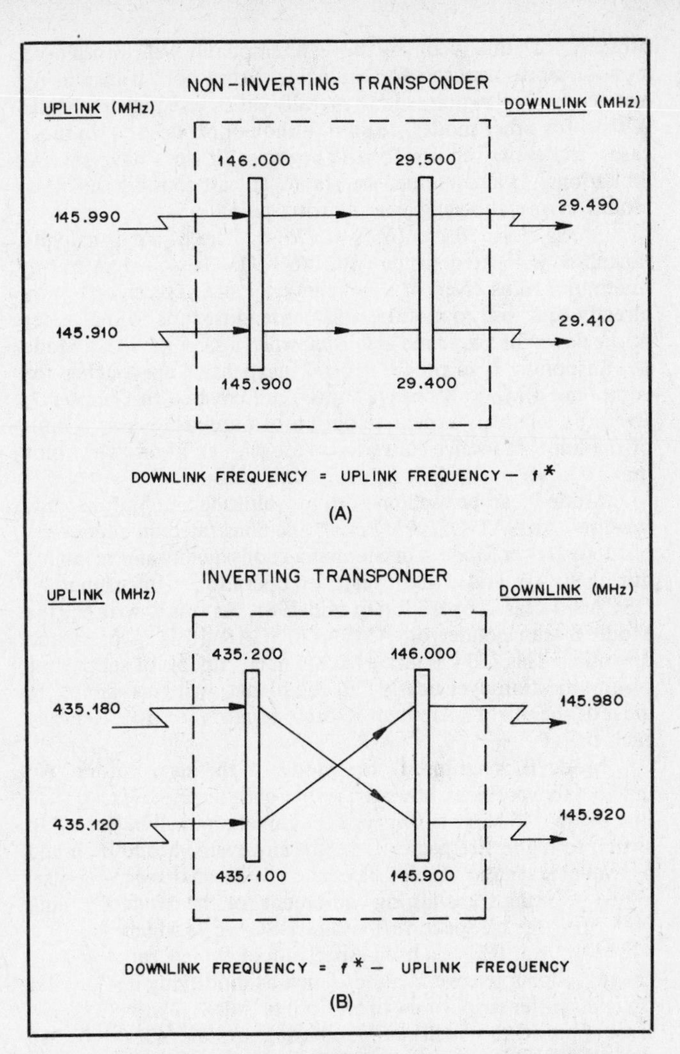

Fig. 4-2 — Examples of a non-inverting transponder (Fig. 4-2a) and an inverting transponder (Fig. 4-2b). Values of f* for each transponder currently in orbit are included in Appendix A.

that those who enjoy satellite activity will eventually want to step up to one of the more suitable modes. When the 10-m band is open for terrestrial communications, Mode A downlink signals from a spacecraft are frequently reflected back into space; nothing is heard on earth. The propagational shortcomings of Mode A are especially noticeable near peaks in the sunspot cycle. For this and other technical reasons Mode A will be limited to low-altitude (Phase II) spacecraft. Since low-altitude satellites are only in range for four to six short passes (10 to 25 minutes each) per day, restricting your operating to Mode A will limit your total daily access time unless several Phase II satellites are active. Another limiting factor of Phase II satellites is that their maximum communications range is usually under 5000 miles.

If you do opt to start on Mode A you'll find that it's important to be able to predict those times when a satellite will be in range. This aspect of tracking is covered in Chapter 5. If you're a little lazy, however, it is possible to ignore tracking temporarily. Just tune your receiver between 29.300 MHz and 29.500 MHz on upper sideband every now and then when you're in earshot. With several low-altitude satellites transmitting in this band (a total of six — RS-3 through RS-8 — at this writing), the probability of hearing a spacecraft is on the order of 0.4 or 40%. After listening to several passes and learning to predict when the spacecraft will be in range, you'll probably begin to consider methods, such as preamps and special antennas, to improve your receiving system's performance (Chapters 6 and 7). Once the

Mode A downlink is coming through reasonably well, you'll need a cw or ssb transmitter for 146 MHz. From both transmitting and receiving viewpoints it's generally easier to equip for Mode A than for other modes. In sum, although Mode A is, in most cases, the easiest choice for the beginner, it does have several limitations. As a newcomer you should at least consider the other modes before choosing your starting point.

Mode B — 70 cm Up/2 m Down: The Mode B satellite downlink is located just below 146 MHz. If you own a 2-m multimode transceiver, or a 2-m converter and hf receiver, you're already equipped to monitor this mode. Just tune to the center of the downlink passband and listen when a satellite with a Mode B transponder is in range. Several inexpensive approaches for equipping a Mode B receive station are covered in Chapter 7. Even the relatively expensive option of purchasing a new, top-of-the-line 2-m receive converter to use with an hf receiver won't make too serious a dent in your bank account.

Mode B can be used on both low-altitude and high-altitude satellites. AMSAT-OSCAR 7 clearly demonstrated the superiority of Mode B over Mode A in communications quality and reliability during its six and a half years of operation. Unfortunately, OSCAR 7 ceased operation in mid 1981. As this is written, the Mode B transponder on AMSAT-OSCAR 10 is in orbit and operating. This 150-kHz-wide transponder, capable of supporting communication over nearly half the planet, will be available to most of the world's Amateur Radio operators for over 10 hours each day.

Mode B is definitely the mode of the near future. An unavoidable problem, however, is that once the 200-kHz segment just below 146 MHz is fully utilized, amateurs will be forced to switch to higher frequency links. In any event, Mode B should be available on a continuing basis and it will remain very popular. When selecting transmitting equipment for the Mode B uplink (435 MHz for all spacecraft except OSCAR 7, which was near 432 MHz) you'll have a broad spectrum of choices ranging from cheap, labor-intensive methods (such as modifying a 450-MHz fm transmitter strip for cw operation) to quick, relatively expensive approaches (such as purchasing a new 435-MHz ssb transceiver or transverter). See Chapter 7 for details. If your primary goal is to operate with a Phase III spacecraft, equipping for Mode B is the logical first step.

Mode J — 2 m Up/70 cm Down: Mode J uses the same bands as Mode B but with the uplink and downlink assignments switched. In the mid 1970s there was considerable discussion of which mode was preferable. Each mode offered clear advantages and clear disadvantages. Feelings on both sides were so strong that early plans for Phase III included both Mode B and Mode J transponders. Users would then have been able to compare the two and choose the future path. Technical problems and weight constraints that surfaced as Phase III-A was being developed forced AMSAT to abandon this approach. Mode B was chosen.

Meanwhile, a group of radio amateurs in Japan (JAMSAT) built a Mode J transponder for AMSAT-OSCAR 8 so that users could gain practical experience with this frequency combination. Because there are no solid plans to include Mode J on future spacecraft, and OSCAR 8 is no longer active, this mode cannot be recommended as the entry point for beginners. Nonetheless, should a future spacecraft carry a Mode J transponder, it does offer several features that make it very attractive to amateurs who already have satellite capabilities. In fact, many experienced users considered the Mode J transponder aboard OSCAR 8 one of the best transponders amateurs have orbited to date.

Before you can fully appreciate Mode J's performance, however, you must take several subtle points into account. These include the need for (1) a very-high-performance receive preamp and a moderate-gain receive antenna and (2) careful feed-line filtering (details appear in Chapter 7) at the ground station to prevent receiver desensitization by the third harmonic of your transmitter. Although Mode J is not currently appropriate as a first step for the beginner, anyone with Mode A capabilities needs only acquire a good 435-MHz receive converter to add this mode. Morever, as we'll note shortly, the 435-MHz receive system will serve in conjunction with the Mode L transponder.

Mode L — 23 cm Up/70 cm Down: The 200-kHz-wide segment of the 2-m band set aside for satellite operation becomes saturated quickly. When this occurs, amateurs are forced to turn to higher frequencies such as the 1269 MHz to 435 MHz combination. Because international treaties limit use of 1269 MHz to uplinking, L-band transponders will be designed so that ground stations will transmit on 1269 MHz and receive on 435 MHz. AMSAT-OSCAR 10 carries an 800-kHz-wide Mode L transponder.

When amateurs first heard of plans for the Mode L transponder the most common reaction was, "Oh no, how am I ever going to put a transmitter on 1269 MHz?" The truth is that it's not a big deal. The equipment prospects for this mode are much brighter today than they were for Mode B back in 1970 when Mode B was first proposed for OSCAR 7. Power requirements for accessing the transponder at 1269 MHz are modest, and most of the technology already exists. Although designing efficient microwave equipment is challenging, duplicating a well-engineered unit can be relatively simple, especially when broadband stripline techniques are used; this approach eliminates the use of many discrete inductors and capacitors by having their equivalents built into the pc-board artwork. Critical wiring and adjustments are thereby minimized. As a result, building a piece of microwave gear can be almost as simple as assembling an audio-frequency kit. And, when it becomes apparent that the market for 1269 MHz transverters has grown, commercial units should become more readily available and construction articles more common.

The need for the Mode L transponder is clear, and it's certainly within the financial and technical grasp of the average amateur. But from an operational viewpoint, what can it provide? An 800-kHz-wide transponder can handle nearly as many cw and ssb users as the 10, 14, 18 and 21 MHz hf bands combined. Alternatively, if amateurs so choose, 20 channels spaced 15 kHz apart could be devoted to fm and 500-kHz to cw and ssb. In any event, this mode will increase the effective spectrum available for long-distance satellite communications by a factor of 15. The AMSAT-OSCAR 10 Mode L transponder by itself will accommodate roughly five times the number of amateurs who use the Mode B transponders aboard Phase III spacecraft. AMSAT groups in Canada and California, currently working on SYNCART (*Syn*chronous *A*mateur *R*adio *T*ransponder), are planning to use the L band for their primary transponder, and Phase III-C will also carry one.

Now that the first transponder of this type is in orbit it makes sense to consider it as a possible option. This illustrates how the trade-offs involved in starting out in space communications are constantly changing as new satellites are placed in orbit, the prices of solid-state microwave components drop, and new commercial microwave gear becomes available. A beginner in the mid 1980s might conceivably elect to start out on this mode.

UoSAT-OSCAR 9

UoSAT-OSCAR 9 is a radio amateur scientific satellite built by a group of radio amateurs, educators and scientists at the University of Surrey in England. It contains several instruments designed to be of use to radio amateurs who are interested in studying propagation, and a camera that is optimized for recording land/water transitions (see Chapter 3) and Appendix A). UoSAT-OSCAR 9 does *not* carry a transponder. Our purpose in mentioning it here is to suggest that radio amateurs who are primarily interested in the scientific aspects of space or propagation may choose it as a starting point. Tracking techniques are similar to those used for any low-altitude, circular-orbit spacecraft.

The receiving equipment needed to monitor the telemetry and picture data from UoSAT-OSCAR 9 is very simple: a 2-m fm receiver or public service scanner that can be tuned to frequencies just below 146 MHz. Radio amateurs may also want to monitor the propagation beacons on the 7-, 14-, 21- and 28-MHz hf bands and on several microwave frequencies. Although receiving signals from this satellite is relatively easy, recovering the scientific information or photos from the telemetry requires additional circuitry (a microcomputer and a special interface between the receiver audio output and computer, or other dedicated circuits for specific functions). See Appendix A for additional information.

Planning the Ground Station

The experienced hf operator will soon discover that the operating and technical trade-offs involved in optimizing the performance of a satellite ground station are novel. Old approaches and habits acquired over years of hf operation will have to be reevaluated constantly. Higher power and bigger antennas do not necessarily lead to better performance. For example, a ground station using a very high gain antenna may find it nearly impossible to track a rapidly moving Phase II satellite properly during a contact.

One significant advantage of satellite communications over hf links is the ability to monitor your own downlink signal in the receiver while you're transmitting (duplex operation). On the hf bands, testing a new antenna often means collecting signal reports over several months of operation and then guessing which antenna, new or old, works better. Ground station transmitting antennas for satellite operation can be compared directly by switching back and forth between them while listening to the downlink: an infinitely faster and more accurate approach.

Duplex operation, however, is more than mere convenience; it's a necessity. If your ground station isn't set up for duplex operation, you'll never know during a satellite pass just what your downlink frequency is, or whether your signal is so strong that it's overloading the satellite, or so weak that the downlink is unintelligible.

Since the ability to transmit and listen *at the same time* is critical to satellite communications, an hf or vhf transceiver cannot simultaneously be used as a central element in both the TX and RX blocks of Fig. 4-1. If, for example, you already own an hf transceiver and you're interested in trying Mode A ssb operation, you might either (1) use the hf transceiver to monitor the 29.5 MHz downlink and buy a multimode 2-m transceiver to use for the uplink, or (2) use the hf transceiver to drive a 2-m transmit converter and buy a modified ssb CB transceiver for 29 MHz reception. You have many other options, of course, but the point here is that the hf transceiver may be used as part of either the uplink or the downlink system, but *not* both. After years of successful hf and vhf transceiver operation it's all too easy for satellite newcomers to get caught in the *transceiver trap.* Don't let it happen to you. Plan your station for full duplex operation from the beginning.

Satellite Operations

Radio amateurs thrive on competition. We want top-notch equipment and we take pride in our operating skills. Satellite communications probably won't change this outlook, but it may change some of our notions of what makes superior stations and effective operators. Cooperation is the key. For example, the early RS satellites challenged our thinking. Use of too much power only succeeded in distorting all signals or shutting down the transponder. This certainly didn't lead to invitations to join the A1 Operators Club. The first-class operator will be one who uses only a fair share of the transponder's bandwidth and power, efficiently and effectively milking every last iota of performance from the system. Successful stations won't be noted for the biggest antennas or the most powerful transmitters. They'll outperform others because of the attention paid to receiver preamp performance, antenna polarization, adjustable transmitter power to match changing conditions, antenna systems designed for minimal operator intervention, and so on. Conventional hf approaches to improving performance, such as increasing power or adding

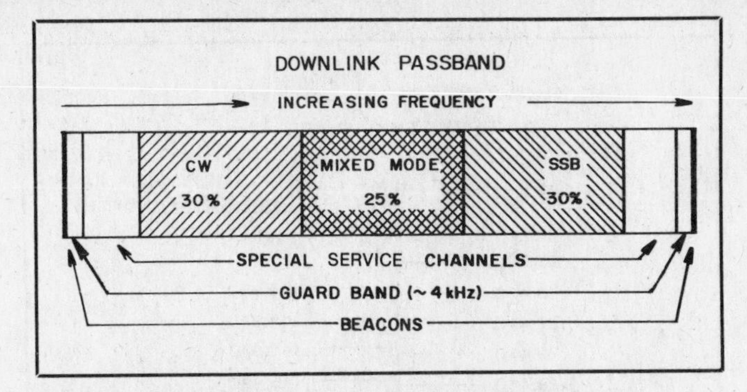

Fig. 4-3 — General downlink bandplan for OSCAR satellites. The special service channels (SSCs) are meant to be used in a coordinated manner for special activities such as bulletins by national societies, code practice, emergency communications and computer networking. Channels are designated H1, H2, . . . (high end), L1, L2, . . . (low end); with H1 and L1 being closest to the beacons. Listen for ARRL W1AW bulletins on AMSAT-OSCAR 10 SSC L2 (145.820 MHz, cw) and H2 (145.962 MHz, ssb).

speech processing, will likely be counterproductive. Experiments on 432-MHz EME, for example, have convinced most operators that with this weak-signal mode any type of speech processing reduces signal intelligibility; a clean ssb signal works best.

Bandplan

Amateurs have voluntarily adopted certain guidelines for use of different types of modulation on amateur satellites. Ssb and cw are preferred because of their efficient use of transponder power and bandwidth. Fm and slow-scan TV are discouraged, except for certain experimental applications, because they use a relatively large share of the transponder's power and bandwidth. The current bandplan is shown in Fig. 4-3, though you can expect small changes to evolve as users' needs change. Note that only the downlink is considered; the bandplan is independent of whether the transponder is inverting or non-inverting.

Operate Effectively

Whether using high-altitude or low-altitude satellites, effective and courteous operators follow a few simple guidelines. To call CQ, select a clear frequency on your receiver. Then, using a chart like the one in Fig. 4-4, set your approximate transmit frequency and send a series of dits while listening for them on the downlink. Adjust your transmitter as necessary (*between* bursts of dits) to bring your signal to the desired downlink frequency. Because of Doppler shift, a little hunting is almost always necessary. The procedure to use for answering a CQ is similar. Good dial calibration and a little experience will minimize the time you spend sending dits and interfering with other stations.

Once contact has been established, adjust only your *transmitter* frequency to compensate for Doppler. This will prevent you from chasing your contacts up or down the band. Although this procedure is especially important with low- altitude satellites where Doppler is more of a problem, it's a good habit to develop with Phase III spacecraft, too.

If you are using the special service channels on Phase III spacecraft, note that they are referenced to the nearest beacon frequency. One of the stations on the special service channel should be designated to check the operating frequency periodically (on the hour and half hour) and adjust it if necessary. All others should tune to this downlink.

Since all stations will be operating duplex (simultaneous transmit and receive), be careful to avoid the horrendous howls and garbled audio that very l-o-n-g feedback loops produce. The most effective (and simple) way is to use headphones. Using a speaker will almost always leave you lunging for your receiver audio gain control and searching for the nonexistent level setting that will provide feedback-free listening. Be considerate; use headphones.

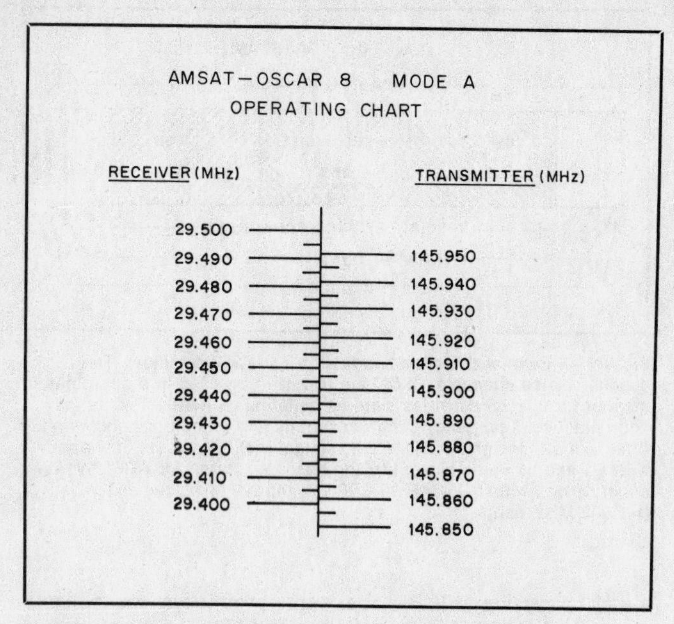

Fig. 4-4 — Ground stations will find it convenient to make an operating chart like the one shown above for each satellite transponder they use. This particular chart was constructed using the translation formula for A-O-8: downlink freq. = uplink freq. − 116.458 MHz. Because of Doppler shift this formula is only approximate (see text).

Most of the tracking methods discussed in the next chapter refer to time after the satellite passes a particular reference point. Using this information in conjunction with a clock that is set either to UTC or local time will require some unnecessary mental arithmetic. Although the addition or subtraction involved is trivial, you'll have to do it many times, often while you've got other operational concerns to contend with. Keeping an extra, inexpensive digital clock at your operating position (one that can be easily reset to zero or some other time) will pay off many times over in operating convenience.

Because DX windows for low-altitude satellites are very brief, you'll note a lot of contest-style short exchanges in the passband. Nevertheless, ragchewing is also common and welcomed on Phase II satellites: The choice is yours.

Contests

A great many amateurs enjoy competing in contests and working toward awards — activities that encourage perfecting operating skills and station performance. Certain features of satellite communication, however, make it very important for us to consider carefully what types of contests and awards are appropriate. Satellites are a shared resource and are most effectively used when everyone cooperates, especially in using the minimum necessary power levels. When communicating via the ionosphere on hf, a low-power operator can use skill and patience to compete against higher-power stations; no one can overload the ionosphere and make it useless for others. Satellites, however, can be overloaded by a few inconsiderate stations, and then no amount of skill or patience will let anyone maintain communication.

Does this mean satellites should not be used for contests or awards? Certainly not! It just suggests that we consider the effects of various activities carefully before endorsing them. Several types of contests can contribute to the advancement and general enjoyment of satellite communication for all. For example, we might wish to encourage emergency preparedness by continuing to support Field Day satellite use. QRP-only DXCC and similar awards might cause operators to improve their ground stations' performance and their operating skills. Contests that encourage the occupancy of underutilized transponders might also prove worthwhile. In any event, remember that AMSAT does not have control; anyone can sponsor a contest or award. Therefore, in a very real sense, the future is up to the user community. Your interest, support and tolerance will determine the future of satellite contests and awards. Be sure to take a stand and make your views known.

A Final Hint

If your personal goals in setting up a satellite ground station include acquiring a firm practical grasp of satellite communications and a thorough understanding of how each element of a ground station contributes to overall performance, you'll find it best to start simple, start cheap, and take one step at a time. If you're mainly interested in getting on the air and communicating, you'll find that all the equipment needed to put together a first-class ground station is available commercially at prices comparable to a modest hf station.

Chapter 5
Tracking Basics

Chapter 5

Tracking Basics

This chapter focuses on basic satellite tracking: what it means, why it's usually necessary, and how to do it. Two simple, widely used techniques for tracking Phase II and Phase III satellites are described in detail. Alternative tracking methods will be presented in Chapter 9 where the detailed mathematical and physical bases of the various techniques will be covered. If you don't care much for mathematics or computers, don't worry. Basic tracking takes only the ability to add, subtract and read a simple map.

Tracking: What, When, Why?

To a scientist, tracking a satellite means being able to specify its position in space. To a radio amateur, tracking more likely refers to practical concerns: When will a satellite be in range and where should the antenna be pointed? Satellites generally are moving targets, so when you use directional antennas, you'll constantly need to update the aiming information. The ability to predict access times is also important because most satellites are in range of a specific ground station for only a part of each day. (Geostationary satellites, which remain over a fixed location on the equator, are an exception we'll discuss later in this chapter.) A low-altitude satellite (such as AMSAT-OSCAR 8, UoSAT-OSCAR 9, RS-3 through RS-8) will generally be in range for less than 25 minutes each time it passes nearby (a *satellite pass*). Four to six passes near a given location usually occur each day. A high-altitude satellite in the elliptical orbit planned for early Phase III missions will generally have two passes each day but with a total access time of (very roughly) 12 hours for Northern Hemisphere stations. The ability to predict when a satellite will be in range will enable you to plan nets and demonstrations, arrange schedules with others in specific locations, and avoid wasted time in front of a silent receiver.

Before we get down to the details of tracking, note that in several situations tracking can be ignored. For example, I spend considerable time in my *radio room* reading (and writing). Sitting here, I often leave a receiver on, tuned to 29.4 MHz so I can monitor the various low-altitude satellites using Mode A (each satellite can be identified by its characteristic telemetry beacon frequency and content) and keep in touch with friends. As omnidirectional transmit and receive antennas are adequate for this type of casual operation, antenna aiming isn't necessary. Users of Phase III spacecraft are also, at times, able to dispense with tracking by simply flipping on a 2-m receiver attached to an omnidirectional antenna such as a ground plane and tuning around 145.9 MHz. If the *band is open* (satellite in range) weak signals will be noticeable. One may then switch to a beam antenna and adjust the rotator controls to peak the receiver S-meter.

The knowledge that tracking can sometimes be avoided might tempt you to skip the rest of this chapter. RESIST! You're bound to encounter situations when you'll wish you knew how to track. The ability to track will add immeasurably to your enjoyment of working with satellites.

We begin by looking at what radio amateurs will need for tracking information. A good satellite tracking aid should enable one to predict:

1) When the satellite will be in range: more specifically, times for *AOS* (*a*cquisition *o*f *s*ignal) and *LOS* (*l*oss *o*f *s*ignal);

2) Proper antenna direction (azimuth and elevation) at any time;

3) The regions of the earth that have access to the spacecraft at any instant.

Often you'll need only one or two of these features. The tracking device should also be simple to construct, easy to use and inexpensive. The *OSCARLOCATOR* (and the *Satellabe*, which is similar) and the *ϕ3 TRACKER* were designed with these needs in mind.

The ARRL OSCARLOCATOR is a simple map with overlays for satellites in circular orbits (fixed height). This includes all currently operating Phase II spacecraft. The ϕ3 TRACKER is an adaptation of the OSCARLOCATOR that permits tracking of satellites in elliptical orbits (height constantly changing) — the type planned for early Phase III missions. Our explanation of the ϕ3 TRACKER will assume that you already know how to use the OSCARLOCATOR; tackle them in order, even if you're only interested in tracking Phase III spacecraft. Geostationary satellites will be treated separately. Since a stationary (geostationary) satellite appears to occupy a fixed position directly over the equator, it presents a very simple tracking problem. If it's in range, antenna aiming parameters have to be determined only once from a given location; they don't change.

Probably the most troublesome hurdle associated with tracking is mastering the new jargon. Take it slow and make sure you understand the informal, practical explanation given with each new italicized term as it is presented. The definitions are summarized in Table 5-1 for your convenience. You're probably already familiar with several of the terms we'll be using. AOS and LOS were just discussed.

The *subsatellite point* (*SSP*) is the point on the surface of the earth directly below the satellite. For most satellites the SSP constantly moves as the satellite moves across the sky. If we were to watch the SSP as the satellite traveled along its orbit it would trace a curve on the surface of the earth called the *ground track* or *subsatellite path*.

A satellite will be in range when the SSP is *close* to your ground station location, and out of range when the SSP is far away from your location. This seemingly obvious statement is the key to using the OSCARLOCATOR and ϕ3 TRACKER. Of course, we have to define how close "close" is. To do this we compute a critical *acquisition distance* associated with each Phase

Table 5-1
Glossary of Tracking Terms

access range (acquisition distance)

acquisition circle: "Circle" drawn about a ground station and keyed to a specific satellite. When SSP is inside circle the satellite is in range.

acquisition distance (access range): The maximum distance between a ground station and SSP, measured along the surface of the earth, at which the satellite is in range (corresponds to 0° elevation).

AOS (Acquisition Of Signal)

apogee: Point on orbit where satellite height is maximum.

ascending node (EQX): Point where ground track crosses equator with satellite headed north.

ascending pass: Satellite pass during which satellite is headed in a northerly direction while in range.

azimuth: Angle in the horizontal plane measured clockwise with respect to North (North = 0°).

coverage circle: Region of earth that is eventually accessible for communication to a particular ground station via a specific satellite.

descending node: Point where ground track crosses equator with satellite headed south.

descending pass: Satellite pass during which satellite is headed in a southerly direction while in range.

elevation: Angle above the horizontal plane.

elevation circle: The set of all points about a ground station where the elevation angle to a specified satellite is a particular value.

EQX (ascending node)

geostationary satellite: A satellite that appears to hang over a fixed point on the equator.

ground track (subsatellite path): Path traced out by SSP over the course of one complete orbit.

increment (longitudinal increment)

longitudinal increment: Change in longitude of ascending node between two successive passes of specified satellite. Measured in degrees West per orbit.

LOS (Loss Of Signal)

node: Point where ground track crosses the equator.

OSCARLOCATOR: A tracking device designed to be used with a satellite in a circular orbit (satellite height fixed).

pass (satellite pass)

PCA (Point of Closest Approach): Point on ground track during orbit of interest where satellite passes closest to specific ground station.

perigee: Point on orbit where satellite height is minimum.

period: The amount of time it takes for a satellite to complete one revolution about the earth.

point of closest approach (PCA)

range circle: "Circle" of specific radius centered about ground station.

reference orbit: First orbit of UTC day.

Satellabe: A tracking device similar to the OSCARLOCATOR but with added features.

satellite pass: Segment of orbit during which satellite passes nearby and in range of ground station.

spiderweb: Set of azimuth curves radiating out from a particular location, and the concentric elevation or range "circles" about the location.

SSP (SubSatellite Point)

subsatellite path (ground track)

subsatellite point (SSP): Point on surface of earth directly below satellite.

TCA (Time of Closest Approach): Time at which satellite passes closest to specific ground station during orbit of interest.

window: Overlap region between acquisition "circles" of two ground stations. Communication between the two stations is possible when SSP passes through window.

ϕ3 TRACKER: A tracking device designed to be used with a satellite in an elliptical orbit (satellite height constantly changing).

Expressions in parentheses are synonyms or acronyms. Note that true circles on the globe are often distorted when transferred to a map. Some minor differences will be found between the definitions in this Table, which focus on the practical aspects of tracking, and those in the Glossary, where more emphasis has been placed on technical precision.

II satellite and with geostationary satellites. When the SSP is closer to you than the acquisition distance, the satellite is in range (see Table 5-2).

The situation is most easily pictured by using a world map and drawing an *acquisition circle* around your station. In Figure 5-1 we show an acquisition circle from AMSAT-OSCAR 8 drawn about Washington, DC. Note how circles on the surface of the earth (roughly a sphere) appear distorted on most flat maps. If the ground track for a specific orbit of a particular satellite passes inside your acquisition circle it will be accessible (in range) during the pass. AOS occurs when the SSP enters the acquisition circle; LOS occurs when SSP leaves the acquisition circle. Determining when a satellite in an elliptical orbit is in range is a little more complicated. We'll look at this problem later in this chapter when we discuss the ϕ3 TRACKER.

The ground track for almost all satellites crosses the equator twice per orbit. The two points where the ground track and equator cross are called *nodes*. The *ascending node* occurs when the SSP crosses the equator headed north; the *descending node* occurs when the SSP crosses the equator headed south. Most tracking approaches use the ascending node, or northward equatorial crossing point (sometimes abbreviated *EQX*) as a reference point.

The amount of time it takes a satellite to go through one complete orbit (revolution of the earth) is called its *period*. The periods of amateur Phase II satellites range from about 95 minutes (UoSAT-OSCAR 9) to 120 minutes (RS-3 through RS-8). The periods of Phase III satellites will probably range from 10 to 24 hours. Knowing a satellite's period (the time for one complete orbit) you can compute the number of orbits per day (about 12 to 15 for Phase II spacecraft, 1 to 2 for Phase III spacecraft).

The OSCARLOCATOR

The OSCARLOCATOR is the most widely used tracking aid for AMSAT-OSCAR 8, RS-3 through RS-8, and UoSAT-OSCAR 9. It consists of two parts:

1) *Map board.* A map centered on the North Pole like the one shown in Fig. 5-1. A full-size map is presented in Appendix

Table 5-2
Distances Between SSP and Ground Station Corresponding to Specified Elevation Angles

Satellite	*Satellite height*	*0° elevation circle (radius) Acquisition distance*	*30° elevation circle (radius)*	*60° elevation circle (radius)*
UoSAT - OSCAR 9	338 miles 544 km	1581 miles 2544 km 22.9°	488 miles 785 km 7.1°	178 miles 286 km 2.6°
AMSAT - OSCAR 8	565 miles 909 km	2000 miles 3218 km 29.0°	741 miles 1192 km 10.7°	280 miles 451 km 4.1°
AMSAT - OSCAR 7	907 miles 1459 km	2456 miles 3952 km 35.6°	1050 miles 1689 km 15.2°	414 miles 666 km 6.0°
RS-3 through RS-8	1050 miles 1690 km	2610 miles 4200 km 37.8°	1161 miles 1868 km 16.8°	464 miles 747 km 6.7°
SYNCART (geostationary)	22,285 miles 35,860 km	5619 miles 9041 km 81.3°	3627 miles 5836 km 52.5°	1774 miles 2854 km 25.7°
Phase III (typical elliptical orbit)				
apogee	22,250 miles 35,800 km	5618 miles 9039 km 81.3°		
perigee	932 miles 1500 km	2485 miles 3955 km 36.0°		

Elevation circle distances are given in miles and kilometers along the surface of the earth and in degrees along a great circle arc. For an explanation of the Phase III elliptical orbit data see the section on the ϕ3 TRACKER.

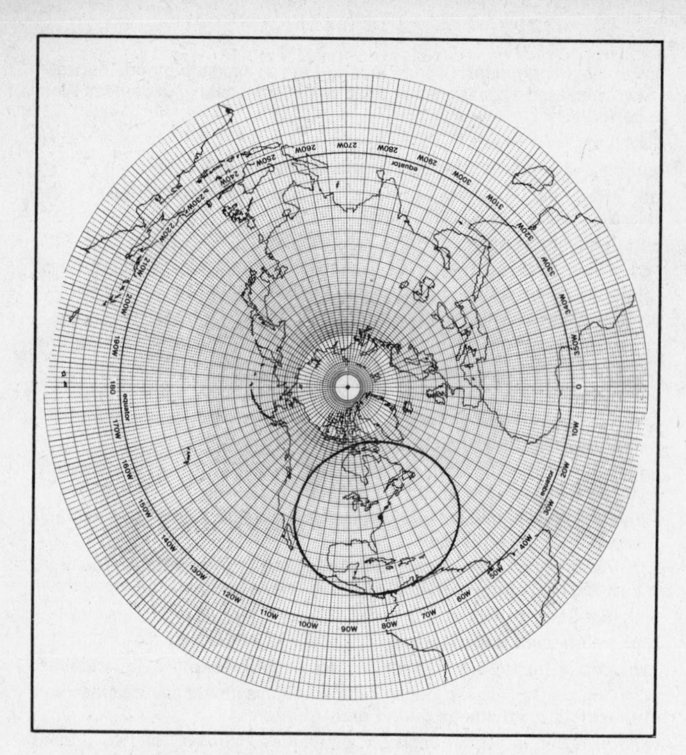

Fig. 5-1 — Typical acquisition circle for a Phase II satellite (OSCAR 8 in this case) drawn about Washington, DC.

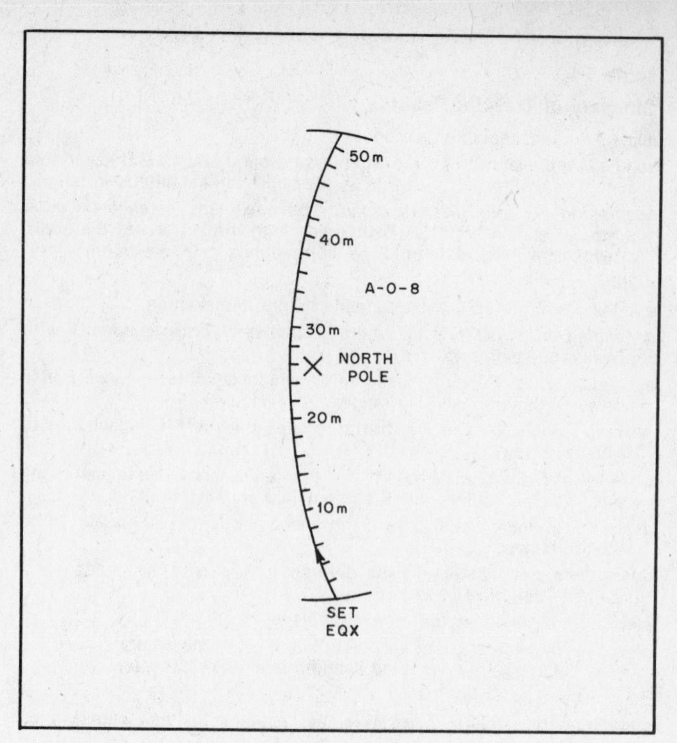

Fig. 5-2 — Typical ground track overlay (OSCAR 8).

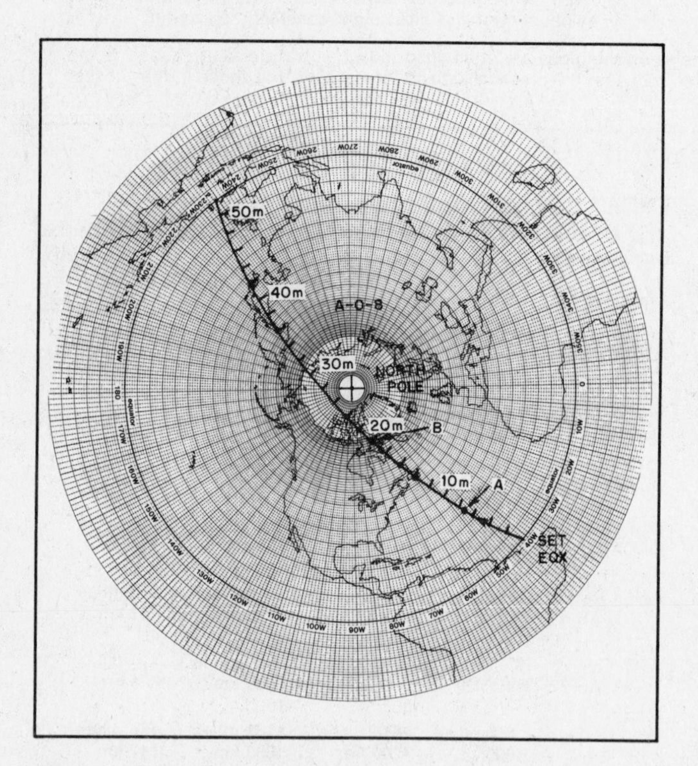

Fig. 5-3 — Ground track superimposed over a polar map with the ascending node set to 41° W longitude.

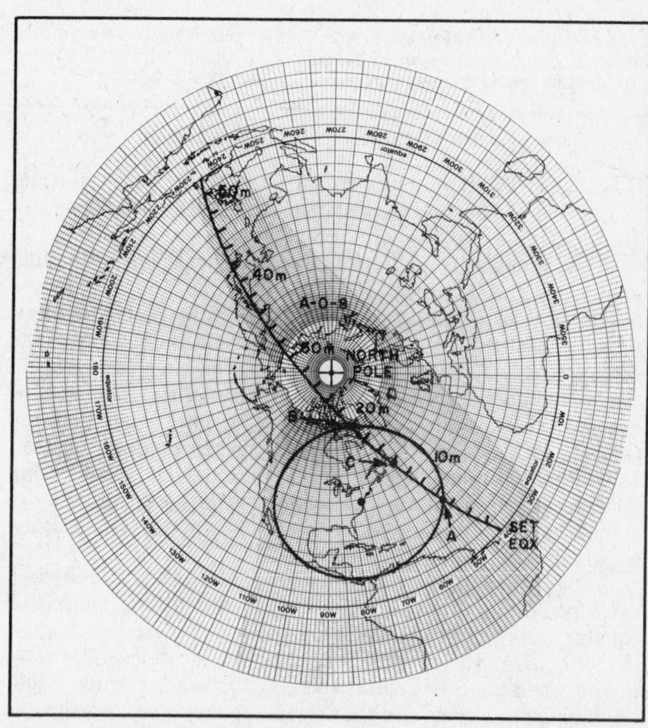

Fig. 5-4 — Acquisition circle (fixed over Washington, DC) and ground track (rotatable around North Pole at center of polar map).

B and as a foldout from the back cover.

2) *Ground-track overlay*. An overlay, usually drawn on transparent material, as shown in Fig. 5-2. The overlay is mounted on the map board so that it can be rotated about the pole.

The OSCARLOCATOR is used in conjunction with an orbit calendar. Table 5-3 shows one day from a three-month calendar for AMSAT-OSCAR 8. The 14 horizontal rows of information correspond to the 14 OSCAR 8 orbits that begin this day.

The calendar provides several pieces of information. The first column contains a reference number that uniquely identifies each orbit. Orbits are numbered consecutively starting at launch, with orbit number one beginning when the first ascending node occurs. The next two columns in the calendar present the time and longitude of the ascending node. (Latitude is zero, obviously, as nodes occur at the equator.) Times here are given in UTC (Universal Coordinated Time) using Hour:Minute:Second (H:M:S) nota-

Table 5-3
One Day From a Complete AMSAT-OSCAR 8 Orbit Calendar
AMSAT-OSCAR 8
21 June 1980 (173) Saturday

Orbit No.	*Time UTC (H:M:S)*	*EQX (°W)*
11695	00:53:37	66.0
11696	02:36:49	91.8
11697	04:20:01	117.6
11698	06:03:14	143.4
11699	07:46:26	169.2
11700	09:29:38	195.0
11701	11:12:50	220.8
11702	12:56:02	246.6
11703	14:39:15	272.4
11704	16:22:27	298.2
11705	18:05:39	324.0
11706	19:48:51	349.8
11707	21:32:03	15.6
*11708	23:15:16	41.4

*Text example.

tion. For our purposes UTC and GMT (Greenwich Mean Time) may be regarded as identical. Longitudes are given in degrees West. Although the calendar is dated June 21, keep in mind that this is a UTC date; the first two orbits would actually be seen by a U.S. East Coast station Friday evening, June 20 EDT; a station on the U.S. West Coast would see the first four orbits begin on June 20 PDT. The number "173" in parentheses following the date indicates that June 21 is the 173rd day of 1980.

The best way to learn how to use the OSCARLOCATOR is to follow an example through from start to finish. For our illustration we've arbitrarily chosen the last OSCAR 8 orbit of the day on Saturday, June 21, 1980. The bottom row of Table 5-3 is the one we need: orbit number 11,708, which has an ascending node at 23:15:16 UTC at 41.4° West longitude. (For most applications, rounding off the time to the nearest minute and the longitude to the nearest degree is acceptable.) To study this orbit the ground-track overlay on the OSCARLOCATOR should be rotated until the ascending node (the "0" end of the ground track) aligns with 41° West longitude on the map board; it will remain set at this point for the entire orbit (see Fig. 5-3). The ground-track overlay contains time ticks (the feathered marks) at 2-minute intervals, which make it possible to tell where the SSP will be at any time during the orbit. At 23:22 UTC (about 7 minutes after ascending node) the SSP will be at point A (Fig. 5-3); at 23:35 UTC (about 20 minutes after ascending node) the SSP will be at point B. You should, by now, be able to locate the OSCAR 8 SSP at any time if you have access to an orbit calendar and an OSCARLOCATOR. We now look at (1) predicting AOS and LOS and (2) obtaining antenna aiming information.

From Table 5-2 we note that OSCAR 8 will be in range whenever the distance between the ground station and SSP (measured along the surface of the earth) is less than 2000 miles. To use this information draw a circle, with a 2000-mile radius centered on your ground station, directly on the OSCARLOCATOR map board:

Whenever the OSCAR 8 SSP is inside your OSCAR 8 acquisition circle the satellite will be in range.

To continue with our example — OSCAR 8, orbit number 11,708 — let's see how a station in Washington, DC, would predict AOS and LOS. Figure 5-4 shows how your map (Fig. 5-3) looks with an OSCAR 8 acquisition circle added about Washington, DC. (Note that circles on the globe become slightly distorted when drawn on most maps.) AOS occurs at point A (about 7 minutes after the ascending node) as the SSP enters the acquisition circle. LOS occurs at point B (about 20 minutes after the ascending node). The distance between the SSP and the ground station is a minimum at point C and is called the *PCA* (*p*oint of *c*losest *a*pproach). On this orbit, the PCA is reached about 13 minutes after the ascending node. The TCA (time of closest approach) is 21:27 UTC. So, OSCAR 8 will be in range of the Washington station during orbit number 11,708 for about 13 minutes starting at 23:22 UTC. Since Washington is on Eastern Daylight Time in June we might prefer to say that the pass begins at 19:22 EDT (7:22 P.M.). (Procedures for converting to or from UTC are reviewed in the last section of this chapter.)

Because the Washington, DC, station sees the SSP when the satellite is headed north, it would call orbit number 11,708 an *ascending pass*. Note that the term ascending pass is relative. On this orbit a station in Japan would see OSCAR 8 come into range from the north and exit heading south. The Japanese station would call orbit number 11,708 a *descending pass*. A ground station generally has access to two or three ascending passes and two or three descending passes each day for OSCAR 8 and the other low-altitude amateur satellites discussed in this book. Now that you can predict AOS, LOS and TCA, we turn to the problem of aiming the antenna.

Aiming for Phase II

To aim a beam antenna we have to determine two angles. *Azimuth* (angle in the horizontal plane, side to side, measured clockwise with respect to North) and *elevation* (angle above the horizontal plane, up and down) are the angles generally used. Each angle can be discussed separately.

Consider elevation first. When the SSP coincides with the ground station location (satellite directly overhead) the antenna should be pointed straight up (90° elevation). At the positions where the SSP crosses the acquisition circle (AOS and LOS) the antenna is set to the horizontal (0° elevation). Between these extremes (SSP inside the acquisition circle) the elevation angle must be set somewhere between 0° and 90°. The method of estimating elevation angles discussed here provides the accuracy generally required. Note that the acquisition circle is composed of all points having 0° elevation. Circles corresponding to various other elevation angles can also be drawn about your location on the map. In Fig. 5-5 we've added elevation circles at 30° and 60° about the Washington, DC, station. (The lines radiating outward from the center relate to azimuth; ignore them for a few moments.) A relatively broad beamwidth antenna that is set to an elevation of 15° will work fine whenever the SSP is between the 0° elevation (acquisition) circle and the 30° elevation circle. Similarly, whenever the SSP is between the 30° and 60° elevation circles, an antenna set to an elevation of 45° should suffice. The key distances for drawing acquisition and elevation circles for a number of popular spacecraft are contained in Table 5-2.

Returning to OSCAR 8's orbit number 11,708 once again (Fig. 5-5), we can estimate the elevation at PCA to be approximately 20°. A reasonable strategy for a Washington, DC, station operating this pass would be to leave the antenna set at 10° elevation for the entire time. Note that normal amateur antenna systems neither permit nor require precise tracking of Phase II satellites; you have a few degrees of tolerance.

Now consider azimuth. Azimuthal directions radiating out from a ground station generally appear as curved lines on a map. Fig. 5-5 shows a set of such curves centered on Washington. Noting the position of the SSP at any time in relation to these curves, we can use "eyeball interpolation" to estimate the correct antenna azimuth heading. Returning again to OSCAR 8's orbit number 11,708, we see that AOS occurs at an azimuth of roughly 110°, PCA at an azimuth of 50°, and LOS at an azimuth of 5°. Taken together, the set of concentric elevation circles and curved azimuth radials is often referred to as a *spiderweb*. Several methods for drawing spiderwebs will be covered in the construction hints section of this chapter.

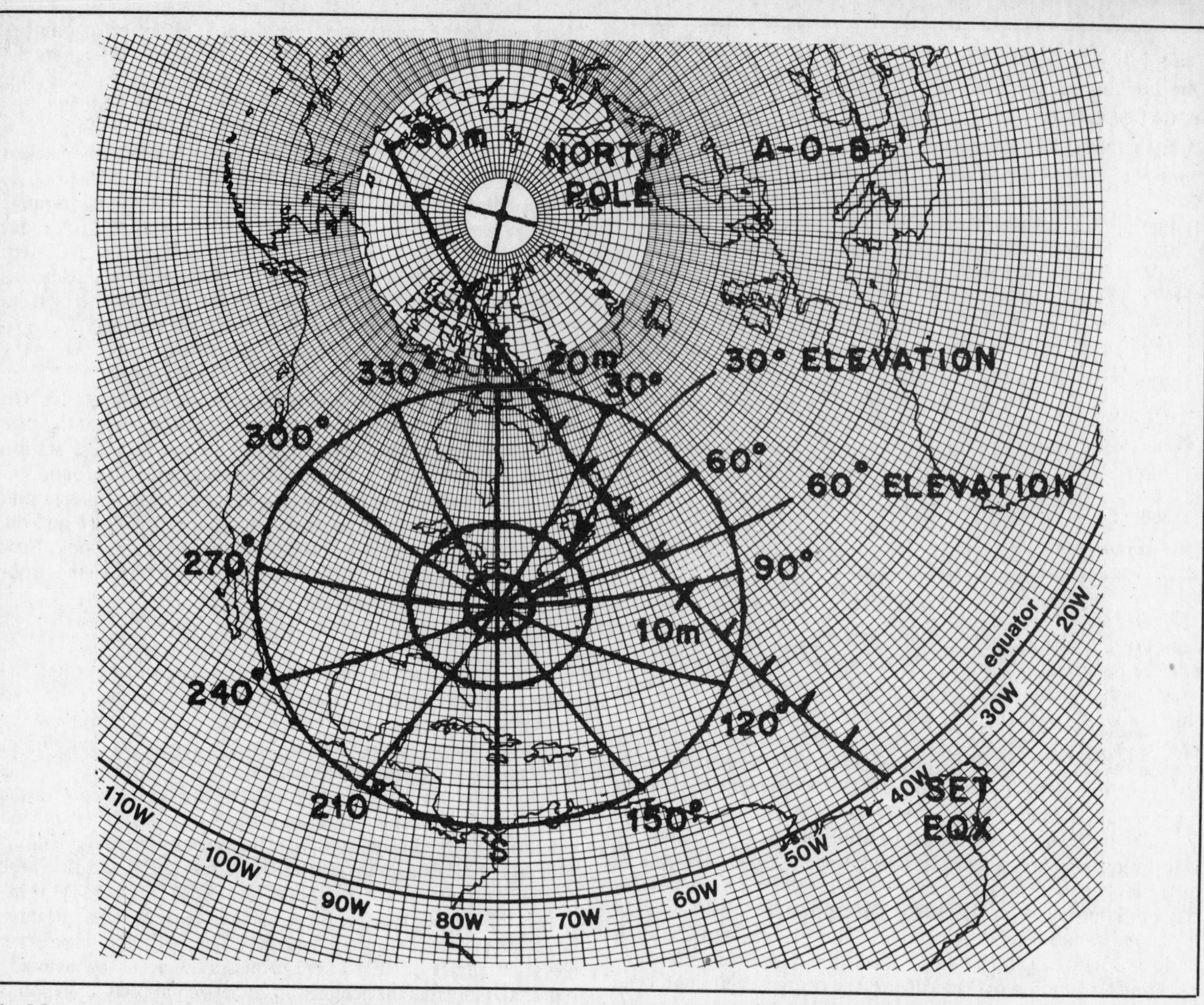

Fig. 5-5 — Added to the previous figure is the "Spiderweb" azimuth radials emanating from the station location, Washington, DC in this case, and concentric elevation circles. These are used for aiming antennas in azimuth (side to side) and elevation (up and down).

Simple extensions of the information just presented make it possible for us to determine (1) those regions of the earth eventually accessible to us via a specific satellite and (2) those orbits suitable for communicating with distant stations. [Note: skipping this paragraph and the following one won't affect your understanding of the remainder of the chapter.] Again take OSCAR 8 as an example. Since the maximum acquisition distance for this satellite (Table 5-3) is 2000 miles, two stations separated by twice this distance (4000 miles) can communicate with one another but only when the SSP is at the midpoint of the great circle path joining them. Just as we drew acquisition and elevation circles on the map board, we can draw a *coverage circle* to show which regions of the earth will eventually be in range of a given ground station. The radius of the coverage circle is twice the acquisition distance. A station in New York who wanted to know if communication with London was possible through OSCAR 8 would only have to check his coverage circle to learn the answer (yes).

To select suitable orbits for communicating over the New York to London path, draw acquisition circles for both stations on the map board as in Fig. 5-6. Whenever OSCAR 8's SSP is in the football-shaped overlapping region (called the *window*), communications between the two stations is possible. To find the best pass, rotate the overlay so the ground track passes through the center of the window. As shown in the figure, when the equatorial crossing is 26° W the mutual window will open approximately 11 minutes after ascending node and last for about 8 minutes. Now rotate the overlay slightly to determine the limits

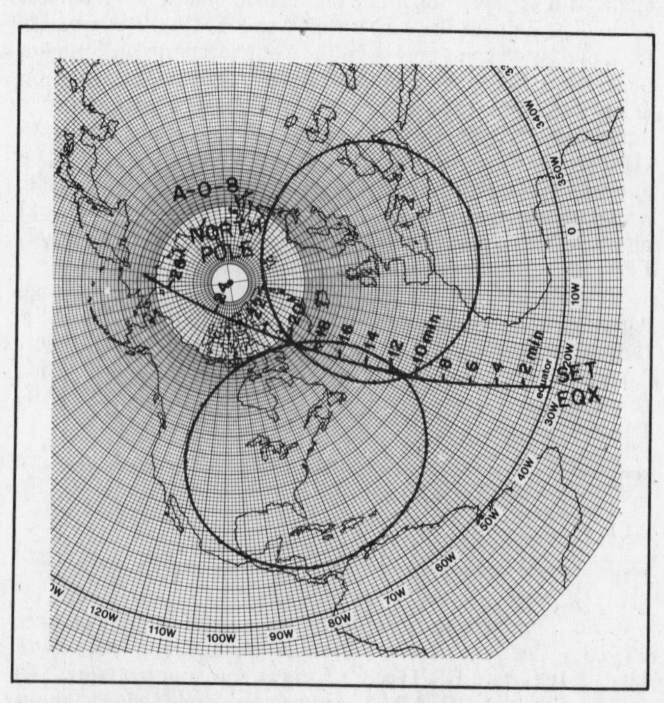

Fig. 5-6 — An AMSAT-OSCAR 8 orbit that passes through the center of the NY-London window.

on equatorial crossings that permit communication. Passes with EQXs between 19° W and 31° W will permit New York to London communication. Descending passes, when the equatorial crossing is between 217° W and 229° W, will also produce ground tracks passing through the window. Once you've got the limits for ascending and descending nodes check through an orbit calendar to locate appropriate orbits.

At this point, we've covered all the basic information needed to operate the OSCARLOCATOR. One further topic related to the orbit calendar must be treated before we proceed to Phase III tracking techniques. A careful analysis of Table 5-3 reveals that each line on the orbit calendar differs from the preceding line by 1:43:12 (1 hour, 43 minutes, 12 seconds in H:M:S notation) or 103.20 minutes (decimal notation) in the time column, and by 25.8° in the longitude column. This means that if, for example, data describing only the first orbit of a day was available, data for all remaining orbits this day could be generated by successively adding 103.20 minutes (time column) and 25.8° (longitude column). The number 103.20 minutes is the *period* (time for one complete orbit) of OSCAR 8. The number 25.8° is the *longitudinal increment* (usually just called the *increment*) for OSCAR 8 and is given in degrees west per orbit. Since a complete calendar is quite lengthy, most magazines carrying this information present data only for the first orbit in each UTC day (called the *reference orbit*) and provide the period and increment values, leaving the reader to compute the remaining orbits. To check your understanding of how this is done, try taking the first row in Table 5-3 and, using the numbers just quoted for the OSCAR 8 period and increment, reproduce the rest of the chart.

Though various orbit calendars and listings available sometimes use different formats, heading abbreviations and time notation, you shouldn't have any trouble using them. Data for each orbit must include time and longitude of the ascending node; all other information is optional. Before you attempt to use an orbit calendar make sure you know the form of time notation being used and whether longitudes are being measured in degrees west of Greenwich. Otherwise you could easily make needless mistakes.

Generating orbital calendar data for several days or weeks from a single reference orbit is possible, but very small inaccuracies in the values for period and increment you're working with will soon produce large cumulative errors in the results. Listing the reference orbit for each day is a convenient compromise. Periods and increments for several satellites are listed in Table 5-4. As we'll see in Chapter 8, the increment and the period are closely related. The increment is very nearly equal to the number of degrees the earth rotates about its axis during one orbit of the spacecraft. The earth revolves about 360° in one day, 15° in one hour and 0.25° in one minute. We now turn to the ϕ3 TRACKER and how it's used to track elliptical orbit Phase III satellites.

Table 5-4

Period and Increment for Several Satellites

These values can be used in conjunction with reference orbit data to predict orbits for up to several days. (See note in text about cumulative prediction errors.)

Satellite	*Period (minutes)*	*Increment (°West/orbit)*
UoSAT-OSCAR 9	95.3	23.8
AMSAT-OSCAR 8	103.20	25.81
AMSAT-OSCAR 7	114.95	28.74
RS-3*	118.46	29.76
RS-8*	119.71	30.07
AMSAT-OSCAR 10	699.5	175.3
SYNCART	approximately 24 hours**	0.0 **

*RS-4 through RS-7: The period and increment lie between these two extremes. See Appendix A for additional information.
**These values are not needed for tracking.

Table 5-5

Color Coding Format for ϕ3 TRACKER Range Circles and Orbit Overlay

Color	*Map board (range circle radius)*	*Orbit overlay (minimum communication range)*
blue	9000 km	9000 km
green	8000 km	8000 km
yellow	7000 km	7000 km
orange	6000 km	6000 km
red	5000 km	5000 km
brown	4500 km	4500 km
black	4000 km	4000 km
not coded	3000 km	—
not coded	2000 km	—
not coded	1000 km	—

ϕ3 TRACKER

The ϕ3 TRACKER is similar in many ways to the OSCARLOCATOR. It must, however, take into account two additional factors associated with elliptical orbits: (1) the constantly changing height of a satellite and (2) slow changes in the shape of the ground-track overlay. The ϕ3 TRACKER consists of:

1) *Map board*. (The same map board can be used for the ϕ3 TRACKER and the OSCARLOCATOR.)

2) *Ground-track overlay*. (This overlay, usually drawn on transparent material, is mounted on the map board so it can be rotated about the pole.)

3) *Elevation angle table*.

In conjunction with the ϕ3 TRACKER you'll need an orbit calendar or reference orbit listing to provide the information for setting the ground-track overlay. Because the height of a satellite in an elliptical orbit is constantly changing, the acquisition distance (also called *access range*) will not be constant. Drawing acquisition and elevation circles about a particular ground station as we did previously is impossible. Instead, a set of *range circles* at fixed distances are drawn about a given location. The range circles are color coded for distance using the format given in Table 5-5. The reason for the color coding will become clear shortly. The ground track overlay for the ϕ3 TRACKER is also color coded (in accordance with the distance specified in Table 5-5) but this time the colors signify the minimal access range during each segment of the orbit. During the green section of the orbit, for example, the access range will be *at least* 7000 km. This means that any ground station located within 7000 km of the SSP will have access to the spacecraft. The color code is the *key* to understanding the operation of the ϕ3 TRACKER:

The green range circle about your ground station is a rough acquisition circle during the green segment of the orbit, the yellow range circle is a rough acquisition circle during the yellow segment of the orbit, etc.

Note that we're interested only in matching colors: It's *not* necessary to memorize the distances associated with each color.

The best way to learn how to use the ϕ3 TRACKER is to follow an example through step by step. For our illustration let's take a look at an imaginary satellite, OSCAR ϕ3*, in an elliptical orbit similar to the one that had been planned for the ill-fated AMSAT-OSCAR Phase III-A mission (see Table 5-6). Other orbit calendars may contain additional parameters for use with other tracking methods; these, however, may be ignored. Take care to note that the time and longitude entries that refer to *apogee* (point of highest altitude) are being used in our ϕ3 TRACKER, *not* those that refer to the ascending node. Orbit reference numbers, however, run from *perigee* (point of lowest altitude) to perigee. In other words, a given orbit begins one-half period (328 minutes for OSCAR ϕ3*) before apogee and ends a half period after apogee. Be sure you understand these important differences before continuing. Because a complete OSCAR ϕ3* orbit takes nearly 11 hours, the orbit calendar will generally contain only two entries each day.

Ground-Track Overlays

For a satellite in a circular orbit, a single ground-track overlay will work year after year. When a satellite is in an elliptical orbit, its ground track will change daily. With most of the elliptical orbits being considered for Phase III, the daily changes will be very slight so a single overlay may be used for a month or longer. With the OSCAR ϕ3* example, a single overlay would serve for about two months. A sample is shown in Fig. 5-7. For satellites in elliptical orbits, updated information for producing new overlays will be presented periodically in *QST, Orbit* and other magazines.

Our example focuses on OSCAR ϕ3* orbit 1607 as seen by a ground station in Washington, DC. Orbit 1607 begins at 02:37 UTC (one half period before apogee) and ends at 13:33 (one half period after apogee). The first step is to rotate the orbit overlay on the map board so the "set apogee" arrow points to the value specified in the calendar: 16° West longitude (See Fig. 5-8). You'll find it much easier to follow the example if you pause at this point to color at least the two outer range circles (blue and green) and the blue and green segments of the ground-track overlay on

Table 5-6

Orbit Calendar Entry For Imaginary Satellite OSCAR ϕ3*

OSCAR ϕ3* — Period = 656 minutes
Increment = 164° West per orbit
Inclination = 57°
1 July 1982 (182) Thursday

Orbit reference number	*Apogee time (UTC)*	*Apogee longitude (°West)*
1607	08:05	16°
1608	19:01	180°

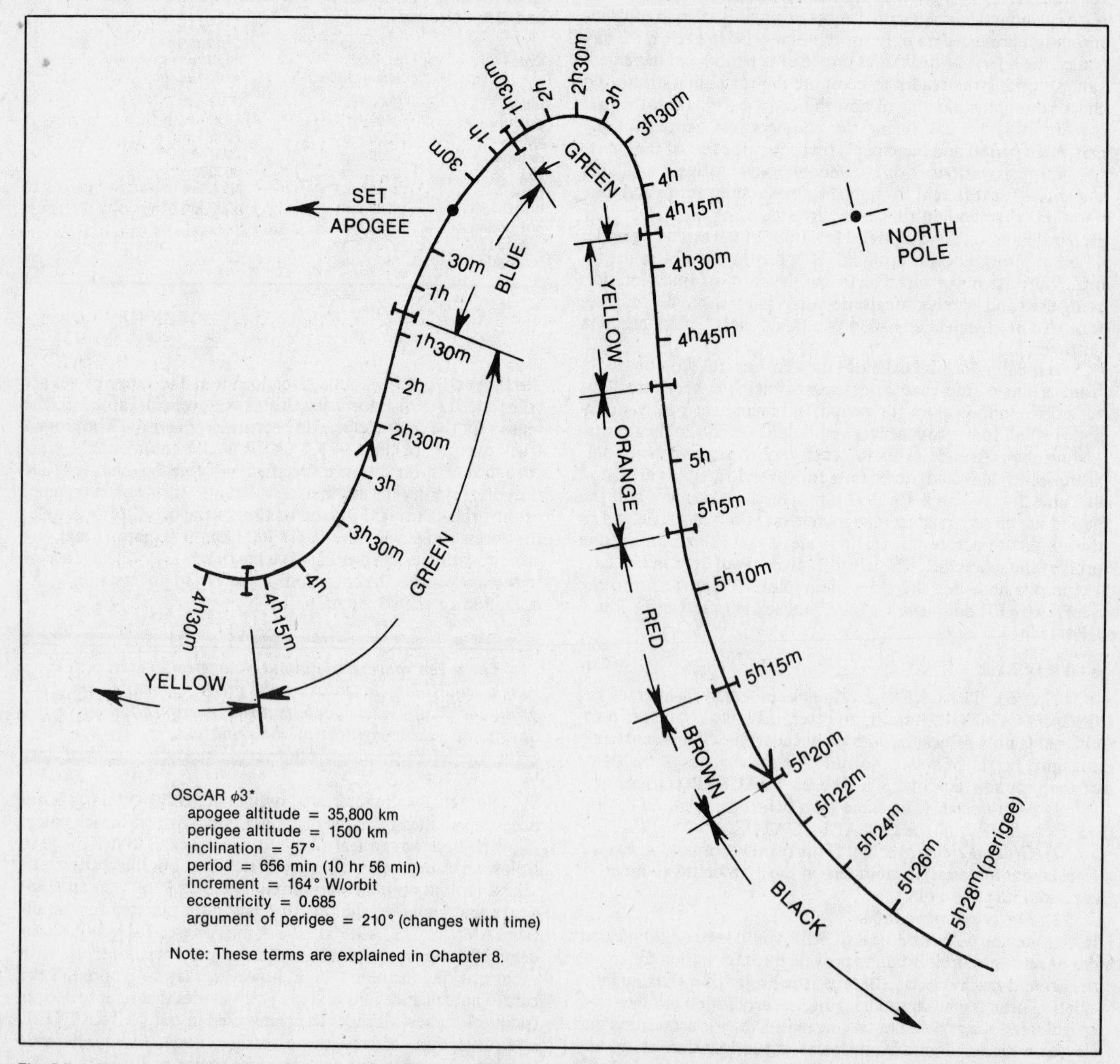

Fig. 5-7 — Typical ground-track overlay for elliptical orbit of the type planned for early Phase III missions.

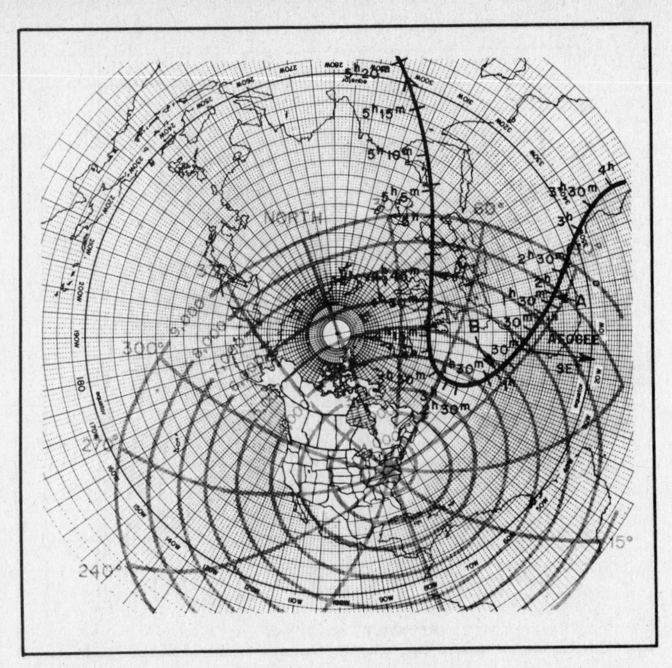

Fig. 5-8 — ϕ3 TRACKER with range and bearing curves drawn about Washington, DC.

Table 5-7
Satellite Elevation Angle (may be used with any spacecraft)

ELEVATION ANGLE (DEGREES)

GROUND TRACK OVERLAY																		
BLUE	85	79	74	69	64	58	53	48	43	38	33	28	23	19	14	9	5	0
GREEN	84	78	72	67	61	55	50	44	39	34	29	24	19	14	9	5	0	
YELLOW	83	75	68	61	55	48	42	36	30	25	19	14	9	5	0			
ORANGE	81	71	63	54	46	39	32	26	20	15	9	5	0					
RED	77	65	54	44	36	28	21	15	10	5	0							
BROWN	73	58	46	35	26	19	13	7	2			OUT OF RANGE						
BLACK	70	52	39	28	20	13	7	2										
	0.5	1	1.5	2	2.5	3	3.5	4	4.5	5	5.5	6	6.5	7	7.5	8	8.5	9

DISTANCE BETWEEN GROUND STATION AND SUBSATELLITE POINT (x 1,000 km)

Fig. 5-8 (as per the directions in Table 5-5 and Fig. 5-7). Our discussion will assume that you have done this.

During orbit 1607, AOS for Washington occurs (roughly) at point A where the green segment on the ground-track overlay crosses the green range circle. The time marks on the ground-track overlay show that this happens about 1 hour and 30 minutes before apogee (at 06:35 UTC). The azimuthal bearing of point A is read directly from the map board (approximately 87°). At AOS the ground station antenna should, of course, be just above the horizon. Now let us look at how the Washington station will determine the position of the satellite at 09:05 UTC (same day). At 09:05 UTC (apogee plus 1 hour) the satellite will be at position B. The ground track color is blue and the SSP is well inside the blue range circle so the satellite is well within range. The azimuthal bearing of the spacecraft is again read directly from the map board (about 83°). The elevation of the satellite cannot be obtained directly from the map; you must refer to Table 5-7 in the following manner. As the color of the ground-track overlay at point B is blue, locate the blue row of Table 5-7. The closest range circle on the map to point B is 5000 km, so we locate the 5000-km column of Table 5-7. The elevation angle, 38° in this case, is contained in the box where the blue row intersects the 5000-km column. LOS for orbit 1607 will occur at position C, about 4 hours and 45 minutes past apogee (at 12:50 UTC) when the yellow orbit overlay segment crosses the yellow range circle. On this typical orbit the Washington station would have an opening that lasts over six hours. Near apogee the satellite will simultaneously be available to stations in North America (except for Alaska and the West Coast), Central America, Europe, Africa, the Middle East, a large part of Asiatic Russia, and South America (except for the southernmost tip). The next apogee (orbit 1608) would occur one period (10 hours and 56 minutes) later at 19:01 UTC at a longitude that is one increment (164°) farther west (179° W). Given a ϕ3 TRACKER and an orbit calendar you should now be able to track any elliptical-orbit Phase III satellite.

Simplified ϕ3 TRACKER

It's likely that early Phase III missions will be placed in orbits similar to the one used for our OSCAR ϕ3* example. A satellite in such an orbit will spend about 80% of its life in the blue and green segments of the ground-track overlay. As it happens, because of the geometry involved, ground stations will find the satellite in this region more than 90% of the time that it is in range. Certain operational difficulties (resulting from rapid satellite motion, spin modulation and Doppler shifts on the radio links) that aren't of concern near apogee will become more pronounced as the satellite approaches perigee. As a consequence of all these factors the majority of users will probably confine their operation to an 8.5-hour window centered about apogee. Now suppose that a simplified ϕ3 TRACKER was constructed with only two range circles — the blue and green ones — and that only the blue and green segments of the ground-track overlay were color coded. This bare-bones ϕ3 TRACKER would tell us if the satellite were in range during the 8.5-hour interval centered on apogee and would provide information on antenna bearing at any time. An amateur using this approach would probably not use Table 5-7 and would peak received signals by scanning the antenna in elevation.

Refining the Approximations

Focusing again on the full featured ϕ3 TRACKER we can refine our approximations even further. Since the color of the overlay segment represents *minimum* access distance during that segment of the orbit, the *true* acquisition distance at any time will actually lie between the range circle of matching color and the next larger range circle. In other words, the satellite will be in range a little before the color-coded ϕ3 TRACKER implies; actual AOS will occur a little before the ground-track segment crosses the acquisition curve of like color.

Figure 5-9 shows acquisition distance as a function of time from apogee (before or after) for OSCAR ϕ3* and how acquisition distance and altitude are related. Using this graph we see that when OSCAR ϕ3* is 4 hours and 15 minutes from apogee (green segment of orbit) the actual acquisition distance is about 8100 km. The green (8000-km) range circle is therefore a good approximation to the true acquisition circle at this time. At 2 hours from apogee (still in the green segment of the orbit) the satellite altitude is greater and the actual acquisition distance is about 8900 km — closer to the blue (9000-km) range circle. Referring back to OSCAR ϕ3* orbit 1607 we see (looking very closely) that AOS occurs approximately two hours before apogee, instead of our 1.5-hour initial estimate. Whether this degree of refinement is warranted in the real world, where the elevation of one's radio horizon often delays one's actual AOS or LOS, is questionable. In the blue and green sections of the orbit especially, the satellite elevation angle generally changes very slowly. This points up a related fact: Although ground station antenna height, by itself, is not important for satellite communications, raising the antenna can have a noticeable effect on AOS and LOS if it reduces your radio horizon by a few degrees.

Stationary Satellites

Tracking a stationary satellite (one that remains over a fixed spot on the equator) involves determining whether the spacecraft is in range and, if so, acquiring antenna aiming data.

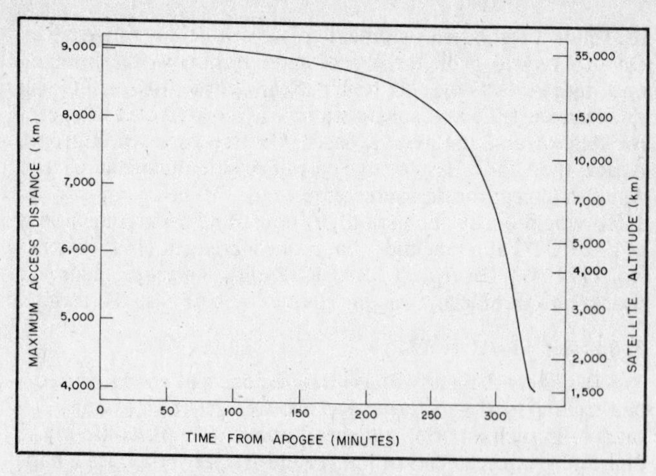

Fig. 5-9 — This curve can be used to determine the acquisition distance at any time for the sample AMSAT Phase III orbit — refer to horizontal and left-hand vertical scales. The relation depends on orbital eccentricity and apogee altitude, so it holds for any inclination. The correspondence between acquisition distance and altitude (left- and right-hand vertical scales) holds for any satellite.

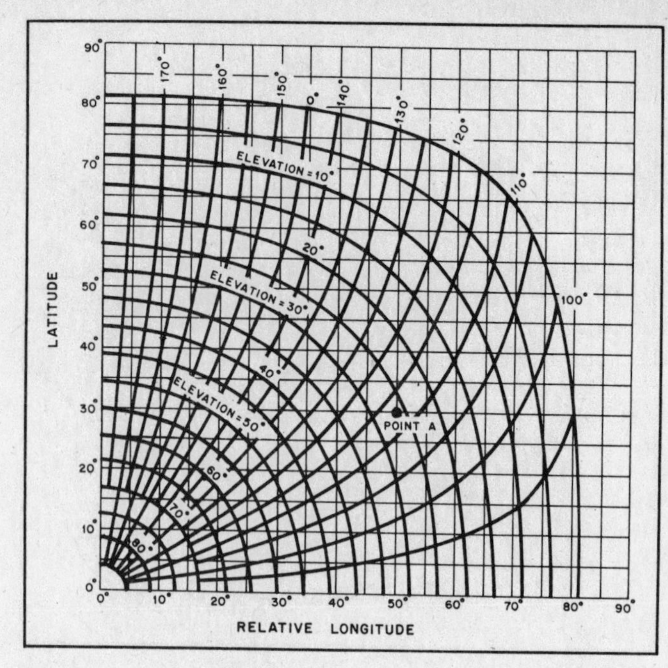

Fig. 5-10 — Chart for obtaining azimuth and elevation directions to geostationary satellite. See text for directions.

Using Fig. 5-10 (a full-page chart is presented at the end of Appendix B) it's easy to find the needed information for *any* stationary satellite if you know its location (longitude). This technique is of practical concern since work on SYNCART, an amateur transponder designed for synchronous orbit, is well under way.

An example illustrating how to use Fig. 5-10 follows. Suppose a ground station in New Orleans (latitude = 30° N, longitude = 90° W) is interested in accessing a stationary satellite located at 40° W longitude.

Step 1. The position of the ground station relative to the satellite is obtained by subtracting the ground station's longitude (90° W) from the satellite's longitude (40° W): 40° W − 90° W = −50° W (all longitudes are expressed in degrees West). Plot the point consisting of the absolute values of (a) the longitude just computed and (b) the latitude of the ground station. See point A (relative longitude = 50°, latitude = 30°) on Fig. 5-10.

Step 2. If the point is inside the 0° elevation circle the satellite is in range of the ground station. The location of point A indicates that the satellite is in range of New Orleans.

Step 3. Noting intersections of elevation circles and radial curves near point A, use "eyeball interpolation" to estimate the antenna elevation and radial values. In our example the radial value and elevation of the satellite, as seen from point A, are approximately 113° (radial value) and 27° (elevation). For Northern Hemisphere stations the azimuth is equal to either (a) the radial value or (b) 360° − (radial value), depending on whether the satellite is east or west of the ground station. For Southern Hemisphere stations the azimuth is equal to either (a) 180° + (radial value) or (b) 180° − (radial value), depending on whether the satellite is east or west of the ground station.

Step 4. The antenna should be aimed as per the results of step 3 and then the direction varied slightly to peak received signals. Since the satellite appears stationary the resulting antenna azimuth and elevation should remain fixed. (An alternative technique for finding azimuth and elevation of stationary satellites using spherical trigonometry is covered in Chapter 8.)

Tracking Hints

Now that you know how to use the OSCARLOCATOR and ϕ3 TRACKER we turn to some practical questions. Do you buy or build one? If you build, where can you obtain maps? Where do you find templates for ground-track overlays? How do you draw spiderwebs? We'll be looking at all these and other questions in this section.

Buying vs. Building

Factors that affect the decision whether to buy or build tracking devices include the cost and availability of raw materials and finished trackers, and the construction effort. Table 5-8 lists some commercial Phase II tracking devices that have been available in recent years. The variety and availability of Phase III satellite tracking devices will probably increase as AMSAT-OSCAR 10 use grows. In late 1983 the only tracker on the market was part of ARRL's *OSCARLOCATOR* package. Check recent periodicals to determine what's available. The rest of this section provides some insight into the effort required to "roll your own" tracking aids.

Table 5-8
Phase II Tracking Devices

OSCARLOCATOR
Description. Kit consisting of two multicolored map boards and several transparent ground-track overlays and spiderwebs, historical and spacecraft data. (Also included is a map board spiderweb for AMSAT-OSCAR 10; updated ground track curves for A-O-10 will have to be traced on the acetate included in the package from published templates.)
Source. Produced by ARRL (also available from AMSAT)
Approximate cost. $8.50 (1983)

Satellabe III
Description. Basically similar to OSCARLOCATOR but includes several additional convenience features and full instructions. Complete and ready to use.
Source. Communications Technology, Greenville, NH 03048
Approximate cost. $7 (1981)

W2GFF Plotter
Description. Somewhat similar to OSCARLOCATOR. Features a real-time readout of position, which makes it very convenient to use. Detailed instructions included.
Source. R. Peacock; 9 Andrea Dr., Setauket, NY 11733.
Approximate cost. $6 (1982)
Note: Separate Plotters are required for satellites in different orbits so be sure to specify the satellite you're interested in and your ground station latitude and longitude.

Check *QST* and *Orbit* for current availability and price.

Building (General). Most of the options in building a tracking device involve trade-offs such as accuracy vs. construction effort, accuracy vs. cost and accuracy vs. convenience. There is no "best" method; select the approach that meshes most closely with your needs. Evaluating the trade-offs, however, can be dif-

Table 5-9

Sources for Polar Maps

1) North pole stereographic* projection, multicolor, extends to equator, "USAF Physical-Political Chart of the World"; GH-2A (41-cm diameter) $0.50; GH-2 (82-cm diameter) $1. Source: Department of Commerce, Distribution Division (C-44), National Ocean Survey, Riverdale, MD 20840.

2) North pole stereographic* projection, 2-color, 105 cm/100 cm, stock no. DOD WPC xx032004, $0.85. Source: same as 1.

3) South pole stereographic* projection, 2 color, 105 cm/100 cm, stock no. DOD WPC xx032007, $0.85. Source: same as 1.

4) North pole azimuthal equidistant projection*, black/white, extends to 30° South latitude, 61-cm diameter. Single copies available at no charge from: APT Coordinator, Department of Commerce, NOAA, National Environmental Satellite Center, Suitland, MD 20233. Request "APT Plotting Board" and reference this book.

5) Plain polar graph paper also makes a very effective map board if political boundaries are not needed (azimuthal equidistant projection*).

6) Black and white North polar azimuthal equidistant projection on heavy stock (same size as maps in this book) available for $1 from ARRL Headquarters, 225 Main St., Newington CT 06111. All tracing templates in ARRL publications will be scaled to be drawn on this map board.

*Azimuthal equidistant polar projection maps are characterized by equally spaced latitude circles. Stereographic polar projection maps are characterized by increasing spacing between latitude circles as they get farther from the center (see Chapter 8 for additional details). Check on current prices before ordering.

ficult if you haven't had practical experience working with satellites, so I will state my personal preferences.

Some people prefer to use a single OSCARLOCATOR to track all Phase II satellites. This is easily accomplished by limiting the spiderweb on the map board to (1) azimuth curves (which are satellite independent) and (2) acquisition circles for particular satellites. Ground tracks may be either (a) placed on separate overlays and changed as needed, or (b) all drawn on a single overlay with ascending nodes spaced out around the perimeter. This approach provides almost no information on antenna elevation. It's mainly suited to stations using omnidirectional antennas or broad beamwidth antennas set at a compromise fixed elevation such as 25°. I use separate OSCARLOCATORs for each Phase II satellite, even though I usually run omnidirectional antennas.

Map Sources and Selection

Large maps can be used to obtain greater tracking accuracy, but only if they're mounted carefully and the overlays and ground tracks are drawn with great precision. Several sources of polar maps are listed in Table 5-9. The map in Appendix B of this book, the identically sized version on the back cover, and extra black and white maps on heavy stock available from ARRL, are my first choice. The tracing templates for spiderwebs and ground tracks provided in Appendix B are all drawn to match these maps, simplifying the construction process for you. Though I've prepared several large, precision OSCARLOCATORs and ϕ3 TRACKERS for lectures to large groups, they usually end up hanging on the wall while the more conveniently sized tracking aids are put through their paces.

Preparing Ground-Track Overlays

Templates for tracing ground-track overlays for Phase II satellites are included in Appendix B. All are scaled to match the most recent ARRL 8.5" × 11" polar maps. The tabular values in Appendix B can be used to construct ground-track overlays for other maps by following the directions accompanying the tables. Additional tracing templates and tabular data will be provided for each new Phase II satellite. Check *QST* and *Orbit* for timely details.

Tentative ground-track templates and data for Phase III satellites will be published before launch. After launch, when the orbit is accurately determined, information on periodically updating ground tracks, will be provided.

A trip to a drafting, art supply or large stationery store will provide you with the raw materials needed for preparing overlays and other parts of the tracker: transparent plastic, Mylar or Plexiglas; a set of colored pencil-crayons that mark well on plastic; round-head paper fasteners to attach the overlay to the map board; and some nylon filament tape to reinforce the overlay and map board where they pivot at the poles. When drawing overlays I often lay Scotch® #810 tape over the sections of the plastic where lines will be drawn. It's much easier to write on and shows colors more vividly than most transparent plastics. This tape also helps to shield all markings against smudging.

Spiderwebs

Drawing a spiderweb can be the most time-consuming part of producing a tracking device. In the "old days" many radio amateurs drew circles on a globe and then transferred the circle, point by point, to a map — an extremely tedious process. Though we'll be discussing a number of shortcuts, the accurate methods still require considerable effort.

OSCARLOCATOR Spiderwebs. We focus first on satellites in low-altitude circular orbits. Several tracing templates for spiderwebs are included in Appendix B. Samples are given for stations at 30° N latitude and 46° N latitude for each active Phase II satellite. All are scaled to the ARRL polar map. If you're using the ARRL map and live between the equator and 60° N latitude, the template closest to your latitude should work well.

If you live further north, or are using a different polar map (as long as it has equally spaced latitude circles), or you're willing to accept a little less accuracy to use a quick and simple method for drawing spiderwebs, try the following "two minute" approach. This technique, ignoring the distortions that occur when a circle is tranferred from a spherical globe to a flat map, approximates acquisition and elevation circles with true circles, and azimuth curves with straight lines. Consider OSCAR 8 as an example. Referring to Table 5-2 we see that the acquisition distance is 2000 miles, which corresponds to a 29° arc measured along a longitude line. The degree measure makes it easy to determine where the acquisition circle will intersect your longitude. For example, a station in New Orleans (30° N, 90° W) would add 29° to his latitude to obtain the point (59° N, 90° W) which would be on the acquisition circle. Once you know the center of a circle (30° N, 90° W) and one point on the circumference (59° N, 90° W) it can be quickly sketched with a drawing compass.

A similar procedure can be used to sketch elevation circles. From Table 5-2 we can determine that the 30° elevation circle for our New Orleans station would include the point (40.7° N, 90° W), and the 60° elevation circle would include the point (34.1° N, 90° W). Approximating true azimuth curves by using a protractor to draw straight line azimuths is consistent with the accuracy of this approach. The errors produced by this technique are worse as we get further from the pole.

Methods for drawing spiderwebs on stereographic polar projection maps (characterized by increasing spacing between latitude circles as they get further away from the center) are covered in Chapter 9.

ϕ3 TRACKER Spiderwebs. Drawing an accurate spiderweb on a ϕ3 TRACKER is more difficult than on the OSCARLOCATOR because the larger range circles on the map board become significantly distorted. A template (scaled to the ARRL polar map), for a station at 40° N latitude is included in Appendix B. It can be used by stations between 25° N and 55° N but accuracy suffers as you get further from 40° N. More accurate spiderwebs can be drawn on *any map* by using the tabular data in Appendix B; the closer you are to the specified latitudes the more accurate the spiderweb will be. The procedure is somewhat tedious but it sure beats trying to lift the information from a globe. A computer algorithm based on spherical trigonometry was used to prepare the tables. The technique is outlined in Chapter 9. Tables keyed to several additional latitudes are available for an s.a.s.e. to ARRL Satellite Programs, 225

Table 5-10
Time Conversion Chart

Time zone	EST	EDT	CST	CDT	MST	MDT	PST	PDT	AK/HI ST	AK/HI DT
Time difference	5	4	6	5	7	6	8	7	10	9

To convert from UTC to ________ (time zone) subtract ________ (time difference) hours.

To convert from ________ (time zone) to UTC add ________ (time difference) hours.

Table 5-11
This table may be used to convert from day/month notation to the day of year notation often used in satellite scheduling and computer programing.

Month	*Day of year = day of month + number listed (regular year)*	*(leap year)*
January	0	0
February	31	31
March	59	60
April	90	91
May	120	121
June	151	152
July	181	182
August	212	213
September	243	244
October	273	274
November	304	305
December	334	335

Main St., Newington, CT 06111 (specify your latitude).

Up-To-Date Data

Certain tracking data, such as the time and longitude of ascending node for Phase II satellites and the time and longitude of apogee for Phase III satellites, can only be predicted accurately a few months in advance. As a result, new orbit calendars, reference orbit listings and ground-track updates for Phase III spacecraft in elliptical orbits must be produced several times each year. Operating schedules (what features and modes can be expected at what times) for the various active satellites are also updated periodically in response to the satellites' health and users' needs. Active satellite users should check *QST* and *Orbit* frequently for this information. Up-to-the-minute details on satellite operations can be obtained by tuning in to the AMSAT nets. Net schedules will be found in *QST* and *Orbit*. A comprehensive calendar listing all orbits for currently active Amateur Radio satellites is being distributed by Project OSCAR, P.O. Box 1136, Los Altos, CA 94022. An s.a.s.e. will bring details. A monthly updated orbit schedule listing all orbits chronologically is available for ARRL members only; send a 4- × 9-in s.a.s.e. with your call sign (2 units of postage for each envelope, please) for each month you're interested in, to ARRL Hq.

Time Zone and Day-of-Year Conversions

Table 5-10 will enable ground stations in the United States to convert to UTC from local time, or to local time from UTC. For reference, the Uniform Time Act of 1966 specifies that Daylight Savings Time in the U.S. will be observed for six months each year, beginning the last Sunday in April and ending the last Sunday in October. Arizona, Hawaii and Michigan choose not to conform. Since satellite scheduling and computer programs for satellite tracking are often based on day of year notation, it's sometimes necessary to convert to or from the more common day/month notation. This can be done using Table 5-11.

Chapter 6
Antennas

Chapter 6

Antennas

Ground station performance is affected by many factors, but one stands out as being critically important: antennas. Although there are no intrinsic differences between antennas for satellite use and those for terrestrial applications, some designs are clearly better suited for satellite work. Properties that make a certain type of antenna desirable for hf operation may make it a poor performer on a satellite link, and vice versa. Before we list the characteristics that make an antenna suitable for satellite operation, we'll review the techniques used to specify antenna properties.

This chapter provides the information you need to select and build satellite ground station antennas. It's divided into three main parts. In the first section we focus on antenna characteristics in general and relate them to satellite links. In the second section we discuss several basic antennas that are suitable for satellite ground station use and provide either detailed plans or, for the more common antennas, references where construction information may be found. In the last section, we discuss a number of related practical topics.

Part 1: Antenna Characteristics

Simply stated, an antenna for monitoring downlink signals should be chosen to provide an adequate signal-to-noise (S/N) ratio at the receiver output; an antenna for transmitting on the uplink should be chosen to provide the desired signal level at the satellite. While pursuing these goals we also try to keep costs down and minimize the complexity associated with large mechanical structures and high aiming accuracy.

The antenna system characteristics we'll focus on in Part I include:

1) Directional properties (gain and pattern)
2) Transmitting vs. receiving properties
3) Efficiency
4) Polarization
5) Link effects (spin modulation, Faraday rotation)

One basic concept we'll refer to time after time is the *isotropic antenna:* an array that radiates power equally in all directions. No one has ever been able to build a practical isotropic antenna but the concept is still very useful as a "measuring stick" against which other antennas can be compared. Closely related is the *omnidirectional antenna,* one that radiates equally well in all directions in a specific plane. Practical omnidirectional antennas are common; the ground plane is one example. Any antenna that tends to radiate best in a specific direction (or directions) may be called a *beam antenna*. Several beams (the Yagi, quad, loop-Yagi and helix) are shown in Fig. 6-1. Even the common dipole can be regarded as a beam since it has preferred directions. The "first law" of antennas is: You don't get something for nothing. A beam can only increase the power radiated in one direction by borrowing that power from someplace else. In other words, a beam acts by concentrating its radiated energy in a specific direction. To quantify how well it accomplishes this task we compare it to the isotropic antenna, our measuring stick.

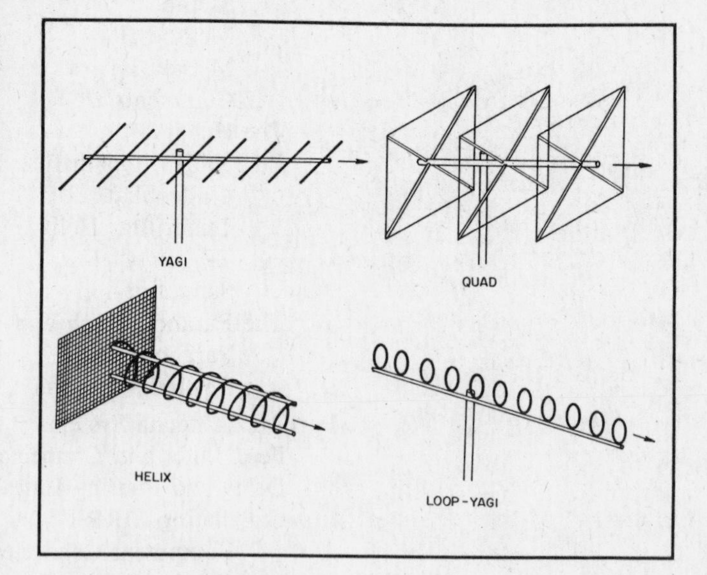

Fig. 6-1 — Four beam antennas.

Gain and EIRP

An imaginary radio link with two stations, A and B, as shown in Fig. 6-2 will help illustrate how the properties of a beam are specified. We'll discuss the transmit characteristics first since they're easier to grasp. Later we'll see how transmitting and receiving properties are related. As the type of antenna at Station B (the receiving station) isn't important for the comparison, a dipole is assumed. Station A (the transmitting station) has a choice of two antennas, a beam whose properties we wish to determine and an isotropic radiator. Our "thought experiment" begins with A using the beam antenna and some convenient power (P). A adjusts his antenna's orientation until B records the strongest signal and notes the level. A then switches to the isotropic antenna

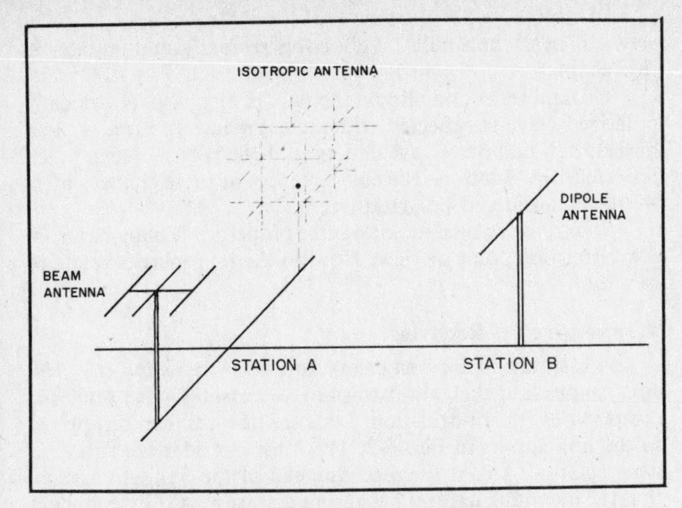

Fig. 6-2 — Radio link involving two stations, A and B.

and adjusts the power (P_i) until B reports the same signal level as noted earlier. The gain (*G*) of the beam is given by the formula

$$G = \frac{P_i}{P} \quad \text{(Eq. 6.1)}$$

For example, if 25 watts to the beam produced the same signal level at B as 500 watts to the isotropic, the gain of the beam would be

$$G = \frac{500}{25} = 20$$

This is roughly what would be expected from a well-designed Yagi with a boom length of 2 wavelengths.

Now suppose that B is the satellite and A is your uplink system. Aha! The satellite sees exactly the same signal whether you run 500 watts to an isotropic radiator or 25 watts to the beam. In either case we'd say the ground station *EIRP* (*E*ffective *I*sotropic *R*adiated *P*ower) is 500 watts. EIRP and the quantity P_i in our "thought experiment" are identical. We can rewrite Eq. 6.1 as $P_i = GP$ (EIRP is equal to the product of "gain" and "power being fed into the beam"). An EIRP of 500 watts can also be produced by a beam with a gain of 4 that is fed 125 watts, a beam with a gain of 10 that is fed 50 watts and so on. The definition of EIRP we've been using just depends on power fed to the antenna and gain. Later we'll see how this can be generalized to include transmitter output power, feed-line losses and even the effects of a misaimed antenna.

To simplify certain calculations, gain is often expressed in decibels (dB).

$$\text{G [in dB]} = 10 \log \frac{P_i}{P} \text{ or,} \quad \text{(Eq. 6.2a)}$$

$$\text{G [in dB]} = 10 \log G \text{ or,} \quad \text{(Eq. 6.2b)}$$

$$G = 10^{G/10} \quad \text{(Eq. 6.2c)}$$

Since we refer to G and *G* as "gain" it's important to note the units. If gain is simply a number (a ratio), we're talking about *G* (Eq. 6.1); if gain is given in decibels we're referring to G.

Eq. 6.1 and Eq. 6.2 clearly depend on what antenna is used for comparison (the reference antenna); it's the isotropic. At times, a half-wave dipole is used for this purpose. The half-wave dipole has a gain of 1.64 (2.14 dB) over an isotropic radiator. As a result, the gain of a specific beam looks better when the reference antenna is an isotropic than when it's a dipole. Eqs. 6.3a and 6.3b describe how the figures can be translated.

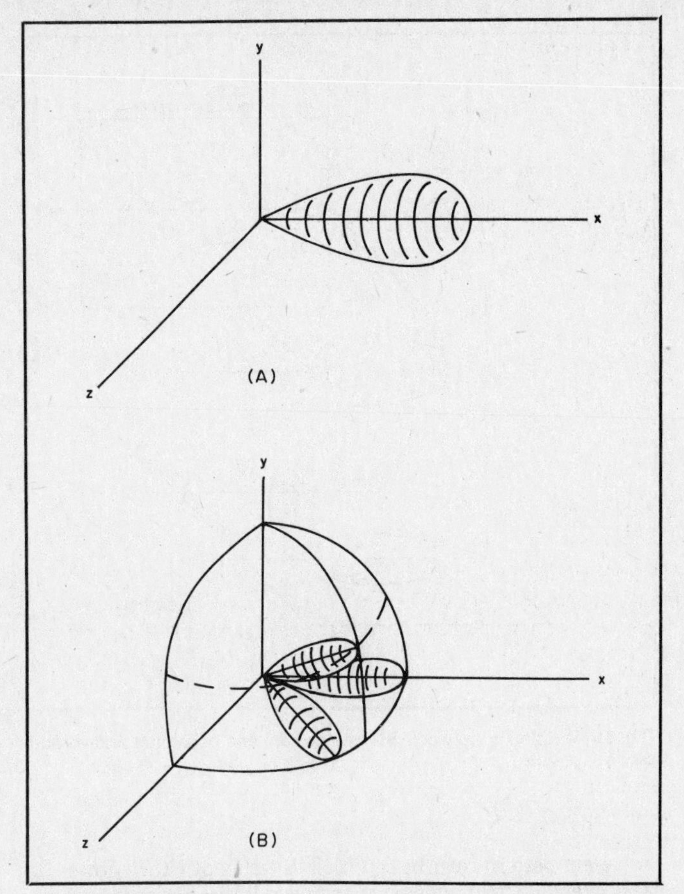

Fig. 6-3 — Three dimensional illustrations of beam patterns. (A) single lobe and (B) multi lobe.

$$G \text{ [isotropic reference]} = (1.64)(G \text{ [dipole reference]}) \quad \text{(Eq. 6.3a)}$$

$$\text{G [isotropic reference]} = \text{G [dipole reference]} + 2.14 \text{ dB} \quad \text{(Eq. 6.3b)}$$

Obviously, it's very important to specify the nature of the reference. This is sometimes done by expressing gain in either dB_i or dB_d, where the last letter describes the reference antenna as *i*sotropic or *d*ipole. Note that so far we've looked only at the one direction in which the maximum signal is radiated.

Gain Patterns

We've seen how one very important antenna characteristic, gain, is specified. Gain tells us nothing, however, about the three-dimensional radiation pattern of an antenna. A beam with a given gain might have one broad lobe as shown in Fig. 6-3A, or several sharp lobes as shown in Fig. 6-3B. A single broad lobe is generally more desirable because it makes the antenna easier to aim and is usually less susceptible to interfering signals. Because drawing quantitative three-dimensional pictures, like those in Fig. 6-3, is difficult, the directional properties of an antenna are more often pictured using one or two two-dimensional cross-sections drawn to include the direction of maximum radiation. In Fig. 6-4 we show two common cross-sections (gain patterns) used to describe a Yagi. When beams are installed for terrestrial communications the cross-sections may conveniently be referred to as horizontal plane (azimuth plane) and vertical plane (elevation plane) gain patterns. When working with antennas that can be aimed upward or those using circular polarization it's important to clearly specify the relation between any two-dimensional pattern pictured and physical orientation of the antenna. Before we continue, note that the gain pattern of an isotropic antenna is a circle in any cross-sectional plane, and the gain pattern of the omnidirectional antenna is a circle in one specific plane.

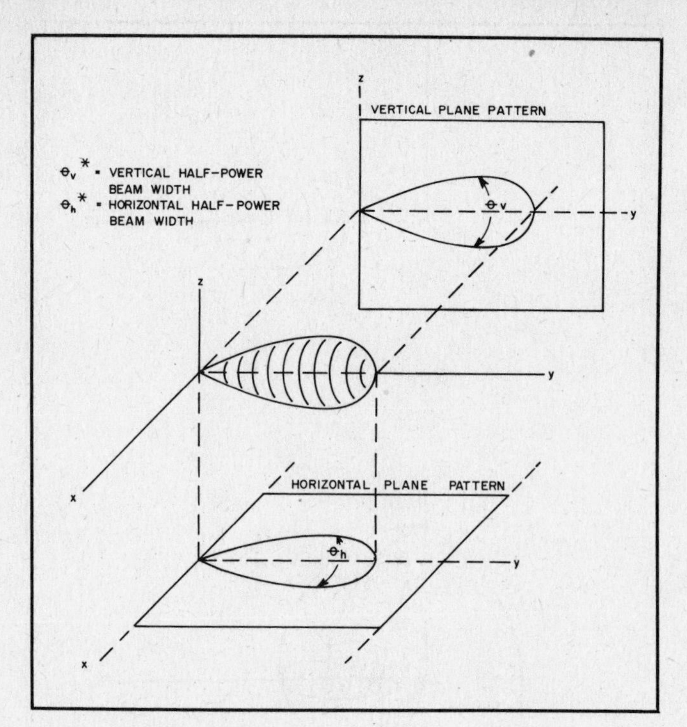

Fig. 6-4 — Relation between 3-D gain pattern and horizontal and vertical cross-sections.

Gain patterns can be specified in terms of either power or field strength (field intensity). As power is directly proportional to the square of field intensity, translating back and forth between the two descriptions is relatively simple. For example, when field intensity drops to 0.707 of its maximum value, power will have dropped to 0.5 of its maximum since $(0.707)^2 = 0.5$.

One important characteristic of the gain pattern of an antenna is the *beamwidth:* the angle between the two straight lines that start at the origin of the pattern and that go through the points where the radiated power drops to one half its maximum value. See Fig. 6-4. Since high antenna gains are obtained by concentrating the radiated power in a specific direction it's clear that beamwidth and gain must be related. High gains can only be obtained by sacrificing beamwidth. It's possible and sometimes desirable, to design an antenna so that two gain cross-sections taken at right angles are shaped significantly differently. For many familiar beam antennas, however, the two patterns are very similar. In cases where an efficiently designed antenna produces a symmetrical pattern (one that's independent of cross-section orientation), the maximum beamwidth, θ^*, for a given gain is given approximately by Eq. 6.4

$$\theta^* \text{ [in degrees]} = 10 \sqrt{\frac{400}{G}} \qquad \text{(Eq. 6.4)}$$

For example, an antenna with a gain of 20 would have a beamwidth of roughly 45°. Tracking a slowly moving target like an elliptical-orbit Phase III satellite near apogee with such an antenna would pose no problem. Tracking a speedy, low-altitude satellite, however, might be difficult.

Most discussions of antennas begin with a *free space* model that represents an antenna as if it were in outer space with no nearby objects to affect its performance. In the real world, rf reflections can have a large impact on an antenna's behavior. This is especially true for vertical plane patterns. At low elevation angles, for example, a receiving antenna will see two signals: a direct signal and a ground-reflected signal. Depending on the phase difference, these signals might add, giving up to 3 dB of ground reflection gain, or destructively interfere to produce a null. The real-world vertical pattern will therefore often consist of a series of peaks and nulls. We'll cover several examples later in this chapter.

Ground reflection also has an impact on phase. A vertically polarized wave is reflected without any phase change, while a horizontally polarized wave undergoes a 180° phase change when reflected. We'll look at the consequences of these changes after we have considered polarization.

So far, we've been looking at the properties of antennas from a transmitting point of view. How do these properties relate to reception?

Transmitting vs. Receiving

A basic law of antenna theory, known as the *reciprocity principle,* states that the gain pattern of an antenna is the same for reception as for transmission. Let's see how this can be applied to the link shown in Fig. 6-2. This time consider the situation where station A is at the receiving end of the link and assume that the incoming natural background noise at A is independent of direction — a reasonable assumption at the frequencies of interest. If station A measures the noise power arriving at the receiver with both a beam antenna and an isotropic antenna of the same efficiency, he'll obtain the same result. The beam actually picks up more noise than the isotropic from the primary direction and less noise than the isotropic from other directions, but the overall result is that both antennas capture the same total noise power. Now let's see what happens when station B transmits a reference signal at any convenient fixed power. The total amount of signal power reaching station A is fixed but the power is arriving from a particular direction. A beam antenna pointed toward station B will provide station A with more signal power than would an isotropic antenna. As a result, when signal *and* noise are present, the beam produces a better S/N power ratio at the input to A's receiver. For well-designed antennas, the improvement in the S/N power ratio over a communications link will be the same whether the antenna switch — beam for isotropic — is made at the transmitting end or the receiving end.

The reciprocity principle does *not* state that a particularly desirable receiving antenna is consequently also desirable as a transmitting antenna, or vice versa (though this is often the case). In transmitting, the objective is to produce the largest possible signal level at the receive point. High efficiency and gain are therefore very important. When receiving, the objective is to obtain the best possible S/N ratio. Though high efficiency and gain may contribute to this goal, the shape of the gain pattern and the location of nulls may have a larger impact on S/N ratio by reducing noise and interfering signals.

Efficiency

A transmitting antenna that is 100% efficient radiates all the power reaching its input terminals. Reduced efficiency causes an antenna to radiate less power in every direction; it has *no* effect on antenna pattern. A transmitting antenna that is 50% efficient only radiates half the power appearing at its input terminals. Since building high-efficiency vhf and uhf antennas is relatively easy, and producing transmit power at 146 MHz and higher frequencies is difficult, inefficient transmitting antennas should never be used at a satellite ground station.

The trade-offs are somewhat different for receive systems. A receiving antenna that's 50% efficient passes along only half the signal *and* half the noise power it intercepts. If the receive chain's S/N ratio is limited by atmospheric or cosmic noise arriving at the antenna, poor antenna efficiency may not affect the observed S/N ratio. In practice, the only situation where relatively inefficient receive antennas may be considered is on the 29-MHz Mode A downlink where a compact half-size Yagi (or crossed-Yagi array) may provide the same performance as a full-size model.

Low antenna efficiencies are usually caused by (1) physically small (relative to design wavelength) elements that require inductive loading, (2) lossy power-distribution and matching systems, especially in multi-antenna arrays, or (3) poor ground systems (when the ground is an integral part of the antenna). Remember, efficiency does not affect the pattern of either transmit or receive antennas.

Polarization

Our treatment of polarization is divided into two parts: (1) a technical description of the term polarization as it's applied to radio waves and antennas and (2) a link comparison that examines how polarization affects satellite radio link performance.

Technical Description

Radio waves consist of electric and magnetic fields, both of which are always present and inseparable. Since most amateur antennas are designed to respond primarily to the electric field, it is possible to limit our discussion to it. When a radio wave passes a point in space, the electric field *at that point* varies cyclically at the frequency of the wave. When we discuss the *polarization of a radio wave* we're focusing on how the electric field varies.

The electric field can vary in *magnitude,* in *direction* or in both. If, at a particular point in space, the magnitude of the electric field remains constant while the direction changes we have *circular polarization* (CP). (Note: All changes referred to here are cyclic ones at the frequency of the passing wave, and the direction of the electric field is confined to a plane perpendicular to the direction of propagation.) If, on the other hand, the direction of the field remains constant, while the magnitude changes, we have *linear polarization* (LP). If both magnitude and direction are varying we have *elliptical polarization.*

Two of the several approaches to describing polarization are of practical interest to radio amateurs. One pictures an elliptically polarized radio wave as consisting of a linear component and a circular component. If the magnitude of the electric field varies only slightly in the course of each cycle, the circular component dominates. If the magnitude of the electric field decreases to nearly zero during each cycle, the linear component dominates. From this viewpoint, circular polarization and linear polarization are simply special cases of elliptical polarization.

The second approach to describing polarization also treats the elliptically polarized wave as having two components; but this time each component is linearly polarized with the two components at right angles physically and 90° out of phase electrically. If the maximum values of both electrical field components are the same, we have circular polarization. If one electrical field component is always zero, we have linear polarization. Both of the approaches to describing polarization just outlined are helpful in understanding antennas and radio link performance.

The polarization characteristics of a radio wave depend on the transmitting antenna; the transmitter itself has absolutely nothing to do with polarization. Like radio waves, antennas can be assigned a polarization label: the polarization of the wave that they *transmit in direction of maximum gain.* The common terms "linearly polarized antenna" and "circularly polarized antenna" can be confusing if you aren't aware that "in the direction of maximum gain" is meant, even though it's not stated explicitly.

The following example illustrates this potential source of confusion. Consider the transmitting antenna and the three observers shown in Fig. 6-5. The antenna pictured, known as a turnstile, consists of two crossed dipoles fed 90° (a quarter cycle) out of phase. Observer A sees a circularly polarized wave; observer B sees an elliptically polarized wave; observer C sees a linearly polarized wave. Since observer A, located in the direction of maximum gain, sees a CP wave, the turnstile is called a circularly polarized antenna, even though observers off the z axis, like B and C, see something quite different.

Most of the circularly polarized antennas that we'll be looking at, such as the helix and crossed Yagis fed 90° out of phase, are similar to the turnstile in that only observers in the direction of maximum gain actually see circular polarization. When using such an antenna at a ground station it's important to keep the array pointed at the spacecraft to reap the benefits of CP. If an antenna of this type is being used on a spacecraft, the only time

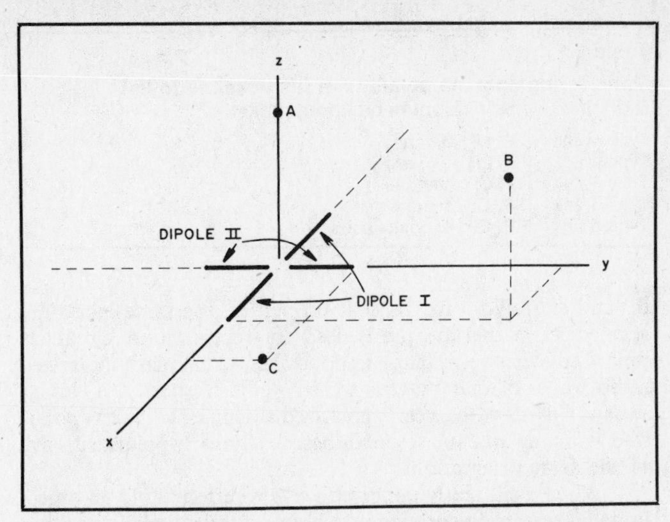

Fig. 6-5 — Turnstile transmitting antenna (two dipoles fed 90° out of phase). Observers measuring the polarization of the radiated signal are positioned at points A, B and C.

Fig. 6-6 — To determine the sense of a circularly polarized antenna the observer is imagined as being in back of the antenna looking in the direction of maximum gain. Because of the possible biological hazards associated with rf radiation (see chapter 7), one should *never* stand this close to a transmitting antenna.

you'll see CP is when the satellite antenna is pointed in your direction. Some antennas produce circularly polarized radiation patterns over most of the beam. We'll look at two such antennas, the quadrifilar helix and Lindenblad, later in this chapter.

Our description of a circularly polarized radio wave emphasized that at a particular point in space, the constant-magnitude electric field rotated at the frequency of the source. It's important to be able to specify whether the *sense* of rotation is *clockwise* or *counterclockwise.* For historical reasons, physicists and electrical engineers specify polarization sense in opposite ways, a fact that can often cause confusion. The IEEE (Institute of Electrical and Electronic Engineers) standard is the one used in most recent radio amateur literature. To specify the sense of a circularly polarized wave using the IEEE standard, picture yourself behind the transmitting antenna (Fig. 6-6) looking in the direction of maximum radiation. Pick a specific point on the main axis (any point will do) and note the position of the electric field at a particular instant and again a fraction of a cycle later. The center of a driven element is a convenient point and a quarter of a cycle is a convenient time interval. If you observe the electric field rotating clockwise, the wave is *right-hand circularly polarized* (RHCP). If the electric field appears to be rotating counterclockwise, the wave is *left-hand circularly polarized* (LHCP). As we obviously cannot "see" the transmitted electric field, we'll discuss shortly how you can determine the sense of a helix or crossed-Yagi array by inspecting the antenna.

Although polarization-sense labels attached to an antenna

Table 6-1

Considering antenna polarization it's possible to list 5 distinct types of communications links.

Type 1 link	(LP, LP, matched)
Type 2 link	(LP, LP, random)
Type 3 link	(LP, CP, random)
Type 4 link	(CP, CP, same sense)
Type 5 link	(CP, CP, opposite sense)

depend entirely on its transmit properties, the same labels are applied when the antenna is used for reception. A circularly polarized receiving antenna responds best to circularly polarized radio waves of matching sense. As we'll see shortly, a similar situation exists with linearly polarized antennas; a linearly polarized receiving antenna responds best to a linearly polarized wave of the same orientation.

When a circularly polarized wave is reflected off an object (a metal screen, the ground or a house, for example), the sense of its polarization is changed. A RHCP signal aimed at the moon returns as a LHCP signal; a feed horn irradiating a parabolic reflector with LHCP produces a main beam that's RHCP.

Link Comparisons

We now look at how polarization effects a communications link involving two stations: T (the transmitting station) and R (the receiving station). Each station can choose from among antennas that provide RHCP, LHCP or LP. The orientation of the LP antennas can be varied by rotating them about the line joining T and R. All antennas are assumed to have the same gain and each is aimed at the other station. Various possible link combinations can be characterized by the polarization at T, the polarization at R, and the relative orientation (linear polarization) or sense (circular polarization) of the antennas used. For example, (LP, CP, random) in Table 6-1 can mean either T has an LP antenna and R a CP antenna, or vice versa, and that the orientation of the LP antenna is random. The ambiguity is intentional; since the reciprocity relation previously mentioned states that system performance will be the same in both cases, there is no need to distinguish between them. (Random orientation means vertical, horizontal or anywhere in between and refers to the elements; the antenna is aimed at the other station.) Only five distinct combinations need be considered. See Table 6-1.

Arbitrarily choosing the Type 1 link as a reference, we compare the performance of the other four combinations.

Type 1 link (LP, LP matched). The received signal level is constant. This link is our reference.

Type 2 link (LP, LP random). The received signal strength varies monotonically from a maximum equal to the reference level when the two antennas are parallel down to zero (theoretically) when the two antennas are perpendicular. In practice, the attenuation is rarely more than 30 dB for the perpendicular situation.

Type 3 link (LP, CP, random). The received signal strength on this link is constant at 3 dB down from the reference level and is independent of the orientation of the LP antenna and the sense of the CP antenna.

Type 4 link (CP, CP, same sense). The received signal strength on this link is constant and equal to the reference level.

Type 5 link (CP, CP, opposite sense). A simple theoretical model predicts infinite attenuation compared to the reference signal link, but in practice attenuations greater than 30 dB are rare.

Having looked at the polarization choices of the five basic links, we return to comparing the performance of various ground station antennas when operated in conjunction with a specific satellite antenna. If the satellite antenna is linearly polarized, our choice of ground station antenna is equivalent to choosing a Type 1, 2 or 3 link. Of the three, the Type 1 may appear best since it provides the strongest signals. From a practical viewpoint, however, it is nearly impossible since the orientation of the incoming wave is continually changing. In reality our choice is limited to a Type 2 or Type 3 link. Although the Type 2 link will sometimes provide up to 3 dB stronger signals (matched orientation), the Type 3 link will equally likely provide 30 dB stronger signals (perpendicular orientation). Of the two, the Type 3 link is preferable.

We can perform a similar analysis for a satellite antenna that is circularly polarized. The choice of ground station antenna here is equivalent to choosing a Type 3, 4 or 5 link. A Type 4 link is clearly perferable. But, it should be noted that the Type 3 link results in signals that are only 3 dB weaker with none of the severe fading problems of Type 2 links. As a result, on links where the S/N ratio is generally good, many ground stations trade a little performance for the mechanical simplicity of LP antennas.

Signals arriving from an actual satellite usually are elliptically polarized. As we've noted, an elliptically polarized wave can be thought of as having linear and circular components; moreover, a CP ground station antenna produces the best performance whether the signal from a spacecraft is CP or LP. Therefore, we can conclude that a CP ground station antenna will provide the best results in the general case where elliptically polarized signals are arriving from a satellite.

So far, we've focused on the effects of matching polarization at the ends of a link. The choice of polarization for a ground station antenna may also be influenced by other factors: spin modulation, Faraday rotation and mechanical constraints.

Spin Modulation

Since a satellite antenna and its gain pattern are firmly anchored to the spacecraft, a ground station's position relative to the pattern will change moment by moment. As we've noted, both the polarization and gain of an antenna vary with the observer's location. A ground station will therefore see cyclical gain and polarization changes on a downlink signal resulting from satellite rotation. These changes are called *spin modulation*. The spin modulation frequency depends on the spacecraft's rotation which, in turn, depends on the attitude stabilization technique employed. OSCARs 5, 6, 7 and 8 rotated at frequencies on the order of 0.01 Hz (about one revolution every four minutes) after a few weeks in orbit. Spin modulation at 0.01 Hz sounds much like a slow fade. Its effect on intelligibility is minor unless the signal drops below the noise level.

The attitude stabilization scheme used on Phase III elliptical orbit missions differs considerably. The spacecraft is spun at roughly 60 revolutions per minute (rpm) about an axis that is parallel to the line joining the apogee and the center of the earth. Because of the tri-star shape of the early Phase III satellites, gain and polarization variations on the links occur three times the spin rate (180 rpm or 3 Hz). When a ground station is located on the fringes of the satellite's antenna pattern, it may observe gain variations that exceed 10 dB. To a user, spin modualtion at a frequency of 3 Hz resembles rapid airplane flutter. It can be very annoying and have a severe impact on intelligibility.

How serious a problem is spin modulation? It is mainly of concern with spin-stabilized spacecraft of the type planned for Phase III elliptical orbits. Even with these spacecraft the effects only become annoying when the ground station is looking at the spacecraft from off to its side, a situation that occurs a small percentage of the time. A ground station can't do much to alleviate true gain variations, but variations caused by polarization mismatch can be minimized by using a circularly polarized antenna.

Faraday Rotation

As a linearly polarized radio wave passes through the ionosphere, the direction of the electric field rotates slowly about the direction of propagation. This rotation, known as the *Faraday effect* (see Chapter 10), is most noticeable at lower frequencies such as 29 MHz and 146 MHz. Its effects can be observed by ground stations that use linearly polarized antennas; slow fades will occur as the angle between the linear component of the incoming wave and the ground station antenna changes during a pass. Faraday rotation is especially noticeable on the 29-MHz downlink since all amateur satellites have used linearly polarized antennas for Mode A. The use of a CP antenna at the ground station would eliminate these effects but very few ground stations employ CP at 29 MHz. It's important to note that circular polarization won't cure all Mode A fading since much of it arises from the constantly changing orientation of the gain pattern of the antennas aboard the spacecraft as it spins. Other factors, such as absorption in the ionosphere, can also contribute to fading.

General Comments

We have discussed a great many technical factors in the selection of ground station antennas in the earlier part of this chapter. Let's restate and integrate the major ones.

1) The receiving antenna gain needed at a ground station is determined by the height, transmitter power and antennas of the specific satellite of interest. Typical receiving antenna gain values for the 146-MHz Mode B downlink are 0 dB_i for OSCAR 7 and 13 dB_i for AMSAT-OSCAR 10. To minimize tracking concerns, one should use the least gain needed to produce an acceptable S/N ratio at the ground station.

2) The EIRP needed by a ground station for uplinking is determined by the height, receiver sensitivity and antennas of the specific satellite of interest. The estimated EIRP value for accessing the 435-MHz Mode B uplink for AMSAT-OSCAR 10 is 500 watts. To obtain this EIRP level we can trade off ground station antenna gain and transmitter power.

3) Antenna beamwidth and gain are closely related: higher gain can be obtained only at the expense of narrower beamwidth.

4) Narrow-beamwidth antennas require greater precision in aiming and consequently increase the difficulty of tracking. The problem is most severe with rapidly moving satellites such as Phase II spacecraft, or elliptical-orbit Phase III spacecraft near perigee.

5) In all cases considered, circular polarization of the correct sense provides better performance than linear polarization. In most instances, however, the differences are small.

Several additional factors may affect your antenna selection. These include the desired performance standards, the acceptable level of mechanical complexity, costs and the difficulty in getting proper performance from certain types of homemade antennas.

Let's look at how some of the selection factors interact. Consider the 50-watt Mode B transponder flown on AMSAT-OSCAR 10. A straightforward prelaunch path-loss calculation by AMSAT (see Chapter 10) suggests that a ground station receive antenna with about 13 dB should provide a respectable S/N ratio. Prelaunch values should always be regarded as estimates though they do provide a reasonable starting point. A good antenna with 13 dB_i gain will have a beamwidth of about 45°. This, in turn, sets the accuracy requirements for our aiming system (rotators and readouts). For maximum convenience you would generally use a transmitting antenna with the same beamwidth and gain. The suggested EIRP value for accessing this spacecraft (see Appendix A) is 500 watts. Earlier in this chapter we computed that 25 watts to an antenna with a gain of 20 (corresponding to 13 dB_i) produces an EIRP of 500 watts. If transmitting antenna feed-line losses amount to 3 dB, a transmitter that put out 50 watts would be needed to provide 25 watts at the antenna terminals; this, in turn, would produce the desired 500-watt EIRP. What might we do if the only available transmitter produced 25 watts output? There are several options: (1) Mount the transmitter directly at the antenna to eliminate the feed-line losses; (2) add a small linear amplifier that would provide 3-dB gain to the system; (3) modify the transmitter to produce an extra 3 dB of output; (4) use a 16-dB-gain transmitting antenna; (5) learn to live with a slightly weaker-than-average signal. The user must select the most desirable (or least objectionable) option. If the 16-dB-gain transmitting antenna were chosen, the aiming requirements would become more stringent. If you're willing to accept these requirements, choosing the 16-dB-gain receive antenna option is reasonable.

The question of whether to use circular or linear polarization still remains. Circular polarization clearly performs better, but there are some barriers. The cumbersome physical structure or complex matching adjustments may deter you from going this route. We'll look at the severity of these barriers in Part II of this chapter; let's temporarily sidestep this question until we've examined a number of practical antennas.

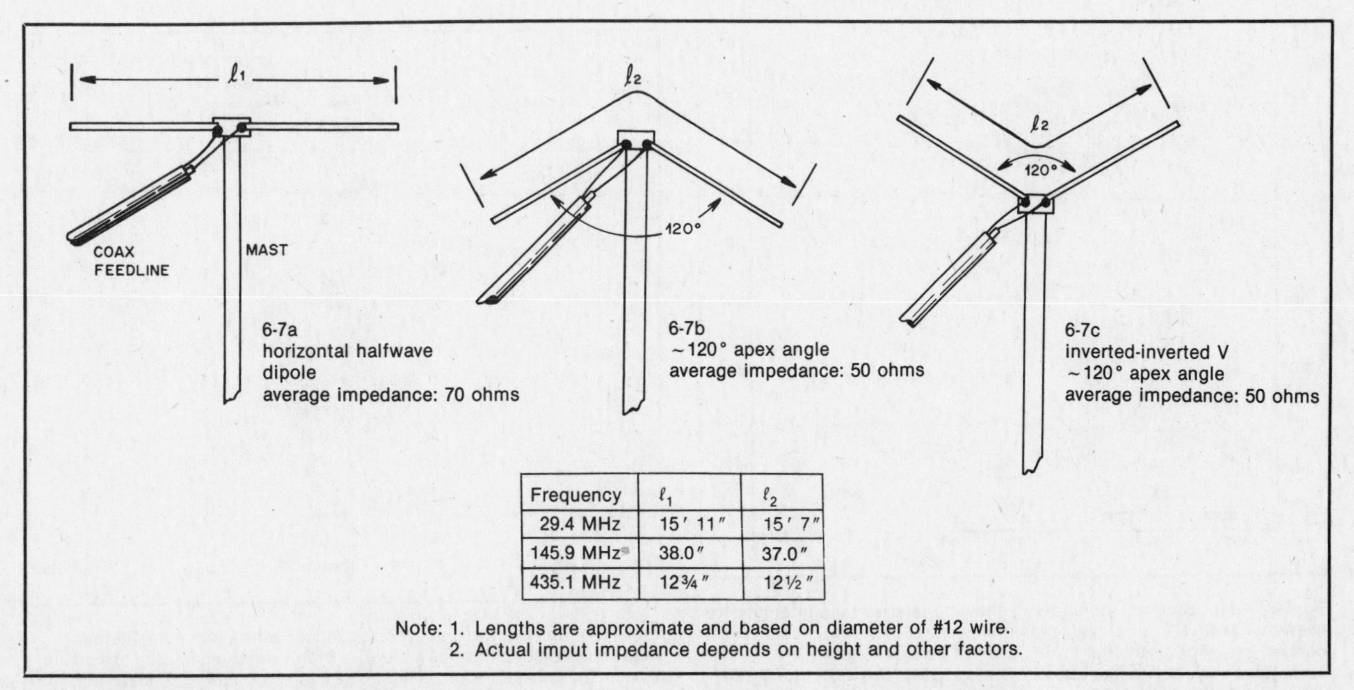

Frequency	ℓ_1	ℓ_2
29.4 MHz	15′ 11″	15′ 7″
145.9 MHz	38.0″	37.0″
435.1 MHz	12¾″	12½″

Note: 1. Lengths are approximate and based on diameter of #12 wire.
2. Actual imput impedance depends on height and other factors.

Fig. 6-7 — Three variations of the halfwave dipole.

Part II: Practical Ground Station Antennas

This section focuses on several practical antennas that may be used at a satellite ground station. You'll no doubt recognize many, as they're also popular for terrestrial hf and vhf communication. We'll point out the advantages and disadvantages of each for accessing low- and high-altitude spacecraft, for construction difficulty, and for general utility as part of an overall antenna system. Construction details are provided for the more unusual models, and references to readily available sources of information are provided for the popular types.

The Dipole and Its Variations

The horizontal half-wave dipole (Fig. 6-7A) is a familiar antenna that can be used at satellite ground stations. Two offshoots of the dipole, the inverted V (Fig. 6-7B) and the somewhat less familiar inverted-inverted V (Fig. 6-7C), have also been used. Be sure not to confuse the V antennas discussed here with the V-beam, which is radically different in construction and performance. Our discussion of the V will focus on the inverted V since it has been investigated thoroughly. Nonetheless, it's safe to assume similar characteristics for the inverted-inverted V.

The dipole and Vs are usually mounted fixed in the same configuration for both satellite and terrestrial applications (as in Fig. 6-7). It therefore makes sense to label patterns as vertical and horizontal. Gain patterns in the horizontal plane for the dipole and inverted V are shown in Fig. 6-8. Note how the horizontal dipole has higher gain broadside and deeper nulls off the ends. Their low gain renders the dipole and V suitable mainly for use with low-altitude satellites. Their broad beamwidth provides reasonably good coverage when the antennas are fixed mounted. Dipoles are most often used to receive the 29-MHz Mode A downlink. A few amateurs have tried them successfully on 146-MHz uplinks and downlinks in conjunction with low-altitude spacecraft on Modes A, B and J, but this has mainly been for experimental, not for general, communication.

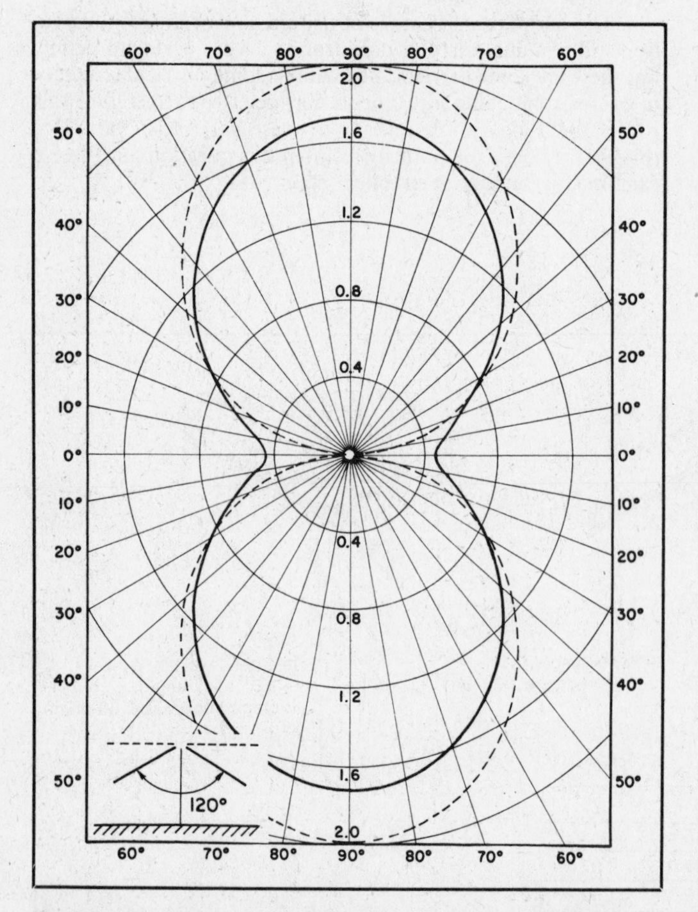

Fig. 6-8 — Horizontal plane patterns showing relative field intensity for inverted V with 120° apex angle (solid line) and horizontal dipole (dashed line). For additional information on inverted V see: D. W. Covington, "Inverted-V Radiation Patterns," *QST*, Vol. XLIX, no. 5, May 1965, pp. 81-84.

Let's look at some practical applications of dipoles and Vs at 29 MHz. Given the patterns in the horizontal plane, most amateurs who are constrained to using a single fixed antenna choose the V to eliminate the deep nulls associated with the dipole. Slightly better overall performance can be obtained by using two totally independent dipoles mounted at right angles to one another. If feed lines for both are brought into the operating position, switching between them to find the dipole that produces the best received signals is a simple matter. Another application, offering even better performance, consists of mounting a 10-m dipole behind a small 2-m beam, using a light-duty azimuth rotator to turn the whole array (Fig. 6-9). If the 2-m beam is inclined at roughly 25° above horizontal, we won't have to use an elevation rotator; moreover, azimuth aiming requirements will be lax. You'll note that for all three examples just presented, improved performance seems to go hand-in-hand with increased cost and complexity.

The free-space gain pattern of the dipole in the vertical plane really isn't of much interest to us because ground reflections change it drastically. As it turns out, the gain pattern depends on the height of the dipole. Look at the patterns in Fig. 6-10 for three specific heights: 1/4, 3/8 and 1-1/2 wavelengths above an infinite, perfectly conducting ground. The pattern in Fig. 6-10C is very poor for satellite work since signals will fade sharply each time the satellite passes through one of the nulls. In reality, the nulls are not as severe as shown because the ground is not a perfect conductor and signals (reflected off nearby objects) often arrive at the ground station receiving antenna from several directions. The pattern in Fig. 6-10B is most desirable since gain variations tend to balance out changes in signal level as the distance between spacecraft and ground station varies. In other words, the gain pattern of Fig. 6-10B is high toward the horizon where signals are weak (large satellite to ground-station distances), and low in the overhead direction where signals are strong (small satellite to ground-station distances). The pattern in Fig. 6-10A is acceptable, though not as good as the one in Fig. 6-10B. Gain patterns

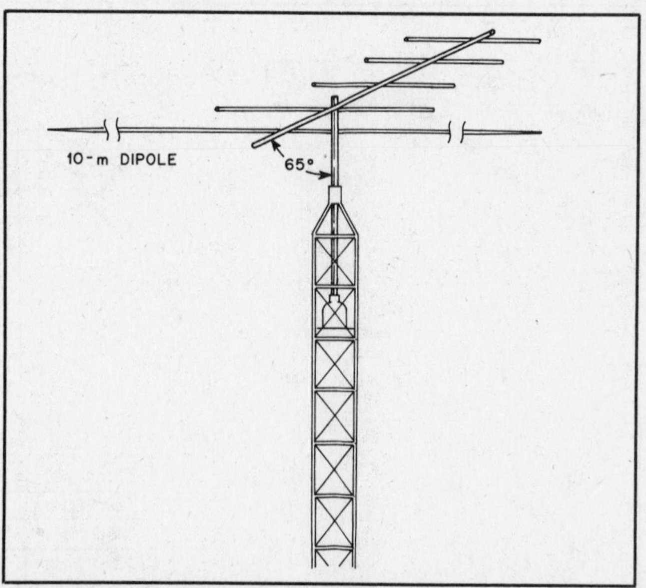

Fig. 6-9 — An effective linearly polarized antenna system for operating mode A consisting of a halfwave 10-m dipole mounted in back of a small 2-m beam. The main boom is inclined at approximately 25° above horizontal (65° from vertical) and only an azimuth rotator is used.

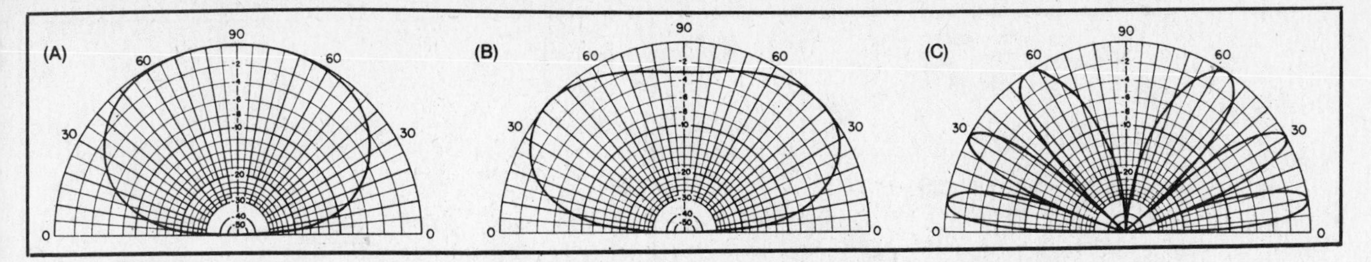

Fig. 6-10. Vertical plane gain patterns showing the relative field intensity for halfwave dipole above perfectly conducting ground. Pattern at right angles to dipole. (A) is for height of 1/4 wavelength, (B) is for height of 3/8 wavelength, and (C) is for height of 1.5 wavelength.

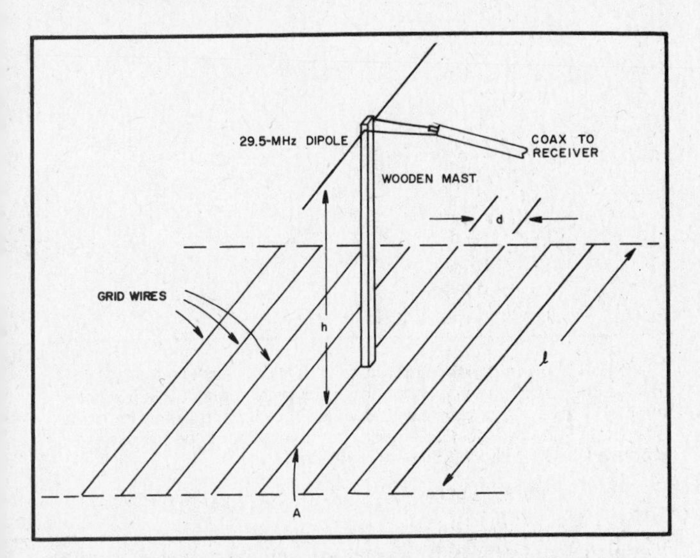

Fig. 6-11 — Dipole mounted above reflecting screen. Best results are obtained when h = 3/8 wavelength, d is less than 0.1 wavelength and ℓ = 0.6 wavelength. Note that it is not necessary to physically connect the grid wires to the dipole or feed line.

for the V antennas are similar when height is measured from the feed point to the conducting surface.

As the effective *electrical* ground does not generally coincide with the *actual* ground surface, you can't simply measure height above ground to figure out which pattern applies to a given antenna. Many dipole users just orient the antenna with regard to the horizontal pattern and mount it as high and as clear of surrounding objects as possible. Although this does not always produce the best system performance, the results are usually adequate. Some users have tried to obtain the desired vertical patterns (Figs. 6-10A or 6-10B) by simulating a ground with a grid of wires placed beneath the dipole as shown in Fig. 6-11. Subjective reports suggest that even a single wire (the one labeled A) placed beneath a dipole or V may improve 29-MHz Mode A reception. At 146-MHz and higher frequencies, a reflecting screen can be used for the ground so that a vertical pattern similar to the one of Fig. 6-10B can be achieved with the antenna mounted in a desirably high location. That the ground screen is not infinite tends to reduce gain at take-off angles below about 15°.

The basic half-wave dipole can also be mounted vertically. In this orientation the horizontal plane pattern is omnidirectional while the actual vertical plane pattern, which depends on mounting height, is likely to have one or more nulls at high radiation angles. Although the characteristics of this antenna appear suitable for work with low-altitude satellites, there is a hitch: the feed line must be routed at right angles to the antenna for at least a half wavelength if one hopes to obtain the patterns described. As a result, it's usually easier to use a ground-plane antenna (see next section), which has similar characteristics. One novel configuration that has proved effective for working DX on Mode A consists of a vertical dipole for 29 MHz hung at the end of a tower-mounted 2-m beam. When the tower-to-dipole distance is set at roughly 6 feet the tower will tend to act as a reflector and the resulting 29-MHz pattern will be similar to that obtained with a vertically mounted 2-element beam.

In truth, we've paid considerably more attention to the dipole and V than their actual use justifies. Nevertheless, they clearly illustrate many of the trade-offs between effective gain patterns and system complexity that a ground station operator is faced with.

The Ground Plane

The ground plane (GP) antenna, familiar to hf and vhf operators alike, is sometimes used at satellite ground stations. Physically, the GP consists of a 1/4- or 5/8-wavelength vertical element and three or four horizontal or drooping spokes that are roughly 0.3 wavelengths or longer. At vhf and uhf frequencies sheet metal or metal screening is often used in place of the horizontal spokes. The GP is a low-gain, linearly polarized antenna. The gain pattern in the horizontal plane is omnidirectional. Because of its low gain the GP is not generally suitable for operating with high-altitude satellites, though it may be used in special cases. We'll focus on its possibilities with respect to low-altitude spacecraft.

Gain patterns in the vertical plane for 1/4-wavelength GP antennas are shown in Fig. 6-12. Although the vertical plane patterns suggest that performance will be poor when the satellite is overhead, stations using the GP report satisfactory results. The reasons are most easily explained in terms of reception. Downlink signals usually arrive at the ground station antenna from several directions after being reflected off nearby objects. These reflected signals can either help (when the direct signal falls within a pattern null) or hinder (when interference between the main and reflected signals results in fading). In practice, the good effects appear to far outweigh the bad; the GP is a good all-around performer for working with low-altitude spacecraft. The particular mode and satellite of interest, and your transmit power, however, do affect the performance you can expect. Most stations receiving the OSCAR 8 Mode J downlink (435 MHz) prefer a beam, but signal levels are generally acceptable when a GP with a good low-noise preamp (1.2-dB noise figure) mounted directly at its base is used. Similarly, stations running low-power transmitters (10 to 20 watts) often find that a small beam is needed for reliable uplinking to OSCAR 8.

A GP may work for receiving signals from high-altitude satellites under certain situations. For example, although the downlink S/N ratio using a GP generally will not be adequate for communication, it should be sufficient for spotting (determining if the spacecraft is in range). The omnidirectional (horizontal plane) pattern of the GP makes it especially suitable for this purpose. Also, the GP may be useful near perigee of elliptical-orbit missions if the spacecraft height is less than a few thousand miles. Spin modulation of the downlink signals, however, may make a circularly polarized ground station antenna mandatory in this situation; only experience will tell. Suitable broad-beamwidth, circularly polarized antennas are discussed later in this chapter. We now turn to some practical GP antennas.

GP antennas designed for the 27-MHz CB market are inexpensive and widely available. For Mode A downlink operation the 1/4-wavelength GP usually out performs the "bigger and better" 5/8-wavelength CB model. To modify a 1/4-wavelength

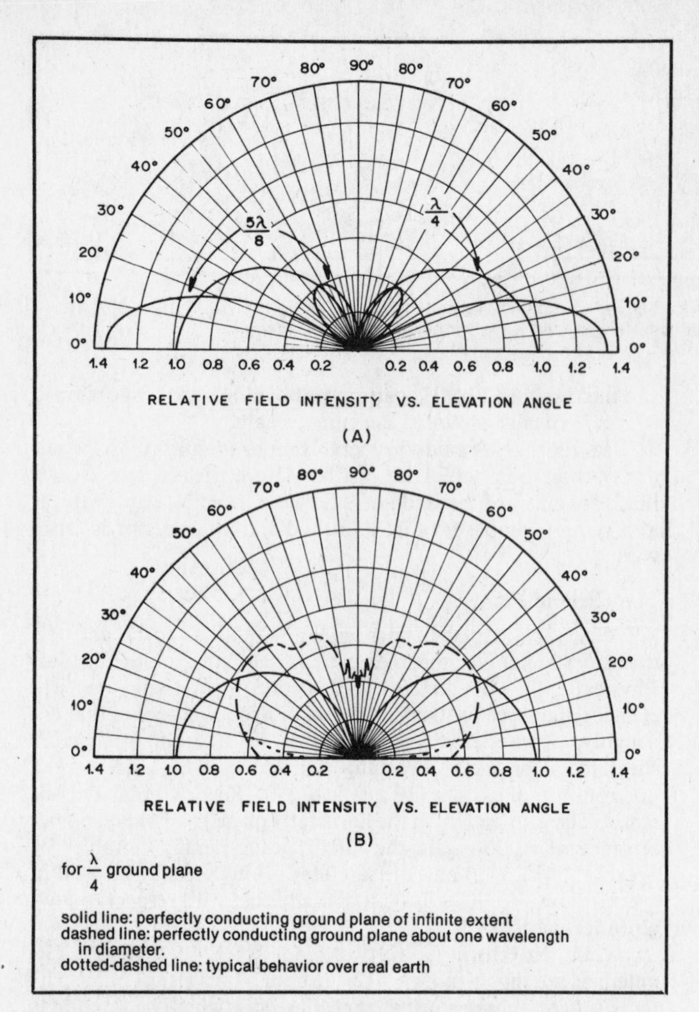

Fig. 6-12 — (A) Vertical plane gain patterns showing relative field intensity for 1/4 wavelength and 5/8 wavelength ground-plane antennas over ideal earth (perfect conductivity and infinite extent). (B) The effects of a limited ground plane and/or resistive ground on the 1/4-wavelength ground-plane antenna.

CB antenna for 29.5-MHz, you need only to shorten the vertical element by about 9%.

GP antennas designed for the 146-MHz and 435-MHz amateur bands are available commercially at moderate cost. Once again, the 1/4-wavelength models produce good results. Some users, however, prefer a 5/8-wavelength GP when the satellite is at low elevation angles, and a different type of antenna when the satellite is at higher elevation angles. A vhf or uhf 1/4-wavelength GP can be assembled at extremely low cost (see the illustration in Fig. 6-13).

Tilting the vertical element of a 1/4-wavelength GP produces a gain pattern in the vertical plane shown in Fig. 6-14B. Note how the overhead null has been eliminated. The horizontal pattern is slightly skewed, but remains essentially omnidirectional. Tilting also tends to reduce the already low input impedance of the GP. One way to compensate for this reduction is to use a folded element as shown in Fig. 6-14A. As in the folded dipole, the folded 1/4-wave element in a tilted GP steps up the input impedance and gives a broader bandwidth antenna.

Yagi and Quad

In many situations a beam will be the preferred antenna at a ground station. In this section we'll look at two familiar beams, the Yagi and the quad, and some of their variants: the quagi, loop-Yagi and delta-loop, all of which produce linearly polarized signals. The next section will focus on techniques for combining pairs of these antennas to produce a circularly polarized signal and the section following that will introduce a beam that produces a circularly polarized wave directly.

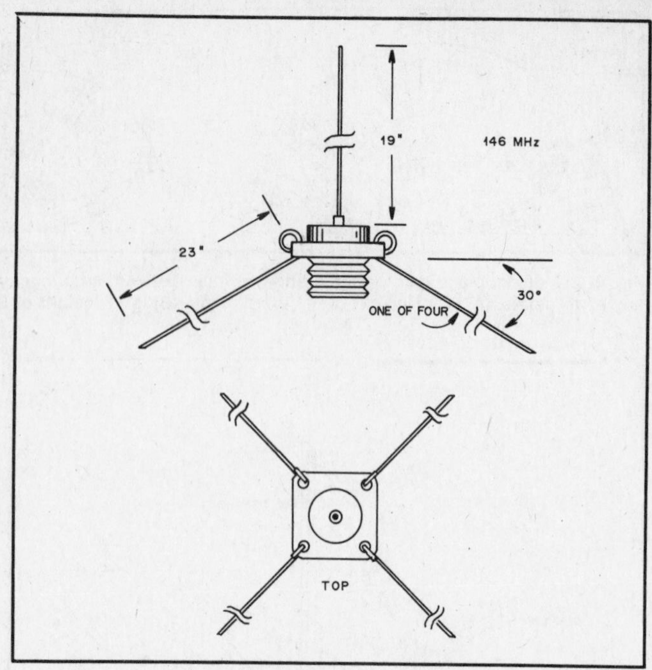

Fig. 6-13 — A ground-plane antenna for 146 MHz is easily constructed using a chassis mount coax connector. A type N connector is preferred but a UHF type is acceptable at 146 MHz. Drooping the radials increases gain slightly at low elevation angles and raises input impedance to produce a better match to 50-ohm feed line.

In certain situations beams may be appropriate for working with low-altitude spacecraft. A beam could be used for (1) an uplink antenna when available ground station transmitter power is very low, (2) a downlink antenna when a very high downlink S/N ratio is desired or (3) both link antennas when one is attempting to contact stations when the spacecraft is near or below the radio horizon. Superior performance has its costs: the financial cost of the rotator(s) needed, and the operational cost of having to "ride" the rotator(s) during a pass.

If several passes of a low-altitude satellite are previewed on an OSCARLOCATOR, you'll see that the satellite often just grazes the outskirts of your acquisition circle. During these horizon passes the satellite elevation angle will generally be between 0° and 15°, and azimuth changes, though larger, will usually be less than 90°. Readers who are already equipped for terrestrial vhf or uhf operation with a beam mounted on an azimuth rotator and aimed at the horizon will find that their setup provides good satellite access on horizon-grazing passes. Before the pass begins, the antenna can be set to an azimuth about 20° past AOS. A single azimuth update will usually suffice for the entire pass.

For general operation with high-altitude satellites, beams are necessary for obtaining an adequate downlink S/N ratio and cost-effective for obtaining the desired uplink EIRP. The burden of keeping the antennas properly aimed during a pass is not as severe with elliptical orbits of the Phase III type because spacecraft motion near apogee will appear very slow. With stationary satellites the antenna will simply be aimed during the initial setup. Once you have committed to using a beam and rotators for the downlink, it's almost always least expensive to use a beam with similar gain on the uplink.

The performance of well-designed Yagis, quads, quagis, loop-Yagis and delta-loops of equal boomlength is very similar. Free-space gain patterns are roughly symmetrical (all cross sections look the same) with a shape somewhat like Fig. 6-4 (sometimes called a pencil-beam). The relation between half-power beamwidth and gain, given earlier in Eq. 6.4, has been

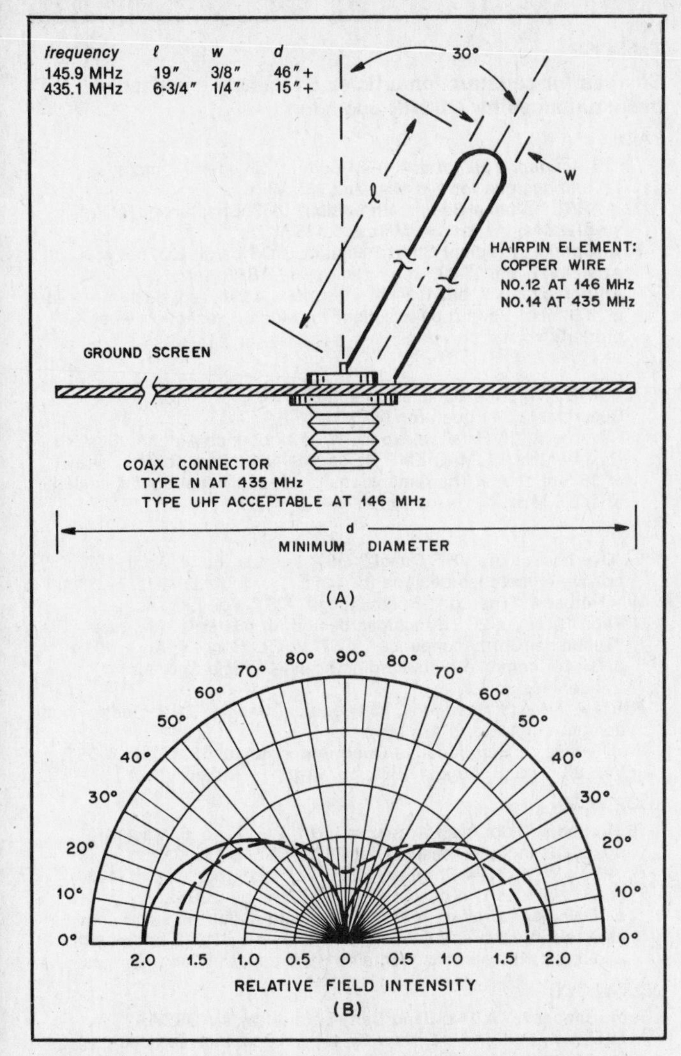

Fig. 6-14 — (A) 1/4-wavelength ground-plane antenna with tilted vertical element. Groundplane may be square or circular, solid or mesh. (B) Vertical plane relative field intensity for 1/4-wavelength groundplane; solid line — element vertical, dashed line — element tilted 30° from vertical.

used to prepare Table 6-2; approximate half-power beamwidths are listed for several gains. Since performance of the Yagi, quad, quagi, loop-Yagi and delta-loop are so similar, selection will depend on difficulty of construction, mounting ease, commercial availability, and suitability for later use as part of a circularly polarized system. We'll consider each type of beam in terms of these criteria shortly; but first, some comments on ground effects.

Ground reflections affect all the beams under discussion similarly. The vertical gain pattern of a beam mounted with its boom parallel to the surface of the earth does *not* look like the clean, free-space pattern shown in Fig. 6-4. Instead, it breaks up into several lobes interspaced with nulls, the number and position depending on the antenna height (in wavelengths). An example can be seen in Fig. 6-10C. These lobes and nulls result from constructive and destructive interference between the direct and ground-reflected signals, as discussed earlier in this chapter. In contrast, when the same beam is pointed significantly above the horizon, the ground-reflected signal contains only a relatively low proportion of the total power; interference effects (both constructive and destructive) become very small. As a result, the tilted beam does produce a clean pattern resembling that in free space.

To illustrate the practical implications of ground effects on vertical patterns, consider a typical ground station antenna for working with Phase III satellites. It gives 13-dB$_i$ gain and 45° beamwidth. Let's focus on the downlink and look at the satellite

Table 6-2

Half-power beamwidth as a function of gain for well designed, symmetric pattern, beam antennas.

G *gain*	G *gain*	θ* *half-power beamwidth*
6 dB$_i$	4.0	100°
8 dB$_i$	6.3	80°
10 dB$_i$	10.0	63°
12 dB$_i$	16.0	50°
14 dB$_i$	25.0	40°
16 dB$_i$	40.0	32°
18 dB$_i$	63.0	25°
20 dB$_i$	100.0	20°
22 dB$_i$	159.0	16°
24 dB$_i$	251.0	13°

near apogee. Assume that both the satellite and the antenna are initially at an elevation angle of 40°. Suppose that one hour later the elevation rotator has not been touched though the satellite has climbed to an elevation angle of 62.5°, a change of one half our antenna beamwidth. With the antenna set at 40° elevation, very little ground-reflected power reaches the antenna and the pattern can be thought of as a clean pencil-beam. When the satellite is at 62.5° elevation, it is at a point 3-dB down on the ground station antenna pattern; we'd expect the downlink signals to have decreased by 3 dB. Practical experience would confirm these expectations.

Now consider a similar situation with the same satellite near apogee and same antenna, but this time let the initial elevations of both satellite and the antenna be 5°. Assume that one hour later the satellite elevation increases to 15° while the antenna elevation remains at 5°. What happens to the link? A prediction based on free-space patterns would yield an almost trivial 1- or 2-dB decrease in signal level since the 10° change in elevation is far less than the 22.5° (half-beamwidth) change it takes to reduce signals by 3 dB. But predictions based on the free-space model are totally inadequate at low antenna elevations where ground reflections play a very pronounced role. In reality, it's nearly impossible to predict the outcome, but changes in the downlink amounting to a decrease of 30 dB, an increase of 3 dB, or anything in between wouldn't be surprising. Even though the outcome can't be predicted, understanding the situation is important: At low satellite elevation angles, aiming the antenna in elevation becomes more critical. With a broad-beamwidth antenna it's very easy to ignore a small, seemingly insignificant change in satellite elevation. While this oversight is safe at high elevation angles, it can be disastrous at low angles.

Our discussion has focused on the downlink. The uplink is analogous except for one fact. Even if uplink and downlink antennas have identical free-space patterns and are mounted at the same physical height, their actual vertical patterns will not be the same; their electrical heights (measured in wavelengths) will be different. This again points out the importance of careful elevation angle control when the satellite is close to the horizon. For reliable operation at low elevation angles it's critically important to monitor your downlink and adjust antenna elevation as often as necessary.

We now turn to some of the practical concerns involved in choosing among the Yagi, quad, quagi, loop-Yagi and delta-loop. The chief advantages of the Yagi are its simple structure, light weight, and low wind-load for a given gain. The overwhelming majority of commercially available vhf and uhf beams are Yagis. The high-gain Yagi, however, is intrinsically a narrow-band device. Dimensions and matching are critical. As a result, home builders working from a published design must be very careful to duplicate all dimensions and spacings exactly as in the original if they want optimum performance. Some Yagi designs use a log-periodic type feed consisting of several driven elements. This produces a broader bandwidth antenna whose dimensions are less

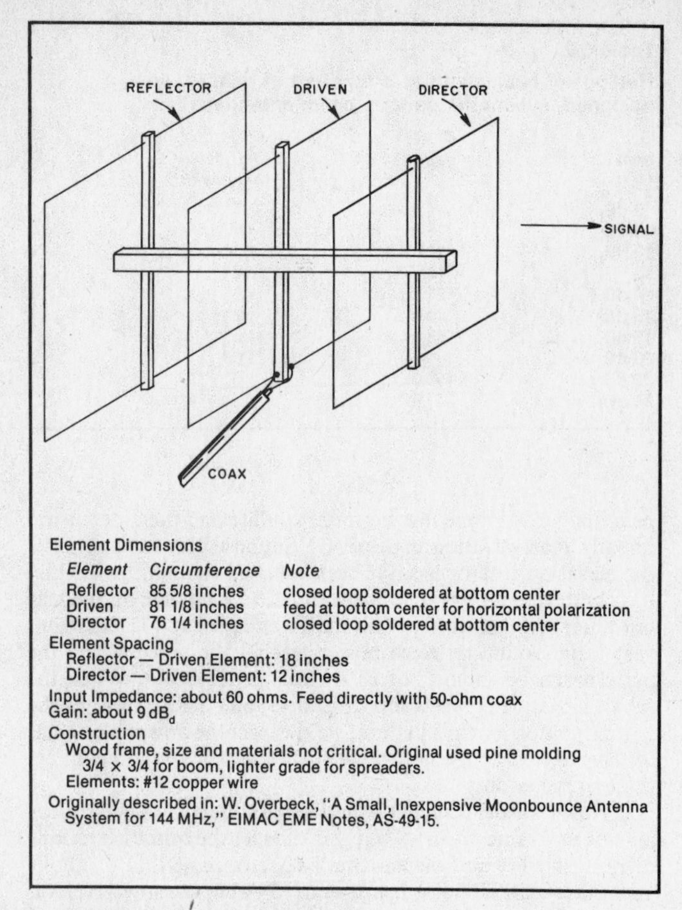

Fig. 6-15 — A 3-el quad for 145.9 MHz.

critical and for which the detuning effects of nearby objects are less severe. Of course, the extra elements add to the weight and windload.

Though the quad antenna is structurally awkward, this is compensated, to some degree, by the ease of matching and the noncritical dimensions. Very few designs for vhf and uhf quads have been published. It's fair to say that this attests to their lack of popularity. Designs for hf quads of four to six elements have been studied by scaling their dimensions to uhf and measuring the performance of the uhf model on an antenna range. Construction details for an easily duplicated 146-MHz, 3-el quad are given in Fig. 6-15.

The quagi is a cross between quad and Yagi. It uses a quad reflector and driven element for easy, efficient matching, and Yagi directors for good gain, low windload and simple structure. Since its introduction in 1977, the quagi has quickly become popular with new vhf and uhf operators who want a simply constructed homemade antenna that can be put on the air without any specialized test equipment and that will perform up to expectations.

The loop-Yagi (Fig. 6-1) and the delta-loop are close relatives of the quad and Yagi. Both have been used for satellite communication. Recently, the loop-Yagi has received considerable attention. Since its structure is mechanically awkward at vhf and lower frequencies, it hasn't seen much use in this part of the radio spectrum. It is gaining in popularity at 435 MHz, 1260 MHz and higher frequencies where a very straightforward mechanical design has evolved. Each loop is formed into a circle from a strip of flat, springy conductor. A single screw holds the loop in shape and secures it to an aluminum boom. Good loop-Yagi designs appear at least to equal, and perhaps exceed, Yagis of the same boom length. As the bandwidth of a loop-Yagi is about five times that of a comparable Yagi, construction tolerances are considerably relaxed. The delta-loop is similar to the quad in mechanical complexity and performance and, like the quad, has not been widely used at vhf or uhf.

Table 6-3

Sources for construction articles on linearly polarized beam antennas for OSCAR operation.

YAGI

Radio Amateur's Handbook, 1984 edition, Chapter 21. Includes several designs for 146 MHz and 435 MHz.

The ARRL Antenna Book, 14th edition, 1982, Chapter 11. Includes several designs for 146 MHz and 435 MHz.

R. J. Gorski, "Efficient Short Radiators," *QST*, Vol. LXI, no. 4, April, 1977, pp. 37-39. Reprinted in *The ARRL Antenna Anthology*, 1978, pp. 112-114. Describes a 2-el Yagi design tested at 100 MHz. Should be excellent for Mode A reception when properly scaled.

QUAD

The ARRL Antenna Book, 14th edition, 1982, pp. 11-13. Describes a 2-el quad for 144 MHz.

W. Overbeck, "A Small, Inexpensive Moonbounce Antenna System for 144 MHz," EIMAC EME Notes, AS-49-15. Describes an array of 16 3-el quads. The dimensions of the individual quads, scaled to 145.9 MHz, are given in Fig. 6-15.

QUAGI

W. Overbeck, "The VHF Quagi," *QST*, Vol. LXI, no. 4, April, 1977, pp. 11-14. Includes designs for 144.5, 147 and 432 MHz.

W. Overbeck, "The Long-Boom Quagi," *QST*, Vol. LXII, no. 2, Feb., 1978, pp. 20-21. Includes design for 432 MHz. Also see "Technical Correspondence," *QST*, Vol. LXII, no. 4, April, 1978, p. 34, for comments concerning scaling quagis to other frequencies.

Radio Amateur's Handbook, 1984 edition, Chapter 21. Includes quagi designs for 146 and 435 MHz.

W. Overbeck, "Reproducible Quagi Antennas for 1296 MHz," *QST*, Vol. LXV, no. 8, August 1981, pp. 11-15.

LOOP-YAGI

R. Harrison, "Loop-Yagi Antennas," *HR*, Vol. 9, no. 5, May 1976, pp. 30-32. Includes designs for 28.5, 146 and 435 MHz.

B. Atkins, "The New Frontier," *QST*, Vol. LXIV, no. 10, Oct. 1980, p. 66. Includes two designs for 1296 MHz by G3GVL: a 38-element array on a 10-ft. boom with about the same gain as a four-ft. dish and a 27-element array on a 7.5-ft. boom with about 1.5 dB less gain. Contains good construction diagrams.

DELTA-LOOP

A. A. Simpson, "A Two-Band Delta-Loop Array for OSCAR," *QST*, Vol. LVIII, no. 11, Nov. 1974, pp. 11-13. Includes designs for 146 and 435 MHz.

A list of relevant construction articles featuring the Yagi, quad, quagi, loop-Yagi and delta-loop is contained in Table 6-3. Before selecting one of these homebuilt designs or a fully assembled commercial model, check the results of recent vhf, uhf and EME contests in *QST* to see which antennas are favored by the "big guns." Each year at vhf/uhf conferences around the U.S., antenna test ranges are used for careful comparisons among antennas. Consistent top performers are quickly adopted by serious contesters and EME buffs. As a rule of thumb, if an array of eight brand-X Yagis is popular with EME operators, one brand-X Yagi is a good bet to use with a Phase III satellite, or a pair might be used to obtain circular polarization, as described in the next section.

Circular Polarization From Linearly Polarized Antennas

There are several techniques for producing a circularly polarized wave from linearly polarized antennas (Table 6-4). The first two methods have been used widely by radio amateurs. EME buffs have had success using the third method at frequencies above 1 GHz as a feed for parabolic antennas. As the remaining approaches do not appear suitable for amateur applications at satellite ground stations, only the first three will be covered here.

We'll look at Methods 1 and 2 in detail. Each requires a pair of matched, linearly polarized antennas. We'll use two identical 2-element Yagis, carefully adjusted to provide a 50-ohm resistive input impedance, to illustrate each method, although two dipoles,

Table 6-4

Methods for producing circular polarization using linearly polarized antennas.

For information on methods not covered in this text see: H. Jasik, *Antenna Engineering Handbook*, McGraw Hill, NY, 1961, Chapter 17.

1. Pair of similar antennas fed 90° out of phase.
2. Pair of similar antennas fed in phase.
3. Dual-Mode horn
4. Combination of electric and magnetic antennas
5. Transmission-type polarizers
6. Reflection-type polarizers

two multielement Yagis, two quagis, and so on, could also serve. (With adjustments in the phasing/matching harnesses, other impedance antennas would also work.)

Method I

In this method the two antennas are mounted as shown in either Fig. 6-16A (single-boom array) or Fig. 6-16B (dual-boom array). Both configurations produce the same results when aimed properly, though the off-axis performance of the single-boom array is slightly better. This advantage is balanced to some extent by the ease of mounting the two-boom array while keeping the rotators well out of the radiation pattern. When employing quads or quagis the two-boom configuration should be used to lessen the interaction between the two antennas.

The feed system is critical to the performance of these arrays. A phasing/matching harness that produces the correct power division, matching and delay parameters is shown in Fig. 6-17. Only when the two antennas are fed *90° out of phase* with *equal power* will the array produce a circularly polarized wave. The effects of various errors in power division and phase difference are described in Table 6-5. For the feed system to perform its function, each antenna must be carefully adjusted to provide an unbalanced, 50-ohm, purely resistive input impedance before it's incorporated into the array. Small adjustment errors in each antenna, even though they may be identical, can have a large effect on power division and phasing and thereby produce an elliptical wave with a large, linear component. This may occur even though the SWR in the main feed line remains acceptable.

Although each array in Fig. 6-16 shows one Yagi mounted vertically and the other horizontally, this particular configuration needn't be employed as long as the Yagis are mounted at right angles to one another. The tilted arrangement shown in Fig. 6-18 is commonly used so that interaction with the cross boom or rotators is balanced. There's little difference in performance between the horizontal-vertical and skewed arrangements at high elevation angles. At low elevation angles, however, where ground reflections have a pronounced effect on the radiated signal, the skewed design may be preferable. Horizontal and vertical signal components undergo different phase changes when reflected off the ground, as discussed earlier in this chapter.

How do we determine the polarization sense of the antenna in Fig. 6-16A when it's fed with the harness in Fig. 6-17A? We could measure the polarization using a technique that will be described shortly, or we could figure out the sense analytically as follows. Imagine yourself standing behind the single-boom array looking in the direction of maximum gain. Focus your attention on the electric field at the point P located at the center of the driven elements. The field at P results from the sum of two components: one component that is parallel to AA′ (contributed by element AA′), and a second component that is parallel to BB′ (contributed by the BB′ element). Because of the 90° phasing, one component will be a maximum when the other one is zero. We wait until the field at P points toward 12 o'clock (parallel to AA′, pointing in the direction of the element connected to the center conductor of the feed line). Exactly one quarter cycle (90°) later, the rf currents at the end of the delay line will produce an electric field parallel to BB′, pointing toward 3 o'clock since element B is connected to the center conductor of the delay line. From your observation position in back of the antenna, you see the electric field at P rotate from 12 o'clock to 3 o'clock (90° clockwise) during this quarter cycle. This configuration therefore produces right-hand circular polarization (RHCP).

How can we change the sense of polarization? We can switch from RHCP to LHCP by interchanging *either* (1) the connections at B and B′ *or* (2) the connections at A and A′. Switching both sets of connections will *not* change the polarization sense.

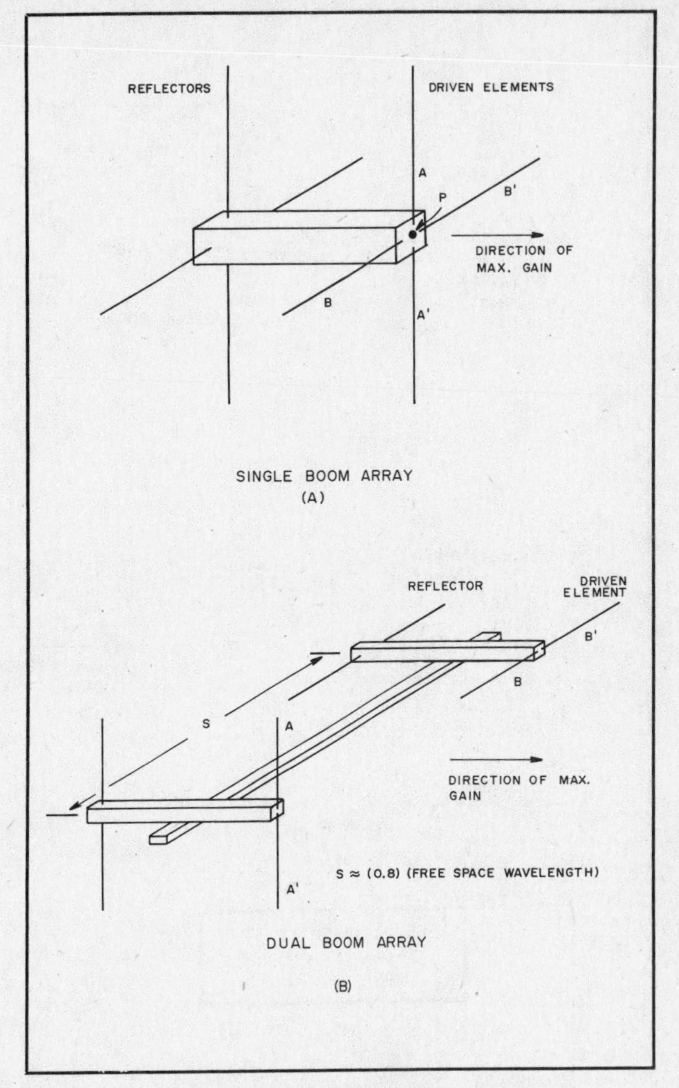

Fig. 6-16 — Yagi placement for Method I production of circular polarization. See Fig. 6-17 for required phasing/matching harness.

Table 6-5

The effects of feed phase and/or power division errors on the performance of the Yagi arrays shown in Fig. 6-16A and Fig. 6-16B.

Phase difference (θ)	*Power Division*	*resulting wave along major axis*
90°	equal	circular polarization
90°	unequal	elliptical polarization
0°	equal	linear polarization in plane midway between planes of two Yagis
0°	unequal	linear polarization plane depends on power division
$0° < \theta < 90°$	equal	elliptical polarization
$0° < \theta < 90°$	unequal	elliptical polarization

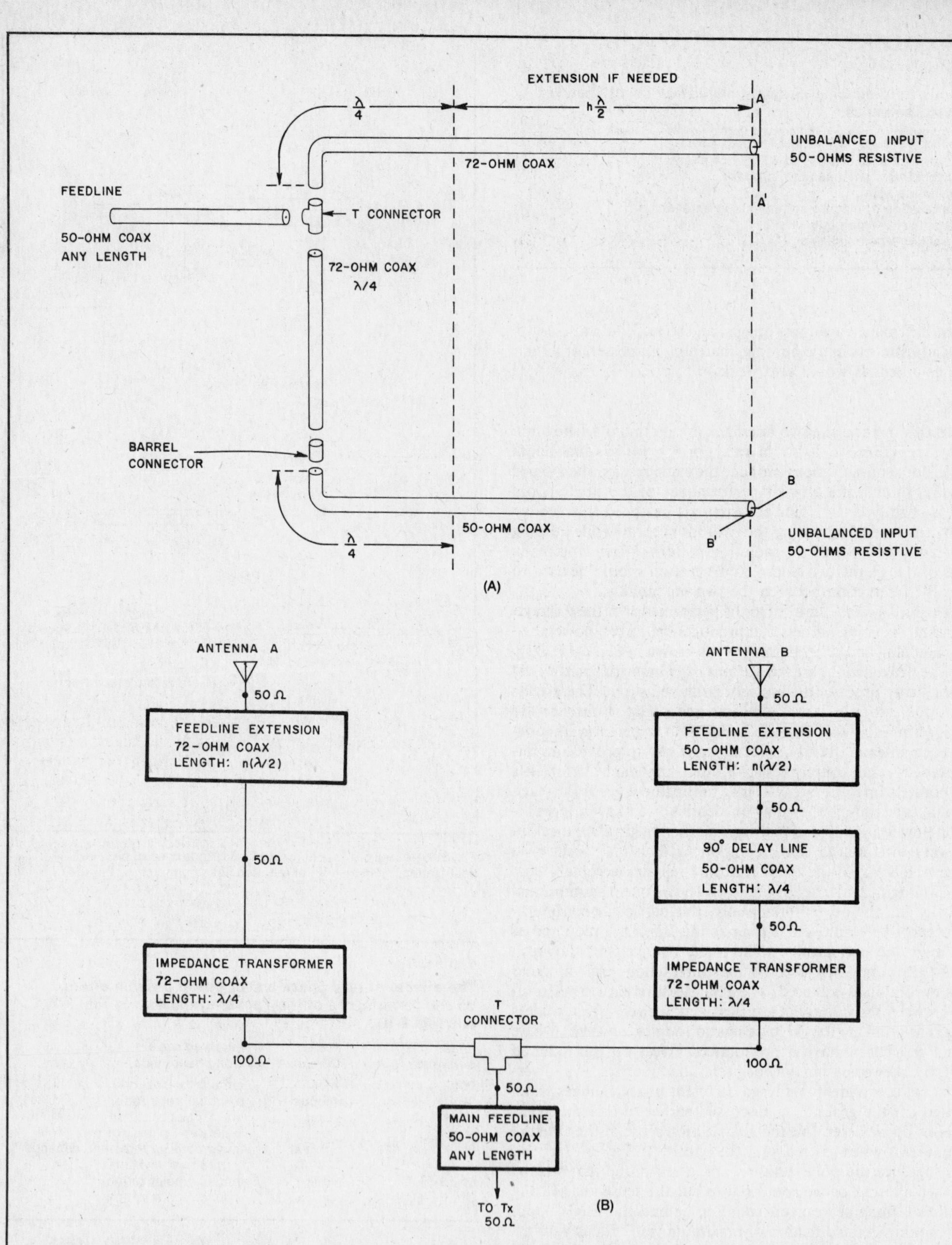

Notes:

1. n may be either 0, 1, 2, etc. depending on how many half wavelength extensions are needed (see text).
2. All cable dimensions refer to electrical length λ = (velocity factor of cable) (free space wavelength)
Values for the velocity factor of 0.66 (regular cable) and 0.81 (foam filled cable) are often used. However, large variations are common. Because small errors in the matching harness can significantly degrade performance one should either (1) measure the electrical length of each cable or (2) measure the actual velocity factor of the cables being used. Methods for doing this are discussed later in this chapter.
3. Since the antenna impedance repeats every half wavelength along the feedline, regardless of the feedline impedance, we can use 72 ohm coax to feed antenna A and eliminate one splice in the matching harness.

Fig. 6-17 — Matching/Phasing harness for arrays shown in Fig. 6-16; (A) physical design, (B) function block diagram.

Fig. 6-18 — Crossed Yagis mounted in skewed orientation.

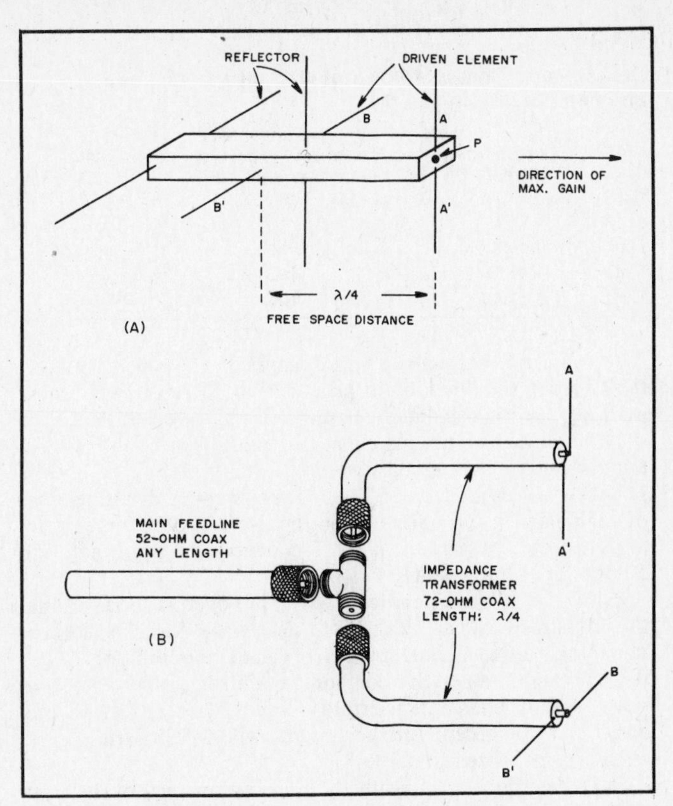

Fig. 6-19 — Illustration of Method II for producing circular polarization from linearly polarized antennas. (A) shows antenna, (B) the matching harness for 50-ohm resistive, unbalanced antennas.

Switching to LHCP can also be accomplished by modifying the matching section of Fig. 6-17 so that the two extensions differ by an odd number of electrical half wavelengths. Any of the techniques just mentioned can, of course, also be used to switch from LHCP to RHCP.

The polarization sense of the dual-boom array of Fig. 6-16B can be predicted by imagining that the two antennas are slipped together (sideways motion only) until the booms overlap and form an array like the one in Fig. 6-16A. As both arrays will have the same sense of circular polarization, and as we already know how to determine whether a single-boom array is RHCP or LHCP, analyzing the sense of a dual-boom array is simple.

As stated earlier, the 90° phase difference and the equal power division are critical to achieving circular polarization. The phasing/matching harness of Fig. 6-17A was designed to work with antennas having an unbalanced, 50-ohm input impedance that's purely resistive. Using the functional block diagram of Fig. 6-17B as a guide, we'll step through its operation. Temporarily ignore the feed-line extensions. Since the array's operation depends on a 90° phase difference between the two sets of elements, the first thing we incorporate is a delay line. A piece of coax that's electrically 1/4-wavelength long does the job. (One wavelength is 360°; one-quarter wavelength is 90°.) The coax delay line would also act as an impedance transformer if the characteristic impedance of the line didn't match the antenna. To obtain the proper power division we don't want any impedance transformation here. Therefore, we use 50-ohm cable. Next, we could connect the two branches in parallel at a coaxial T connector and have both equal power division and correct delay, but the feed-point impedance would be 25 ohms, a value that's awkward to match. Instead, we install two identical impedance transformers consisting of 1/4-wavelength sections of 72-ohm coax to step up the impedance of each branch to 100 ohms. When the two 100-ohm lines are connected in parallel at a coaxial T connector, we obtain a good match to 50-ohm feedline. The two impedance transformers do, of course, also act as delay lines. But, since we've used a pair of equal lengths, the phase *difference* between the two Yagis isn't affected.

The harness is now complete except for one mundane consideration: The two branches may not be long enough to reach the antennas. Two identical pieces of 50-ohm coax (any length) would work as extensions; we can save a coax connector and its consequent losses, however, by using electrical-half-wavelength sections cut from coax of different impedances as shown. Note that adding an extra half-wavelength to one of the extensions will reverse the sense of polarization.

This completes our discussion of Method I. We now turn to a second technique for obtaining circular polarization.

Method II

To illustrate this method we again use two 2-element Yagis. They can be mounted on a single boom as in Fig. 6-19A, or on two separate booms as described earlier. A 90° phase difference is again the key to the operation. This time it's obtained by

Table 6-6
Data Used to Compute Polarization Sense of Antenna and Feed Shown in Fig. 6-19

	Field at center of BB′ (from element BB′ only)	*Field at center of AA′ (from element AA′ only)*	*Total field at P*
Time 1	9 o'clock	12 o'clock	xxx
Time 2	zero magnitude	zero magnitude	9 o'clock
Time 3	3 o'clock	6 o'clock	6 o'clock

physically offsetting one Yagi a quarter of a wavelength along the boom in the direction of propagation. With this approach, no delay line is needed in the feed harness. The feed system need only take into account matching and equal power splitting. An appropriate matching harness is shown in Fig. 6-19B. The 1/4-wavelength sections of 72-ohm coax step-up the impedance of each Yagi to 100 ohms. When the two 100-ohm impedances are connected in parallel at the T connector, a good match to 50-ohm feed line results.

Method II has a significant advantage over Method I in that the adjustment of each Yagi is not nearly as critical. As long as both Yagis are identical, small errors in the input impedance or the presence of a reactive component will *not* disturb the equal power split or phasing; the errors will only affect SWR. As long as the SWR is acceptable the antenna will produce the desired circularly polarized pattern.

To determine analytically the polarization sense of the array shown in Fig. 6-19, imagine yourself standing behind it looking in the direction of maximum gain. Focus your attention on the electric field at the center of the front driven element. The field at the center of AA′ (point P) results from the sum of two components: a contribution from element AA′ in the vertical direction, and a contribution from element BB′ in the horizontal direction. Note that the contribution of the element BB′ to the total field at P was actually produced by element BB′ a quarter-cycle earlier. Because of the quarter-wavelength offset it takes a quarter-cycle for the field to travel from the center of BB′ to point P.

We're going to compute the direction of the electric field at P by combining the fields produced at the center of each element at three different times. Table 6-6 will help us keep track of all the needed information.

We start our observations at *Time 1* when the rf current in the feed line is producing a maximum field at each element. *Time 2* occurs after a quarter cycle has elapsed. *Time 3* occurs after an additional quarter cycle has passed. In the second column of Table 6-6 we describe the field at the center of element BB′, from this element only, at each of the three times. In the third column we describe the field at the center of element AA′, from this element only, at each of the three times. Finally, we fill in the last column of Table 6-6 for each time, by adding together the field at the center of AA′ and the field that was produced at the center of BB′ a quarter of a cycle earlier and that is just reaching P.

The dashed lines in Table 6-6 will help you follow the additions. Note that the last column opposite *Time 1* has been left blank since we didn't compute the contribution of BB′ a quarter cycle earlier. From our observation position in back of the antenna we see the electric field at point P rotate from 9 o'clock to 6 o'clock as a quarter cycle elapses. The wave is therefore counterclockwise (LHCP).

Yagis using a balanced driven element such as a folded dipole are particularly well suited to Method II. An efficient matching harness using open wire balanced line of an appropriate impedance and a 1:1 or 4:1 balun as needed can be designed and constructed easily.

Method III

At frequencies above 1 GHz the horn antenna is both convenient and effective as a feed for parabolic reflectors. Amateurs have learned from experience that a surprisingly efficient horn

Table 6-7
Variable polarization antenna systems are usually limited to a few discrete choices. Three examples are listed.

Polarization	*2-option system*	*4-option system*	*6-option system*
RHCP	X	X	X
LHCP	X	X	X
LP-vertical		X	X
LP-horizontal		X	X
LP-45°			X
LP-135°			X
max. loss due to polarization mismatch	less than 3 dB	less than 3 dB	less than 2 dB

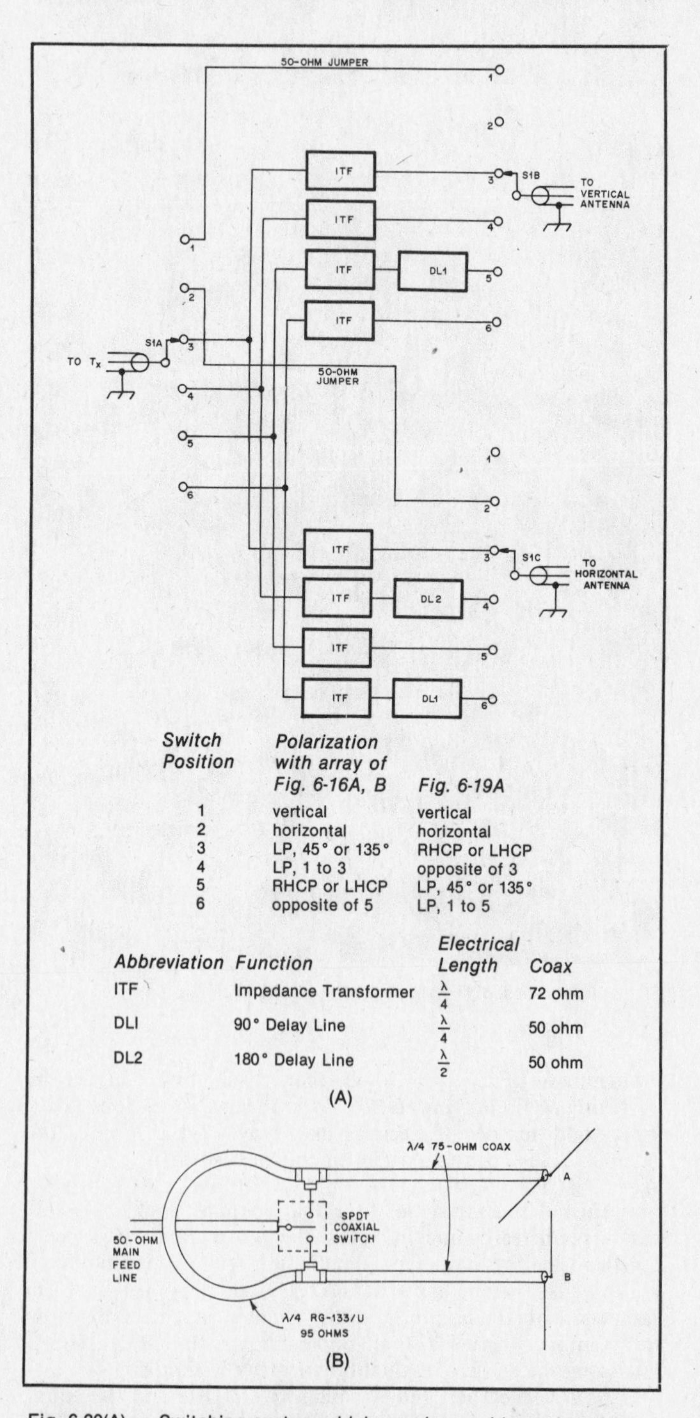

Fig. 6-20(A) — Switching system which may be used in conjunction with two 50-ohm Yagis configured as in Figs. 16A, 16B, 19A. To provide a choice of RHCP, LHCP, or linear polarization at 0°, 45°, 90°, and 135°. (B) — Switching system providing choice of RHCP or LHCP with two 50-ohm Yagis configured as in Fig. 6-16A or 6-16B.

can be built from a tin can in which a quarter wave monopole is soldered to a coax chasis connector and mounted on the inside curved surface. The dimensions of the can and placement of the probe depend on your operating frequency and the shape of your dish. Antenna tuning, phasing and circularity adjustments are accomplished with a few stragetically placed brass screws protruding into the can.

As amateur satellites begin to use links at 1.2 GHz and 2.4 GHz, antenna arrays made of a dish and feed horn will become popular. Since parabolic dishes are passive reflectors, a linearly polarized feed will produce a linearly polarized array and a circularly polarized feed will produce a circularly polarized array. For a detailed description of a 2.3-GHz circularly polarized feed horn see *The ARRL Antenna Book,* 13th Ed., 1974, pp. 259-260. Additional information on parabolic dish antennas is contained later in this chapter.

Comments

The analytic procedures outlines for determining the polarization sense of a crossed Yagi or similar array may leave your brain feeling like it's been spin modulated. Don't worry; you're in good company — the telecommunications engineers setting up the first satellite transatlantic TV broadcast via TELSTAR, for example, built a link with RHCP at one end and LHCP at the other! The calculation approach for determining whether an antenna is RHCP or LHCP can be sidestepped. You can, for example, determine the polarization sense of an array simply by testing it on a link with a CP antenna of known sense at the other end. It's also possible to design an array whose polarization can be switched from the operating position. You then select the sense of circular polarization that produces the strongest signals without worrying about what it happens to be. Let's look at both possibilities.

Setting up a terrestrial test link with an antenna of known polarization at one end isn't a problem, though you may need to build a small, 3-turn helix to lend to a cooperative local amateur. As we'll see shortly, the sense of a helix is determined easily. The array being studied should be mounted temporarily in an easily accessible location so that its sense can be reversed several times while you are monitoring signal strength at the receiving end of the link. S-meter readings should be several units higher when the polarization of the array under test matches the helix.

Being able to switch polarization from the operating position is very desirable. Using a pair of identical linearly polarized antennas mounted as in Figs. 6-16A, 6-16B or 6-19A, it's theoretically possible to obtain any polarization — linear (any orientation), circular (RH or LH) or elliptical (any combination of linear and circular) — by adjusting the power division between the two antennas and the relative phasing. In practice, systems providing a continuous range of choices are very complex; selection is usually restricted to several discrete options.

Three common systems that provide 2 options, 4 options, and 6 options are listed in Table 6-7. A suggested switching diagram for the 6-option system is shown in Fig. 6-20A. To obtain a 4-option switch design just omit the unused paths. A very simple 2-option switching system is shown in Fig. 6-20B. Note that the feed/phasing systems of Fig. 6-20 will work only with 50-ohm antennas. The 6-option switch is generally located at the operating position and separate feed lines are run to each antenna. The feed lines *must* be of equal electrical length; they should be pruned carefully using the dip-meter techniques discussed later in this chapter.

Although the worst-case link performances of the 2-option and 4-option systems are identical (3-dB below matched polarization), the 4-option system will, on the average, be about 1-dB better. The 6-option system will, in turn, provide about a 1-dB advantage over the 4-option system. Because these performance differences are so small, the 2-option system (RHCP or LHCP) will provide excellent results in all but the most demanding situations.

The 6-option switching system can be mechanical or electrical. In either case, RG-58/U and RG-59/U can be used for transformers and delay lines at the moderate power levels used for satellite work. Each coax section should be pruned carefully to length using the dip-meter technique. A mechanical switch will work reasonably well at 146 MHz and marginally at 435 MHz if care is taken in construction, miniature low-loss rotary switch

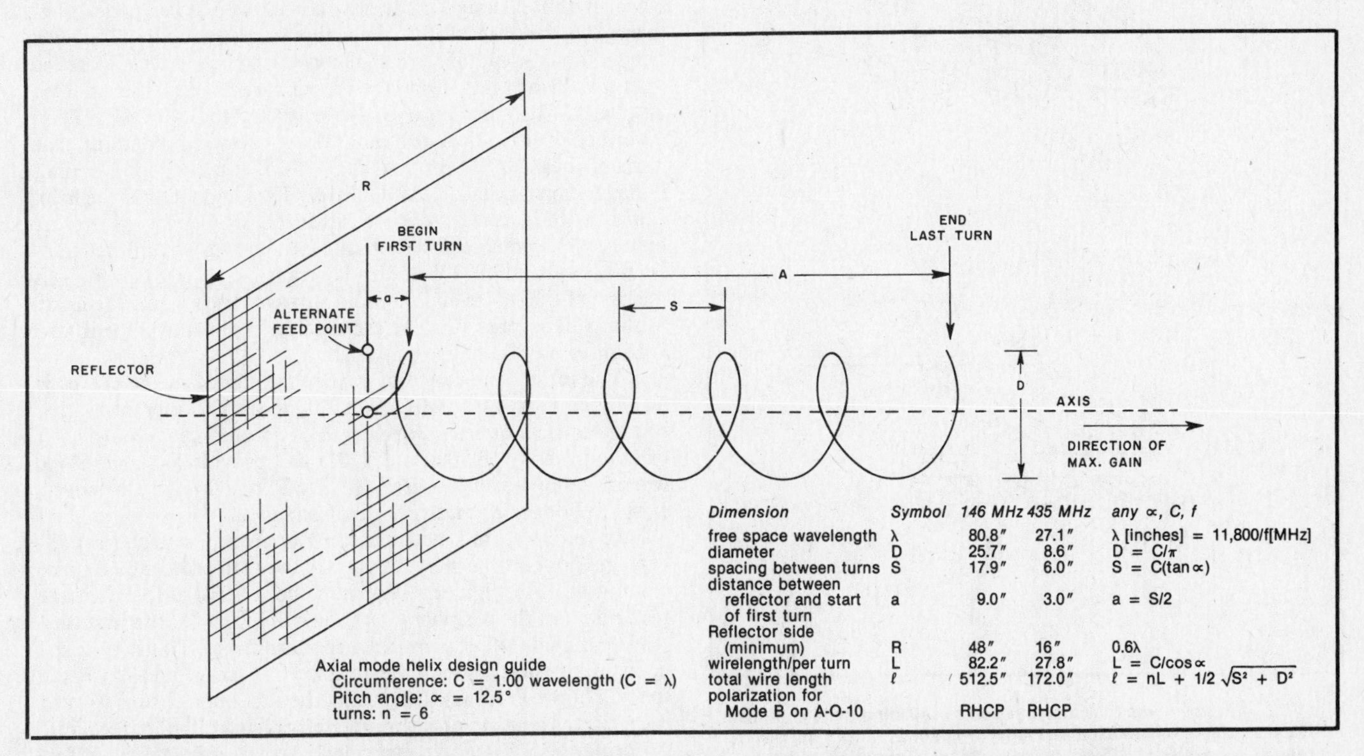

Dimension	*Symbol*	*146 MHz*	*435 MHz*	*any ∝, C, f*
free space wavelength	λ	80.8″	27.1″	λ [inches] = 11,800/f[MHz]
diameter	D	25.7″	8.6″	$D = C/\pi$
spacing between turns	S	17.9″	6.0″	$S = C(\tan\alpha)$
distance between reflector and start of first turn	a	9.0″	3.0″	$a = S/2$
Reflector side (minimum)	R	48″	16″	0.6λ
wirelength/per turn	L	82.2″	27.8″	$L = C/\cos\alpha$
total wire length	ℓ	512.5″	172.0″	$\ell = nL + 1/2\sqrt{S^2 + D^2}$
polarization for Mode B on A-O-10		RHCP	RHCP	

Fig. 6-21 — Dimensions for axial mode helix. For additional information see: J. D. Kraus, *Antennas*, McGraw-Hill, NY, 1950, Chapter 7; H. E. King and J. L. Wong, "Characteristics of 1-8 Wavelength Uniform Helical Antennas," *IEEE Transactions on Antennas and Propagation*, Vol. AP-28, no. 2, March 1980, pp. 291-296.

decks are used, and shields are placed between switch decks.

Note. It's widely believed that providing for switchable polarization by running two feed lines between the operating position and the antenna adds 3 dB loss to the system. *This is a fallacy.* Consider two identical ground stations using crossed Yagi antennas. Let station A use a single, 100-ft feed line and a power splitter at the antenna. Let station B use two feed lines, each 100 ft, and a power splitter at the operating position so that switchable polarization can be employed. Each station has a transmitter putting out 200 watts and we assume that 100 ft of feed line has 3 dB loss at the frequency of interest.

Table 6-8
Helix Characteristics†

No. of turns (n) [*1]	Gain (G) [*2]	Gain (G)	Half-power beamwidth [*3]	Approx. boom length 146 MHz [*4]	435 MHz
3	10.0	10.0 dB_i	64°	5.0 ft	2.0 ft
4	13.3	11.0 dB_i	55°	6.5 ft	2.5 ft
5	16.6	12.2 dB_i	49°	8.0 ft	3.0 ft
6	20.0	13.0 dB_i	45°	9.5 ft	3.5 ft
7	23.3	13.7 dB_i	42°	11.0 ft	4.0 ft
8	26.6	14.2 dB_i	39°	12.5 ft	4.5 ft
9	30.0	14.8 dB_i	37°	14.0 ft	5.0 ft
10	33.3	15.2 dB_i	35°	15.5 ft	5.5 ft
11	36.6	15.6 dB_i	33°	17.0 ft	6.0 ft
12	40.0	16.0 dB_i	32°	18.5 ft	6.5 ft

[*1] for n less than 3 the helix pattern changes radically
[*2] Theoretical values: Measurements suggest these values are 1 or 2 dB too high. Gain (G) ~ 15 n tan α (Note: α pitch angle = 12.5°)
[*3] Half-power beamwidth = $52° / \sqrt{n \tan\alpha}$ †
[*4] Boomlength = $\lambda (n + 0.5) \tan \alpha$

†Based on 1-wavelength circumference ($C = \lambda$) and 12.5° pitch angle ($\alpha = 12.5°$)

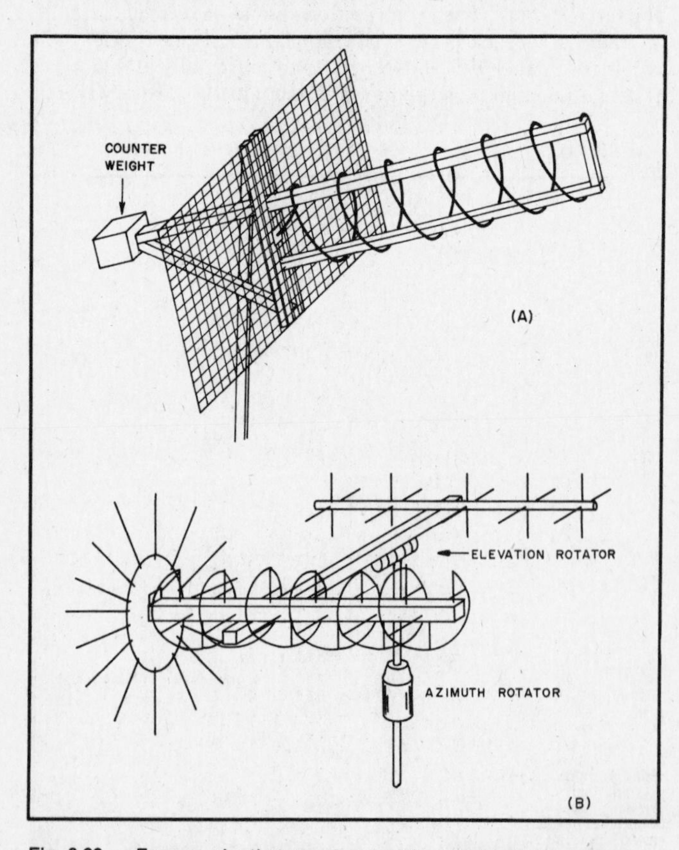

Fig. 6-22 — Frameworks that may be used for building helix antennas. A lattice structure, often used at 146 MHz, is shown at (A); the structure in (B) is popular at 435 MHz. For practical information on helix structures see: D. Jansson, "Helical Antenna Construction for 146 and 435 MHz," *Orbit*, Vol. 2, no. 3, May/June 1981, pp. 12-14.

Consider station A

Transmitter output: 200 watts
Power at antenna end of feed line: 100 watts (3 dB loss)
Power reaching each Yagi: 50 watts (after power splitter)

Consider station B

Transmitter output: 200 watts
Power fed into each feed line: 100 watts (after power splitter)
Power reaching each Yagi: 50 watts (3 dB loss on each feed line)

Running two feed lines in this instance does *not* add any extra loss to the system.

The Helix

Imagine a beam antenna that (1) produces a circularly polarized wave without a complex feed harness, (2) operates over a wide bandwidth and (3) is very forgiving with respect to dimensions and construction techniques. Unlike the imaginary isotropic antenna, this one's for real. Called an axial mode helix (*helix* for short), it's an excellent choice for satellite users. Before we get carried away describing the advantages of the helix, note that it does have a few shortcomings, which we'll also discuss.

A helix is characterized by three basic parameters:

C, the circumference of the imaginary cylinder on which the helical element is wound (usually expressed in terms of wavelength so that it's frequency independent)

α, the pitch angle, essentially a measure of how closely the turns of the helical element are spaced (also frequency independent)

n, the total number of turns

When these parameters lie in these ranges,

$$0.8\,\lambda \geq C \geq 1.2\,\lambda$$
$$12° \leq \alpha \leq 14°$$
$$n \geq 3$$

the helix will produce a beam pattern similar to the Yagi and quad. Dimensions are given in Fig. 6-21. A 6-turn helix suitable for use with AMSAT-OSCAR 10 is shown but the number of turns may be scaled up or down (see Table 6-8) to change the gain and beamwidth.

When a helix is built with the circumference equal to the wavelength it is designed for, it will work well at frequencies between 20% below and 30% above the design frequency. The wide bandwidth is advantageous: It allows you to be a little less precise than usual when measuring the proper antenna dimensions. This bandwidth also makes it possible to use the 146-MHz helix described in Fig. 6-21 for monitoring scientific satellites that transmit near 137 MHz, and the 435-MHz model for listening to navigation satellites near 400 MHz. The bandwidth of the helix can contribute to receiver desensitization problems, however, if high-power commercial stations are located nearby. Unfortunately, megawatt EIRP TV and radar transmitters are common in the part of the spectrum that radio amateurs use for satellite links. A sharp band-pass filter at the receiver input may help if you encounter any trouble.

The input impedance of a helix that is fed at the center is usually close to 140 ohms. A matching transformer consisting of an electrical quarter wavelength of 75-ohm coax (RG-11/U) or 80-ohm coax (Belden no. 8221) will provide a decent SWR when 50-ohm feed line is used. The SWR improvement, however, exists only over a relatively small bandwidth.

In recent years a new matching approach with several advantages has become increasingly popular with professional space communication engineers. When the helix is fed at the alternate feed point on the periphery, as shown in Fig. 6-21, the first turn may be thought of as an impedance transformer. To use this feed point, dimension a should be doubled (i.e., set a equal to S, the spacing between turns). Displacing the first quarter turn toward the reflector tends to produce a better match to 50-ohm feed line. To bring the SWR down even closer to 1:1, increase the effective wire diameter of the first quarter turn by soldering a strip of thin brass shim stock or copper flashing (width roughly 5 times

Table 6-9
Comparison of Three Circularly Polarized Beam Antennas

	Crossed Yagis with delay line	*Crossed Yagis offset 1/4 λ in direction of max. gain*	*Single helix*
Length for 12 dB_i gain	1.0 λ	1.25 λ	1.4 λ (plus boom for counter-weight if needed)
Bandwidth	~ 2% of center frequency	~2% of center frequency	From 20% below 30% above center frequency
Matching/ phasing system	Highly complex	Moderately complex	Relatively simple
Adjustment Procedure	Complex	Complex	Simple
Are dimensions and construction materials critical?	Yes	Yes	No
Relative size, weight, mounting complexity	Small, light, simple	Small, light, simple	Moderately large, heavy, complex
Can polarization sense be exter-nally switched?	Yes (see note 1)	Yes (see note 1)	No

Note 1: Although this appears to be a strong plus for crossed Yagis, it requires the use of a complex switching system. If the switching system is located at the operating position, a receive preamp cannot be mounted at the antenna.

the wire diameter) to it. This technique is described in detail by J. D. Kraus in "A 50-ohm Input Impedance for Helical Beam Antennas," *IEEE Transactions on Antennas and Propagation,* Vol. AP-25, No. 6, Nov. 1977, p. 913, and J. Cadwallader, "Easy 50-ohm Feed for a Helix," *QST,* June 1981, pp. 28-29. With this matching technique the SWR remains below 2:1 over a range of about 40% of the center frequency.

The helical element must be supported by a nonconductive structure. Two common approaches to building such a frame are illustrated in Fig. 6-22. Lightweight woods with good weathering properties, such as cedar or redwood, are preferred for large 146-MHz lattice structures, while varnished pine or oak dowels may be used for the smaller 435-MHz model. The construction of the reflector is not critical as long as it meets the minimal size requirements. Square or round sections of hardware cloth for 435-MHz helices, or 2″ × 4″ welded wire fencing for 146-MHz helices are suitable. A small aluminum hub with 18 or more evenly spaced spokes radiating outward can also be used. At 146 MHz the helical element may be wound from 1/4-in. flexible copper tubing or from a length of old coaxial cable (impedance is not important) with the inner conductor and outer braid shorted together. At 435 MHz and higher frequencies, no. 12 wire is acceptable.

The main problem with using a helix is its cumbersome physical structure. Comparing a well-designed crossed-Yagi array and a helix with the same gain, you'll find that generally the Yagi array will be considerably shorter and have less than half the weight and windload (see Table 6-8). Several serious EME operators, experimenting with arrays of helices, have concluded that helices are not suitable for providing the very large gains required for EME communication. If, however, you want an inexpensive, easily reproducible, moderate-gain antenna for satellite operation, consider the helix.

To determine the polarization sense of a helix, picture yourself standing in back of the reflector looking out along the frame in the direction of maximum gain. If you were to place your index finger on the feed point and slide it forward along the surface of the helical element, you would see it trace out either a clockwise pattern or a counterclockwise pattern. Clockwise corresponds to an RHCP helix; counterclockwise corresponds to an LHCP helix. As mentioned earlier, a 3-turn helix makes an excellent reference antenna for determining the polarization sense of a crossed Yagi or similar array.

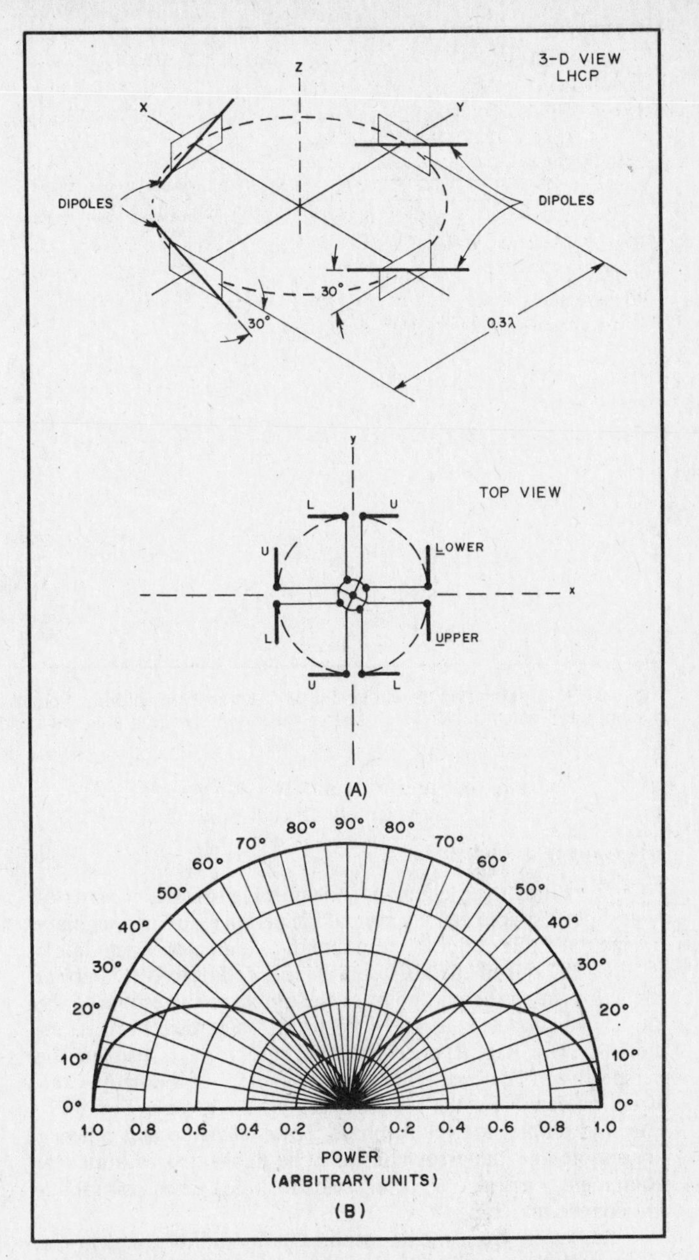

Fig. 6-23(A) — The Lindenblad antenna consists of four λ/2 dipoles oriented as shown in the 3-D view and fed as illustrated in the top view. (B) Free space vertical plane power vs. elevation angle for Lindenblad antenna. Ground reflections will decrease gain at very low elevation angles and introduce nulls.

Arrays of four helices have been popular on 1.2 GHz and 2.3 GHz for many years (see *The ARRL Antenna Book,* 13th Edition, pp. 261-263, for construction details). The noncritical nature of the helix makes it an excellent moderate gain antenna at these frequencies, where the test equipment needed for optimizing performance is often difficult to come by. When gains above 20 dB_i are needed, however, the parabolic dish, or variations on the Yagi, are more appropriate antenna choices. EME operation at 1.2 GHz is rapidly growing in popularity as has 2.3-GHz EME and terrestrial activity recently, probably as a result of the improved availability of equipment for this part of the spectrum. As a result, designs for effective 1.2-GHz and 2.3-GHz Yagis, loop-Yagis, quagis and disc-Yagis will probably be refined before amateur satellites begin to use these frequencies.

Lindenblad, Quadrifilar Helix and TR-Array

Three other circularly polarized antennas are also worthy

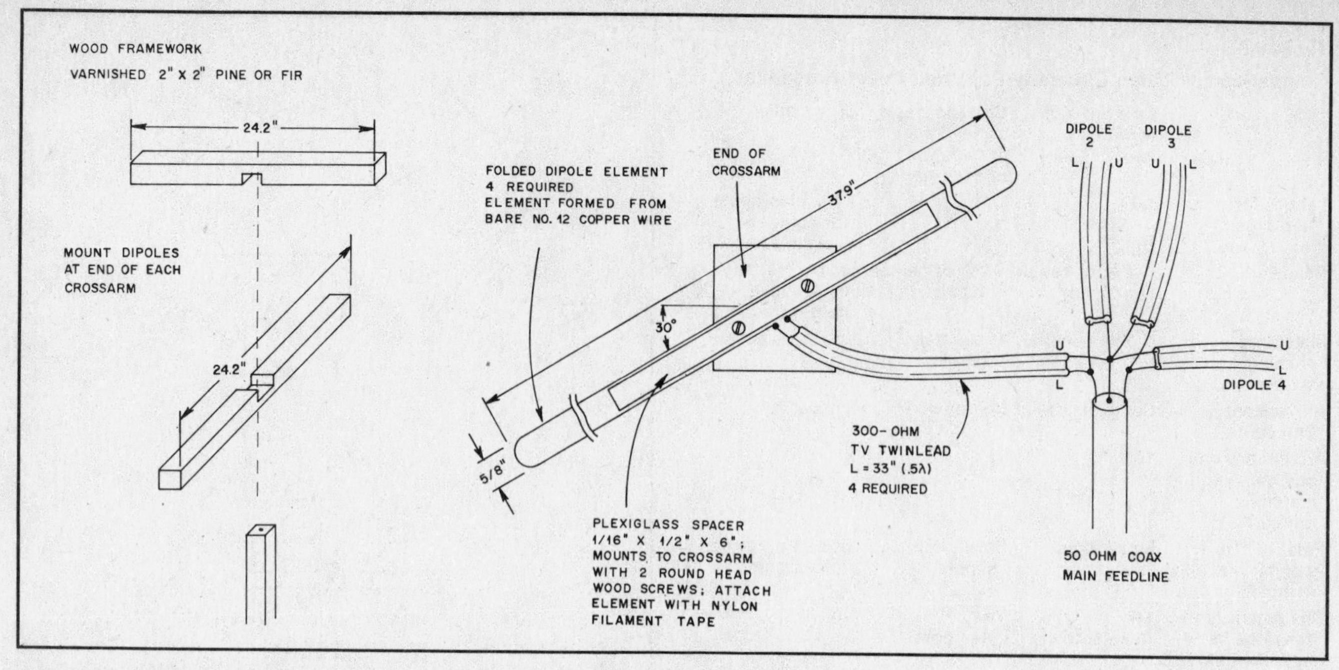

Fig. 6-24 — Construction details for 146 MHz Lindenblad antenna. Folded dipoles have been used to simplify matching. If desired, a 75-ohm to 50-ohm transformer and/or balun may be inserted between the antenna and main feedline.

of consideration. All are low-gain, broad-beamwidth designs primarily suited for use with low-altitude spacecraft.

Lindenblad

The Lindenblad antenna, shown in Fig. 6-23A, consists of four dipoles spaced equally around the perimeter of an imaginary horizontal circle about 0.3 wavelength in diameter. Each dipole is tilted 30° out of the horizontal plane; rotation (tilt) is about the axis joining the mid point of the dipole to the center of the circle. All four dipoles are tilted in the same direction: either clockwise (for RHCP) or counterclockwise (for LHCP) from the perspective of an observer located at the center of the array. Construction details for the 146-MHz version are given in Fig. 6-24. Since all dipoles are fed in phase, power division and phasing are simple and the array can easily be duplicated without test equipment. Furthermore, using folded dipole elements simplifies impedance matching.

Radiation from the Lindenblad is omnidirectional in the horizontal plane and favors low elevation angles in the vertical plane (see Fig. 6-23B). When used with low-altitude, circular-orbit satellites, the increased power at low elevation angles compensates somewhat for increased satellite-ground station distance; signal levels therefore remain fairly constant over a considerable range of elevations. The radiated signal is nearly circularly polarized in all directions, a very desirable characteristic. As mentioned earlier, the polarization sense is determined by the direction in which the dipoles are rotated (tilted) out of the horizontal plane. Polarization can't be reversed by modifying the feed harness; if you want to change from RHCP to LHCP, or vice versa, you must change the antenna structure.

Quadrifilar Helix

The quadrifilar helix (Fig. 6-25A) consists of four 1/2-turn helices (A, A′, B, B′) equally spaced around the circumference of a common cylinder. Opposite elements (A and A′, B and B′) form a bifilar pair; the two bifilars must be fed equal amounts of power but 90° out of phase. As with other antennas requiring a 90° phase difference and equal power division, problems arise in designing an adequate feed system. The solution favored by professional antenna engineers is to build one small bifilar to resonate slightly above the operating frequency (input impedance has a capacitive reactance component) and one large bifilar designed to resonate slightly below the operating frequency (in-

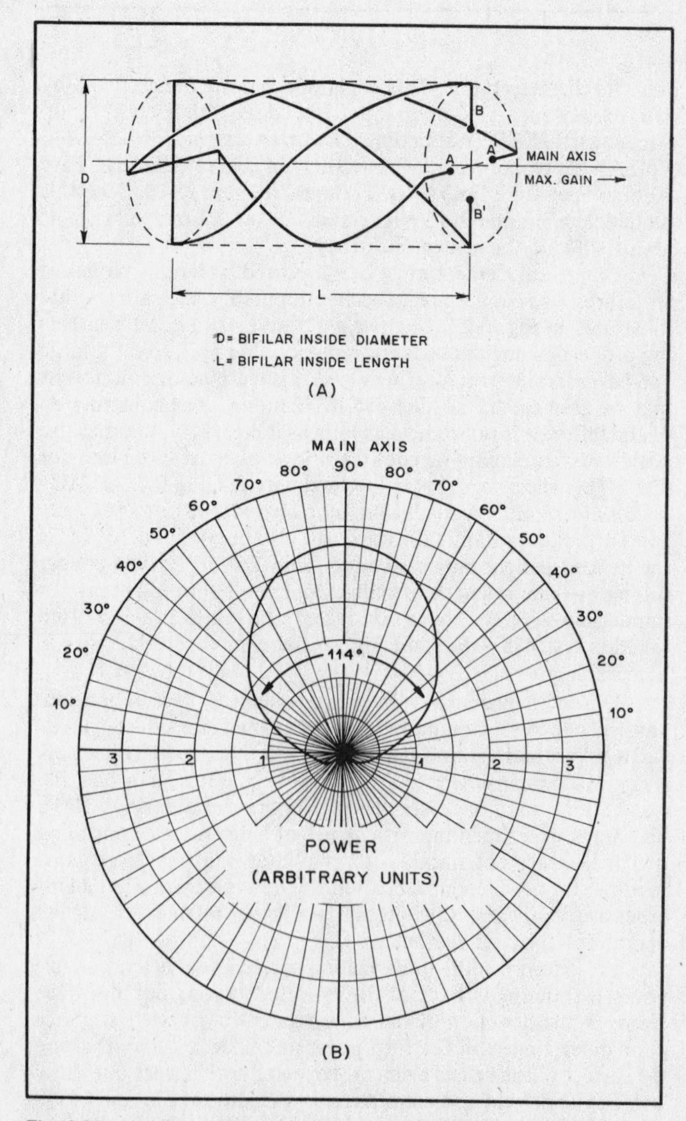

Fig. 6-25 — (A) The quadrifilar helix antenna (B) Power gain pattern of quadrifilar helix.

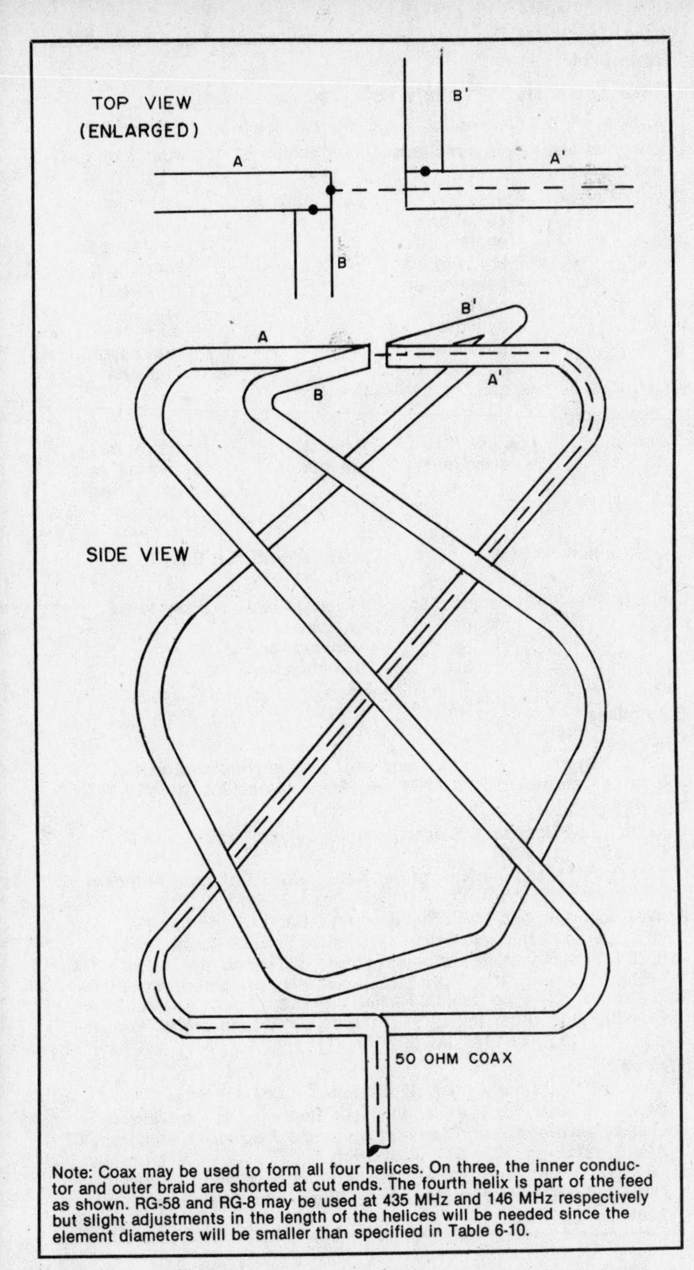

Fig. 6-26 — Quadrifilar helix employing self-phasing and infinite balun.

Table 6-10
Design Data for Quadrifilar Helix

	Small Bifilar			*Large Bifilar*			
	D (in.)	*L (in.)*	*Length A-A' (in.)*	*D (in.)*	*L (in.)*	*Length B-B' (in.)*	*Wire diameter (in.)*
146 MHz	12.62	19.25	82.19	13.99	21.03	90.60	0.71
435 MHz	4.23	6.46	27.57	4.69	7.05	30.39	0.24
Any frequency	0.156 λ	0.238 λ	1.016 λ	0.173 λ	0.260 λ	1.120 λ	0.0088 λ

Note: Dimensions should be regarded only as a guide. Special thanks to Walter Maxwell, W2DU, for providing this information.

put impedance has an inductive component). If the diameters of the bifilars are adjusted so that the magnitudes of all resistive and reactive components of the two input impedances are equal, the current in the small bifilar will lead the input by 45° while the current in the large bifilar lags by 45°. This yields the desired 90° phase difference and a purely resistive input impedance of about 40 ohms when the two bifilars are fed in parallel. In effect, matching and phasing are built into the antenna itself. An "infinite balun" is conveniently used in conjunction with the "self-phased" quadrifilar.

The radiation pattern of a quadrifilar helix is omnidirectional in the plane perpendicular to its main axis. In a plane containing the main axis (Fig. 6-25B) the maximum gain is about 5 dB_i and the beamwidth 114°. Radiation is nearly circularly polarized over the entire hemisphere irradiated. In many situations an antenna with these characteristics is ideal for a ground station. For example, it could be used as part of an unattended automated command or data retrieval station. The quadrifilar helix also makes an excellent spacecraft antenna. On AMSAT- OSCAR 7 one was used in the downlink for the 2.3-GHz microwave beacon.

Because small changes in the dimensions and dielectric properties of the quadrifilar support structure, and the presence of nearby objects, can have a large effect on power division and phasing, the radio amateur without sophisticated test equipment will have difficulty duplicating the desired performance. Nevertheless, the intrepid experimenter will find construction details for 146-MHz and 435-MHz quadrifilars in Fig. 6-26 and Table 6-10. Dimensions, scaled from a 2-GHz model, should only be regarded as a guide. Phasing and balun details are also included.

TR-Array

The TR-array (*t*urnstile-*r*eflector array) shown in Fig. 6-27A consists of a pair of dipoles mounted above a reflecting screen

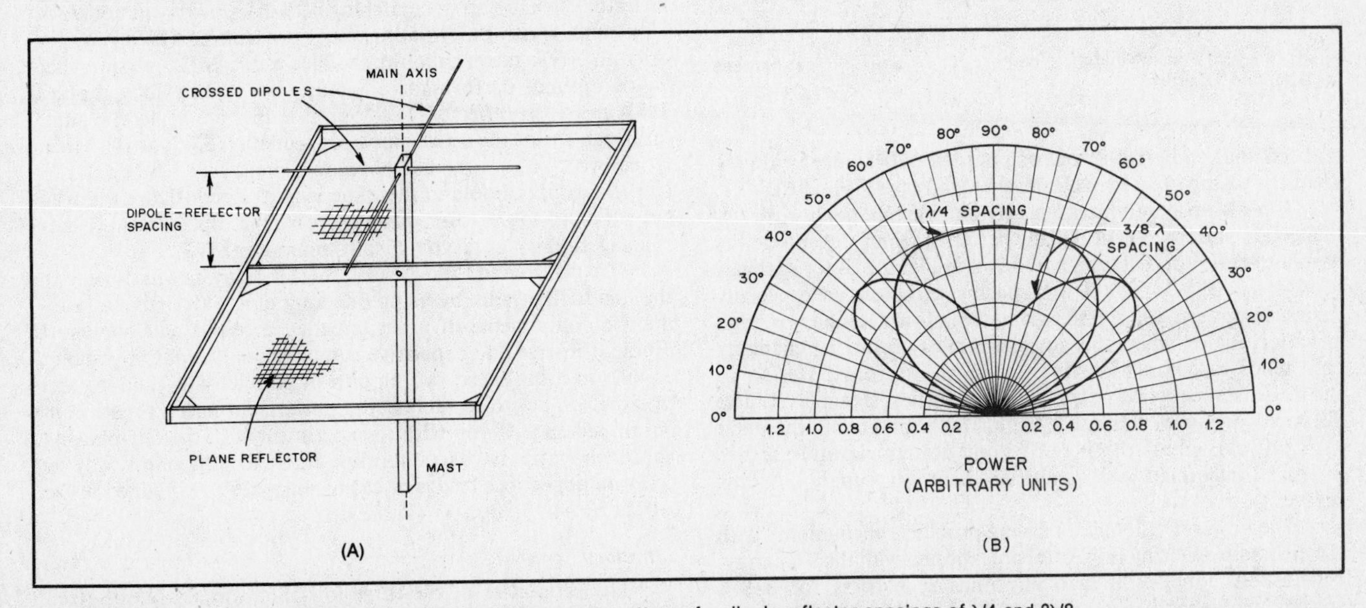

Fig. 6-27 — (A) Turnstile-Reflector array. (B) Vertical plane power patterns for dipole-reflector spacings of λ/4 and 3λ/8

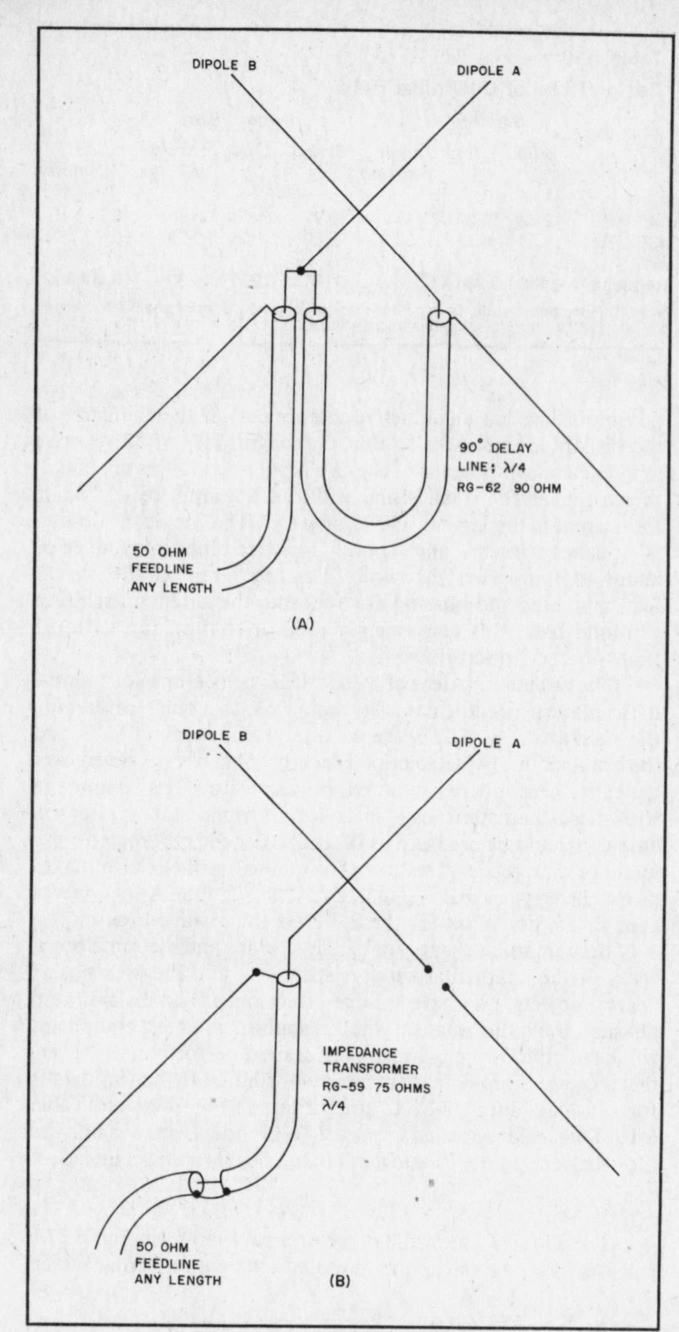

Fig. 6-28 — (A) Phasing/matching harness for TR-array (B) Test harness for adjusting TR-array.

and fed equal power, 90° out of phase. Performance is almost identical to the crossed 2-element Yagi array (Fig. 6-16).

The TR-array produces a nearly omnidirectional horizontal-plane gain pattern. Vertical plane patterns, which depend on the dipole-to-reflector distance, are shown in Fig. 6-27B for spacings of λ/4 and 3/8 λ. The 3/8 λ spacing produces an especially desirable pattern for a fixed ground station antenna. At high elevation angles, where this antenna is most useful, the changing gain tends to compensate for variations in ground station to satellite distance, yielding a relatively constant signal level. The TR-array produces a circularly polarized signal along the main axis. Off-axis circularity is fairly good at high elevation angles but the Lindenblad and quadrifilar helix are superior in this regard.

The power division and phasing problems encountered with the crossed-Yagi array (Fig. 6-16) are repeated with the TR- array. Fig. 6-28A contains a matching/phasing harness for 3/8 λ spacing. Note that the impedance of the dipoles varies with dipole-reflector distance, so the matching network shown will not work with other spacings. An adjustment procedure, which requires only an SWR meter, should produce a 146-MHz version that yields optimal performance. Set up two slightly long dipoles 3/8 λ above the reflector. Feed one as in Fig. 6-28B; let the other one float. Prune the active dipole for minimum SWR at 146 MHz. Don't worry about the actual value as long as it's below 1.5:1. Cut the second dipole to the same length. Reconfigure the feed system as in Fig. 6-28A. Then increase the dipole- to-reflector spacing slightly until you obtain minimum SWR.

It is possible to "self-phase" the TR-array as was done with the quadrifilar helix by using one long dipole (resistive and inductive components of input impedance equal) and one short dipole (resistive and capacitive components of input impedance equal). Feeding these two dipoles in parallel will yield correct phasing, an approximately equal power split and a resistive input impedance. If you wish to experiment with the self-phasing approach, you'll have to determine dipole lengths empirically by using an impedance bridge or calculate values as explained in the article by M. F. Bolster (Table 6-11).

Table 6-11

Three Low-Gain, Circularly Polarized Antennas.

	Lindenblad	*Quadrifilar Helix*	*TR-Array*
Horizontal plane gain pattern	Omnidirectional	Omnidirectional	Omnidirectional
Vertical plane gain pattern	Favors low elevation angles, gain tends to compensate for changing satellite-groundstation distance	Favors main axis	Favors high elevation angles, gain tends to compensate for changing satellite-groundstation distance
Half-power beamwidth	NA	114°	140°
Circularity	Excellent in all directions	Excellent in all directions	Falls off away from main axis, good over most of pattern
Construction	Easy to build	Moderately difficult to build	Easy to build
Adjustment	No adjustment required	Specialized test equipment required for adjustment	Easy to adjust
Bandwidth	± 8%	± 2%	± 4%

References

Lindenblad

G. H. Brown and O. M. Woodward, Jr., "Circularly-Polarized Omnidirectional Antenna," *RCA Review*, Vol. 8, June 1947, pp. 259-269.

Quadrifilar Helix

C. C. Kilgus, "Resonant Quadrifilar Helix Design," *Microwave Journal*, Dec. 1970, pp. 49-54.

C. C. Kilgus, "Resonant Quadrifilar Helix," *IEEE Trans. on Antennas and Propagation*, Vol. 17, May 1969, pp. 349-351.

R. W. Bricker, Jr. and H. H. Rickert, "An S-Band Quadrifilar Antenna for Satellite Communications," Presented at 1974 International IEEE/P-S Symposium, Georgia Institute of Technology, Atlanta, GA. Authors are with RCA Astro-Electronics Div., Princeton, NJ 08540.

C. C. Kilgus, "Shaped-Conical Radiation Pattern Performance of the Backfire Quadrifilar Helix," *IEEE Trans. on Antennas and Propagation*, Vol. 23, May 1975, pp. 392-397.

TR-Array

M. Davidoff, "A Simple 146-MHz Antenna for OSCAR Ground Stations," *QST*, Sept., 1974, pp. 11-13. Reprinted in *Specialized Communications Techniques for the Radio Amateur*, Newington, CT: ARRL (1975), pp. 173-174 (out of print).

M. F. Bolster, "A New Type of Circular Polarizer Using Crossed Dipoles," *IRE Trans. on Microwave Theory and Techniques*, Sept., 1961, pp. 385-388.

Summary

The properties of the three low-gain, circularly polarized antennas suitable for working with low-altitude satellites are sum-

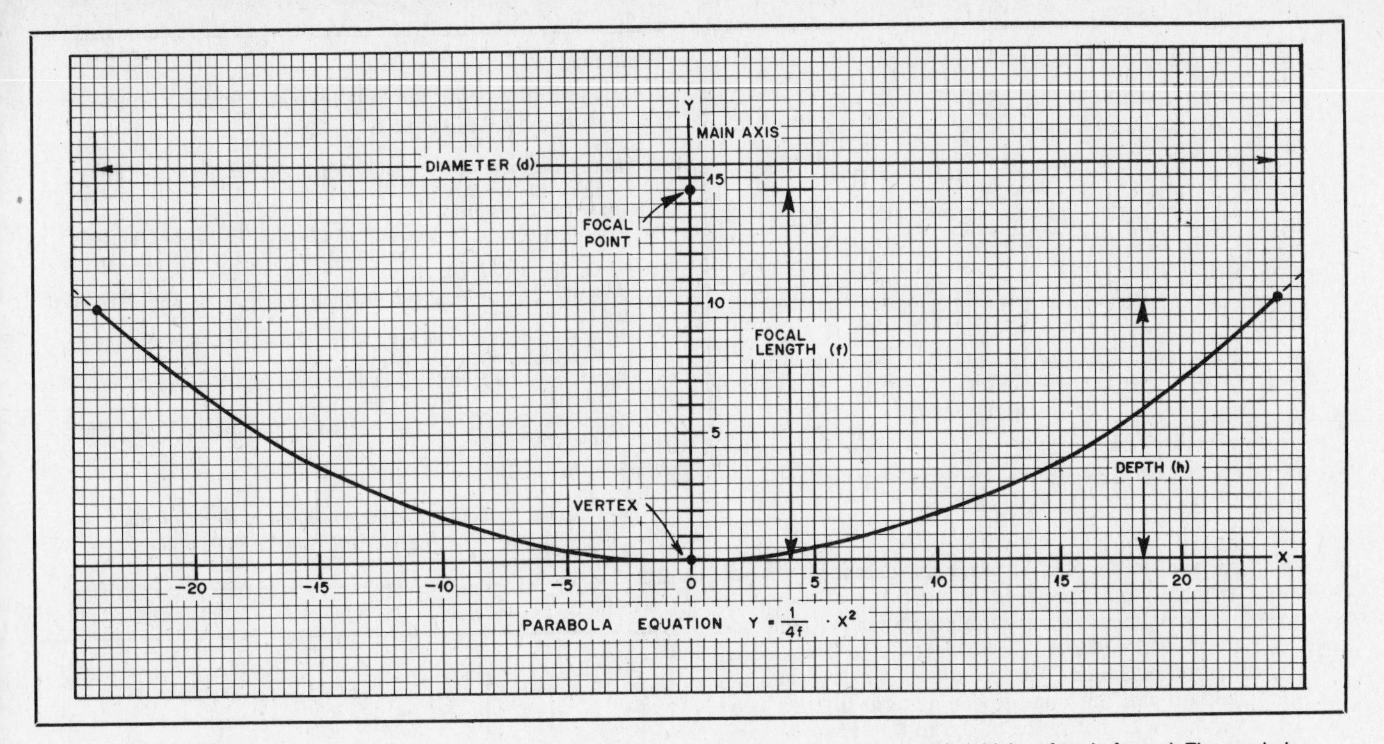

Fig. 6-29 — Parabola geometry and definitions. When a 2-D parabola is rotated about main axis a paraboloidal surface is formed. The parabola shown has an f/d ratio of 0.3.

marized in Table 6-11. The table also includes references to literature discussing each of them.

The Parabolic Dish and Related Antennas

The parabolic dish antenna is a high-gain beam that will grow in popularity as the Amateur Satellite Service begins to use frequencies above 1 GHz. It belongs to a family of antennas that employ a large reflecting surface and a separate feed antenna. The corner reflector, spherical dish and TR-array are members of the same family.

Parabolic Dish

To understand how the parabolic dish antenna operates we have to look at dish geometry and feed systems, and the relationship linking these two factors. The reflecting surface is formally known as paraboloidal, but following common usage we'll refer to both the reflector and the entire antenna as a dish or parabolic dish. The three dimensional dish surface is formed by rotating a two dimensional parabolic curve (see Fig. 6-29) about its main axis. The property that makes the dish interesting is that incoming signals that arrive parallel to the main axis, but which are spread over a large area are reflected off the dish and concentrated at a *focal point*. Similarly, a signal source located at the focal point that illuminates the dish will produce a beam parallel to the main axis, in much the same way that a good flashlight focuses the light emitted by its bulb.

The location of the focal point depends on dish geometry. It's usually specified in terms of *focal length* (f): the distance between the vertex (center) and the focal point. If you need to determine the focal length of a surplus dish, measure the edge-to-edge diameter (d) and the depth (h). The focal length can be determined from the following relationship:

$$f = \frac{d^2}{16h} \qquad \text{(Eq. 6.5)}$$

For example, suppose we find a 4-foot-diameter dish (d = 48″) with a depth of 10 inches (h = 10″). Substituting in Eq. 6.5 we obtain

$$f = \frac{(48)^2}{16(10)} = 14.4''$$

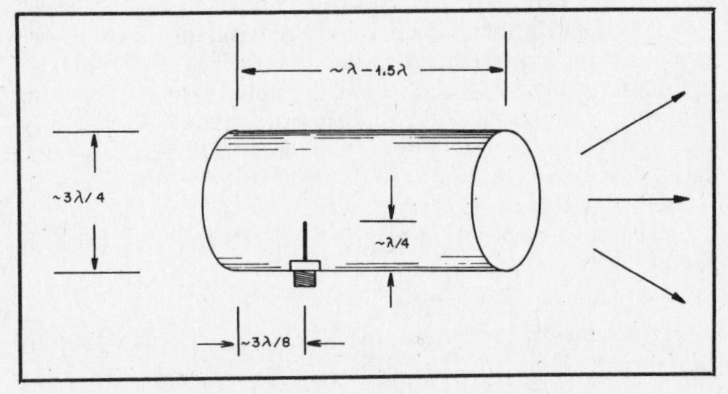

Fig. 6-30 — Feed horn suitable for illuminating a parabolic dish with an f/d ratio of 0.5 to 0.6. Horn has about 10 dB_i gain.

Note that modifying the diameter of a dish by sawing off the outer rim or adding extensions does *not* change the focal length. The easiest dishes to feed properly are those having an f/d ratio (focal length to diameter ratio) of 0.5 to 0.6. The example we just looked at had an f/d ratio of 14.4/48 or 0.3, making it a poor choice for our application.

The feed antenna is placed at the focal point and aimed toward the vertex of the dish. Although a wide variety of feed systems can be used, the cylindrical horn (Fig. 6-30) is the most popular above 1 GHz among radio amateurs. Tin cans of various sizes make surprisingly efficient horns: 1 gallon motor oil cans at 1.2 GHz (~7″ diameter) and 1 pound coffee cans at 2.3 GHz (~4″ diameter) work well. The diameter, not the original contents, is the important parameter. A quarter-wave monopole soldered to a coax connector typically is used to excite the horn. The dimensions shown in Fig. 6-30 are only approximate. By varying the diameter and length of the feed horn (size of the can), the spacing between the monopole and the closed end of the can and the dimensions of the monopole element, we can shape the beamwidth and adjust matching to some extent. The horn shown in Fig. 6-30 will produce a linear wave. By inserting several brass screws into the can at strategic points, circular polarization can be obtained. At 432 MHz, where a horn would assume the size

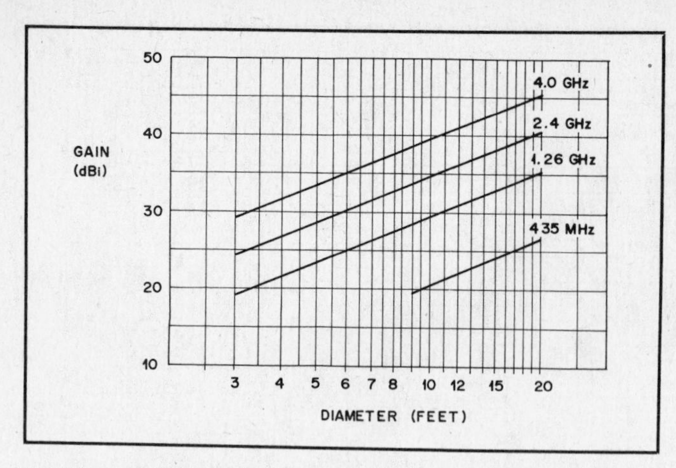

Fig. 6-31 — Parabolic dish gain vs. diameter for several frequencies of interest.

of a small garbage can, the NBS standard-gain antenna is generally used as a linearly polarized feed.

To obtain maximum gain from a parabolic dish antenna, the feed pattern at the edge of the dish should be down about 10 to 12 dB from maximum. The feed horns just described and the NBS standard-gain antenna provide patterns that are good matches to dishes with f/d ratios of 0.5 to 0.6. From a transmitting point of view, if the reflector f/d is too high, much of the feed power will spill over the edge and be wasted; if the reflector f/d is too low the outer rim of the reflector will essentially go unused. Similar problems exist from a receiving point of view.

The polarization of the parabolic dish antenna depends entirely on the feed system: A circularly polarized feed results in a circularly polarized signal; a linearly polarized feed results in a linearly polarized signal. Note that the sense of a circularly polarized signal is reversed when it bounces off the reflecting surface; we must use an LHCP feed horn to produce or receive an RHCP signal and vice-versa.

The gain and beamwidth of an efficiently fed parabolic dish are given by

$$G = 7.5 + 20 \log d + 20 \log F \quad \text{(assumes 55\% feed efficiency)} \qquad \text{(Eq. 6.6)}$$

and

$$\theta^* = \frac{70}{(d)(F)} \qquad \text{(Eq. 6.7)}$$

where

G = gain in dB_i
d = dish diameter in feet
F = frequency in GHz
θ^* = 3-dB beamwidth in degrees.

The gain (Eq. 6.6) is plotted in Fig. 6-31.

When we think of dishes we may envision discouraging images of large commercial monsters, or the mini-monsters used for EME. In reality, the dishes used for satellite work will probably be comparatively small and lightweight. A good part of the complexity of dish installations for EME is related to the necessary precision of aiming and readout systems. A dish set up to access a geostationary satellite needs none of these. At 1.2 GHz an efficiently illuminated 6-foot-diameter dish will provide about 25 dB_i gain. Homebrew dishes, modeled after the stressed rib design detailed by Dick Knadle, K2RIW, are lightweight and simple to build. A 6-footer can be constructed from materials available at a local hardware store — 1/4-inch oak dowels and window screen — in an afternoon. For anything larger, it's best to stay with an aluminum framework.

Shape errors amounting to 1/8 wavelength or less in a dish have very little effect on gain. Errors do, however, have an important impact on the size and shape of sidelobes and thereby limit the ultimate S/N ratio and the ability to receive very weak signals. Commercial dishes designed to minimize sidelobes must therefore have very stiff, accurate structures. A much simpler, lightweight framework will suffice for most amateur satellite applications.

Table 6-12

Sources of Information on Parabolic and Related Antennas

Parabolic Reflectors and Feed Systems

General

D. S. Evans and G. R. Jessop, *RSGB VHF-UHF Manual*, 3rd Ed., 1976, pp. 8.50-8.70.

Stressed Rib Design

R. T. Knadle, Jr., "A Twelve-Foot Stressed Parabolic Dish," *QST*, Aug. 1972, pp. 16-22. Reprinted in *The ARRL Antenna Book*, 14th Ed., 1982, pp. 254-260.

A. Katz, "Simple Parabolic Antenna Design," *CQ*, Aug. 1966, p. 10.

Feed Design

N. J. Foot "Cylindrical Feed Horn for Parabolic Reflectors," *Ham Radio*, May 1976, pp. 16-20.

N. J. Foot, "Second Generation Cylindrical Feedhorns," *Ham Radio*, Vol. 15, no. 5, May 1982, pp. 31-35.

N.B.S. Standard-Gain Antenna, *The ARRL Antenna Book*, 14th Ed., 1982, Ch. 15.

Spherical Reflector

General

A. W. Love, "Spherical Reflecting Antennas with Corrected Line Sources," *IRE Trans. on Antennas and Propagation*, Vol. AP-10, pp. 529-539, Sept. 1962.

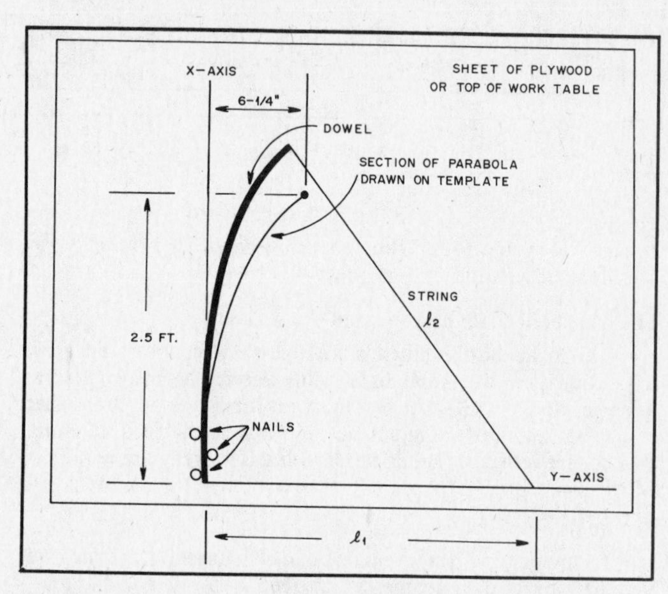

Fig. 6-32 — Template used to determine appropriate lengths for ℓ_1 and ℓ_2 when constructing a stressed-rib parabolic reflector.

A ground station can use a single dish for duplex uplink and downlink operation by mounting two feed horns side by side. Though both will be offset slightly from the focal point and mutually block a small part of the main beam, the effect on gain is negligible (less than 1 dB). Once again, the main impact is on the sidelobes.

It is possible to set up a fixed parabolic reflector and steer the beam by moving the feed away from the focal point. Departures of several beamwidths from the main axis may be made before gain decreases by 1 dB.

Warning: To avoid building a dish from scratch, it's often tempting to buy a small surplus model with the intention of increasing its size by adding to the outer rim. This almost never works. Extending a dish always decreases the f/d ratio. Since most surplus dishes have low f/d ratios to begin with, feeding the resulting reflector efficiently will be almost impossible.

References to construction articles that focus on stressed rib reflectors, feed horns and the NBS standard-gain antenna are given in Table 6-12.

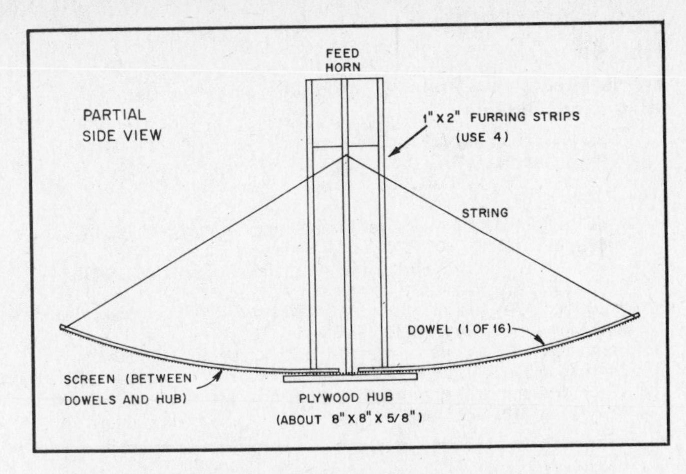

Fig. 6-33 — Construction of a wood-framework parabolic antenna.

An Example: A step-by-step design example suitable for uplinking on Mode L (1269 MHz) follows. The tentative uplink EIRP (subject to change) suggested for AMSAT-OSCAR 10 Mode L is 5000 watts. Suppose we have 20 watts of rf at the antenna. Eq. 6-2 tells us that 24 dB_i of antenna gain is needed.

$$G = 10 \log \frac{5000}{20} = 24 \text{ dB}_i$$

From Fig. 6-31 we deduce that a 5-ft-diameter dish (23.5-dB_i gain) will be adequate. The dish will have a 3-dB beamwidth (Eq. 6.7) of 11°: narrow but usable without great difficulty. The next question is f/d ratio. Anything between 0.5 and 0.6 will work fine. We'll choose 0.6 because it results in a shallower dish, minimizing the unavoidable practical departures from a true parabola. Now that d (5 ft) and f/d (0.6) have been chosen we can compute the focal length, f = 3.0 ft.

Since we're mainly interested in testing a prototype, a wood structure is appropriate. For a permanent, tower-mounted antenna, aluminum would be more suitable. A quick stop at a hardware store for some aluminum screening (a 12-ft length of 36-in.-wide material will do), aluminum wire and 16 quarter-inch oak dowels provides the materials that aren't in our scrap box.

Before construction begins we build a template as shown in Fig. 6-32. Lay out the axes accurately and then use the equation in Fig. 6-29 to draw the section of the parabola $y = (1/12)x^2$. Drive three nails to form a snug channel that will hold a dowel firmly over the x-axis. Tie some Dacron fishing line to the end of the dowel and pull it toward the y-axis. Vary the distances ℓ_1 and ℓ_2 and/or the point where the string is attached to the dowel until you get a good match to the desired parabolic section. Errors can be kept below one-half inch, which amounts to better than 1/16 wavelength. Finally, put it all together as shown in Fig. 6-33. To feed the dish use either a horn or a 3-turn helix.

Related Antennas

Several other reflector-type antennas may turn out to be useful at radio amateur satellite ground stations. We'll briefly mention two: the spherical reflector and the toroidal reflector.

Spherical reflectors are nothing more than sections of a sphere. A spherical reflector does not focus incoming signals at

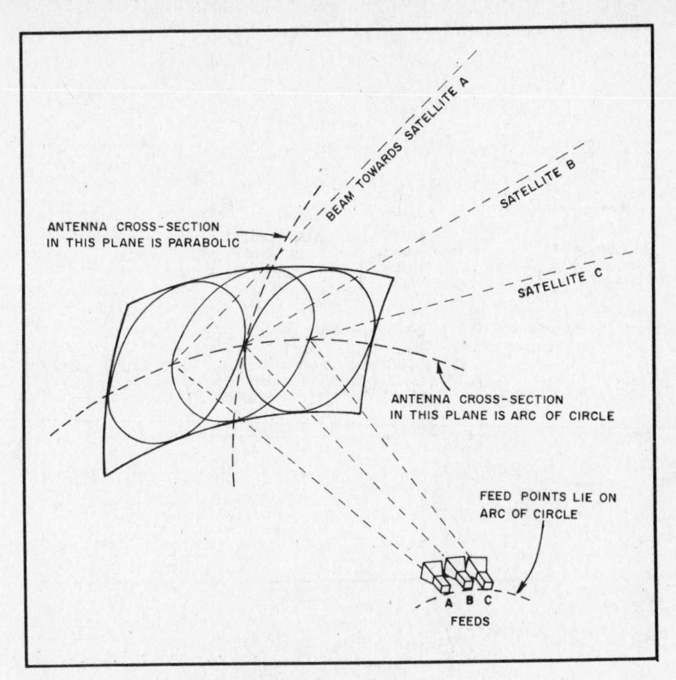

Fig. 6-34 — The multiple-beam torus antenna

a point, so it makes little sense to talk about focal length. Instead, spherical reflectors are characterized by the radius (r) of the sphere they're cut from, rim-to-rim diameter (d) and r/d ratio. The gain of a spherical reflector is about 2 dB less than that of a parabolic reflector of the same diameter, but the spherical reflector does offer other advantages beyond its simpler geometric shape. Notably, moving the feed antenna up to about 45° off axis is possible before gain begins to decrease substantially. As a result, it's possible to use a fixed reflector in conjunction with either a single feed on a movable mount for tracking a satellite over a considerable region of the sky, and/or multiple feeds to access several spacecraft simultaneously. A special type of feed is generally required for spherical reflectors. A number of homebrewers wrestling with direct reception of commercial 4-GHz satellite TV signals are experimenting with spherical reflectors, so information on practical feed systems should be available soon. As a side note, the famous 1000-foot-diameter radio telescope at Arecibo, Puerto Rico, uses a spherical reflector covering 20 acres and having an accuracy of 1/8 inch! [A slide-tape presentation on the Arecibo "monster" is available for loan from the ARRL.]

Another reflector geometry of possible interest is the torus. This configuration has been examined carefully by COMSAT Laboratories because its properties make it especially well suited for simultaneously receiving signals from several stationary satellites that are spaced along the geosynchronous arc above the equator (Fig. 6-34). Commercial toroidal reflectors and matched feeds for 4-GHz satellite TV downlinks were first marketed in 1980, and a large number are now used at cable TV earth stations. As yet, it's not clear whether the toroidal reflector's properties will make it appropriate for use with future radio amateur satellites.

Part III: Antenna Systems

Designing an antenna system involves several additional concerns. We'll look at them briefly.

Feed Lines and Connectors

Satellite radio links generally use vhf and higher frequencies. Anyone setting up a ground station should become familiar with the special techniques used at these short wavelengths to minimize the rf power losses associated with both feed lines and connectors.

Almost always coaxial feed lines are used at ground stations.

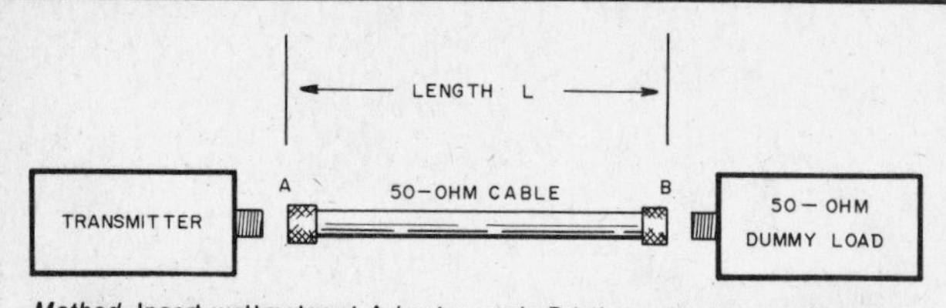

Method. Insert wattmeter at A (meter reads P_A) then adjust transmitter for proper operation. Now move meter to B and read P_B without touching transmitter adjustments. Cable attenuation, in dB, will equal $10 \log(P_A/P_B)$. This value can be scaled to 100 feet as follows:

$A_o = (100/L)(A_L)$

where

A_o = attenuation per 100 feet at test frequency

A_L = measured attenuation

Compare the value obtained to the value listed in Table 6-13 to determine if cable is performing up to specifications. Although it's best to make measurements at the satellite link frequency you'll be using a test setup at 2 m will give a good indication of cable quality.

Fig. 6-35 — Experimental setup for measuring attenuation of a random length of cable. This approach is not designed for high accuracy but it is useful for rough estimates and comparitive measurements.

Table 6-13

Approximate Attenuation Values for Coaxial Lines

	Power Loss Per 100 Feet (dB)			
Cable	*29.5 MHz*	*146 MHz*	*435 MHz*	*1260 MHz*
RG-58 series	2.5	6.5	12	22
RG-58/U foam	1.2	4.5	8	15
RG-8/M foam	1.3	3.2	7.2	13
RG-8 and RG-213	1.2	3.1	5.9	11
RG-8/U foam	0.9	2.1	3.7	6.3
RG-17/U		1.0	2.3	
1/2" Hardline	0.4	1.0	1.8	3.4
3/4" Hardline	0.3	0.8	1.6	3.0
7/8" Hardline	0.3	0.7	1.3	2.5

Note: Attenuation values for old or bargain cable may be much higher.

All coaxial cable produces some attenuation. Typical losses for 100-foot runs of some common cables are shown in Table 6-13. The values quoted are for new, high-quality line; losses increase with age and exposure to the elements. The attenuation of bargain cable is often significantly greater. Measuring cable loss is relatively simple if you have access to a wattmeter and doing so is good insurance! (See Fig. 6-35).

Radio-frequency power attenuation is directly proportional to the coaxial cable's length. Doubling the length doubles the attenuation. To compute the loss expected from a given length of cable at a particular frequency use Table 6-13 (or your own measured attenuation value per hundred feet) and the formula

$$A_L = \frac{L}{100} A_o \qquad \text{(Eq. 6.8)}$$

where

A_L = attenuation [in dB] of cable of length L

L = length [in feet] of cable

A_o = attenuation [in dB] of 100 feet of cable

Coaxial connectors may also cause losses. Amateurs working at hf often use the so-called UHF series of connectors (PL-259 plug and SO-239 recepticle) with RG-8/U and RG-213 cable. *UHF* connectors should never be used at uhf frequencies; they produce intolerable losses. In fact, this misnamed series shouldn't even be used at 146 MHz unless losses are of little concern. At 146, 435 and 1260 MHz, the Type N series of connectors (UG-21 plug and UG-58 recepticle) may be used with RG-8/U sized cable. Though RG-58 can be used for very short jumper cables, there are several better choices, short pieces of which are often available through surplus channels. These include semi-rigid Uniform Tubing UT-141; RG-142/U that features Teflon dielectric, double shielding, and a silver-plated center conductor; and the more common RG-141/U, RG-223/U and RG-55/U. All these cables can be used with BNC, TNC and SMA connectors which give excellent results up to 4 GHz at low power levels. E. F. Johnson produces a widely available series of low-cost SMA-compatible connectors (JCM type) which are justifiably popular with amateur microwave experimenters. Most Hardline cables have matching low-loss connectors that mate to the Type N series. Since Hardline connectors are relatively expensive, some amateurs have devised makeshift connectors by combining Type N connectors and standard plumbing fittings. Table 6-14 lists several references that contain practical information on interfacing Hardline at amateur stations.

Table 6-14

Articles Containing Practical Information on Interfacing Hardline

C. J. Carroll, "Matching 75-Ohm CATV Hardline to 50-Ohm Systems," *Ham Radio*, Sept. 1978, pp. 31-33.

J. H. Ferguson, "CATV Cable Connectors," *Ham Radio*, Oct. 1979, pp. 52-55.

M. D. Weisberg, "Hardline Coaxial Connectors," *Ham Radio*, April 1980, pp. 32-33.

G. K. Woods, "75-Ohm Cable in Amateur Installations," *Ham Radio*, Sept. 1978, pp. 28-30.

D. DeMaw, P. O'Dell, "Connectors for CATV 'Hardline' and Heliax," Hints and Kinks, *QST*, Sept. 1980, pp. 43-44.

D. Pochmerski, "Hardline Connectors and Corrosion," Technical Correspondence, *QST*, May 1981, p. 43.

L. T. Fitch, Matching 75-ohm hardline to 50-ohm systems, *Ham Radio*, Vol. 15, no. 10, Oct. 1982, pp. 43-45.

Delay and Phasing Lines

Short sections of coaxial cable are often used as delay lines or matching transformers in antenna systems. Numerous examples were given earlier in this chapter. In many antenna systems the electrical length of these devices is critical. Because signals travel slower in a cable than in free space, the measured and electrical lengths of a section of cable are not equal. They are related by the formula

$$\text{(measured length)} = \text{(velocity factor)(electrical length)} \qquad \text{(Eq. 6.9)}$$

where the velocity factor is generally given as 0.66 (regular cable) or 0.80 (foam dielectric cable). Random measurements, however, show that these values vary by as much as 10% from cable to cable, or up to a few percent along the length of a given piece of cable. Although an error of a few percent may not be important, a 10% error can have a drastic effect on antenna system performance. Therefore, it's best to cut all delay and matching lines about 10% long and then prune them to frequency using the dip-meter approach (Fig. 6-36).

For example, suppose we need a half-wavelength (electrical length) section of foam-dielectric coax line for 146 MHz. The free-space wavelength is given by

$$\lambda \text{ [inches]} = 11{,}810/f \text{ [MHz]} \qquad \text{(Eq. 6.10)}$$

At 146 MHz, λ = 80.9 inches. A free-space half wavelength is therefore 40.5 inches. The estimated measured (physical) length of the piece of coax we need is (0.80) (40.5 inches) = 32.4 inches. A piece about 10% longer than this value is cut and trimmed to length using one of the methods shown in Fig. 6-36.

Calculating EIRP

Here is how EIRP is calculated.

Equipment Characteristics

Transponder to be used: AMSAT-OSCAR 10 Mode L (1269-MHz uplink)

Recommended ground-station EIRP: 5000 watts (*tentative*)

Ground Station Transmitting System:

Power Output = 50 watts (P_o)

Antenna gain = 23.5 dB_i (5-foot-diameter parabolic dish — see Fig. 6-31)

Feed-line loss = 3 db (50 feet of RG-8/U foam — see Table 6-31)

Coax connector loss = 0.5 dB (two sets of Type N connectors)

Calculation

Step I: Find G (gain, or loss, of entire feed and antenna system expressed in dB_i)

$G = 23.5\ dB_i - 3\ dB - 0.5\ dB = 20\ dB_i$

Step II: Convert gain G (in dB_i) to gain ***G*** (pure number)

$\boldsymbol{G} = 10^{G/10} = 10^{20/10} = 10^2 = 100$

Step III: Calculate EIRP

$EIRP = \boldsymbol{G}\ P_o = (100)\ (50\ watts) = 5000\ watts$

Comments

The transmitting system provides the correct EIRP for the transponder of interest when the antenna is aimed correctly. Since the antenna described has a 3-dB beamwidth of 10°, an aiming error of 5° will reduce the uplink signal reaching the satellite by 3 dB. This is equivalent to reducing the EIRP to 2500 watts. In certain situations aiming errors may be purposely introduced to lower EIRP.

Closing Hints

Having come this far, some brief final suggestions considering antenna systems seem in order. First, start simple and then make improvements where they most affect your operating needs. For example, with Phase II satellites try a ground plane or a Lindenblad before you decide that a circularly polarized beam with azimuth and elevation rotators is necessary. With Phase III satellites, listen to the 146-MHz downlink with a homebuilt, linearly polarized quagi before deciding that you need full circular polarization on both links.

Second, don't get caught in the trap of thinking that you need one ultimate array. Often, it's more convenient and effective to have access to several simple antennas set up so that you can quickly switch to the one that produces the best results. Consider Phase II satellites again. The multiple antenna approach is most effective when the antennas are complementary in either (1) azimuth response (e.g., two horizontal 29-MHz dipoles at right angles), (2) elevation response (e.g., a 2-m TR-array for high elevation angles and a beam aimed at the horizon for low elevation angles) or (3) polarization (e.g., dipoles and a ground plane for 29 MHz).

Third, be sure to consider whether the satellite of interest and your particular location affect antenna selection. A station at 50° N latitude that is interested in working with Mode B on AMSAT-OSCAR 10 might, after studying typical passes on a ϕ3 TRACKER, decide that rotators are an unnecessary expense. A fixed-elevation array set at 20° and a manually adjustable azimuth control might be perfectly satisfactory. In most cases the operator could set azimuth prior to a pass and not need to adjust it any further.

The well-equipped station working with Phase III satellites will eventually benefit from circular polarization. If asked to guess which antennas will prove most popular with the serious Phase III user, I'd predict (1) the crossed-Yagi array at 146 MHz, (2) the helix at 435 MHz, where its awkward mechanical structure is less of a problem and (3) the parabolic dish or quad-helix (array of four helices) at 1269 MHz.

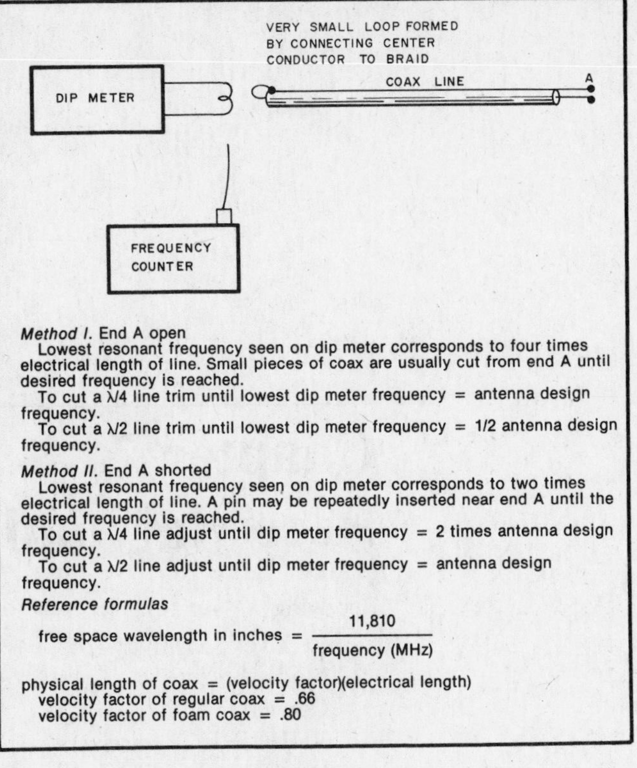

Fig. 6-36 — Two methods for using a dip meter to prune a section of coaxial line to a specific electrical length.

Chapter 7
Receiving and Transmitting

Chapter 7

Receiving and Transmitting

This chapter focuses on the ssb and cw equipment needed to work with the communications transponders on radio amateur satellites. Though SSTV, RTTY, fm and other modes are sometimes employed for experimental purposes or special tests, their use is not widespread. Since most digital encoding schemes (including SSTV and RTTY) are based on cw/ssb equipment, our focus really isn't limited. Topics covered include receiving equipment for 29.5, 146 and 436 MHz, and transmitting gear for 146, 436 and 1269 MHz. We'll also discuss safety and FCC regulations pertaining to transmitting. Later chapters will introduce receiving techniques for weather satellites at 137 MHz and 1.691 GHz, and commercial TV at 4 and 12 GHz.

In most cases, to receive telemetry (Chapter 12) you'll want the same basic receive equipment, though some special processing devices may be needed for the high-speed modes. The amateur/scientific satellite UoSAT does not fit the general pattern, so anyone interested in using this spacecraft should check Chapter 12 and Appendix A carefully.

Ssb and cw equipment designed to access satellites is essentially indistinguishable from gear built for terrestrial use at the same frequencies and power levels. Thus, the existing literature on vhf and uhf construction is directly applicable to our needs. References to useful articles in various radio amateur publications will be given throughout this chapter. Our main concern will be to describe several practical approaches to building a ground station, systematically evaluating the trade-offs involved in various choices.

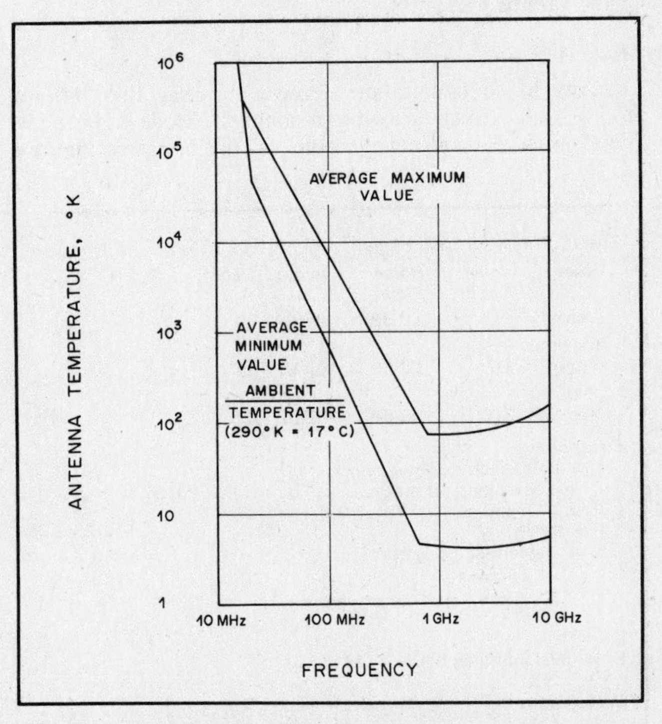

Fig. 7-1 — The sky noise arriving at an earth-based antenna depends on several factors including (1) the portion of the galaxy being observed, (2) the elevation angle of the antenna, and (3) to a lesser extent the water-vapor content of the atmosphere. Average values of the upper and lower limits on sky noise are shown in the graph. For details see: J. D. Kraus, *Radio Astronomy,* McGraw-Hill, NY, 1966, p. 237.

Receiving

A good hf cw/ssb "communication receiver" must meet certain minimal criteria with respect to sensitivity, stability, selectivity and freedom from overload or spurious responses. Anyone with hands-on hf experience knows roughly what these terms mean, and for our purposes we won't need to quantify most of them. Sensitivity, however, deserves special attention.

Receiver Sensitivity

At all radio frequencies, noise arriving by way of the antenna ultimately limits our ability to receive weak signals. Over a considerable range of the vhf, uhf and microwave spectra, the dominant source of external noise is cosmic in origin. Fig. 7-1 shows the background noise levels observed at various frequencies. The dip in the central section shows that the absolute level of this noise is very low at 146 and 435 MHz, making it possible, theoretically, to discern very weak signals. In practice, noise generated in the receiving system itself often masks these weak sources. Our ultimate goal in designing a ground-station receiver is to reduce the internally generated noise to a level below that of the incoming cosmic noise. In reality we usually don't reach this goal but new receiver technology continually makes it easier and less expensive for us to approach it.

A receiver can be depicted as a chain of individual stages (Fig. 7-2), each characterized by two properties: *gain* and *noise factor* (or noise figure), a quantity related to the amount of noise the stage introduces. Noise factor is a dimensionless number greater than or equal to one (noise figure is given in decibels). The lower the noise factor, the better the performance. In a receiver, the noise contribution of each stage acts to reduce the overall system signal-to-noise ratio. The impact of a particular stage on total receiver noise factor depends on the gain prior to

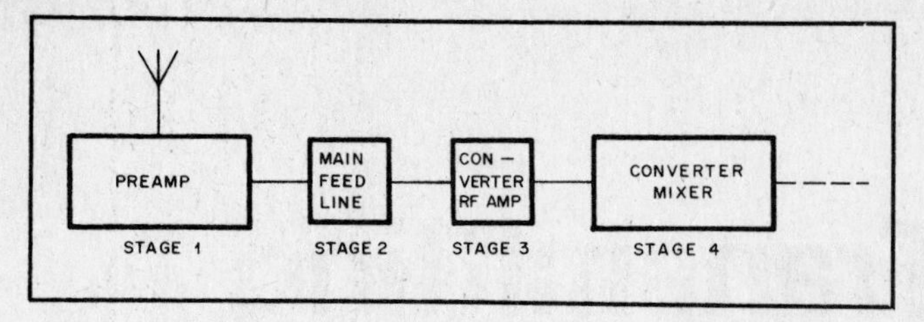

Fig. 7-2 — A receiver is thought of as consisting of a chain of individual stages in order to analyze its weak signal performance.

the stage and the noise factor of the stage. An analysis of the mathematics (Fig. 7-3) shows that the first few stages in a receiver dominate the overall receiver noise factor. Therefore, using very-low-noise devices in the first stage or two of a receiver and avoiding runs of lossy feed line in front of active devices is very important. Once you grasp this basic point you're well on your way to designing an effective receive system. Though many readers may prefer to skip the computations at this point, anyone interested in putting together a high-performance system should look at the sample calculations in Fig. 7-3 to see the consequences of the various trade-offs.

Mode A

Any hf communications receiver covering the 29.0- to 29.5-MHz range can be used to monitor the Mode A downlink (Fig. 7-4). A good low-noise preamp will, in many cases, improve reception significantly. This is true even with expensive receivers because they've been designed to satisfy the less-stringent sensitivity requirements (noise factor) for terrestrial communication. Mounting the preamp at the antenna always provides the best performance. At 29 MHz, however, the improvement is small when using less than 100 feet of RG-8/U feedline; most operators give in to convenience and place the preamp at the receiver. Many amateurs with receivers having "hot" 10-m front ends find reception satisfactory without a preamp. You may too, so give it a try without the preamp — you can always add one later.

Converters

A crystal-controlled converter that uses an hf receiver as a tunable i-f is usually used for 146-MHz and 435-MHz cw/ssb reception (Fig. 7-5). Using this approach, anyone with a good hf receiver can acquire state-of-the-art vhf or uhf receive capabilities at moderate cost. Most modern converters and receivers are designed to use a 28.000- to 30.000-MHz i-f range. With a converter, tuning 435.000 to 435.200 MHz becomes as easy as tuning 28.000 to 28.200 MHz on the hf receiver.

Characteristics that distinguish a good converter from a mediocre performer include low noise figure, freedom from spurious responses, low susceptibility to IMD and gain compression, high frequency stability and low susceptibility to burnout. Although noise figure is important, it shouldn't be of overriding concern. Once you commit to placing a good 1.0-dB-noise-figure, 16-dB-gain preamp at the antenna, whether the converter noise figure is 3 dB or 4 dB makes little difference. The next characteristic, spurious responses, often results from poor interstage

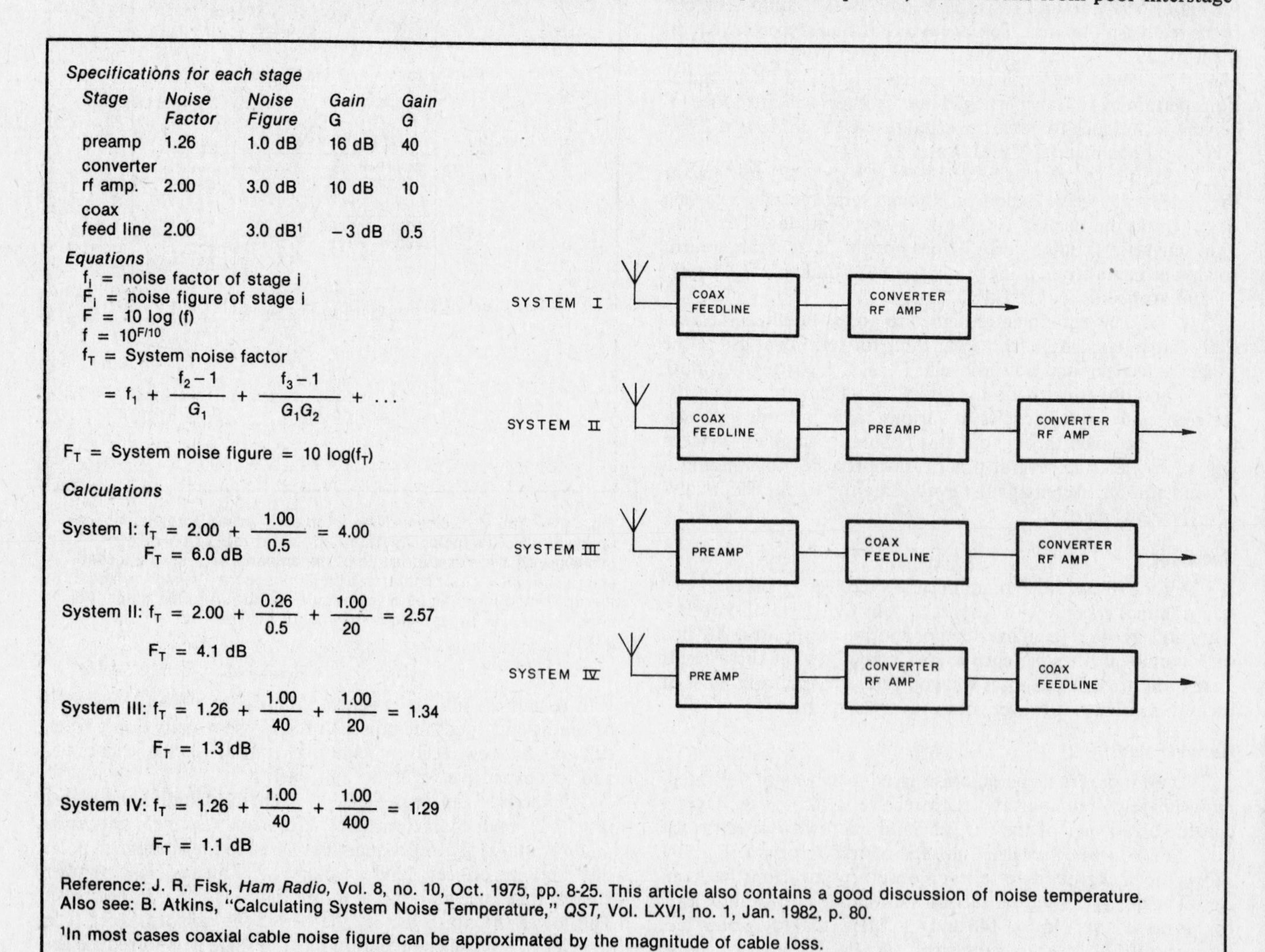

Specifications for each stage

Stage	*Noise Factor*	*Noise Figure*	*Gain* G	*Gain* G
preamp	1.26	1.0 dB	16 dB	40
converter rf amp.	2.00	3.0 dB	10 dB	10
coax feed line	2.00	3.0 dB[1]	−3 dB	0.5

Equations

f_i = noise factor of stage i

F_i = noise figure of stage i

$F = 10 \log(f)$

$f = 10^{F/10}$

f_T = System noise factor

$$= f_1 + \frac{f_2 - 1}{G_1} + \frac{f_3 - 1}{G_1 G_2} + \ldots$$

F_T = System noise figure $= 10 \log(f_T)$

Calculations

System I: $f_T = 2.00 + \frac{1.00}{0.5} = 4.00$

$F_T = 6.0$ dB

System II: $f_T = 2.00 + \frac{0.26}{0.5} + \frac{1.00}{20} = 2.57$

$F_T = 4.1$ dB

System III: $f_T = 1.26 + \frac{1.00}{40} + \frac{1.00}{20} = 1.34$

$F_T = 1.3$ dB

System IV: $f_T = 1.26 + \frac{1.00}{40} + \frac{1.00}{400} = 1.29$

$F_T = 1.1$ dB

Reference: J. R. Fisk, *Ham Radio*, Vol. 8, no. 10, Oct. 1975, pp. 8-25. This article also contains a good discussion of noise temperature. Also see: B. Atkins, "Calculating System Noise Temperature," *QST*, Vol. LXVI, no. 1, Jan. 1982, p. 80.

[1]In most cases coaxial cable noise figure can be approximated by the magnitude of cable loss.

Fig. 7-3 — Comparing noise figures of four systems.

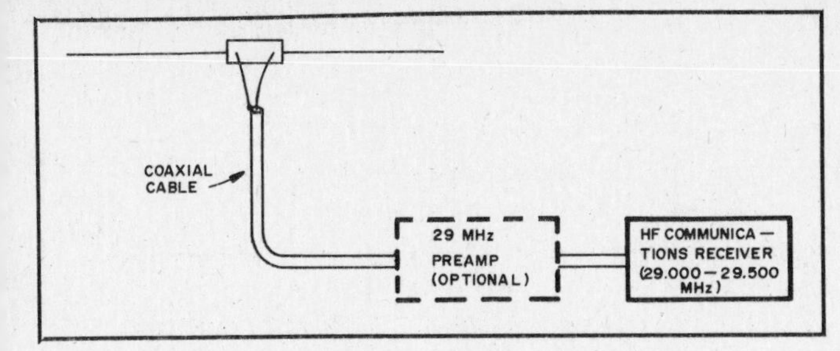

Fig. 7-4 — Basic receive system for Mode A reception.

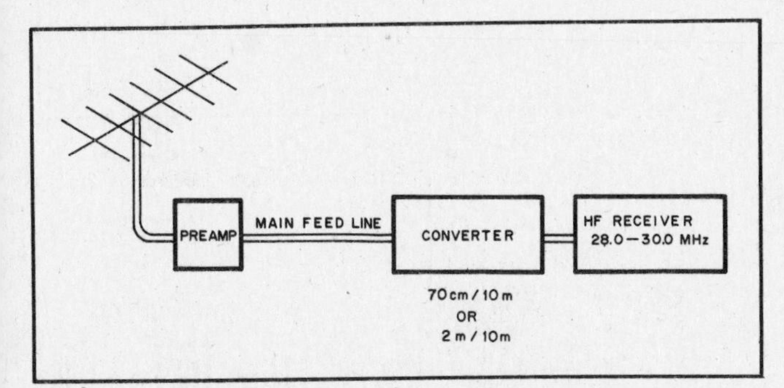

Fig. 7-5 — A crystal controlled converter used in conjunction with an HF receiver acting as a tuneable i-f can provide state-of-the-art VHF/UHF receive capabilities.

filtering in the rf and local-oscillator chain circuits or instability (undesired oscillations) anyplace in the converter. IM distortion and gain compression can arise from poor choices of mixer injection power, bias levels or gain distribution. Poor stability is usually the result of cost-cutting shortcuts such as using cheap crystals and heavily loaded oscillators. To overcome the problems just mentioned, you may need to find special test equipment, develop good diagnostic skills and redesign the circuitry — in other words, these shortcomings are often difficult to cure.

Burnout is primarily associated with the first rf stage in solid state converters or preamps. High-power transmitters and transients related to lightning, relay switching or other sources can burn out expensive rf devices even when extensive precautions are taken. If you build a solid-state preamp, don't skimp by omitting the recommended protection circuits. For a detailed analysis of converter design and hints for improving preformance, see J. Reisert, Jr., "VHF/UHF Receivers — How to Improve Them," *Ham Radio*, March 1976, pp. 44-48.

If you don't want to invest in a modern, well-engineered, low-noise-figure converter, you can often obtain excellent performance using an older, well-engineered one even if it falls short in the noise-figure department; you need only add a good preamp. Several well-designed converters manufactured in the mid and late 1960s used Nuvistors®, miniature tube-type devices typified by the 6CW4, in the front end. The better units had noise figures in the 3.0- to 3.5-dB range at 2 m and in the 4- to 5-dB range at 70 cm. You can often find them very inexpensively at hamfests. A number of operators actually prefer these older converters in situations where IM distortion, gain compression or burnout have been a problem. In any event, avoid poorly engineered units, no matter how impressive the noise-figure specifications may seem.

Many operators report that their receiver performance degrades (desenses) whenever their transmitters are keyed. This problem is especially prevalent when operating Mode J since the third harmonic of the transmitter is very close to the receive frequency. Spotting the downlink frequency and evaluating the uplink performance become difficult, if not impossible, under these conditions. To alleviate the problem amateurs have tried (1) filters between the transmitter and its feed line, and between the receiver and its feed line, (2) separating the transmit and

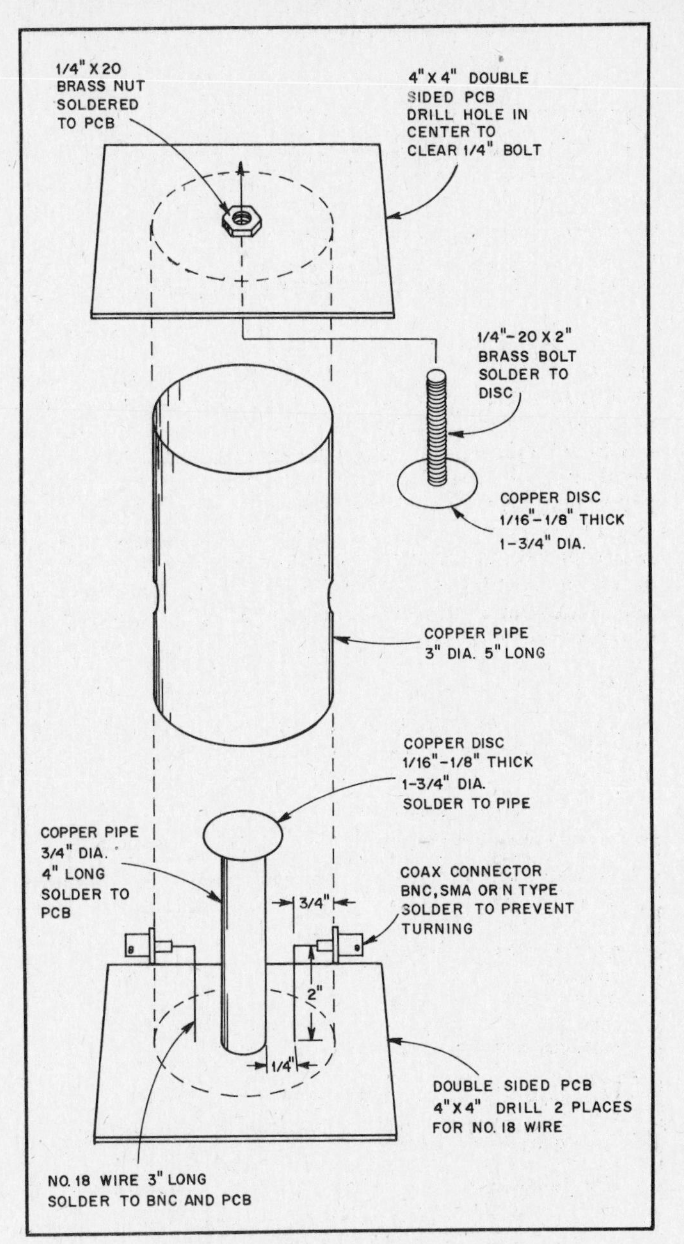

Fig. 7-6 — A 435-MHz cavity filter can significantly improve reception when front end overload by out of band signals is a problem. Don't expect this filter to have much effect on the 3rd harmonic of a 2-m transmitter when operating Mode J since the 3-dB bandwidth is generally about 15 MHz. The 3rd harmonic should be removed at the transmitter using a bandpass filter. Suitable circuits may be found in the *1984 Radio Amateur's Handbook* (p. 14-12), *FM and Repeaters* (p. 56), *ARRL VHF Manual*, 3/E, (pp. 334-337).

receive antennas and feed lines physically and (3) using older type converters in place of modern, solid-state units. Several users have reported cases where serious overload, intermodulation or other receive problems didn't respond to filters or physical displacement but were cured by switching to a tube-type converter. Nonetheless, good filtering and adequate antenna and feed-line separation should also be pursued. One receive filter is shown and several others referenced in Fig. 7-6.

Another feature that you may want in a converter is coverage of more than a 2-MHz segment of a band, e.g., 144 to 148 MHz on 2 m, or 432 to 434 MHz and 435 to 437 MHz on 70 cm. If so, look for provisions for switching crystals in the injection chain.

Many modern converters have very good noise figures (often under 2 dB), suggesting the possibility of eliminating the preamp. To take advantage of the converter's noise figure, however, it should be mounted remotely at or near the antenna using a very

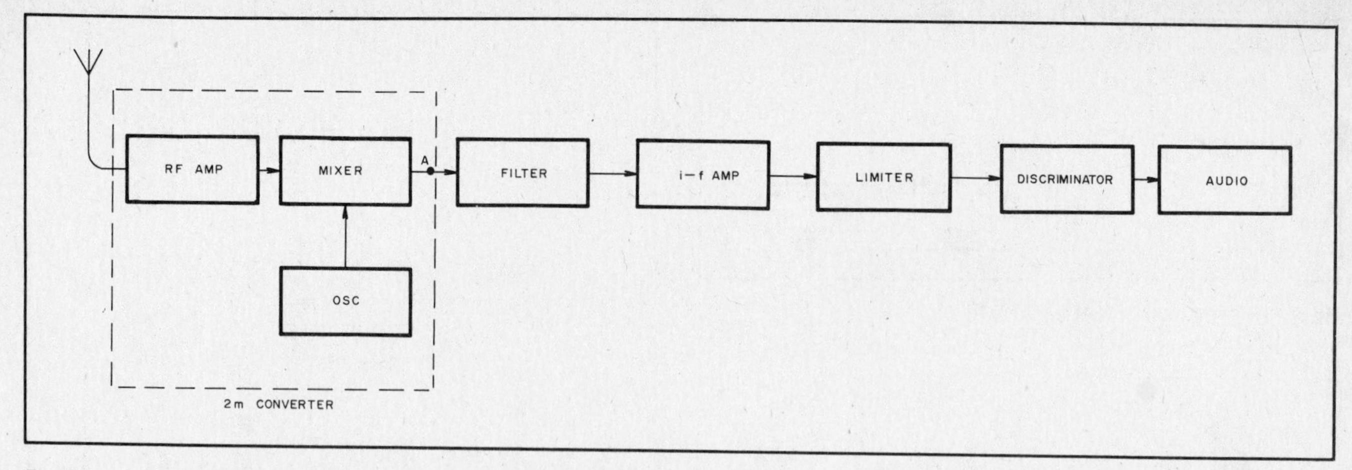

Fig. 7-7 — A block diagram for a typical 2-m fm receiver is shown. The section enclosed in the dotted lines forms a 2-m converter. It's usually possible to pick off a portion of the rf signal at point A without impairing 2-m fm operation. The frequency at point A varies from receiver to receiver, e.g., ICOM 211 (10.7 MHz), Clegg FM-28 (16.9 MHz), G.E. Progline (8.7 MHz).

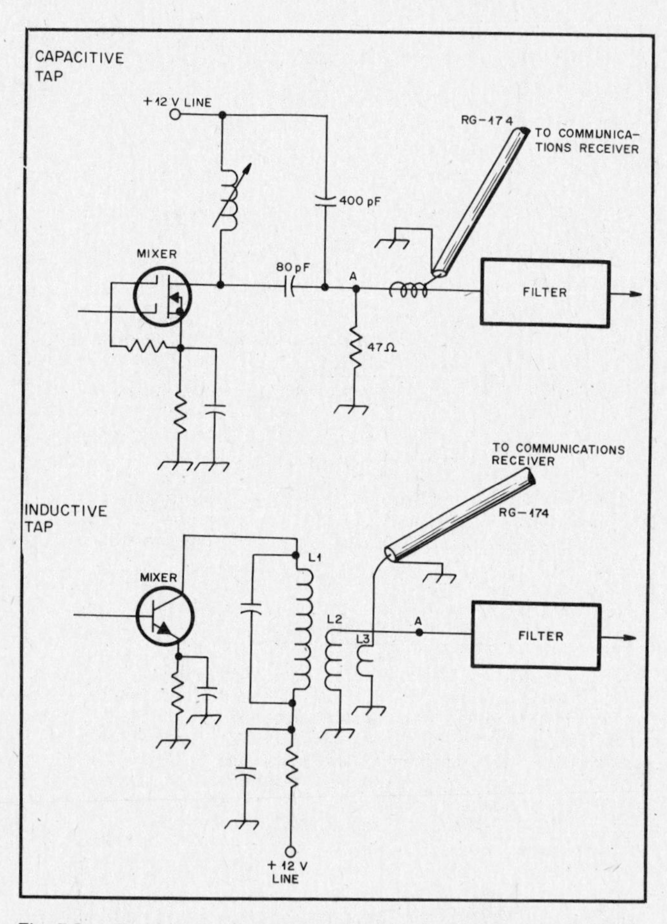

Fig. 7-8 — Techniques for tapping off a little rf from a 2-m fm receiver.

short run of low-loss feed line. Most feed-line losses will then occur after that point in the overall system where the noise figure is established. The problems this approach generates relate to weatherproofing, environmental temperature extremes, oscillator drift, switching the frequency range and adjusting the converter gain, some of which can be difficult to overcome. If you're considering remote converter mounting be sure to check whether the converter is suited to the planned environment. In most cases you'll have fewer headaches by placing a good preamp at the antenna and keeping the converter in the shack.

Many hf operators don't realize that they already have a good 2-m converter at the operating position: the receiver front end in a 2-m fm transceiver (Fig. 7-7). It's usually a simple matter to "steal" a little of the signal from the fm transceiver and feed it into the communications receiver that is being used as a tunable i-f. This gives full cw/ssb capabilities. If you're considering this approach dig out the instruction manual for the fm transceiver and find the frequency of the first i-f (point A in Fig. 7-7). Next, determine whether the hf receiver following the converter can be adjusted to tune this range. With some receivers there's no problem (with the Drake R4 series one can order an appropriate crystal or, in many cases, a standard crystal can be utilized by tuning the preselector to an image frequency). Adjusting some receivers to cover the proper range may involve more work or expense than is justified. If everything checks out to this point, the next step is to pick off the signal. Check the schematic for an easily accessible low-impedance point between the mixer and filter (or between the first mixer and second mixer in double-conversion units) and then try either a capacitive or inductive probe (Fig. 7-8).

VHF/UHF Transceivers

Before turning to preamps, one additional receive alternative will be mentioned. In the past few years a number of multimode ssb/cw and ssb/cw/fm transceivers for 2 m and 70 cm have been marketed. Though expensive, these units are convenient to use and often provide excellent performance. If you need all the features provided, these transceivers may prove to be cost-effective. Some manufacturers make receiver upconverters that can be used with a 2-m transceiver to monitor the 0.1 to 30 MHz hf spectrum. A futuristic amateur satellite ground station, capable of operating on Modes A, B, J and L, may be configured something like the one shown in Fig. 7-9.

Preamps

The late 1970s brought a revolution in preamp performance. Noise figures of 2 dB at 146 MHz and 436 MHz, considered state of the art just a few years before, are now considered ho-hum. Noise figures under 1.0 dB using solid-state devices costing less than $10 are becoming commonplace. The most promising device is the GaAs FET (gallium-arsenide field-effect transistor). A single-stage preamp using a GaAs FET typically yields a noise figure under 0.8 dB with a gain of 15 to 18 dB at vhf and uhf frequencies. One modern circuit for 435 MHz is shown in Fig. 7-10, and references to several other designs are given in Table 7.1. Although recent attention has focused on the spectacular performance of the GaAs FET, silicon bipolar and FET transistors needn't be totally ignored. Several such devices that sell for under $3 can provide noise figures in the 1.5-dB range. See, for example, the 146-MHz and 435-MHz preamps in Figs. 7-11 and 7-12. Changes in the availability and cost of these devices occur so rapidly that anyone who intends to build a preamp should be sure

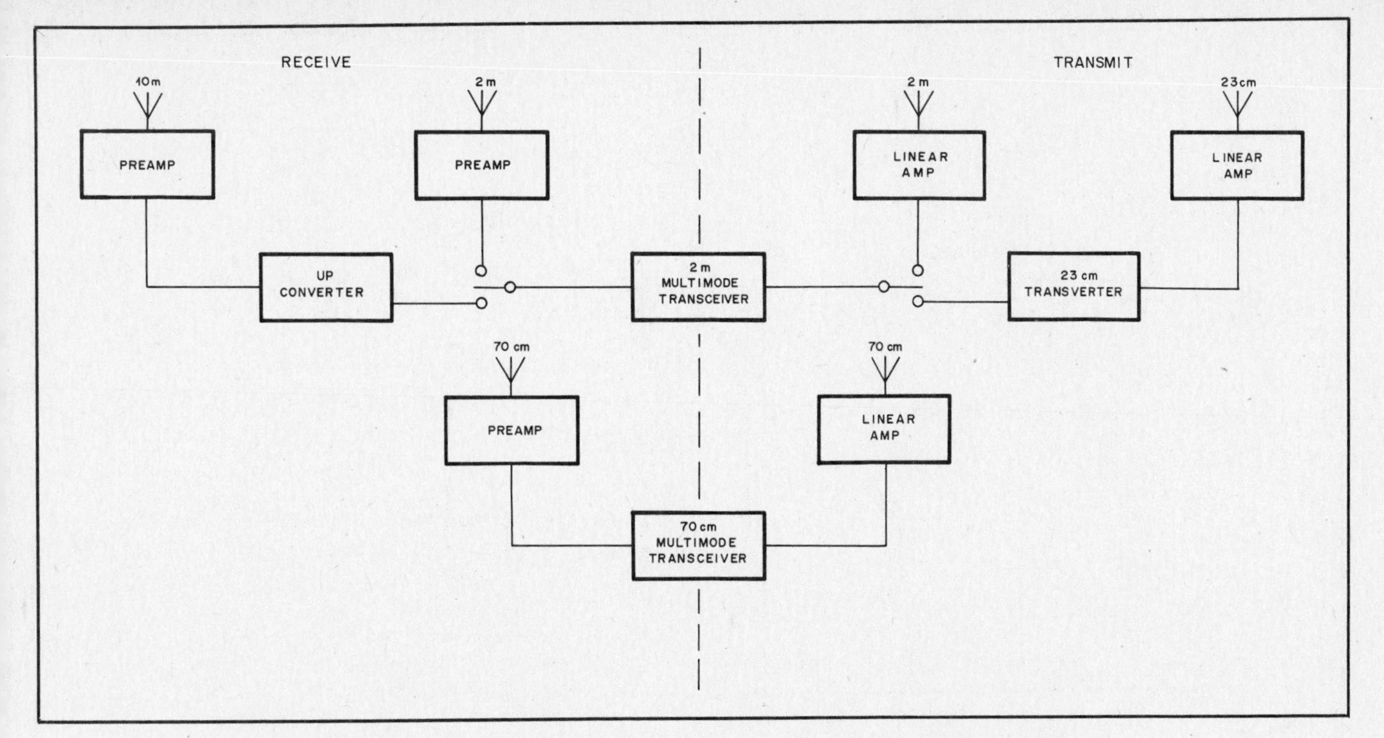

Fig. 7-9 — One possible configuration for a satellite ground station capable of using all presently planned modes.

to check recent issues of *QST, Orbit* and *Ham Radio* for the latest information. A list of manufacturers producing converters and preamps is given in Table 7.2.

Two practical problems are encountered when mounting a preamp at the antenna: supplying power and weatherproofing the installation. Let's look at power first. Most solid-state units require a single positive supply of about + 12 volts dc. Because the power drain is considerable and the location is often inaccessible, placing batteries at the preamp is out of the question. (Batteries are sometimes included in GaAs preamps to provide gate bias; the current drawn in this instance is negligible and the battery should last for its shelf life.)

The most direct method of supplying power to the preamp is to run a separate lead up to the antenna from a + 12-volt power supply in the shack and use the outer braid of the coaxial cable as the power-supply ground return. A more elegant solution (Fig. 7-13) completely eliminates the need for any extra wires by using the coax feed line to carry both dc power and rf signals. Measurements reveal no discernible rf losses when this technique is used. Although two junction boxes are shown, one (or both) is usually eliminated by enclosing the circuitry with the preamp or converter. Capacitors C1 and C2 are needed only when the rf signal preamp-exit or converter-entry point is at dc ground potential. Note how C1 has been eliminated in the preamps of Figs. 7-11 and 7-12.

Weatherproofing can be simple. One common technique (Fig. 7-14) is to mount the preamp on the cover of an inverted plastic food container using a liberal amount of RTV sealer. Buy a good quality refrigerator or freezer container; avoid cheap plastics that tend to become brittle and crack after a few months of exposure to the sun.

To crudely judge the reliability of antenna-mounted preamps I tried the following test. A microwave converter was placed in a double plastic bag and set up on the deck of my house where it was fully exposed to the weather. No weatherproofing was used and no attempt was made to seal the unit. The plastic bags were simply gathered together using a twist tie and the cables were routed out the bottom to discourage seepage. After two years of operation absolutely no problems have occurred. Although this approach is definitely *not* recommended, it does suggest that the difficulty of weatherproofing externally mounted equipment may be overestimated.

Table 7.1

Sources of 30- to 435-MHz Preamp Construction Information

Latest edition of *The Radio Amateur's Handbook.* Check chapters on vhf and uhf receiving techniques, and fm and repeaters.

G. Krauss, "Low-Noise, Low-Cost 10- To 60-MHz Preamp," *HR*, Vol. 14, no. 5, May 1981, pp. 65-68. Excellent designs for use either as a Mode A preamp or as a post amp following a low-gain converter.

G. Krauss, "VHF Preamplifiers," *HR*, Vol. 12, no. 12, Dec. 1979, pp. 50-60. This article contains an extensive summary of recent work by the author and others on preamp design and performance over the range of 30 to 435 MHz. It's a must for anyone building their own. Also contains an extensive bibliography of earlier articles.

G. Krauss, "VHF and UHF Low-Noise Preamplifiers," *QEX*, Vol. 1, no. 1, Dec. 1981, pp. 3-8.

P. Wade and A. Katz, "Low-Noise GaAs FET UHF Preamplifiers," *QST*, June 1978, pp. 14-15. An introduction to GaAs FET preamplifiers for 432 MHz and 1296 MHz. Later testing at 432 MHz showed that lowest noise figure usually coincided with zero gate voltage. As a result, the source bias circuit shown in the *QST* article can generally be omitted by grounding the junction of L1 and the gate feedthrough capacitor.

S. Sando, "Very Low-Noise GaAs FET Preamp for 432 MHz," *HR*, Vol. 11, no. 4, April 1978, pp. 22-27. Uses NE 24406 (2SK85), gain: 18 dB, NF: 0.7 dB.

Also see references listed in Figs. 7-10, 7-11 and 7-12.

Transmitting

All radio amateurs using the transponders on OSCAR satellites must share the available power and bandwidth. Cooperation is essential. Stations employing too high an eirp will use more than their share of spacecraft energy and may even activate the transponder automatic-gain-control circuitry making it impossible for low-power ground stations to be heard. For general communications, cw and ssb are recommended. The high peak-to-average power characteristic of ssb, and the low duty cycle of cw, use satellite energy effectively and efficiently. Users should generally not use fm, SSTV (slow-scan television), a-m or ssb with

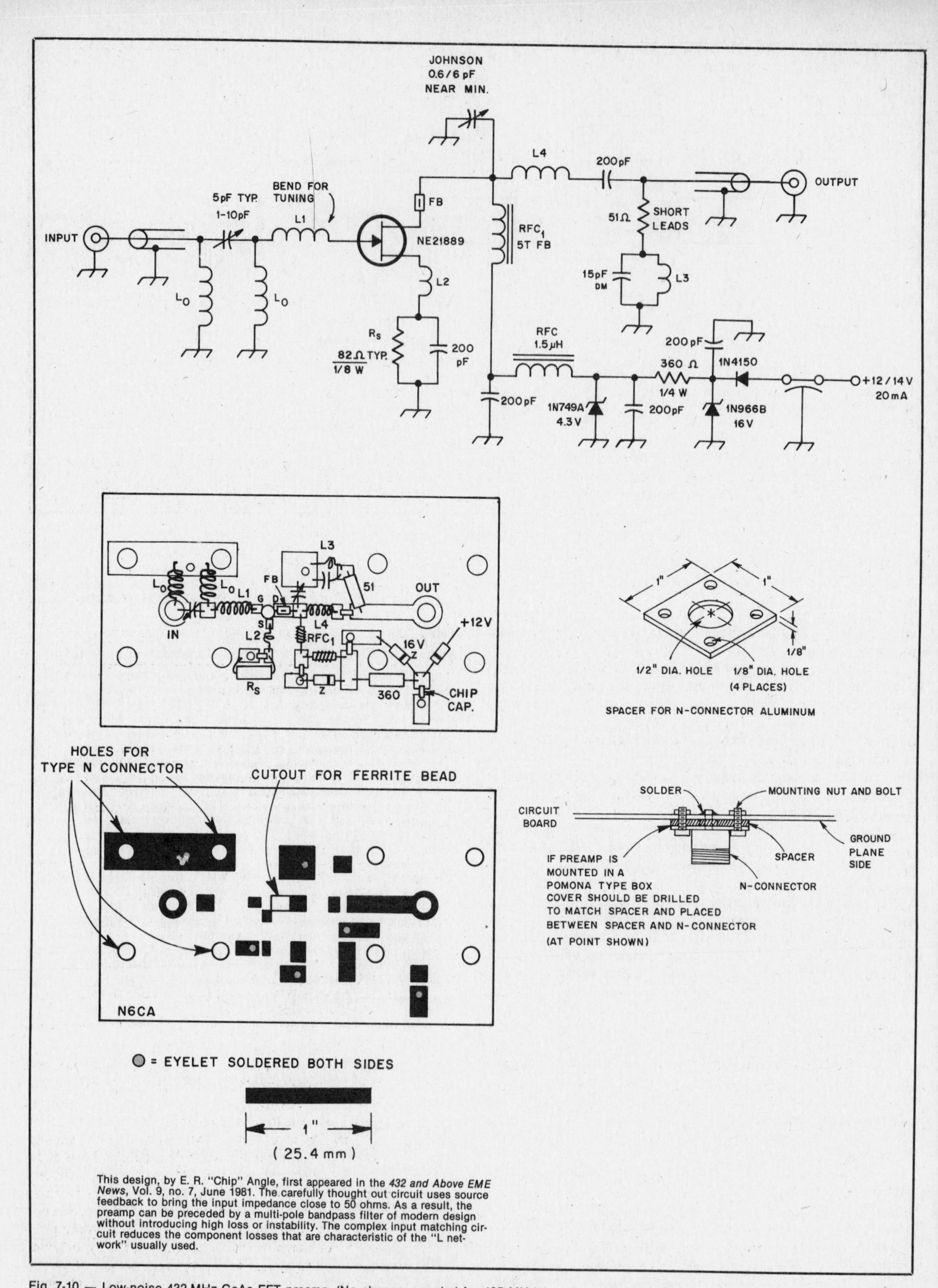

This design, by E. R. "Chip" Angle, first appeared in the *432 and Above EME News*, Vol. 9, no. 7, June 1981. The carefully thought out circuit uses source feedback to bring the input impedance close to 50 ohms. As a result, the preamp can be preceded by a multi-pole bandpass filter of modern design without introducing high loss or instability. The complex input matching circuit reduces the component losses that are characteristic of the "L network" usually used.

Fig. 7-10 — Low-noise 432-MHz GaAs FET preamp. (No changes needed for 435 MHz)
Noise figure: 0.5 dB
gain: ~16 dB

Q1: NE21889 specified but inexpensive MGF 1200 should perform nearly as well.

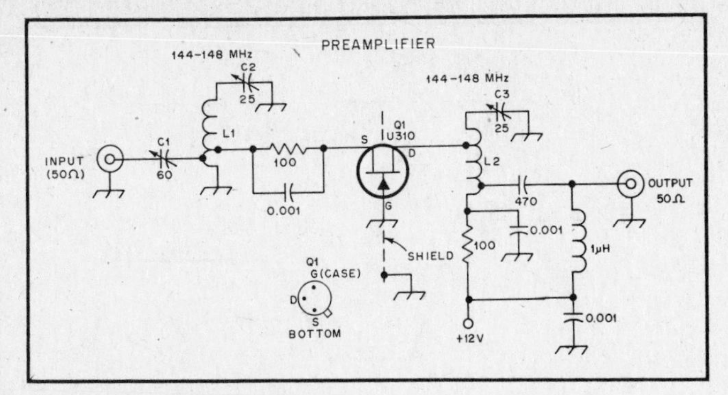

Fig. 7-11 — Low-noise 2-m preamp
Noise figure: ~1.5 dB
gain: ~10 dB
Q1: Siliconex U310 or 2N5397

See 1984 *ARRL Handbook*, p. 13-10 for additional information. A similar design by C.E. Scheideler with detailed construction information appeared in "A Preamplifier for 144-MHz EME," *EME Notes*, AS-49-9, Eimac Division of Varian Associates, 301 Industrial Way, San Carlos, CA.

speech processing, because these forms of modulation drain an excessive amount of satellite energy. These modes may be used for special experimental purposes, however.

Cooperation

Recommended eirp levels are listed in Appendix A: Spacecraft Profiles for each transponder aboard each spacecraft. AMSAT has calculated these levels to ensure that the transponders provide reliable communications when *all* users cooperate. Though there's little chance of overloading the satellite when you abide by the eirp guidelines, it can happen, especially when a Phase II spacecraft is directly overhead. In certain circumstances modestly higher eirp levels may be used, particularly when path losses are exceptionally high (on passes below your radio horizon, when beaming through a thick stand of trees or when the ionosphere is disturbed). Courteous, experienced users monitor their downlink signals to ensure that they are not overloading or monopolizing the spacecraft transponder.

Cooperation also means taking care to produce a *clean* signal. Key clicks and ssb splatter will degrade transponder performance. In fact, ssb splatter can raise the noise floor of the transponder, making it impossible for any users to copy medium-strength stations that would otherwise be perfectly readable. This problem is potentially a serious one, because users of the wide, underpopulated terrestrial (non-satellite) 70-cm and 23-cm bands haven't had much incentive to worry about spectral purity. After all, it's difficult enough to put an ssb signal on 1.26 GHz; if you're only 10 dB above the noise at your closest neighbor's shack, no one will know whether or not your splatter is down 15 dB or 30 dB. As a result, not enough attention has been paid to the spectral purity of ssb signals at 435 and 1260 MHz in the past, but amateurs are finally beginning to focus on this problem. See, for example, the article "Solid-State VHF Linear Amplifiers," by C. F. Clark, *Ham Radio*, Jan. 1980, pp. 48-50, in which vhf solid-state amplifier design is discussed from the viewpoint of assuring linearity. Before turning to transmitting equipment, we'll consider a very important aspect of generating rf energy at vhf and uhf frequencies: rf power hazards.

RF Power — Hazardous?

Amateur Radio is basically a safe activity but accidents can always occur if we don't use common sense. Most of us know enough not to place an antenna where it can fall on a power line, insert our hand into an energized linear amplifier, or climb a tower on a windy day. We also should not venture overexposure to rf energy. Large amounts of rf energy can cause damage in people by heating tissues. The effects depend on the wavelength, incident energy density of the rf field, and on other factors such as polarization.

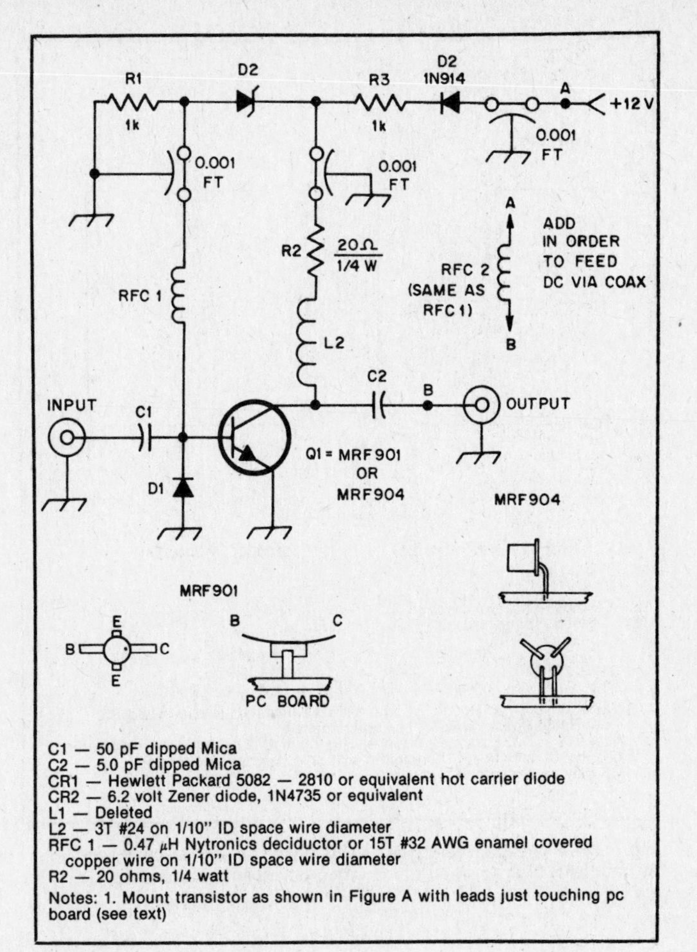

Fig. 7-12 — Low-noise 435-MHz preamp
Noise figure: 1.5-2.0 dB
Gain: 12-14 dB
Q1: MRF 901 or MRF 904

This design first appeared in: J. Reisert, "An Inexpensive AMSAT-OSCAR Mode J Receive Preamplifier," *AMSAT Newsletter*, Vol. X, no. 2, June 1978, pp. 10-11. For additional information on this preamp see: J. Reisert, "Ultra Low-noise UHF Preamplifier," *HR*, Vol. 8, no. 3, March 1975, pp. 8-19. The complete absence of tweaking adjustments makes this design a pleasure to replicate.

Table 7.2
Some Manufacturers of Preamps and Converters

Advanced Receiver Research, Box 1242, Burlington, CT 06013.
Hamtronics, Inc., 65K Moul Rd., Hilton, NY 14468.
Janel Laboratories, 33890 Eastgate Circle, Corvallis, OR 97330.
Lunar Electronics, 2785 Kurtz St., Suite 10, San Diego, CA 92110.
Microwave Modules, Brookfield Drive, Aintree, Liverpool L9 7AN, England. Distributed in U.S. by Spectrum International, Inc., P.O. Box 1084, Concord, MA 01742.

At the frequencies of interest to satellite users — 146, 436 and 1269 MHz — large power densities may be accessible. The most susceptible parts of the body, the tissues of the eyes and gonads, don't have heat-sensitive receptors to warn us of the danger before the damage occurs. Symptoms of overexposure may not appear until after irreversible damage has been done. Though the problem should be taken seriously, with reasonable precaution operation at 146, 436 and 1269 MHz can be safe.

The primary aim of this section is to show where protection from rf energy may be necessary. Our emphasis is on the practical problems encountered by satellite operators; for a more comprehensive technical treatment see the references that follow this section. The rf-protection problem consists of two parts: (1) determining safe exposure levels and (2) estimating the local rf levels produced by a given power and antenna at a particular location.

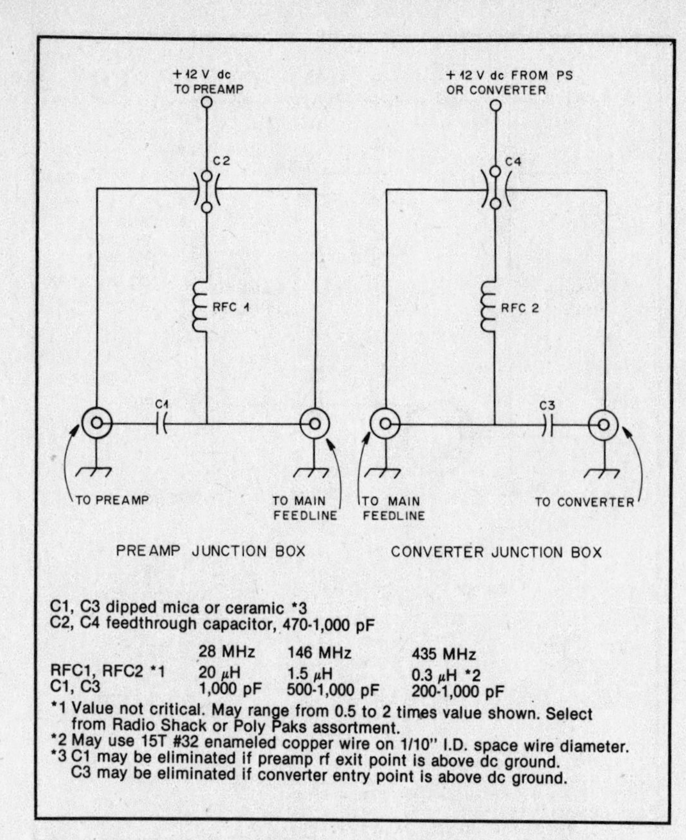

Fig. 7-13 — The main feed line can be used to carry dc for the preamp as well as rf signals when the junction boxes shown are used.

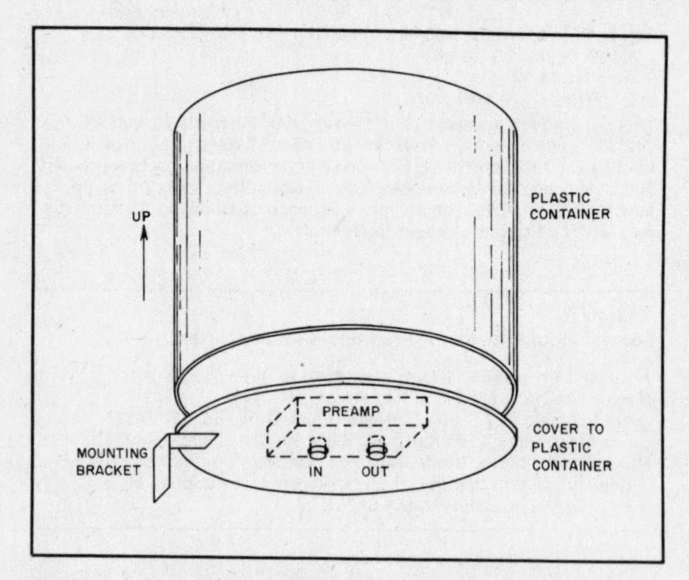

Fig. 7-14 — A preamp can be mounted in a sealed plastic container for waterproofing. Liberal use of RTV sealer and a couple of plastic bags over the whole shebang will help insure integrity.

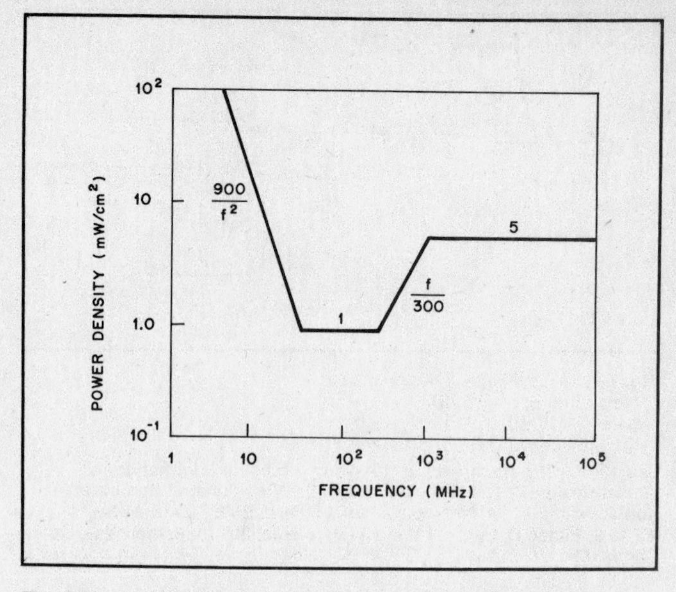

Fig. 7-15 — ANSI RFPG for whole-body exposure of human beings.

If the actual rf power density levels are greater than or even roughly equal to the safe levels, protection or precaution is required by limiting access or some other means. We begin with the question of safe exposure levels.

Safe Exposure Levels. In recent years scientists have devoted a great deal of effort to determining safe rf-exposure limits. As the problem is very complex, it's not surprising that some changes in the recommended levels have occurred as more information has become available. The American Radio Relay League believes that the latest "Radio Protection Guide of the American National Standards Institute (ANSI)" is the best available protection standard; it took nearly five years to formulate and has undergone repeated critical review by the scientific community. This 1982 guide recognizes the phenomenon of whole-body or geometric resonance and establishes a frequency-dependent maximum permissible rf exposure level.

Resonance occurs at frequencies for which a body's long axis, if parallel to the ambient field, is about 0.4 wavelength long. Because of the range of human heights the resonant region spans a broad range of frequencies. The most stringent maximum permissible exposure level, the bottom of the "valley" (see Fig. 7-15), is 1 mW/cm^2 for frequencies between 30 and 300 MHz. On either side of those "corner" frequencies the rise is gradual. At 3 MHz the maximum permissible exposure level is 100 mW/cm^2; at 1500 MHz and above, 5 mW/cm^2. The valley region includes some active amateur bands (10, 6 and 2 meters) as well as all fm and TV broadcasting. The rationale for specifying a constant 5 mW/cm^2 above 1500 MHz takes into consideration that with the extremely short wavelengths there is very little penetration into tissue. Until the Environmental Protection Agency (EPA) promulgates a general population standard, the ANSI Guide will likely be the most commonly accepted one. The EPA began its promulgation process in early 1982, so several years may yet pass before its standard is in place.

Estimating Power Density. Our task is to estimate the power density levels that could be produced at a given location by a specified antenna and power. Since most amateurs do not have the special equipment needed to measure rf electric fields accurately, power density will have to be estimated by calculation. The estimates involve certain approximations but ought to be upper-bound estimates. While our estimates will always tend toward the conservative side when there's a choice, keep in mind that the results should be used only as a guideline for pointing out situations to avoid. The results should *never* be taken as *proving* that a particular setup is safe. For example, we only consider radiation from an antenna. Radiation can also take place directly from a power amplifier (if operated without proper shielding), from transmission lines (if poorly shielded or connectors are improperly installed) and in other situations. Take care to see that the only radiation from your station is at the antenna. Also, we'll be using a free-space propagation model to get a first estimate of power density. You may and should allow up to a 4- or 6-dB margin to provide for cases where a reinforcing reflection might occur.

Generally, antenna engineers divide the region around an antenna into a *far-field* and a *near-field*. At large distances from an antenna located in free space, power density can be estimated by applying Eq. 7.1.

$$\rho = \frac{P\,G}{4\,\pi\,R^2} \qquad \text{(Eq. 7.1)}$$

where

ρ = estimated power density at distance R from antenna (units will be W/m^2 if P is in watts and R is in m)

R = distance from observation point to closest point on antenna (in m)

P = *average* power at antenna feedpoint terminals (watts)

G = gain as a power ratio, that is, the numerical gain (do *not* use gain expressed in decibels)

Note that Eq. 7.1 holds only if both of the following requirements are met: The free space model is appropriate and we're at a sufficiently large distance from the antenna to be in the far field. The far field in fact is defined as the region where Eq. 7.1 is valid.

How far away from the antenna must the observer be for Eq. 7.1 to hold? The answer depends on the type of antenna and on our accuracy requirements.

For several types of antennas likely to be used in space communication activity, we list the minimum distance at which Eq. 7.1 may be applied with some confidence to provide an upper-bound (conservative) estimate.

To assure an upper-bound estimate, the free-space antenna gain should be used if the actual value is not known. A textbook value will generally be useful.

Antenna type	*Dimension*	*Minimum distance, R, for Eq. 7.1*
Parabolic dish	diameter, D	$(1/2)(D^2/\lambda)$
Broadside array	max. linear, L	$(1/2)(L^2/\lambda)$
End-fire types:		
Yagi	$L = \lambda/2$	$2L^2/\lambda$
Loop (quad) Yagi	L = max. width of loop (quad)	$2L^2/\lambda$
Axial-mode helix	L = diameter of turn	$2L^2/\lambda$

Example. For an example consider a station with the following characteristics. A Yagi antenna with 13-dB_i gain ($G = 20$) is mounted atop a 33-ft (10-meter) tower. Wavelength at 430 MHz is approximately 0.7 meter. Average power reaching the antenna terminals is 50 watts.

To determine the minimum R at which the far-field power density formula (Eq. 7.1) can be applied, note that for a Yagi the maximum linear dimension "L" is roughly the length of an element, or $\lambda/2$.

$$R_{min} = 2L^2/\lambda = \frac{2(0.35\ m)^2}{0.7\ m} = 0.35\ m$$

Assume the interest is in the power density likely to occur at a point 50 feet on axis (15 meters) away. Since this distance is greater than R_{min} we can use Eq. 7.1.

$$\rho = \frac{P\,G}{4\,\pi\,R^2} = \frac{50 \times 20}{4\,\pi\,(15)(15)} = 0.35\ \text{watts/m}^2$$

Since 1.0 watt/m^2 equals 0.1 milliwatts/cm^2 this yields

ρ = 0.035 milliwatts/cm^2

ANSI 1982 has a protection level of (430/300) milliwatts/cm^2 at 430 MHz, or 1.43 mW/cm^2, for far-field exposure. Our result in the example is less than 3% of the ANSI level.

Comments. The ARRL through its Committee on the Biological Effects of RF Energy will continue to keep the Amateur Radio community informed on current protection issues and knowledge of e-m bioeffects. Even though the preponderance of Amateur Radio operation, because of its fundamentally intermittent nature and relatively low power, poses little rf-protection requirements on antenna proximity, hams should keep abreast of developments and always follow recommended rules of good practice on the management and uses of rf energy. Consult *QST* and other technical sources for articles dealing with safe operation.

Our thanks to Dr. David Davidson, W1GKM, member of the ARRL Committee on the Bio-Effects of RF Energy, for his assistance with this section. For further details in the areas of the biological effects of rf energy, refer to the following.

ANSI C95.1-(1982). Safety Levels with Respect to Human Exposure to Radio Frequency Electromagnetic Fields (300 kHz to 100 GHz). New York: American National Standards Institute.

Balzano, Q., O. Garay, K. Siwiak, "The Near Field of Dipole Antennas, Part I: Theory." *IEEE Trans. Vehicular Technology* (VT) 30, p. 161, November 1981). Also "Part II; Experimental Results," same issue, p. 175.

Guy, A. W., C. K. Chou, "Thermographic Determination of SAR in Human Models Exposed to UHF Mobile Antenna Fields," Paper F-6, Third Annual Conference, Bioelectromagnetics Society, Washington, DC, Aug. 9-12, 1981.

Lambdin, D. L. "An Investigation of Energy Densities in the Vicinity of Vehicles with Mobile Communications Equipment and Near a Hand-Held Walkie Talkie," *EPA Report ORP/EAD 79-2*, March 1979.

R. J. Spiegel, "The Thermal Response of a Human in the Near-Zone of a Resonant Thin-Wire Antenna," *IEEE Trans. Microwave Theory and Technology* (MTT) 30(2), pp. 177-185, Feb. 1982.

Transmitting Equipment

There are four basic approaches to obtaining rf power at 146, 435 and 1269 MHz:

1) Purchase Amateur Radio cw/ssb equipment (new or used).
2) Convert commercial fm or military surplus equipment.
3) Modify amateur fm equipment.
4) Build transmitters or transmitting converters.

We'll discuss each of these approaches in general terms and then go to a band-by-band survey of desired eirp levels and examples of suitable gear. But first, we'll consider two common devices used to produce rf at 146 MHz and higher frequencies: the transmitting converter (transverter) and the varactor multiplier.

Transverter. A transverter works very much like a receive converter. In a transverter a mixer is used to combine energy from an ssb or cw signal at one frequency with energy from a local oscillator chain at the second frequency to produce a signal at a sum or difference frequency. Receive converters and low-power transverters are often very similar in design. In fact, two modular, low-power ssb/cw transceivers for 1.26 GHz that are referenced later in this chapter use the same mixer and oscillator chain on both transmit and receive. Transverter mixers, however, are sometimes optimized to work with signals of several watts (high-level mixers). Since signal levels in a transverter are generally better defined than in a receive converter (in which input signals can vary by a factor of a million), transverters are generally easier to design. Most transverters for 146 and 435 MHz have been standardized to be driven by a 28 to 30 MHz input (usually at a fraction of a watt). Output power is usually in the 0.5- to 10-watt range. At 1269 MHz, however, it's better to use a 144-MHz input to eliminate the need for extensive image filtering.

Varactor Multiplier. A power varactor is a type of semiconductor diode whose properties make it an efficient frequency multiplier in the 1- to 100-watt range. Although varactors have been used as doublers, triplers, quintuplers and higher-order

Table 7.3

Varactor Multipliers

H. H. Cross, "Frequency Multiplication with Power Varactors at U.H.F.," *QST*, Oct. 1962, pp. 60-62. This pioneering article describes a 144/432-MHz tripler that uses conventional inductors. The unit yields 40% efficiency at 20-watts input with a Microwave Associates MA-4060A diode. The article gives lots of good, practical advice.

D. Blakeslee, "Practical Tripler Circuits," *QST*, March 1966, pp. 14-19. Contains a practical tripler that incorporates a strip-line output filter. The unit yields 60% efficiency at 20-watts input with an Amperex H4A (1N4885) diode. The basic design was reprinted in several editions of *The Radio Amateur's Handbook* in the late '60s and early '70s.

D. DeMaw, "Varactor Diodes in Theory and Practice," *QST*, March 1966, pp. 11-14. Contains a thorough and understandable discussion of basic varactor doubler and tripler design considerations.

The Radio Amateur's VHF Manual, ARRL, 3rd edition, 1972. 144/432 tripler using H4A, pp. 289-290; 432/1296 tripler using MA4062D, pp. 292-293 (out of print).

D. S. Evans and G. R. Jessop, *VHF-UHF Manual*, 3rd edition, RSGB, London, 1976 (available from ARRL). Contains general information (pp. 5.20-5.21), a 145/435-MHz tripler (pp. 5.21-5.23) that uses 1N4387 (40 W in/25 W out) or BAY 96 (15 W in/9 W out) and a 384/1152-MHz tripler (pp. 5.70-5.71) using BXY 35A (30 W in).

FM & Repeaters, 2nd edition, ARRL, 1978. Contains a practical design for a 145/435-MHz tripler that uses an H4A (pp. 49-50).

D. R. Pacholok, "Microwave-frequency Converter for UHF Counters," *Ham Radio*, July 1980, pp. 40-47. Describes how transistor collector-base junction can be used as varactor. As a result, a bipolar transistor can be used simultaneously to amplify at the input frequency and multiply using efficient varactor effect.

Complete triplers for 145/435 MHz and 420/1260 MHz are available from Microwave Modules at several power levels.

Note: In most cases the varactors specified are interchangeable as long as power dissipation is taken into account. A summary of the varactors used in various amateur construction projects and the maximum rf power input follows.

Device	*Max.recommended input power*	*Manufacturer*
MA-4060A	20 watts	Microwave Associates
H4A (1N4885)	20 watts	Amperex
BAY 66	12 watts	Mullard
BAY 96	40 watts	Mullard
1N4387	40 watts	Motorola
BXY35A	30 watts	Mullard
MA4062D		Microwave Associates

Table 7.4

Sources for Used Commercial FM Equipment

Gregory Electronics Corp.
246 Rte. 46
Saddle Brook, NJ 07662

Spectronics
1009 Garfield Ave.
Oak Park, IL 60304

QST Ham Ads
Hamfests

Sources for manuals and schematics

General Electric Co.
Marketing Communications Production
Box 4197
Lynchburg, VA 24502

Motorola Communications and Electronics, Inc.
1301 E. Algonquin Rd.
Schaumburg, IL 60172

RCA Corporation
Customer Technical Information Service
Meadow Lands, PA 15347

FM Schematic Digest
Contains schematics and notes on Motorola land mobile equipment manufactured in 1950s. Includes T44, 30D, 80D, 140D and Dispatcher transceivers. (136 p.) Available from S. Wolf, Box 535, Lexington, MA 02173.

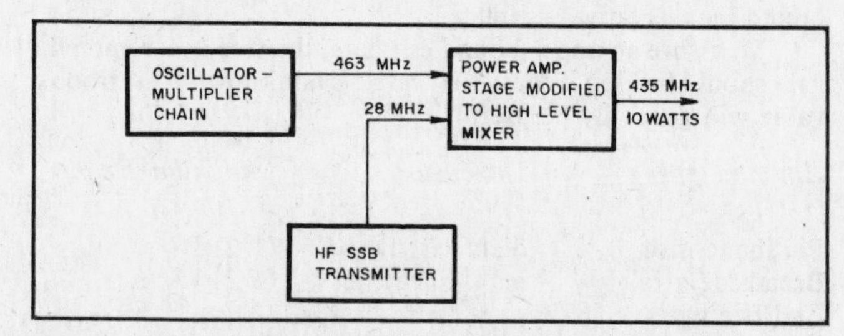

Fig. 7-16 — Block diagram showing how a uhf commercial fm transmitter strip can be converted to 435 MHz. See R. Stevenson, *QST*, Hints and Kinks, Vol. LX, no. 3, March 1976, p. 40; and R. Stevenson, "SSB on Mode B, Using Modified FM Equipment," *AMSAT Newsletter*, Vol. VII, no. 4, Dec. 1975, p. 10.

multipliers, their most common application is in tripling 145 MHz to 435 MHz, or 420 MHz to 1260 MHz. Efficiencies of 50% to 60% (output rf power × 100%/input rf power) are common in the tripler configuration. A varactor multiplier does *not* require any dc power for operation so it's relatively simple to mount one remotely at the antenna. The greatest shortcoming of varactor multipliers is their inability to work on ssb. They're only suitable for cw input signals or as part of a local-oscillator chain leading to a high-level mixer. Table 7.3 contains an extensive list of articles on varactor operation and construction.

Purchase of Amateur Equipment. It's now possible to purchase a complete 146-MHz or 435-MHz ssb/cw transmitter suitable for satellite operation "off-the-shelf." At 1.26 GHz we're almost at this point. Transverters for 1.26 GHz are available but at power levels that are a little below what is desirable. And, a varactor tripler capable of producing 18 watts when driven with 30 watts at 420 MHz is available for under $150. Users will likely be able to purchase a complete transmitter for 1.26 GHz uplinking within a year after the Mode L transponder aboard AMSAT-OSCAR 10 has been operational.

Transmitters for 146 and 435 MHz are available as part of 10 to 30 watt ssb/cw or ssb/cw/fm transceivers by ICOM, Kenwood, KLM, Yaesu and others. A 10-watt transceiver coupled with a 10-watt/80-watt linear amplifier makes a very effective combination for uplinking. Be careful when shopping, though, as many of the commercial amplifiers in this power range are designed for fm and cw only. If you're interested in ssb make sure you acquire an amplifier that's truly *linear*.

Transverters have proved very popular with current satellite operators. Units for 146, 435 and 1269 MHz are being marketed by Lunar, Microwave Modules, SOTA and others. Hamtronics produces transceiver kits for 146 and 435 MHz. Most yield from 0.5- to 10-watts output when driven with a fraction of a watt. When comparing prices, be sure to take into account power levels and the fact that some units include a receive converter. Because of the low output of transverters they're usually used in conjunction with linear amplifiers.

Commercial varactor triplers for producing up to 40 watts (145/435 MHz) and 18 watts (420/1260 MHz) are available from Microwave Modules. These comments on the commercial availability of transmitters, transverters and varactor triplers are meant only to introduce you to the ever-growing range of new equipment suitable for satellite operation. Several additional manufacturers are producing fine pieces of equipment as well. If you're interested in purchasing commercial gear you'll certainly want to check the latest ads in *Orbit, QST* and other Amateur Radio publications to see what's available.

Several excellent though discontinued 2-m transmitters and transverters sometimes appear on the used-equipment market at reasonable prices. Desirable items for cw include the AMECO TX-62 40-watt transmitter and gear by Clegg, all originally designed for a-m/cw operation. Also look for the Gonset Sidewinder, a 10-watt ssb transmitter, and its companion 100-watt linear amplifier. Transverters by Drake and Collins have also given good service, but they tend to command premium prices.

Converting Commercial Fm or Military Surplus Equipment. Old tube-type commercial fm gear designed for the land mobile service (130 to 160 MHz and 420 to 460 MHz) is widely available

Table 7.5

Sources of Information on Converting Commercial Fm Transmitting Equipment for Satellite Ground Station Use

FM & Repeaters, 2nd edition, ARRL, Newington, 1978. The chapter on surplus fm equipment contains a great deal of useful general information.

D. P. Clement, "Using the Motorola TU-110 Series Transmitters on 420 MHz," *QST*, Sept. 1971, pp. 39-41, 45. Contains detailed information on converting the TU-110 to a 20-watt-output cw transmitter. Treats such topics as obtaining a stable, chirp-free signal.

F. R. McLeod, Jr., "ATV with the Motorola T44 UHF Transmitter," Part I, *QST*, Dec. 1972, pp. 28-32; Part II, *QST*, Feb. 1973, pp. 36-43. These articles are very useful to anyone wishing to put the widely available T44 on 435 MHz.

R. Stevenson, "SSB on Mode B, Using Modified FM Equipment," *AMSAT Newsletter*, Dec. 1975, p. 10. Shows how an RCA CMU-15 designed for 460 MHz can be converted to a 435-MHz transverter. Conversion involves modifying the 5894 power amplifier to operate as a high-level mixer as shown in Fig. 7-16. This information was also made available in *QST*, Hints & Kinks, March 1976, p. 40.

W. R. Gabriel, "A 70-cm Linear Amplifier from a Motorola T44," *AMSAT Newsletter*, March 1977, pp. 4-5. Illustrates how the 2C39 output stage of a Motorola T44 can be used as a 435-MHz linear amplifier. Specific power levels aren't given but the design should provide 6-10 dB of gain at up to 40-watts output.

at modest prices. (Sources for equipment and schematics are listed in Table 7.4.) Amateurs have successfully converted fm transmitter strips into cw transmitters, transverters for ssb and cw, and linear amplifiers. A block diagram showing how a 460-MHz transmitter strip may be converted into a 435-MHz transverter is given in Fig. 7-16. Conversion usually involves the following basic steps: (1) constructing an appropriate ac power supply, (2) retuning resonant circuits and cavities to the correct frequencies, (3) adding provisions for keying (producing a stable, chirp-free signal at 435 MHz can be challenging) and (4) changing power amplifier biasing to AB_1 and AB_2 linear operation if necessary. Transmitter strips that are rated at 15- to 60-watts output in commercial service can safely provide 50% more power for amateur operation. At 435 MHz, a single crystal can usually be pulled enough to give a 75-kHz tuning range. Table 7.5 gives an annotated list of conversion articles.

Although few pieces of military surplus equipment lend themselves to our needs, it's important to keep an open mind to this approach. One of the most desirable pieces of gear for 1260 MHz is the AN/UPX-6 that can be turned into a linear amplifier that will provide about 50 watts. See the article by R. Stein referenced in Table 7.6.

Modifying an Amateur 146-MHz Fm Transceiver. This approach will appeal to beginners who are interested in gaining temporary access to Mode A. Most amateur 2-m fm transceivers can be modified easily for cw operation on the OSCAR uplink frequencies. Modification may be as simple as plugging in an appropriate crystal, removing the mike element and keying the push-to-talk switch! Of course, it's far better to change the push-to-talk circuitry so the unit can be left in transmit while only the driver and final amplifier are keyed. This is especially true with synthesized rigs that take a fraction of a second to lock onto the transmit frequency.

Construction of Transmitters, Transverters or Varactor Triplers. Though collecting components, building and debugging can involve a relatively large amount of time and total expense, many amateurs still prefer the rewards of building their equipment from scratch. Plans for transmitters, transverters, varactor triplers and other pieces of gear in the power ranges needed for amateur satellite work are referenced in Tables 7.3 and 7.6. We turn now to a band-by-band survey of transmitter approaches.

The Transmitting Station

146 MHz. The 2-m band, 146 MHz, is suitable as an uplink for working both high- and low-altitude spacecraft. For low-altitude satellites the suggested eirp is usually well under 100 watts. Most amateurs that operate on Modes A and J on AMSAT spacecraft, and Mode A on the RS satellites, have employed transmitters that run 10 to 80 watts of actual rf output power. Feed-line loss and antenna gain are manipulated (chosen) to provide the eirp appropriate for the satellites being used. Although there are no firm plans to employ 146 MHz as an uplink on amateur high-altitude spacecraft, doing so is technically feasible. If this is done eventually, an rf power output of 50 to 80 watts into a beam with 13-dBi gain should provide an adequate uplink signal.

There's an extensive choice of equipment for uplinking on 146 MHz. Transverters and transceivers are used widely on ssb. Old a-m/cw equipment, such as the AMECO TX-62 and commercial fm transmitter strips, provide an inexpensive option for cw operators. Another popular approach for cw has been to convert a low-power fm transmitter kit from Hamtronics and to use it to drive an amplifier. Table 7.6 lists references containing information on building transmitting equipment for 2 m.

70-cm Power Restrictions. For many years the FCC has restricted amateurs who use the 420-450 MHz band to 50-watts input power in certain parts of the United States. In response to growing satellite activity and a concurrent increase in requests for Special Temporary Authorizations (often referred to as STAs) to use higher power, the FCC has acted to "ease" the restrictions.

As a result of FCC actions, beginning in April 1981 (1) additional restricted areas were introduced and (2) the power limitations were divided into two categories, one for terrestrial operation and another for satellite communications. In August 1982 the restricted regions were increased in number and size.

The restricted areas now include:

1) Those portions of Texas and New Mexico bounded by latitudes 33° 24′ N., 31° 53′ N., and longitudes 105° 40′ W. and 106° 40′ W.

2) The entire state of Florida, including the Key West area and the areas enclosed within circles of 320-kilometer (200-mile) radius of Patrick Air Force Base (28° 21′ N., 80° 43′ W.) and Eglin Air Force Base (30° 30′ N., 86° 30′ W.).

3) The entire state of Arizona.

4) Those portions of California and Nevada south of latitude 37° 10′ N., and the area within a 320-kilometer (200-mile) radius of the U.S. Naval Missile Center (34° 09′ N., 119° 11′ W.).

5) In the state of Massachusetts within a 160-kilometer (100-mile) radius of Otis Air Force Base (41° 45′ N., 70° 32′ W.).

6) In the state of California within a 240-kilometer (150-mile) radius of Beale Air Force Base (39° 08′ N., 121° 26′ W.).

7) In the state of Alaska within a 160-kilometer (100-mile) radius of Elmendorf Air Force Base (64° 17′ N., 149° 10′ W.).

8) In the state of North Dakota within a 160-kilometer (100-mile) radius of Grand Forks Air Force Base (48° 43′ N., 97° 54′ W.).

The 50-watt input-power limit will continue to apply to stations that are engaged in terrestrial communication in the restricted area. Amateurs engaged in satellite communication on frequencies between 435 and 438 MHz in those areas, however, will be permitted to use 1000 watts eirp provided their antennas' elevations are adjusted so that the half-power points of the radiated pattern remain at least 10° above the horizon. See Section 97.61, 97.421 and 97.422 of the Amateur Rules and Regulations in *The FCC Rule Book*, published by ARRL, for more information.

435 MHz. The 70-cm band, 435 MHz, is suitable as an uplink band on both high- and low-altitude spacecraft. Mode B on AMSAT-OSCAR 7 had a recommended ground-station eirp of 80 watts. Most stations working with this transponder used transmitters that ran from 5- to 40-watts output power — 10 watts being most common — into small beams. For Mode B on AMSAT-OSCAR 10 and Phase III-C, the recommended eirp is 500 watts, or 50 watts of rf energy to a beam having 10 dBi gain.

As at 146 MHz, commercial transverters and transceivers are popular at 435 MHz, but their higher costs have led amateurs

Table 7.6

Construction Information Sources: Transmitting Equipment

General

The Radio Amateur's Handbook, ARRL, Newington. See chapters on vhf and uhf transmitting, and fm and repeaters in recent editions.

The Radio Amateur's VHF Manual, 3rd edition, ARRL, Newington, 1972 (out of print).

FM and Repeaters for the Radio Amateur, ARRL, Newington, 1978. See Chapter 4 on fm transmitters and Chapter 12 on surplus fm equipment. Note that many of the amplifiers described are not suitable for ssb.

VHF-UHF Manual, 3rd edition, by D. S. Evans and G. R. Jessop, RSGB, London, 1976. An excellent source of information, especially for 1296 MHz and higher frequencies.

2 m (146 MHz)

L. Leighton, "Two-meter Transverter Using Power FETs," *Ham Radio,* Sept. 1976, pp. 10-15. Contains modular transmitting converter. Mixer unit: input — 1 mW at 28 MHz; output — 100 mW at 145 MHz. Linear amp. I: input — 100 mW; output — 2 W
Linear amp. II: input — 2 W; output — 10 W

R. S. Stein, "Solid-State Transmitting Converter for 144-MHz ssb," *Ham Radio,* Feb. 1974, pp. 6-18.
Transverter: input — fraction of watt at 28 MHz; output — 6 W
Linear amp.: input — 6 W, output — 30 W
See correction, *Ham Radio,* Dec. 1974, p. 62

70 cm (435 MHz)

J. Buscemi, "A 60-watt Solid-State UHF Linear Amplifier," *QST,* July 1977, pp. 42-45. Describes a two-stage linear amplifier that puts out 60 W for 1-3 W input.

R. R. Eide, "A Solid-State Transverter for 70 cm," *QST,* Sept. 1978, pp. 28-30. Output is about 1 W when driven with a fraction of a watt at 28 MHz.

R. T. Knadle, "A Strip-Line Kilowatt Amplifier for 432 MHz," Part I, *QST,* April 1972, pp. 49-55; Part II, *QST,* May 1972, pp. 59-62, 79. Also see: J. Reisert, "More on the 432-MHz kW Strip-Line Amplifier," *QST,* July 1975, p. 47. This is the standard high-power, 435-MHz amplifier. It will loaf along efficiently at lower-power levels for satellite work. A complete unit and parts kits are available from ARCOS, Box 546, E. Greenbush, NY 12061.

F. J. Merry, "Phase III with a Tetrode UHF Amplifier," *QST,* Aug. 1982, pp. 41-44.

T. McMullen and C. Greene, "A Tramplifier for 432 MHz," *QST,* Jan. 1976, pp. 11-15. Describes a varactor tripler (145/435 MHz) and tube-type amplifier. 7 W in at 435 MHz yields 100-W output; 10 W in at 145 MHz yields 80 W output. This unit was also described in several editions of *The Radio Amateur's Handbook* in the late '70s and early '80s.

C. F. Moretti, "A Heterodyne Exciter for 432 MHz," *QST,* Nov. 1973, pp. 47-50, 95. Describes a tube-type transverter that puts out 10 W when driven by 0.5 W at 28 MHz. Be sure to note the correction: *QST,* March 1974, p. 83.

R. K. Olsen, "100-watt Solid-State Power Amplifier for 432 MHz," *Ham Radio,* Sept. 1975, pp. 36-43.
Output power: 100 W PEP
Gain: 10 dB
Active devices: 2 MRF306s using a 28-V power supply

H. P. Shuch, "UHF Local-Oscillator Chain," *Ham Radio,* July 1979, pp. 27-33. Simple-local oscillator chain that provides excellent spectral purity, stability and calibration tolerance. Provides 5 mW between 380 and 540 MHz. Suitable for receive converters and transmit mixers at 435 and 1269 MHz.

T. Souza, "432-MHz Power Amplifier," *Ham Radio,* June 1977, pp. 10-14. Describes a high-power, grounded-grid linear amplifier that uses strip-line techniques.
Input power: 30 W
Output power: 600 W
Gain: 13 dB
The power level is higher than needed for satellite operation but the design lends itself to scaling down to 2C39-series tubes.

F. Telewski, "A Practical Approach to 432-MHz ssb," *Ham Radio,* June 1971, pp. 6-21. Contains an extensive review of tube-type mixers and linear amplifiers at all power levels for 432 MHz. Since vacuum tube techniques at this frequency have not changed significantly over the last decade, the information is still valuable for anyone interested in 6939 mixers and the 2C39 family of amplifiers.

L. Wilson, "Solid-State Linear Power Amplifier for 432 MHz," *Ham Radio,* Aug. 1975, pp. 30-35.
Output power: 10 W PEP
Gain: 10 dB
Active device: CM10-12 using 12-V power supply

23 cm (1.26 GHz)

Though most of the units referenced below were built for 1296 MHz, they'll work equally well at 1269 MHz by changing oscillator frequencies.

B. Atkins, "1296 MHz Power and SWR Indicator," *QST,* Nov. 1980, p. 69.

D. Bingham, "A Modular Transceiver for 1296 MHz," *QST,* Dec. 1975, pp. 29-35. Uses a solid-state modular approach. Mixer produces about 2 mW at 1296 MHz. Two linear amplifiers bring this up to about 0.5 W.

J. M. Cadwallader, "1296-MHz Transverter," *Ham Radio,* July 1977, pp. 10-17. This transverter uses a 2C39-series tube as a high level mixer to produce 5-15 W PEP at 1296 MHz. Requires 5 W of oscillator injection at 1116 MHz for 1269 MHz operation and 5 W of ssb signal at 144 MHz.

R. E. Fisher, C. W. Schaible, G. W. Schober and R. H. Turrin, "A Power Amplifier for 1296 MHz," *Ham Radio,* March 1970, pp. 43-50. Home-built cavity-type amplifier using two 3CX100s. Produces 10-dB gain and 100-W output at 50% efficiency. (See Laakmann article.)

G. Hatherell, "Double-stub Tuner for 1296 MHz," *Ham Radio,* Dec. 1978, pp. 72-75. If you don't know what a Double-stub tuner is, just think of it as a Transmatch. Requires machine work to duplicate.

J. Hinshaw, "Solid-State Power for 1296 MHz," *Ham Radio,* Feb. 1981, pp. 30-38. Describes two solid-state amplifiers.
AMP-I: input — 2 mW, output — 100 mW, gain — 17 dB, Class A
AMP-II: input — 100 mW, output — 2 W, gain — 13 dB, Class C
Data for transistor used in AMP-II suggests that it may be operated Class B (linear) with appropriate bias, but this was not tried.

P. Laakmann, "Cavity Amplifier for 1296 Mc.," *QST,* Jan. 1968, pp. 17-19, 146. Home-built cavity amplifier using two 2C39As. Produces 6-10 dB gain and up to 100-W output. Anyone working with surplus cavity amplifiers may wish to review this article and the Fisher, et al., article for background information.

H. P. Shuch, "1296-MHz ssb Transceiver," *Ham Radio,* Sept. 1974, pp. 8-23. Describes a solid-state, low-power, modular ssb transceiver. The balanced mixer that uses inexpensive hot-carrier diodes producing 3-mW output and the oscillator-multiplier chain for 1267 MHz are of interest.

H. P. Shuch, "Rat-race Balanced Mixer for 1296 MHz," *Ham Radio,* July 1977, pp. 33-39. Describes a low-level mixer that can be used in transmit and receive applications. Produces about 1 mW at 1296 MHz when driven with 3 mW of ssb at 144 MHz and 5-10 mW near 1152 MHz.

H. P. Shuch, "Improved Grounding for the 1296-MHz Microstrip Filter," *Ham Radio,* Aug. 1978, pp. 60-63. Includes a design for a 70-cm to 23-cm tripler that provides a clean signal suitable for a local-oscillator chain.

R. S. Stein, "Converting Surplus AN/UPX-6 Cavities," *Ham Radio,* March 1981, pp. 12-17. Describes a 3-stage amplifier for 1296 MHz that produces 40-W output for 100-mW input (26-dB gain). At drive levels under 1 mW gain is 28-34 dB. Uses 2C39A tubes operating Class A. Listeners report a clean ssb signal at all power levels.

P. Wade, "A High-performance Balanced Mixer for 1296 MHz," *QST,* Sept. 1973, pp. 15-17. Describes a receiver mixer based on a quadrature-hybrid coupler using hot-carrier diodes. This mixer can be used as a low-level transmit mixer.

P. Wade, "Clean local-oscillator chain for 1296 MHz," *Ham Radio,* Oct. 1978, pp. 42-47. Produces about 0.5-mW output.

to a more thorough investigation of alternatives. The conversion of 420- to 460-MHz commercial land mobile equipment has received considerable attention. In the list of references in Table 7.5 you'll find plans for converting an fm transmitter strip into (1) a cw transmitter producing about 20 watts, (2) a high level transverter providing about 10-watts output (see Fig. 7-16) and (3) a linear amplifier producing about 20 watts. The low-power fm transmitter kits from Hamtronics have also served as popular building blocks, either for a cw transmitter or for the injection chain in a high-level mixer. Several amateurs interested in cw have had excellent results with varactor triplers from 145 MHz, either the commercial units by Microwave Modules that are available in several models putting out up to 40 watts, or the homemade designs listed in Table 7.3.

For the amateur who wants the ultimate in transmitting capabilities for this band, the high power RIW amplifier mentioned in Table 7.6 is available assembled or in kit form from ARCOS (Amateur Radio Component Service, 35 Highland Dr., East Greenbush, NY 12061). This amplifier will loaf along efficiently at low power (100 to 200 watts out), making it possible to access high-altitude satellites with a small antenna.

1.26 GHz. Because of the Doppler effects associated with low orbits, 1.26 GHz is suited mainly to high-altitude spacecraft. The recommended uplinking eirp on AMSAT-OSCAR 10 is in the range of 2000 to 5000 watts. Most amateurs will probably use transmitters that produce from 5 to 50 watts and high-gain antennas.

This is the only uplink band for which, in early 1983, one cannot go out, checkbook in hand, and buy a complete ssb transmitting station. It's also true that a Mode-L transponder hasn't yet operated extensively (September 1983). By no coincidence both situations are likely to change at about the same time: when the Mode L transponder aboard AMSAT-OSCAR 10 is given more "air time." Several pieces of commercial equipment are available for this band, including transverters by Microwave Modules and SOTA that put out several watts, and a varactor tripler by Microwave Modules that generates 18 watts at 1.26 GHz when driven with 30 watts at 420 MHz.

Articles describing easily reproduced transmitting equipment for 1.26 GHz have begun to appear in the last few years. In the early 1970s most designs focused on home-built cavity amplifiers that used the 2C39 family of tubes. Although these designs worked, reproduction called for machining facilities not available to most amateurs. By the late 1970s amplifier and tripler designs using the 2C39 in a strip-line configuration began to appear, greatly simplifying the construction process. Solid-state devices capable of handling appreciable amounts of power at 1.26 GHz are still very expensive, but this will inevitably change.

As of early 1983, solid-state techniques can reasonably take the average amateur to about the 5-watt level (though, with difficulty, power levels of 160 watts or more are technologically possible). Above this point amplifiers employing the 2C39 family of tubes are suitable. When building from scratch, strip-line designs are easiest to duplicate. Using surplus cavities such as the AN/UPX-6, however, may be quicker and cheaper. Anyone traveling the AN/UPX-6 route may want to read the early articles on cavity amplifier construction just to acquire some technical background in this area.

On possible way for the interested user to move up to Mode L from Mode B involves:

Step 1. Build a varactor tripler producing about 5 watts at 1.26 GHz when driven by 10 watts at 420 MHz. This gives you a chance to try low-power cw operation on Mode L.

Step 2. If you wish to adopt ssb, the next step is to build a high-level mixer like the one described by Cadwallader (1977) (see Table 7.6 for full references). Use the varactor tripler, retuned to 1116 MHz, as an oscillator injection chain. The mixer will also need a few watts of ssb rf energy at 2 m. This setup will provide you with about 5 watts of ssb on 1.26 GHz.

Step. 3. Modify an AN/UPX-6 cavity amplifier (Stein, 1981) or build a strip-line amplifier using a 2C39 (RSGB *VHF-UHF Manual).* This will give you about 50 watts of rf energy at 1.26 GHz.

Another possible approach to getting on 1.26 GHz is to use modules based around a low-level mixer. Designs by Shuch (1977) and Bingham (1975) can be reproduced without special machining facilities. The output of these units is in the 1- to 2-mW range. One or two stages of solid-state linear amplification (Bingham, 1975; Hinshaw, 1981) will bring you up to the 200- to 500-mW level where the AN/UPX-6 can be used effectively. Alternatively, amplification to the 2-watt level (Hinshaw, 1981) can be followed by a single strip-line 2C39A amplifier (RSGB *VHF-UHF Manual*) to provide 20+ watts of rf energy.

Several newer, more appropriate circuits will likely have appeared in the periodical literature by the time you read this. The main point here has been to show that you don't have to be a microwave engineer to be capable of transmitting on the 1.26-GHz satellite uplinks. A cw or ssb transmitter can be built using modules constructed from articles listed in Table 7.6 or purchased commercially. The downward trend in the prices of microwave solid-state components, the increasing availability of commercial products and the growing numbers of amateurs developing equipment for 1.26 GHz are sure to simplify the task of generating rf energy at this frequency in the future.

Part III

Chapter 8
Satellite Orbits

Chapter 8

Satellite Orbits

Using the step-by-step techniques of Chapter 5, radio amateurs can track OSCAR spacecraft without needing to know the basic physics of satellite motion or how a satellite moves in space. This chapter is for those amateurs interested in "why" as well as "how." Here we'll examine satellite motion from a more detailed physical/mathematical point of view.

Several of the topics we look at in this chapter are usually found in texts designed for graduate level scientists and engineers. These texts, rigorous and generalized, are often incomprehensible to readers who don't have an advanced mathematical background. Yet most of the ideas and results can be expressed in terms that someone with a solid background in algebra, plane trigonometry and analytic geometry can understand. We'll keep the mathematics in this chapter as simple as possible, but — face it — mathematics is a key element in understanding satellite motion. Study the solved *Sample Problems* scattered throughout this chapter to see how key formulas are applied. As they also form the basis for later work, the problems may be the most valuable part of the chapter.

At several points we had to raise the mathematical level slightly higher than desired to avoid obscuring potentially useful information. Much of the material in this chapter is not serial in nature, however, so you can skip big chunks and still follow later sections. By now you should realize that this chapter is not for the fainthearted. If you elect to plow through, reviewing the tracking material of Chapter 5 before beginning will make the path a little easier. Also note that Table 8.1 summarizes repeatedly used symbols. Good luck!

The objectives of this chapter are:

1) to introduce the satellite-orbit problem from a scientific point of view);

2) to provide the reader with an overview of satellite motion (including both an understanding of important parameters and an ability to visualize the motion in space and with respect to earth);

3) to summarize the important equations needed to compute orbital parameters so that these equations will be easily accessible when needed.

Background

The satellite-orbit problem (determining the position of a satellite as a function of time and finding its path in space) is essentially the same whether we are studying the motion of the planets around the sun, the moon around the earth, or artificial satellites revolving around either. The similarity arises from the nature of the forces affecting an orbiting body that doesn't have a propulsion system. In the early 17th century Kepler discovered some remarkable properties of planetary motion; they have come to be called *Kepler's Laws.*

I) Each planet moves around the sun in an ellipse, with the sun at one focus (motion lies in a plane);

II) The line from the sun to a planet (radius vector, r) sweeps out across equal areas in equal intervals of time;

III) The ratio of the square of the period (T) to the cube of the semimajor axis (a) is the same for all planets in our solar system. (T^2/a^3) is constant.

Table 8.1

Symbols Used in This Chapter

Symbol	Meaning
a	primary: semimajor axis of ellipse (secondary: side of spherical triangle)
A	angle in spherical triangle
b	primary: semiminor axis of ellipse (secondary: side of spherical triangle)
B	angle in spherical triangle
c	primary: distance between center of ellipse and focal point (secondary: side of spherical triangle)
C	angle in spherical triangle
e	eccentricity of ellipse
E	eccentric anomaly (angle)
G	gravitational constant
h	satellite height above surface of earth
h_a	satellite height above surface of earth at apogee
h_p	satellite height above surface of earth at perigee
i	orbital inclination
I	true longitude increment
$\bar{I}$	estimated longitude increment
m	primary: mass of satellite (secondary: abbreviation for meter)
M	mass of earth
m/s	meters per second
r	satellite-geocenter distance
r_a	satellite-geocenter distance at apogee
r_p	satellite-geocenter distance at perigee
R	mean radius of earth = 6371 km
R_{eq}	mean equatorial radius of earth = 6378 km
s	seconds
t	elapsed time since last ascending node (circular orbits) or last perigee (elliptical orbits)
T	period of satellite
v	magnitude of satellite velocity with respect to static earth
w_o	argument of perigee
$\dot{w}_o$	rate of change of argument of perigee
θ	polar angle in orbital plane
ϕ	latitude
λ	longitude
$\dot{\Omega}$	rate of orbital-plane precession about earth's N-S axis

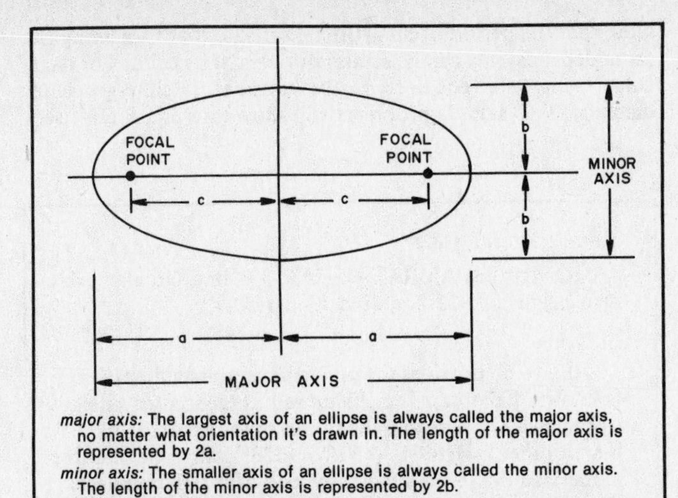

Fig. 8-1 — Geometry of the ellipse.

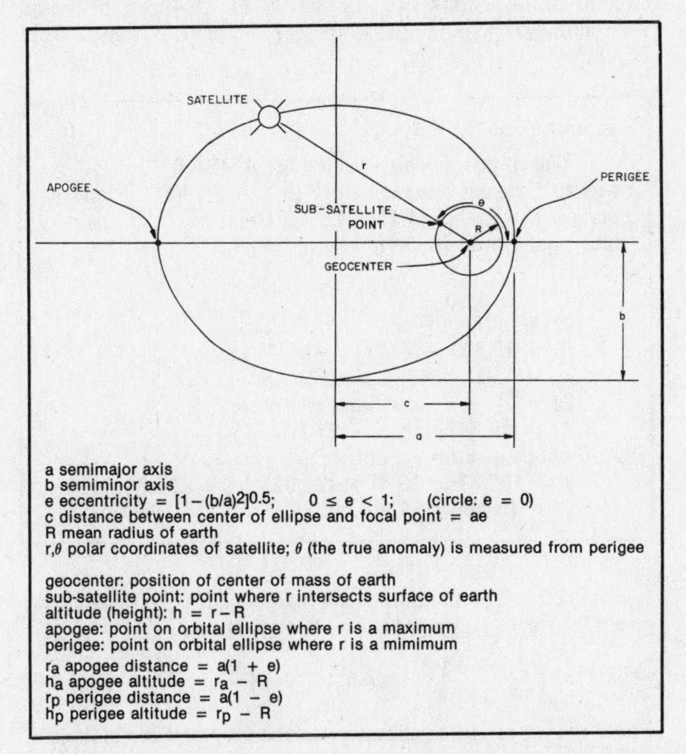

Fig. 8-2 — Geometry of the orbital ellipse for an earth satellite.

These three properties summarize observations; they say nothing about the forces governing planetary motion. It remained for Newton to deduce the characteristics of the force that would yield Kepler's Laws. The force is the same one that keeps us glued to the surface of the earth — good old gravity.

Newton showed that Kepler's Second Law would result if the planets were being acted on by an attractive force always directed at a fixed central point: the sun (central force). To satisfy the First Law, this force would have to vary as the inverse square of the distance between planet and sun ($1/r^2$). Finally, if Kepler's Third Law were to hold, the force would have to be proportional to the mass of the planet. Actually, Newton went a lot further. He assumed that not only does the sun attract the planets in this manner, but that every mass (m_1) attracts every other mass (m_2) with a force directed along the line joining the two masses and having a magnitude (F) given by

$$F = \frac{Gm_1m_2}{r^2} \quad \text{(Universal Law of Gravitation)} \qquad \text{(Eq. 8.1)}$$

where G is the Universal Gravitational Constant.

The Geometry of the Ellipse

As Kepler noted in his First Law, ellipses take center stage in satellite motion. A brief look at the geometry of the ellipse is, therefore, in order (see Fig. 8-1). The lengths a, b and c shown in Fig. 8-1 are not independent. They're related by the formula

$$c^2 = a^2 - b^2 \quad \text{or} \quad c = \sqrt{(a^2 - b^2)} \qquad \text{(Eq. 8.2)}$$

Using Eq. 8.2, any one of the parameters a, b or c can be computed if the other two are known. In essence, it takes two parameters to completely describe the shape of an ellipse. One could, for example, give the semimajor and semiminor axes (a and b), the semimajor axis and the distance from the origin to one focus (a and c), or the semiminor axis and the distance from the origin to one focus (b and c).

There's another convenient parameter, called eccentricity (e), for describing an ellipse. Eccentricity may be thought of as a number describing how closely an ellipse resembles a circle. When the eccentricity is 0 we've got a circle. The larger the eccentricity, the more elongated the ellipse becomes. To be more precise, eccentricity is given by

$$e^2 = 1 - (b/a)^2 \quad \text{or} \quad e = \sqrt{1 - (b/a)^2} \qquad 0 \leq e < 1 \quad \text{(Eq. 8.3)}$$

Because of its mathematical definition, e must be a dimensionless number between 0 and +1. Using Eqs. 8.2 and 8.3 we can derive another useful relationship:

$$c = ae \qquad \text{(Eq. 8.4)}$$

As stated earlier, it always takes two parameters to describe the shape of an ellipse. Any two of the four parameters, a, b, c or e, will suffice.

Fig. 8-2 shows the elliptical path of a typical earth satellite. Since the earth is located at a focal point of the ellipse (Kepler's Law 1), it's convenient to introduce two additional parameters that relate to our earth-bound vantage point: the distances between the center of the earth and the "high" and "low" points on the orbit. Fig. 8-2 summarizes several useful relations and definitions. Note especially

$$\text{apogee distance: } r_a = a(1 + e) \qquad \text{(Eq. 8.5a)}$$

$$\text{perigee distance: } r_p = a(1 - e) \qquad \text{(Eq. 8.5b)}$$

We now have six parameters, a, b, c, e, r_a and r_p, any two of which can be used to describe an ellipse. With the information we've learned so far, many practical satellite problems can be solved. (See Sample Problem 8.1.)

When the major and minor axes of an ellipse are equal, the ellipse becomes a circle. From Eq. 8.2 we see that setting a = b gives c = 0. This means that in a circle both focal points coalesce at the center. Setting a = b in Eq. 8.3 yields e = 0, as we stated earlier.

Since the circular orbit is just a special case of the elliptical orbit, the most general approach to the satellite-orbit problem would be to begin by studying elliptical orbits. Circular orbits, however, are often simpler to work with, so we'll look at them separately whenever it makes our work easier.

Our approach to the satellite-orbit problem involves first determining the path of the satellite in space and then looking at the path the sub-satellite point traces on the surface of the earth. Each of these steps is, in turn, broken down into several smaller steps.

Satellite Path In Space

The motion of an object results from the forces acting on

Sample Problem 8.1

AMSAT-OSCAR 10 has an apogee distance (r_a) of 6.57R and a perigee distance (r_p) of 1.62R. Specify the orbit in terms of the semimajor axis (a) and eccentricity (e). (Note: In studying earth satellites, distances are sometimes expressed in terms of R, the mean radius of the earth).

Solution (Given r_a and r_b, solve for a and e)
Subtracting Eq. 8.5b from Eq. 8.5a we obtain:

$$r_a - r_p = 2ae \text{ or } e = \frac{r_a - r_p}{2a}$$

Adding Eq. 8.5a to Eq. 8.5b gives

$$r_a + r_p = 2a; \; a = 4.10 \text{ R}$$

$$\text{so, } e = \frac{r_a - r_p}{r_a + r_p} = \frac{6.57R - 1.62R}{6.57R + 1.62R} = 0.604$$

it. To determine the path of a satellite in space (1) we'll make a number of simplifying assumptions about the forces on the satellite and other aspects of the problem, taking care to keep the most important determinants of the motion intact; (2) we'll then solve the simplified model; and then (3) we'll add corrections to our solution, accounting for the initial simplifications.

Simplifying Assumptions

We begin by listing the assumptions usually employed to simplify the problem of determining satellite motion in the orbital plane.

1) The earth is considered stationary and a coordinate system is chosen with its origin at the earth's center of mass (geocenter).

2) The earth and satellite are assumed to be spherically symmetric. This enables us to represent each one by a point mass concentrated at its center (M for the earth, m for the satellite).

3) The satellite is subject to only one force, an attractive one directed at the geocenter; the magnitude of the force varies as the inverse of the square of the distance separating satellite and geocenter ($1/r^2$).

The model just outlined is known as the two-body problem, a detailed solution for which is given in most introductory physics texts.[1,2] Some of the important results follow.

Solution to the Two-Body Problem

Initial Conditions. Certain initial conditions (the velocity and position of the satellite at burnout, the instant the propulsion system is turned off) produce elliptical orbits ($0 \le e < 1$). Other initial conditions produce hyperbolic ($e > 1$) or parabolic ($e = 1$) orbits, which we will not discuss.

The Circle. For a certain subset of the set of initial conditions resulting in elliptical orbits, the ellipse degenerates (simplifies) into a circle ($e = 0$).

Satellite Plane. The orbit of a satellite lies in a plane that always contains the geocenter. The orientation of this plane remains fixed in space (with respect to the fixed stars) after being determined by the initial conditions.

Period and Semimajor Axis. The period (T) of a satellite and the semimajor axis (a) of its orbit are related by the equation

$$T^2 = \frac{4\pi^2}{GM} a^3 \qquad \text{(Eq. 8.6a)}$$

where M is the mass of the earth and G is the Universal Gravitational Constant. For computations involving a satellite in earth orbit the following equations may be used (T in minutes, a in kilometers).

$$T = 165.87 \times 10^{-6} \times a^{3/2} \qquad \text{(Eq. 8.6b)}$$

$$a = 331.25 \times T^{2/3} \qquad \text{(Eq. 8.6c)}$$

Note that the period of an artificial satellite that is orbiting the earth depends only on the semimajor axis of its orbit. For a circular orbit, a is equal to r, the constant satellite-geocenter distance. Two sample problems will show how Eq. 8.6 is used.

Sample Problem 8.2

Given that AMSAT-OSCAR 8 is in a circular orbit at a height of 912 km, find its period.

Solution

Eq. 8.6b provides the period when the semimajor axis is known. To obtain the orbital radius (geocenter-satellite distance) we have to add the radius of the earth to OSCAR 8's altitude: $r = 6371 + 912 = 7283$ km. Plugging this value into Eq. 8.6b yields a period of 103.1 minutes:

$$T = 165.87 \times 10^{-6} \times (7283)^{3/2} = 103.1 \text{ minutes.}$$

A graph of period vs. height for low altitude spacecraft in circular orbits is shown in Fig. 8-3. In Fig. 8-4 we plot period vs. semimajor axis. Both of these plots were obtained from Eq. 8.6b.

Sample Problem 8.3

The elliptical orbit planned for AMSAT Phase III-A was to have an apogee height (h_a) of 35,800 km and a perigee height (h_p) of 1500 km. If it had reached this orbit, what would its period have been?

Solution

R = radius of earth = 6371 km
$r_a = 35{,}800 + 6371 = 42{,}171$ km
$r_p = 1500 + 6371 = 7871$ km
$2a = r_a + r_p$ (see Sample Problem 8.1)
$2a = 50{,}042$ km; $a = 25{,}021$ km

Applying Eq. 8.6b we obtain

$$T = 165.87 \times 10^{-6} \times (25{,}021)^{3/2} = 656.5 \text{ minutes}$$
$$= 10 \text{ hours } 56.5 \text{ minutes.}$$

Velocity. The magnitude of a satellite's total velocity (v) generally varies along the orbit. It's given by

$$v^2 = GM\left(\frac{2}{r} - \frac{1}{a}\right) = 3.986 \times 10^{14}\left(\frac{2}{r} - \frac{1}{a}\right) \qquad \text{(Eq. 8.7)}$$

where r is the satellite-geocenter distance, r and a are in meters, and v is in meters/sec (see Fig. 8-2). Note that for a given orbit, G, M and a are constants so that v depends only on r. Eq. 8.7 can therefore be used to compute the velocity at any point along the orbit if r is known. The range of velocities is bounded: The maximum velocity occurs at perigee and the minimum velocity occurs at apogee. The direction of motion is always tangent to the orbital ellipse. For a circular orbit $r = a$ and Eq. 8.7 simplifies to (r in meters, v in m/s)

$$v^2 = \frac{GM}{r} = (3.986 \times 10^{14})\left(\frac{1}{r}\right)$$

(circular earth orbit only)

Note that for circular orbits, v is constant. Sample Problems 8.4 and 8.5 illustrate the use of Eq. 8.7.

Position. Fig. 8-2 shows how the satellite position is specified by the polar coordinates r and θ. Often, it's necessary to know r and θ as a function of the elapsed time, t, since the satellite passed perigee (or some other reference point when a circle is being considered).

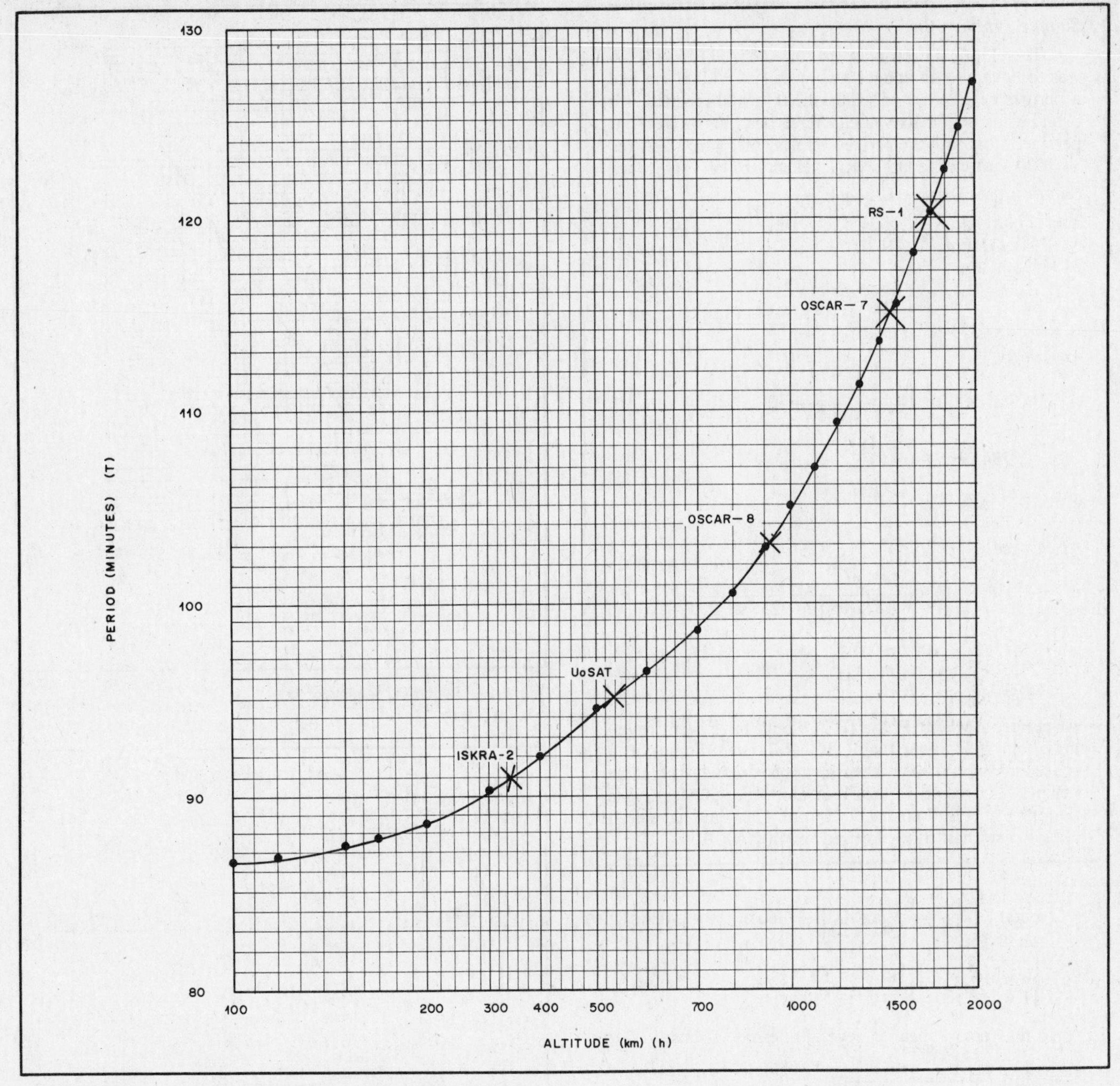

Fig. 8-3 — Period vs. altitude for satellites in low-altitude circular orbits.

Sample Problem 8.4

AMSAT-OSCAR 7 is in a 1460-km-high circular orbit. What is its velocity?

Solution

In a circular orbit we can use the simplified form of Eq. 8.7.

$$v^2 = \frac{GM}{r} = (3.986 \times 10^{14})\ (1/r)$$

$$r = 1460 \text{ km} + 6371 \text{ km} = 7.831 \times 10^6 \text{ m}$$

$$v^2 = \frac{3.986 \times 10^{14}}{7.831 \times 10^6} = 0.5090 \times 10^8 \ (\text{m/s})^2$$

$$v = 0.7134 \times 10^4 = 7134 \text{ m/s}$$

For a satellite in a circular orbit moving at constant speed:

$$\theta \text{ [in degrees]} = \frac{t}{T}(360°); \text{ or } \theta \text{ [in radians]} = 2\pi\frac{t}{T} \qquad \text{(Eq. 8.8)}$$

and the radius is fixed.

The elliptical-orbit problem is considerably more involved. We know (Eq. 8.7) that the satellite moves much more rapidly near perigee. The relation between t and θ, can be derived from Kepler's Law II. Since the derivation is involved we'll skip it and suggest outside references for details.[3,4]

In an elliptical orbit, time from perigee, t, is given by

$$t = \frac{T}{2\pi}[E - e \sin E] \qquad \text{(Eq. 8.9)}$$

where the angle E, known as the eccentric anomaly, is defined by the associated equation

Sample Problem 8.5

The elliptical orbit planned for AMSAT Phase III-A was to have had an apogee height (h_a) of 35,800 km and a perigee height (h_p) of 1500 km. If it had reached this orbit, what would its velocity have been at apogee? At perigee? Compare the perigee velocity of Phase III-A (h_p = 1500 km) to the OSCAR 7 velocity (h = 1460 km).

Solution (See Sample Problem 8.3)

$a = 25{,}021$ km
$r_a = 42{,}171$ km
$r_p = 7871$ km

Use Eq. 8.7: $v^2 = 3.986 \times 10^{14} \left(\frac{2}{r} - \frac{1}{a}\right)$

At Apogee

$$v^2 = 3.986 \times 10^{14} \left(\frac{2}{42{,}171{,}000} - \frac{1}{25{,}021{,}000}\right)$$

$$= 2.9734 \times 10^6 \text{ (m/s)}^2$$

$v = 1724$ m/s

At Perigee

$$v^2 = 3.986 \times 10^{14} \left(\frac{2}{7{,}871{,}000} - \frac{1}{25{,}021{,}000}\right)$$

$$= 85.353 \times 10^6 \text{ (m/s)}^2$$

$v = 9239$ m/s

At perigee AMSAT Phase III-A would have been traveling about 30% faster than OSCAR 7, even though both spacecraft would have been at approximately the same height. As a result, Phase III-A Doppler shift at perigee (Doppler shift is discussed in Chapter 11) would have been 30% worse than that observed on OSCAR 7.

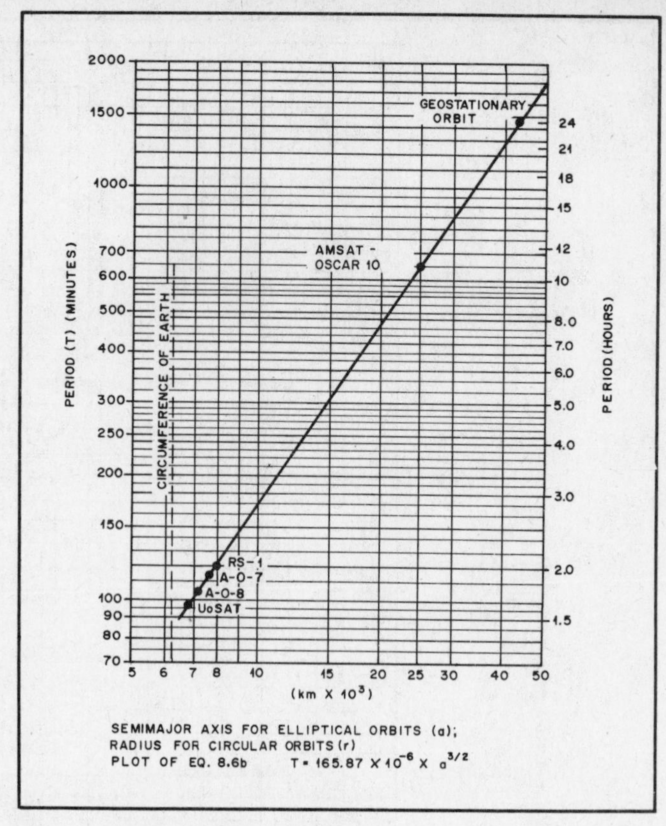

Fig. 8-4 — Period vs. semimajor axis.

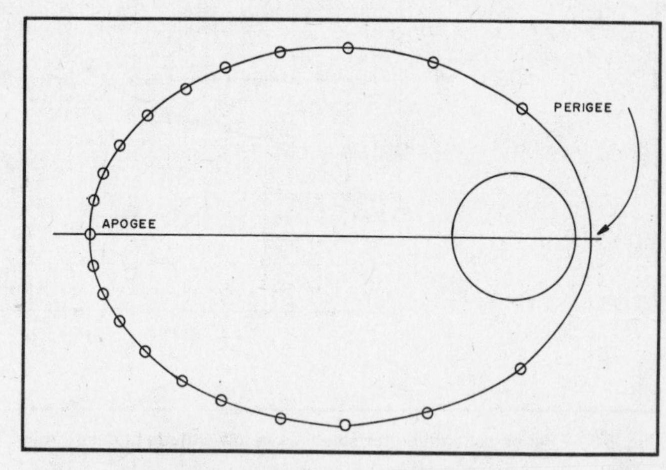

Fig. 8-5 — This orbital plane diagram shows the position of a satellite in a 12-hour elliptical orbit at half-hour intervals. Note that near apogee the satellite moves relatively slow.

$$E = 2 \arctan \left[\left(\frac{1-e}{1+e}\right)^{0.5} \tan \frac{\theta}{2}\right] + 360°n \qquad \text{(Eq. 8.10)}$$

$$n = \begin{cases} 0 \text{ when } -180° \le \theta \le 180° \\ 1 \text{ when } 180° < \theta \le 540° \end{cases}$$

Eq. 8.10 may also appear in several alternate forms:

$$E = \arcsin \left[\frac{(1-e^2)^{0.5} \sin \theta}{1 + e \cos \theta}\right] \text{ or}$$

$$E = \arccos \left[\frac{e + \cos \theta}{1 + e \cos \theta}\right]$$

Note that here, "anomaly" just means angle. Eqs. 8.9 and 8.10, taken together, are commonly referred to as Kepler's Equation. Fig. 8-5 shows the position of a satellite in an elliptical orbit (similar to those of Phase III) as a function of time. It should give Kepler's Equation some physical meaning.

Almost everyone in solving Eqs. 8.9 and 8.10 for the first time makes two mistakes. Eq. 8.9 contains the first pitfall. Since the expression e sin E is a unitless number, the E term standing by itself inside the brackets *must* be given in *radians*. The second pitfall is encountered when working with the various forms of Eq. 8.10. Although all inverse trigonometric functions are multivalued, hand calculators and computers are programmed to give only principle values. For example, if $\sin \theta = 0.99$ then θ may equal 82° or 98° (or either of these two values +/− any integer multiple of 360°) but a calculator only lights up 82°. If the physical situation requires a value outside the principle range, appropriate adjustments must be made. Eq. 8.10 already includes the adjustments needed so that it can be used for values of θ in the range −180° to +540°. If the alternate forms of Eq. 8-10 are used it's up to you to select the appropriate range. A few hints may help: (1) E/2 and $\theta/2$ must always be in the same quadrant; (2) as θ increases, E must increase; (3) adjustments to the alternate forms of Eq. 8.10 occur when the term in brackets passes through ± 1.

We now have a procedure for finding t when θ is known: Plug θ into Eq. 8.10 to compute E, then plug E into Eq. 8.9 to obtain t. The reverse procedure, finding θ when t is known is more involved. The key step is solving Eq. 8.9 for E when t is known. Unfortunately, there isn't any way to neatly express E in terms of t. We can, however, find the value of E corresponding to any value of t by drawing a graph of t vs. θ then reading it "backwards," or by using an iterative approach. An iterative technique is just a systematic way of guessing an answer for E,

Table 8.2
Subroutine KEPLER

Purpose: to give values of θ (angle from perigee) and r (satellite-geocenter distance) as a function of time from perigee.

Language: BASIC

Input:
- P1 3.1416
- t time from perigee in minutes $0<t\leq T$
- T period in minutes
- e eccentricity
- a semimajor axis in km

Output:
- θ true anomaly in radians
- r in km

Other variables:
- E eccentric anomaly
- Z initial guess of E
- Z3 best currently available correction to estimated E

```
6000 REM***KEPLER***
6010 LET Z = 2 * P1 * t/T
6020 LET E = Z
6030 LET Z3 = (E - e*SIN(E) - Z) / (1 - e*COS(E))
6040 LET E = E - Z3
6050 IF ABS(Z3) > .0001 THEN 6030
6060 LET θ = PI
6070 IF E = P1 THEN 6110
6080 LET θ = 2*ATN(SQR((1 - e) / (1 + e))*TAN(E/2))
6090 IF E < P1 THEN 6110
6100 LET θ = 2*P1 + θ
6110 LET r = a*(1 - e*e) / (1 + e*COS(θ))
6120 RETURN
```

Notes

(1) To enable the reader to follow the algebraic steps, the letters a, e, r, t and θ are used in the listing. To run properly on a computer, however, they must be changed to upper case conforming to the version of BASIC in use. When doing so, be sure to differentiate between e and E.

(2) All angles are in radians as required by most versions of BASIC.

Sample Problem 8.6

Consider the orbit planned for AMSAT Phase III-A (see Sample Problem 8.3).

r_a = 42,171 km
r_p = 7871 km
a = 25,021 km
T = 656.5 minutes

(a) Compute the satellite altitude (h) when θ = 108°. (b) How long after perigee does this occur?

Solution

(a) *Step 1:* Solve for the eccentricity (see Sample Problem 8.1).

$$e = \frac{r_a - r_p}{2a} = 0.685$$

Step 2: Solve for r using Eq. 8.11.

$$r = \frac{a(1 - e^2)}{1 - e\cos\theta} = 10{,}960 \text{ km}$$

$$h = r - R = 10{,}960 \text{ km} - 6371 \text{ km} = 4589 \text{ km}$$

(b) *Step 3:* Compute the eccentric anomaly using Eq. 8.10.

$$E = 2\arctan\left[\left(\frac{1-e}{1+e}\right)^{0.5} \tan\frac{\theta}{2}\right]$$

$$= 2\arctan\left[\left(\frac{1-0.685}{1+0.685}\right)^{0.5} \tan 54°\right]$$

E = 61.5° = 1.07 radians

Step 4: Compute t from Eq. 8.9.

$$t = \frac{T}{2}[E - e\sin E]$$

$$= \frac{656.5}{2}[1.07 - (0.685)\sin(61.5°)]$$

= 48.9 minutes

Table 8.3
The Approximate Magnitudes of Various Forces Acting on a Satellite

Source of perturbing force	*Relative force on satellite at specified height above earth*	
	h = 370 km	h = 37,000 km
Sun	7×10^{-4}	3×10^{-2}
Moon	4×10^{-6}	1×10^{-4}
Earth's oblateness	1×10^{-3}	4×10^{-6}

$$\text{Relative force} = \frac{\text{Average force exerted by perturbation}}{\text{Force exerted by symmetrical earth}}$$

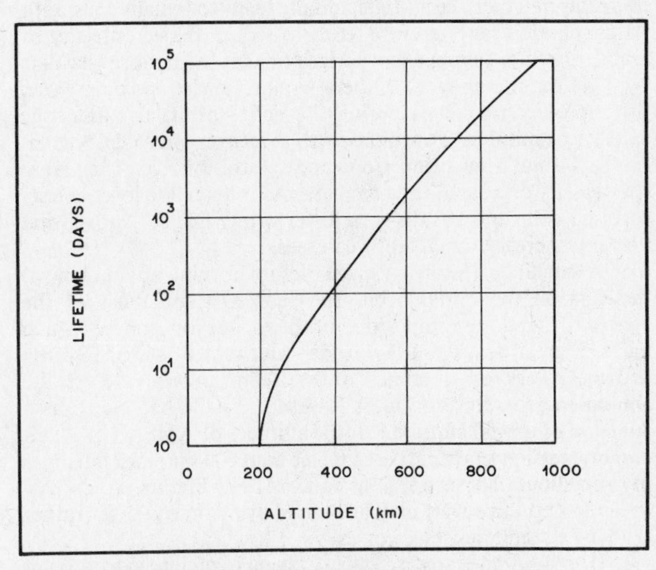

Fig. 8-6 — Satellite lifetime for circular orbit with the satellite geometry and mass similar to AMSAT-OSCARs 7 and 8.

computing the resulting t to determine how close it is to the desired value, then using the information to make a better guess for E. Although this procedure may sound involved it's actually simple. The iterative technique usually employed to solve Kepler's Equation is known as the Newton-Raphson method. A BASIC language subroutine that calculates θ when t is known is shown in Table 8.2.

We now turn to r, the satellite to geocenter distance. Rather than attempt to express r as a function of t, it's simpler and often more useful just to note the relation between r and θ.

$$r = \frac{a(1 - e^2)}{1 - e\cos\theta} \qquad \text{(Eq. 8.11)}$$

Now try Sample Problem 8.6

Corrections to the Simplified Model

Now that we've looked at the solutions to the two-body problem (the simplified satellite-orbit model), let's examine how a more detailed analysis would modify our results.

1) In the two-body problem, the stationary point is the center of mass of the system, not the geocenter. The mass of the earth is so much greater than the mass of an artificial satellite that this correction is negligible.

2) Treating the earth as a point mass implicitly assumes that the shape and the distribution of mass in the earth are spherically symmetrical. Taking into account the actual asymmetry of the earth (most notably the bulge at the equator) produces additional central force terms acting on the satellite. These forces vary as higher orders of 1/r (e.g., $1/r^3$, $1/r^4$, etc.). They cause (i) the major axis of the orbital ellipse to rotate slowly in the plane of the satellite and (ii) the plane of the satellite to rotate about the

earth's N-S axis. Both of these effects are observed readily and we'll return to them shortly.

3) The satellite is affected by a number of other forces in addition to gravitational attraction by the earth. For example, such forces as gravitational attraction by the sun, moon and other planets; friction from the atmosphere (atmospheric drag), radiation pressure from the sun and so on, enter into the system. We turn now to the effects of some of these forces.

Atmospheric Drag. At low altitudes the most prominent perturbing force acting on a satellite is drag caused by collisions with atoms and ions in the earth's atmosphere. Let us consider the effect of drag in two cases: (i) elliptical orbits with high apogee and low perigee and (ii) low-altitude circular orbits. In the elliptical-orbit case drag acts mainly near perigee, reducing the satellite velocity and causing the altitude at the following apogee to be lowered (perigee altitude initially tends to remain constant). Atmospheric drag therefore tends to reduce the eccentricity of elliptical orbits having a low perigee (makes them more circular) by lowering the apogee. In the low-altitude circular orbit case, drag is of consequence during the entire orbit. It causes the satellite to spiral in toward the earth at an ever *increasing* velocity. This is not a misprint. Contrary to intuition, drag can cause the velocity of a satellite to *increase.* As the satellite loses energy through collisions it falls to a lower orbit; Eq. 8.7 shows that velocity increases as height decreases.

A satellite's lifetime in space (before burning up on reentry) depends on the initial orbit, the geometry and mass of the spacecraft, and the composition of the earth's ionosphere (which varies a great deal from day to day and year to year). Fig. 8-6 provides a very rough estimate of the lifetime in orbit of a satellite similar in geometry and mass to AMSAT-OSCAR 7 or 8 as a function of orbital altitude.[5] As the altitudes of AMSAT-OSCAR communication spacecraft are greater than 800 km, their lifetimes in orbit should not be a serious concern. The lifetime of UoSAT (h = 530 km), however, may be limited by its time in orbit rather than by its onboard electronics.

Gravitational Effects. The effects on a satellite's orbit from gravitational attraction by the sun and moon are most prominent when the apogee distance is large. The sun and moon will have a significant effect on the orbit of AMSAT Phase-III satellites. The casual user need not worry about this, but AMSAT scientists must investigate the long-term effects of these forces in detail to ensure that the chosen orbit is stable. Instabilities because of resonant (cumulative) perturbations can cause the loss of a satellite within months. Table 8.3 shows the relative strengths of selected perturbing forces. Now that the motion of the satellite in space has been described, we turn to the problem of relating this motion to an observer on the surface of the earth.

Satellite Motion Viewed From Earth

Terrestrial Reference Frame

To describe a satellite's movement as seen by an observer on the earth, we have to establish a terrestrial reference frame. Once again we simplify the situation by treating the earth as a sphere. The rotational axis of the earth (N-S axis) provides a unique line through the geocenter that intersects the surface of the earth at two points that are designated the *north* (N) and *south* (S) geographic *poles.* The intersection of the surface of the earth and any plane containing the geocenter is called a *great circle.* The great circle formed from the *equatorial plane,* that plane containing the geocenter that also is perpendicular to the N-S axis, is called the *equator.* The set of great circles formed by planes containing the N-S axis are also of special interest. Each is divided into two *meridians* (half circles), connecting north and south poles.

Points on the surface of the earth are specified by two angular coordinates, *latitude* and *longitude.* As an example, the angles used to specify the position of Washington, DC are shown in Fig. 8-7.

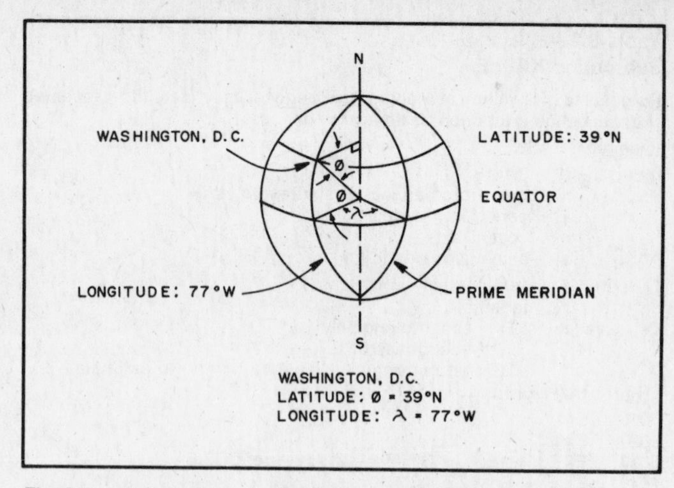

Fig. 8-7 — The location of Washington, DC on the earth can be described by giving its latitude and longitude coordinates.

Latitude. Given any point on the surface of the earth, the latitude is determined by (i) drawing a line from the given point to the geocenter, (ii) dropping a perpendicular from the given point to the N-S axis and (iii) measuring the included angle. A more colloquial but equivalent definition for latitude is the angle between the line drawn from the given point to the geocenter and the equatorial plane. To prevent ambiguity, an N or S is appended to the latitude to indicate whether the given point lies in the northern or southern hemisphere. The set of all points having a given latitude lies on a plane perpendicular to the N-S axis. Although these latitude curves form circles on the surface of the earth, most are *not* great circles. The equator (latitude = 0°) is the only one to qualify as a great circle, since the equatorial plane contains the geocenter. The significance of great circles will become apparent later in this chapter when we look at spherical trigonometry. Better models of the earth take the equatorial and other asymmetries into account when latitude is defined. This leads to a distinction between geodetic, geocentric and astronomical latitude. We won't bother with such refinements.

Longitude. All points on a given meridian are assigned the same longitude. To specify longitude one chooses a reference or "prime" meridian (the original site of the Royal Greenwich Observatory in England is used). The longitude of a given point is then obtained by measuring the angle between the lines joining the geocenter to (i) the point where the equator and prime meridian intersect and (ii) the point where the equator and the meridian containing the given point intersect. For convenience, longitude is given a suffix, E or W, to designate whether one is measuring the angle east or west of the prime meridian.

The Inclination

As the earth rotates about its N-S axis and moves around the sun, the orientation of both the plane containing the equator *(equatorial plane)* and, to a first approximation, the plane containing the satellite *(orbital plane)* remain fixed in space relative to the fixed stars. Fig. 8-8A shows how the orbital plane and equatorial plane are related. The line of intersection of the two planes is called the *line of nodes* since it joins the ascending and descending nodes. The relative orientation of these two planes is very important to satellite users. It is partially specified by giving the inclination. The *inclination,* i, is the angle between the line joining the geocenter and north pole and the line through the geocenter perpendicular to the orbital plane (to avoid ambiguity the half line in the direction of advance of a right-hand screw following satellite motion is used). An equivalent definition of the inclination, the angle between the equator and sub-satellite path as the satellite enters the northern hemisphere, is shown in Fig. 8-8B.

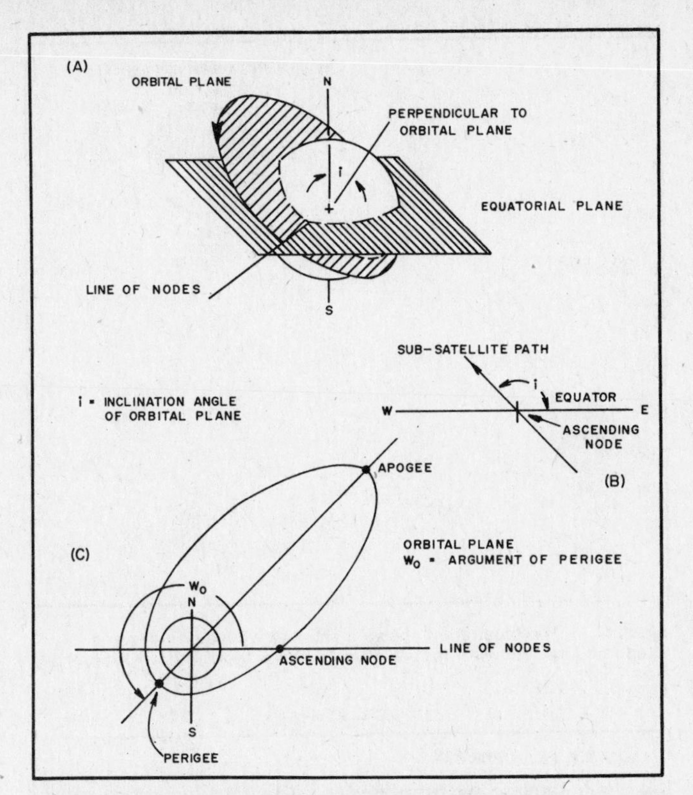

Fig. 8-8 — The orientation of the orbital plane relative to the equatorial plane is given by i, the inclination angle. The position of the perigee in the orbital plane is given by w_o, the argument of perigee.

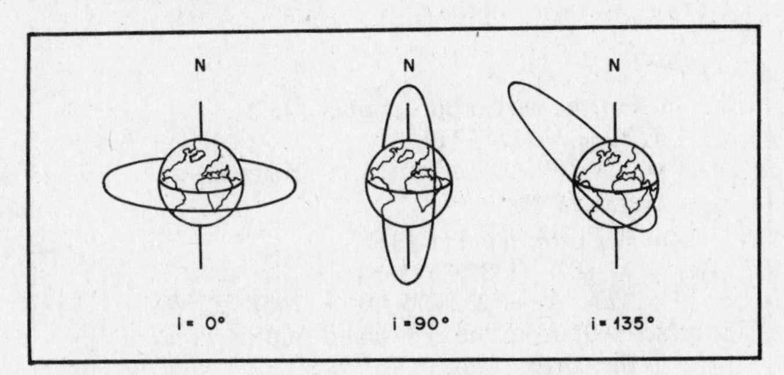

Fig. 8-9 — Satellite orbits with inclination angles of 0°, 90° and 135°.

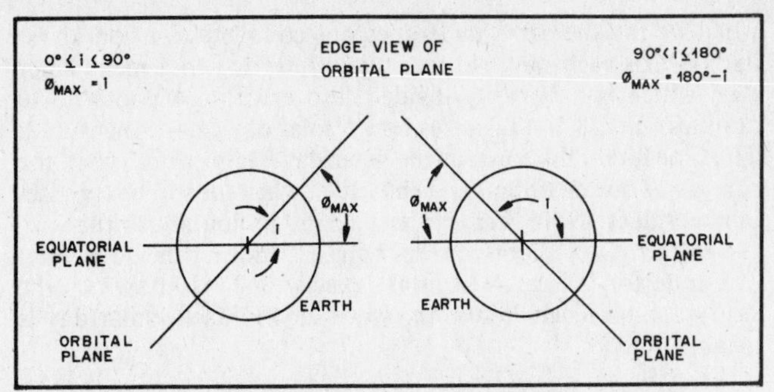

Fig. 8-10 — The maximum latitude reached by the subsatellite point depends only on the inclination angle of the orbital plane. The cross sections shown in the diagram are taken through the geocenter perpendicular to the orbital and equatorial planes.

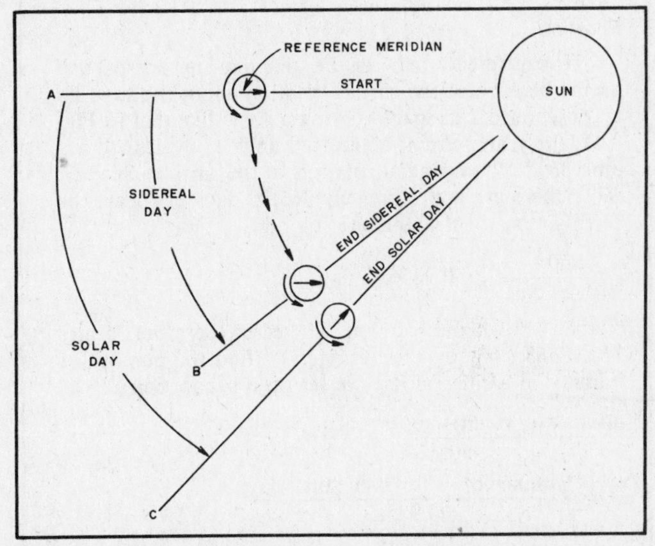

Fig. 8-11 — The figure shows the relation between the solar day and the sidereal day from the vantage point of an observer on the North Star. The measured day begins at A as the reference meridian aligns with the sun. The sidereal day ends at B when the reference meridian rotates 360°. The solar day ends at C when the reference meridian again aligns with the sun.

The angle between the line of nodes (the section joining the geocenter to the ascending node) and the major axis of the ellipse (the section joining the geocenter and perigee) is known as the *argument of perigee*. Fig. 8-8C shows how the argument of perigee serves to locate the perigee in the orbital plane. In the simplified two-body model of satellite motion the argument of perigee is constant. In reality, however, it does vary with time, mainly as a result of the earth's equatorial bulge. The rate of precession (variation) is given by

$$\dot{w}_o = 4.97 \left(\frac{R_{eq}}{a}\right)^{3.5} \frac{(5 \cos^2 i - 1)}{(1 - e^2)^2} \qquad \text{(Eq. 8.12)}$$

where

$\dot{w}_o$ = rate of change of argument of perigee in degrees per day
R_{eq} = mean equatorial radius of earth in same units as a
a = semimajor axis
i = inclination
e = eccentricity

Focusing on the $(5 \cos^2 i - 1)$ term, we see that no matter what the values of a and e, when i = 63.4° the argument of perigee will be constant. The position of the perigee rotates in the same direction as the satellite when $i < 63.4°$, and in the opposite direction when $i > 63.4°$.

The inclination can vary from 0° to 180°. To the first order, none of the perturbations to the simplified model we discussed earlier cause the inclination to change, but higher-order effects result in small oscillations about a mean value. Diagrams of orbits having inclinations of 0°, 90° and 135° are shown in Fig. 8-9. A quick analysis of these three cases yields the following information. When the inclination is 0°, the satellite will always be directly above the equator. When the inclination is 90°, the satellite passes over the north pole and over the south pole once each orbit and over the equator twice, once heading directly north and once heading directly south.

Orbits are sometimes classified as being polar (near polar) when their inclination is 90° (near 90°) or equatorial (near equatorial) when their inclination is 0° (near 0° or 180°). Finally, for other values of inclination, 135° for example, we see that the satellite still passes over the equator twice each orbit but it never crosses above the north or south poles. The maximum latitude (ϕ_{max}), north or south, that the sub-satellite point will reach equals (i) the inclination when the inclination is between 0° and 90° or (ii) 180° less the inclination when the inclination is between 90° and 180°. This can be seen from Fig. 8-10.

Solar and Sidereal Time

Living on earth we quite naturally keep time by the sun. So

when we say the earth undergoes one complete rotation about its N-S axis each day, we're actually referring to a mean *solar day*, which is arbitrarily divided into exactly 24 hours (1440 minutes). Fig. 8-11 illustrates how a solar day can be measured. The time interval known as the solar day begins at A, when the sun passes our meridian, and ends at C, when the sun next passes our meridian. Note that, because of its motion about the sun, the earth rotates slightly more than 360° during the solar day. The time for the earth to rotate exactly 360° is known as the *sidereal day*. When we use the word day by itself, solar day is meant.

Sample Problem 8.7

(a) How many degrees does the earth rotate in one solar day?

(b) How many minutes are there in a sidereal day?

Solution

The difference between the solar day and sidereal day occurs because of the earth's rotation about the sun. To an observer off in space viewing a scene like that in Fig. 8-11, the yearly circuit about the sun is equivalent to adding 360° of extra axial rotation to the earth each year. Since there are approximately 365.25 days per year, the earth's movement about the sun adds

$$\frac{360°}{365.25 \text{ days}} = 0.98563°/\text{day}$$

to the earth's axial rotation. The earth therefore rotates 360.98563°/day on the average. To find the number of minutes in a sidereal day we set up a proportion

$$\frac{\text{number of minutes in sidereal day}}{360°} = \frac{\text{number of minutes in solar day}}{360.98563°}$$

Solving, we obtain approximately 1436.07 minutes for the sidereal day.

Longitude Increment

Now that the difference between solar day and sidereal day has been examined we're prepared to look at the *longitude increment* (I) or, simply, increment. The *increment* is defined as the change in longitude between two successive ascending nodes. In mathematical terms

$$I = \lambda_{n+1} - \lambda_n \qquad \text{(Eq. 8.13a)}$$

where λ_{n+1} is the longitude at any ascending node in degrees west of Greenwich [°W], λ_n is the longitude at the preceding ascending node in °W, and I is in degrees west per orbit [°W/orbit].

There are two ways to obtain the increment: experimentally by averaging observations over a long period of time, or theoretically by calculating it from a model. Though the best numbers are obtained experimentally, the calculation approach is needed; we do, after all, want a value for I before launch and in the early weeks or months in orbit when observations haven't accumulated over a long time period.

Once the increment is known we can compute the longitude of any ascending node, λ_m, given the longitude of any other ascending node, λ_n. The orbit reference integers, m and n, may either be the standard ones beginning with the first orbit after launch, or any other convenient serial set.

$$\lambda_m = \lambda_n + (m-n)\,I \qquad \text{(Eq. 8-13b)}$$

This formula works either forward or backward in time. When

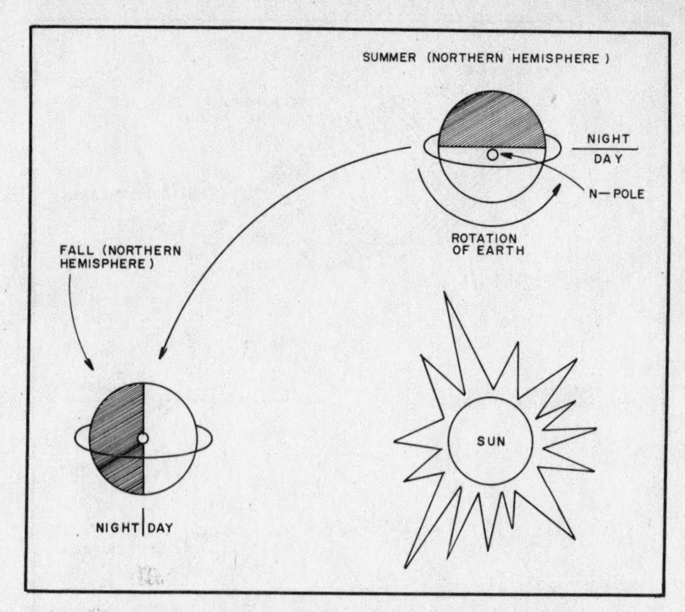

Fig. 8-12 — The illustration shows a satellite whose orbital plane is fixed in space. The view is that of an observer looking down from the North Star.

Sample Problem 8.8

A satellite we're interested in has an increment of I = 28.75 °W/orbit. If the longitude of ascending node on orbit number 1256 is 123 °W, find the longitude of the ascending node on (a) the next orbit and (b) orbit number 1337.

Solution

(a) For the next orbit (number 1257)

$$\lambda_{1257} = \lambda_{1256} + (1257\text{-}1256)\,I$$
$$= 123\ °W + (1 \text{ orbit})(28.75\ °W/\text{orbit})$$
$$\lambda_{1257} = 151.75\ °W$$

(b) For orbit number 1337

$$\lambda_{1337} = \lambda_{1256} + (1337 - 1256)\,I$$
$$= 123\ °W + 2328.75\ °W = 2451.75\ °W$$

Subtract 360° from the right-hand side six times to put λ_{1337} in the correct range.

$$\lambda_{1337} = 291.75\ °W$$

future orbits are being predicted, m > n. The right side of Eq. 8.13b must be brought into the range of 0-360° by successive subtractions or additions of 360° if necessary (see Sample Problem 8.8).

We now examine how I may be calculated. First, let's ignore corrections to the two-body model and only take the earth's rotation about its N-S axis and the motion about the sun into account. In this case the plane of the satellite has a fixed orientation in space and the increment results entirely from the earth's rotation. For a satellite having a period T, we just compute how much the earth rotates (as seen by an observer fixed in space) in an elapsed time equal to T. This can be accomplished by setting up a proportion

$$\frac{\text{angular rotation of earth during one complete orbit } (\bar{I})}{\text{angular rotation of earth during one sidereal day } (360°)} = \frac{\text{number of minutes for one complete orbit (T)}}{\text{number of minutes in one sidereal day (1436.07)}}$$

or,

$$\frac{\bar{I}}{360°} = \frac{T}{1436.07 \text{ minutes}}$$

The bar over the I is to remind us that this is a theoretical value

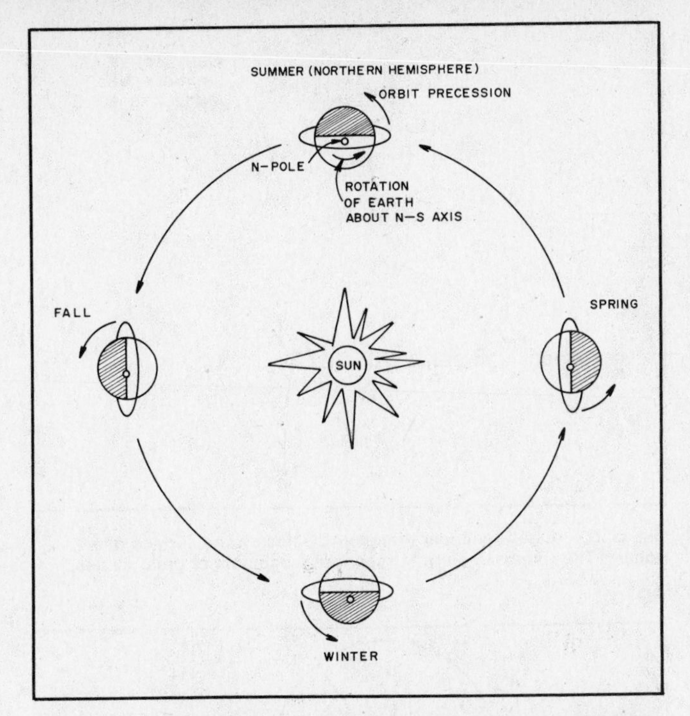

Fig. 8-13 — Sun-synchronous orbit like the one chosen for OSCARs 6, 7 and 8, and UoSAT. The view of the sun-earth-satellite system is from the North Star. Note how the orbital precession can keep the satellite near the twilight line year 'round when total precession for a year is 360°.

that doesn't take some important perturbations into account; we'd expect it to differ slightly from the true I. Solving for $\bar{I}$ we obtain

$$\bar{I} = (0.250684\,°/\text{min})\ T \qquad \text{(Eq. 8.14)}$$

From Eq. 8.14 we see that it's easy to estimate $\bar{I}$ quickly by computing T/4. The difference between I (Eq. 8.13a) and $\bar{I}$ (Eq. 8.14) is because the orientation of the orbital plane does not remain fixed in space. We'll see how the motion of the orbital plane effects $\bar{I}$ in the next section.

Precession: Circular Orbits. Fig. 8-12 shows a satellite whose orbital plane is fixed in space as the earth moves about the sun. In the illustration the satellite closely follows the terminator (day-night line) in summer. As a result, passes accessible to a ground station will be centered near 6 A.M. and 6 P.M. each day. Three months later the satellite passes over the center of the day and night regions. Accessible passes now occur near 3 A.M. and 3 P.M. each day.

Although the two-body model predicts that the orbital plane will remain stationary, we've already noted that when the earth's equatorial bulge is taken into account, the plane precesses about the earth's N-S axis. Fig. 8-13 shows an example of such precession. For circular orbits the precession is given by

$$\dot{\Omega} = 9.95\left(\frac{R_{eq}}{r}\right)^{3.5} \cos i \qquad \text{(Eq. 8.15)}$$

(circular earth orbits only)

where

$\dot{\Omega}$ = orbital plane precession rate in °/day. A negative precession is shown in Fig. 8-13 (Counterclockwise as seen from above N pole).
R_{eq} = mean equatorial radius of earth = 6378 km
r = satellite-geocenter distance in same units as R_{eq}
i = orbital inclination

Once we know $\dot{\Omega}$ we can use it to correct $\bar{I}$ as shown in the following problem.

Sample Problem 8.9

About four weeks after the launch of the Soviet RS-1, observations had yielded the period to six significant digits, T = 120.389 minutes, and had confirmed the TASS and NASA reports that gave i as 82.6° (i is not very critical in $\dot{\Omega}$ computations).
Using this data, compute the longitudinal increment.

Solution

Step 1. Compute r from the period using Eq. 8.6c.
$r = 331.25 \times T^{2/3} = 8076$ km

Step 2. Calculate $\bar{I}$ from the period using Eq. 8.14.
$\bar{I} = (0.250684)(120.389) = 30.1796$ °W/orbit

Step 3. Calculate $\dot{\Omega}$ using Eq. 8.15.

$$\dot{\Omega} = +9.95\left(\frac{6378}{8076}\right)^{3.5} \cos(82.6) = +0.5610\,°/\text{day}$$

Step 4. The number of orbits per day is given by

$$\frac{1440 \text{ min/day}}{120.389 \text{ min/orbit}} = 11.9612 \text{ orbits/day}$$

Step 5. The precession per orbit is therefore

$$\begin{array}{c}\text{precession}\\ \text{in °/orbit}\end{array} = \frac{\begin{array}{c}\text{precession}\\ \text{in °/day}\end{array}}{\begin{array}{c}\text{number of}\\ \text{orbits/day}\end{array}} = \frac{+0.5610\,°/\text{day}}{11.9612 \text{ orbits/day}}$$

$$= +0.0469\,°/\text{orbit}$$

Step 6. Adding the precession per orbit (Step 5) to $\bar{I}$ (Step 2) gives the corrected value for I
$I = \bar{I}$ + (correction factor) = 30.1796° + 0.0469°
= 30.227 °W/orbit

Note: Observations of RS-1 over several months yielded the same value for I.

Sun-Synchronous Orbits. By choosing the altitude and inclination of satellites we can vary $\dot{\Omega}$ over a considerable range of values. Looking at the example of Fig. 8-13 you may have noted that the orbital plane precessed exactly 360° in one year. As a result, the satellite remained above the terminator the entire time. An orbit that precesses very nearly 360° per year is called sun-synchronous. Such orbits pass over the same part of the earth at roughly the same time each day, making communication and various forms of data collection convenient. They also provide nearly continuous sunlight for solar cells and good sun angles for weather satellite photos when the injection orbit is similar to Fig. 8-13. Because of all these desirable features orbits are often carefully chosen to be sun-synchronous.

To obtain an orbital precession of 360° per year we need a precession rate of 0.986°/day (360°/365.25 days). Since precession can be clockwise or counterclockwise (as seen from above the north pole) we must substitute −0.986°/day for $\dot{\Omega}$ in Eq. 8.15 to obtain the correct direction.
Making this substitution and solving for i we obtain

$$i^* = \arccos\left[-(0.09910)\left(\frac{r}{6378}\right)^{3.5}\right] \qquad \text{(Eq. 8.16)}$$

where i* is the inclination needed to produce sun-synchronous circular orbit.

In this form we can plug in values of r and calculate the inclination which will produce a sun-synchronous orbit. Graphing Eq. 8.16 in Fig. 8-14 we see that for low-altitude satellites sun- syn-

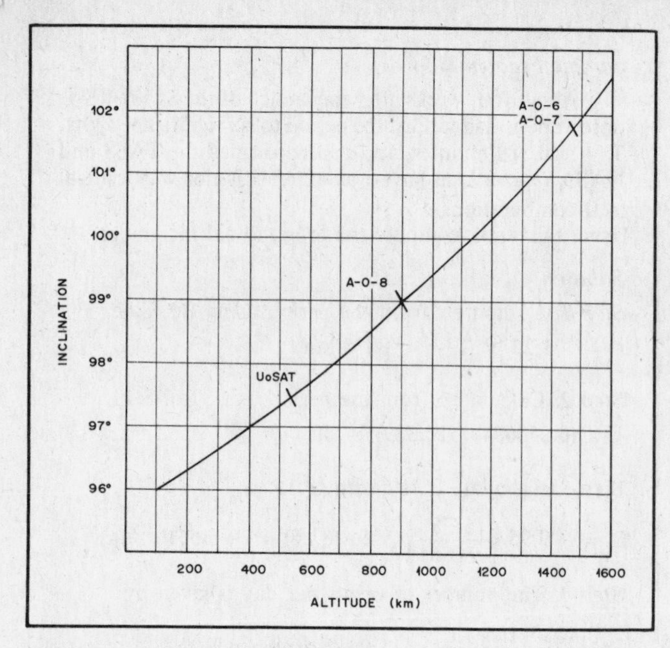

Fig. 8-14 — This graph shows the inclination value which results in a sun-synchronous circular orbit.

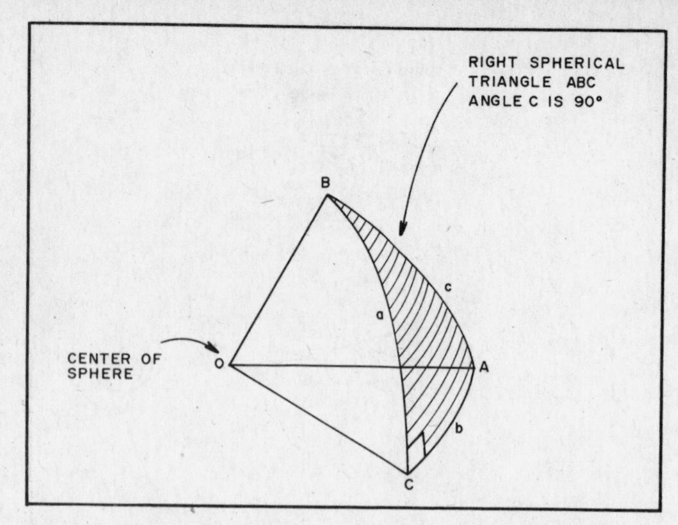

Fig. 8-15 — Right spherical triangle ABC lies on the surface of a sphere. The three sides are formed from segments of great circles.

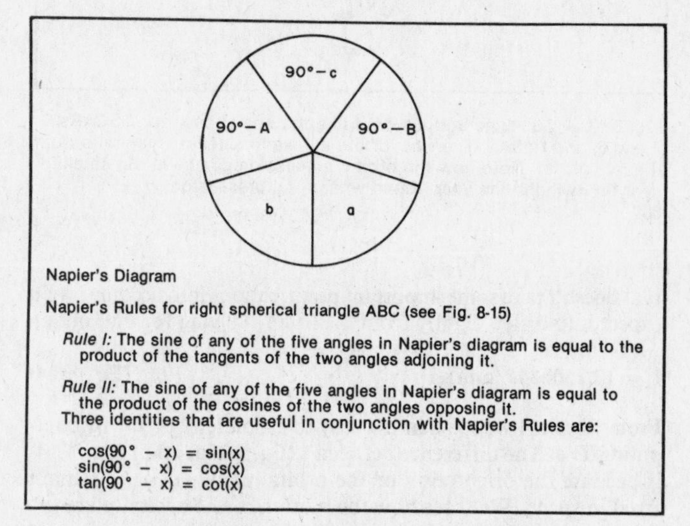

Fig. 8-16 — Napier's Rules and diagram offer an easy way to remember and apply the rules for right spherical triangles.

chronous orbits will be near polar. You may have noted that the 0.986°/day precession rate needed to produce a sun- synchronous orbit exactly corresponds to the amount in excess of 360° that the earth rotates each solar day (Sample Problem 8.7). This is no accident; the precession rate was chosen precisely for this purpose.

Precession: Elliptical Orbits. The precession of the orbital plane about the earth's N-S axis for elliptical orbits is given by

$$\dot{\Omega} = 9.95 \left(\frac{R_{eq}}{a}\right)^{3.5} \frac{\cos(i)}{(1-e^2)^2} \qquad \text{(Eq. 8.17)}$$

If $a = r$ and $e = 0$ (i.e., when the ellipse becomes a circle) Eq. 8.17 simplifies to Eq. 8.15.

Ground Track

This section begins with a bare-bones introduction to spherical trigonometry, a mathematical tool that is then used to derive the ground-track equations for circular orbits. The results for circular orbits are then generalized and summarized. [Readers who just need access to the ground-track equations for programming a computer can skip the spherical trigonometry and derivation sections and jump right to the summary.] We then go on to derive and summarize the ground-track equations for elliptical orbits. This section on spherical trigonometry will also be referred to in the next chapter when we discuss "spiderwebs."

Spherical Trigonometry Basics

A triangle drawn on the surface of a sphere is called a spherical triangle *only* if all three sides are arcs of great circles. A great circle is *only* formed when a plane containing the center of a sphere intersects the surface. Many circles on the sphere, such as latitude lines (other than the equator) and range circles about a ground station, are *not* great circles. A spherical triangle that has at least one 90° angle is called a right spherical triangle (see Fig. 8-15).

Spherical trigonometry is the study of the relations between sides and angles in spherical triangles. The notation of spherical trigonometry closely follows that of plane trigonometry. Surface angles and vertices in a triangle are labeled with capital letters A, B and C, and the side opposite each angle is labeled with the corresponding lower case letter as shown in Fig. 8-15. Note that the arc length of each side is proportional to the central angle formed by joining its end points to the center of the sphere. For example, side b is proportional to angle AOC. The proportionality constant is the radius of the sphere, but because it cancels out in the computations we'll be interested in, the length of a side will often be referred to by its angular measure.

Although the properties of the sine and cosine *functions* of plane trigonometry remain the same in spherical trigonometry, the rules governing the relationships between sides and angles in triangles change. In spherical trigonometry the internal angles in a triangle do *not* usually add up to 180° and the square of the hypotenuse does *not* generally equal the sum of the squares of the other two sides in a right triangle.

Recall how in plane trigonometry the rules for right triangles were simpler than those for oblique triangles. In spherical trigonometry the situation is similar: The rules for right spherical triangles are simpler than those for general spherical triangles. Fortunately, since the spherical triangles we'll be working with have at least one right angle we need only consider the simple laws for right spherical triangles. A convenient method for summarizing these rules, developed by Napier, is shown in Fig. 8-16. Sample problem 8.10 illustrates how Napier's Rules can be applied.

Two major pitfalls await newcomers attempting to apply spherical trigonometry for the first time. The first pitfall, the degree-radian trap, comes from overlooking the fact that angles

Sample Problem 8.10

Given a right spherical triangle like the one in Fig. 8-15, assume that A and c are known. (a) First solve for a in terms of A and c. (b) Then solve for b in terms of a and c.

Solution

(a) To find a, apply Napier Rule II to the indicated segment of the Napier diagram

$\sin(a) = \cos(90° - c)\cos(90° - A)$

Using the identity for $\cos(90° - x)$ we obtain

$\sin(a) = \sin(c)\sin(A)$, or

$a = \arcsin[\sin(c)\sin(A)]$

(b) To find b apply Napier Rule II again, this time to the segment shown

$\sin(90° - c) = \cos(a)\cos(b)$

Using the identity for $\sin(90° - c)$

$\cos(c) = \cos(a)\cos(b)$

$$\cos(b) = \frac{\cos(c)}{\cos(a)}$$

$$b = \arccos\left[\frac{\cos(c)}{\cos(a)}\right]$$

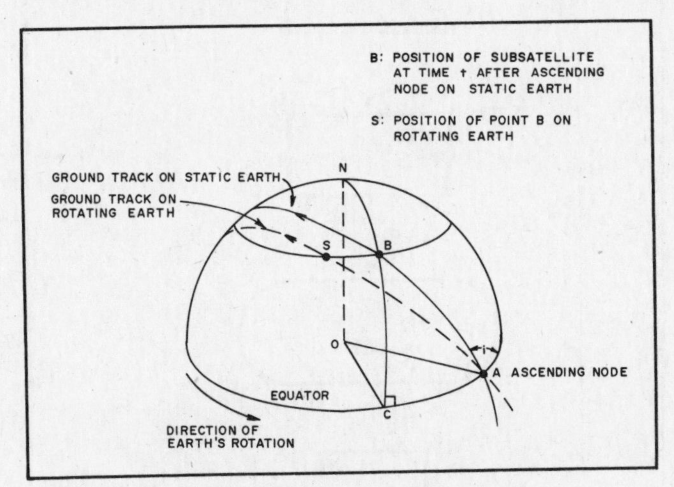

Fig. 8-17 — Illustration for applying the principles of spherical trigonometry to the circular-orbit ground-track problem.

must be expressed in units appropriate to a given equation *and* computing machine. For example, focus on the angle $\theta = 30° = \pi/6$ radians. Consider the machine dependent aspect first. To evaluate $\sin(\theta)$ on most simple scientific calculators you must input "30" since the calculator expects θ to be in degrees. To evaluate $\sin(\theta)$ in BASIC on a microcomputer you must input $\pi/6$ (or 0.52360) because the BASIC language expects θ to be in radians. In some situations, especially in cases where θ is *not* the argument of a trigonometric function, the form of the equation determines whether θ must be in degrees or radians. Consider a radio station at 30° N latitude trying to use the equation $S = R\theta$ to find the surface distance (S) along a meridian (earth radius = R) to the equator. The equation only holds for θ in radians so the input must be $\pi/6$.

The second trap awaiting spherical trigonometry novices is using a latitude line as one side of a spherical triangle. The only latitude line that will serve in this manner is the equator. All other latitude lines do not work since they are not arcs of great circles.

Circular Orbits: Derivation

The most important step in deriving the ground-track equations for circular orbits is drawing a clear picture. In Fig. 8-17 we've chosen to show i between 90° and 180° and a satellite headed north in the Northern Hemisphere. Our object is to compute the latitude and longitude of the subsatellite point (SSP) — ϕ_S and λ_S — when it reaches S, t minutes after the most recent ascending node. We assume that the period T, orbit inclination i, and the longitude of the ascending node, λ_o, are known.

Since arc AS, along the actual ground track, is *not* a section of a great circle we first consider the situation for a static earth (one not rotating about its N-S axis). On such an earth, the SSP would be at point B at t minutes after the ascending node. Triangle ABC is a spherical triangle. Angle A is given by $180° - i$. Arc AB (side c of the spherical triangle) is a section of the circular orbit with

$$c = 2\pi\frac{t}{T} \qquad \text{(see Eq. 8.8).}$$

By definition, the latitude of point B is equal to a.

The problem of finding the latitude of point B, ϕ_B, in terms of i, t and T is identical to the problem of finding a in terms of A and c. This was solved in Sample Problem 8.10 where we found that

$a = \arcsin[\sin(c)\sin(A)]$.

Substituting the variables ϕ_B, T, i, t we obtain

$$\phi_B = \arcsin[\sin(2\pi\frac{t}{T})\sin(180° - i)]$$

Using the symmetry of the sine function this simplifies to

$$\phi_B = \arcsin[\sin(2\pi\frac{t}{T})\sin(i)] \qquad \text{(for a non-rotating earth).}$$

If for computations we wish to specify c in degrees, we would replace

$$2\pi\frac{t}{T} \text{ by } 360°\frac{t}{T}.$$

To solve for the longitude at B, λ_B, we note that $b = \lambda_o - \lambda_B$. So, our problem of solving for λ_B in terms of ϕ_B, t and T is equivalent to solving for b in terms of a and c. This was also done in Sample Problem 8.10 where we found that

$$b = \arccos\left[\frac{\cos(c)}{\cos(a)}\right]$$

Making the appropriate substitutions this yields

$$\lambda_o - \lambda_B = \arccos\left[\frac{\cos(2\pi t/T)}{\cos(\phi_B)}\right]$$

(for a non-rotating earth).

Taking the rotation of the earth into account, point B moves to position S. The latitude does not change so $\phi_S = \phi_B$. The longitude varies only by the angular rotation of the earth in time t. To a first approximation the rotation rate of the earth is 0.25°/minute so, if we measure t in minutes, $\lambda_S = \lambda_B - t/4$. A more accurate figure for the earth's rotation could be used (as we saw when the longitudinal increment, I, was discussed), but our accuracy needs here don't warrant this refinement. It's far simpler to apply the longitude equation for one orbit at a time and then update the ascending node value using the best available increment before beginning the next orbit.

This completes the derivation for the case illustrated. A more complete derivation would consider several additional cases: satellites in the southern hemisphere, i between 0° and 90°, spacecraft headed south, and so on. As the approach is similar, we'll just summarize the results in the next section.

Circular Orbits: Summary

Latitude of SSP: $\phi(t) = \arcsin[\sin(i)\sin(360°\, t/T)]$ (Eq. 8.18)

Note: "$\phi(t)$" should be read "latitude as a function of time"; it does *not* mean ϕ times t.

Longitude of SSP: $\lambda(t) =$

$$\lambda_o - (-1)^{n_2+n_3} \arccos[\frac{\cos(360°t/T)}{\cos(\phi(t))}] - t/4 \quad \text{(Eq. 8.19)}$$

$$n_2 = \begin{cases} 0 \text{ when } 90° \le i \le 180° \\ 1 \text{ when } 0° \le i < 90° \end{cases}$$

$$n_3 = \begin{cases} 0 \text{ when } \phi(t) \ge 0° \text{ (Northern Hemisphere)} \\ 1 \text{ when } \phi(t) < 0° \text{ (Southern Hemisphere)} \end{cases}$$

Sign Conventions

Latitude
North: positive
South: negative

Longitude
East: positive
West: negative

All angles are in degrees and time is in minutes
i inclination of orbit
T period
t elapsed time since most recent ascending node
λ_o longitude of SSP at most recent ascending node

Comments.

1) Please note the sign conventions for east and west longitudes. Most maps used by radio amateurs in the U.S. are labeled in degrees west of Greenwich. This is equivalent to calling west longitudes positive. Because there are important computational advantages to using a right-hand coordinate system, however, almost all physics and mathematics books refer to east as positive, a custom that we follow for computations. When calculations are completed it's a simple matter to relabel longitudes in degrees west. This has been done for all user-oriented data in this book.

2) Eq. 8.19 should only be applied to a single orbit. At the end of each orbit the best available longitude increment should be used to compute a new longitude of ascending node. Eq. 8.19 can then be reapplied.

3) Eq. 8.18 and Eq. 8.19 can be solved at any time, t, if i, λ_o and T are known. In other words, it takes four parameters to specify the location of the SSP for a circular orbit. The four we've used are known as the "classical orbital elements." They were chosen because each has a clear physical meaning. There are several other sets of orbital elements that may also be employed.[6]

4) If you have a programmable hand calculator or microcomputer you can use Eqs. 8.18 and 8.19 to run your own predictions either to follow a particular satellite pass or to produce data for an OSCARLOCATOR ground-track overlay. The flow chart of Fig. 8-18 outlines one simple approach. All sorts of refinements can be added, but it's best to get the basic program running first. You might, for example, input the time increment instead of using a fixed value of two minutes. Or you might add a time delay circuit to the loop to produce a real-time display. The Tables in Appendix B for circular orbit spacecraft were produced using an algorithm based on the flow chart of Fig. 8-18 by inputting $\lambda_o = 0°$. It's a good idea to use these tables to check any program you write.

5) Tracking programs are available in several hand calculator and microcomputer formats. Table 8-4 gives a partial list.

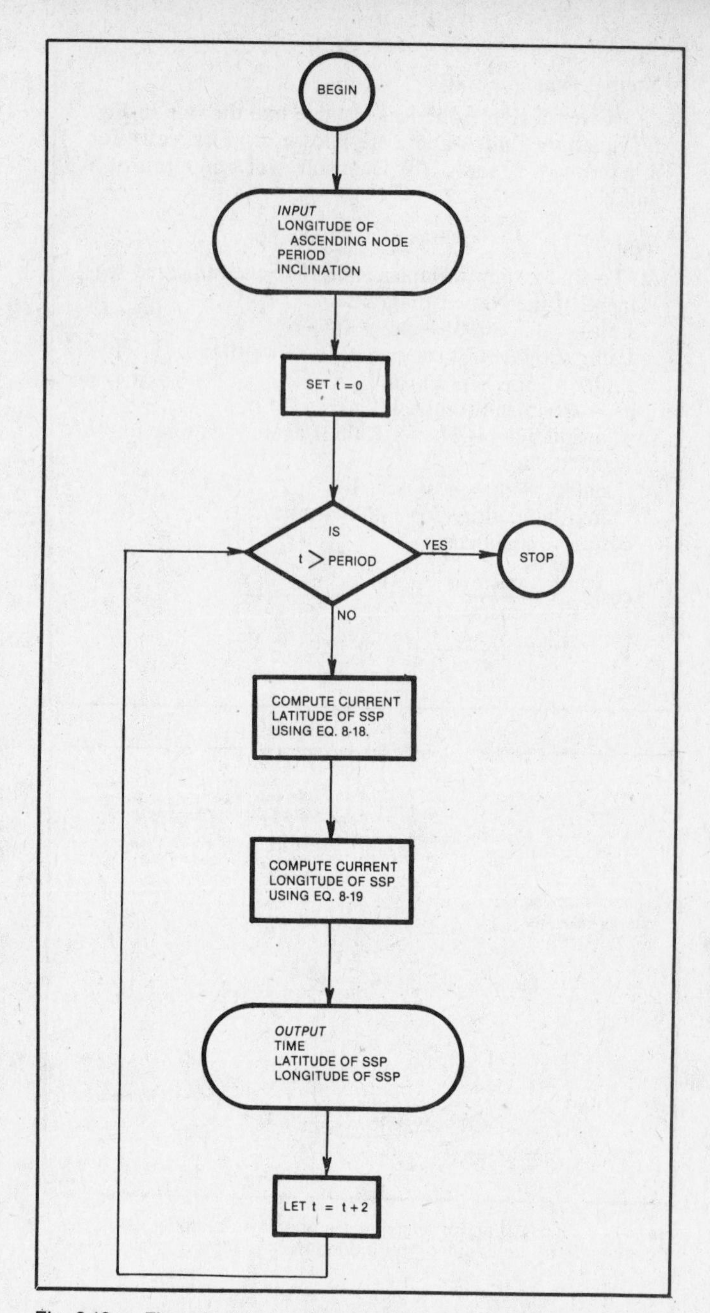

Fig. 8-18 — Flow chart for circular-orbit ground-track program.

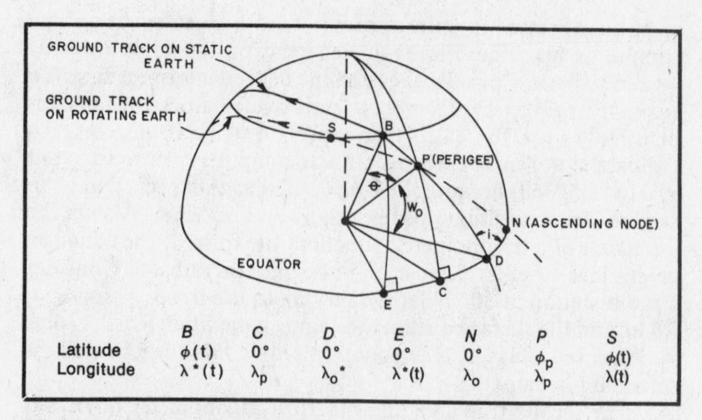

	B	C	D	E	N	P	S
Latitude	$\phi(t)$	0°	0°	0°	0°	ϕ_p	$\phi(t)$
Longitude	$\lambda^*(t)$	λ_p	λ_o^*	$\lambda^*(t)$	λ_o	λ_p	$\lambda(t)$

Fig. 8-19 — Illustration for applying the principles of spherical trigonometry to the elliptical-orbit ground-track problem.

Elliptical Orbits: Derivation

Now that we've seen how the ground-track equations for a circular orbit are derived, we go on to look at the additional parameters and steps required for elliptical orbits. The overall approach uses the same basic principles and is, in many ways, similar. To follow the derivation you should be familiar with all the material in the earlier ground-track section including the introduction to spherical trigonometry. Once again, a clear diagram is essential. In Fig. 8-19 we've chosen an inclination between 90° and 180°, a satellite perigee in the northern hemisphere, and the spacecraft headed north in the northern hemisphere. A diagram of the orbital plane (Fig. 8-20) is also very helpful.

We assume that the following parameters are known:

Table 8.4

Sources for Calculator and Microcomputer Programs Focusing on Orbital Calculations

AMSAT Software Exchange
(Administered by J. Montague, WØRUE)
P.O. Box 27
Washington, DC 20044
The Software Exchange maintains an extensive collection of user-contributed programs. Listings are available at cost, and machine readable formats are available at very reasonable prices. This service is for AMSAT members only. Please send a no. 10 s.a.s.e. for the latest catalog.

HP 67/97 Calculator; Circular Orbits
Circular Orbit Program for OSCAR and RS Satellites
Program no. 02020D by R. Welsh
Available from HP 67/97 Users Group
1000 N. E. Circle Blvd.
Corvallis, OR 97330

HP 67/97 Calculator; Elliptical Orbits
Elliptical Orbit Program for Phase III with Apogee Parameters
Program no. 02457D by R. Welsh
Available from HP 67/97 Users Group
1000 N. E. Circle Blvd.
Corvallis, OR 97330

J. Branegan, *Satellite Tracking Software for the Radio Amateur*, AMSAT-UK, 46 pp., 1982. Covers circular, elliptical and geostationary orbits.

BASIC: Ellpitical Orbits
T. Clark, "Basic Orbits," *Orbit*, Vol. 2, no. 6, March/April 1981, pp. 6-11, 19, 20, 29. This program is the standard prepared for the Phase III command stations. In addition to providing ground-track data it provides azimuth and elevation from any specified ground station (using oblate earth corrections) and provides the instantaneous Doppler shift for any specified frequency. The program is available for any microcomputer (cassette or disk) including such inexpensive models as the Timex Sinclair 1000 or Commodore VIC 20.
[Note that there's a line missing from the program listing in the reference. On page 29 insert the instruction "165 RESTORE 280" between lines 160 and 170.]

BASIC: Circular Orbits
B. Nazarian and D. Mitchell, "Tracker — the Ultimate OSCAR Finder," *73*, Jan. 1981, pp. 88-95. Through numerous flow charts and full listings this article provides good insight into how a user-oriented program is designed.

HP-41C Calculator; Elliptical Orbits
P. Bunnell, "Tracking Satellites in Elliptical Orbits," *Ham Radio*, March 1981, pp. 46-50.

BASIC/FORTRAN; Elliptical Orbits
Sat Trak Package by Sat Trak International. For information on this commercial user-oriented software package see: P. Karn, "A Review of Sat Trak International's Software Products," *Orbit*, Vol. 1, no. 4, Nov./Dec. 1980, p. 31.

TI-59; Elliptical Orbits
J. Molnar, "TI-59 Program Tracks Satellites in Elliptical Orbits," *Electronics*, Oct. 6, 1981, pp. 146-149.

T (period in minutes), i (inclination in degrees), λ_p (SSP longitude at perigee), w_o (argument of perigee), e (eccentricity). Our object is to solve for the latitude and longitude of the SSP — $\phi(t)$ and $\lambda(t)$ — at any time t. We will measure t from perigee. At perigee, $t=0$; after perigee, t is positive; before perigee, t is negative.

The actual ground track is not a great circle so our strategy will again be to focus first on a static-earth model where the principles of spherical trigonometry can be applied. The results will then be adjusted to take into account the rotation of the earth. In Fig. 8-17, which was drawn for circular orbits, we elected to let the static-earth ground track coincide with the true ground track at the ascending node. In Fig. 8-19 we've chosen to let the two ground tracks coincide at perigee.

Step 1. Our object here is to relate our perigee-based parameters to the ascending node. More specifically, we wish to calculate (a) elapsed time as the satellite moves from D to P, (b) the latitude at perigee and (c) the longitude at the ascending node.

1a) Consider the static-earth model and focus on spherical triangle CPD. From Fig. 8-20 we see that arc PD is, by definition, equal to the argument of perigee, w_o. Using Kepler's Equation (Eqs. 8.9 and 8.10) we can plug the value of w_o in for θ and calculate the elapsed time between perigee and the ascending node which we call t_p.

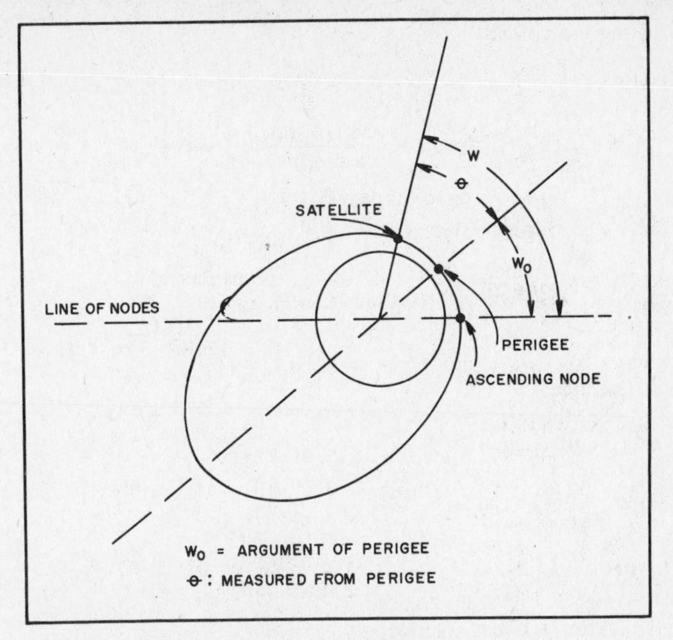

Fig. 8-20 — Satellite position in orbital plane.

1b) The latitude at perigee is, by definition, the length of arc PC. Angle PDC is equal to $180° - i$. Knowing angle PDC and arc PD we use Napier Rule II to solve for arc PC.

$$\phi_p = \arcsin[\sin(i)\sin(w_o)]$$

1c) To obtain the longitude at point D we again apply Napier Rule II.

$$\lambda_o^* = \lambda_p + \arccos\left[\frac{\cos(w_o)}{\cos(\phi_p)}\right]$$

1d) The actual longitude at the ascending node is found by computing how far the earth rotated as the satellite traveled from the ascending node to perigee and adding this to the preceding static-earth result.

$$\lambda_o = \lambda_p + \arccos\left[\frac{\cos(w_o)}{\cos(\phi_p)}\right] + |t_p|/4$$

Step 2. We now turn to the problem of locating the SSP at S, any time, before or after perigee. We again begin by focusing on the static-earth model to find the latitude and longitude of point B. To do this we use spherical triangle BDE.

2a) Comparing Figs. 8-19 and 8-20 we see that arc BD is equal to $(\theta + w_o)$. To emphasize that θ changes with time, we write this term as $(\theta(t) + w_o)$. Using Napier Rule II we obtain the latitude of point B which is also the actual latitude of SSP at S.

$$\phi(t) = \arcsin[\sin(i)\sin(\phi(t) + w_o)]$$

2b) Applying Napier Rule II once again we obtain the longitude of point B.

$$\lambda^*(t) = \lambda^*_o - \arccos\left[\frac{\cos(\theta(t) + w_o)}{\cos(\phi(t))}\right]$$

2c) Finally, correcting for the rotation of the earth we obtain the actual longitude of the SSP at S.

$$\lambda(t) = \lambda_o - \arccos\left[\frac{\cos(\theta(t) + w_o)}{\cos(\phi(t))}\right] - t/4 - |t_p|/4$$

Elliptical Orbits: Summary
(See Fig. 8-19 and Fig. 8-20)

Latitude of SSP:

$$\phi(t) = \arcsin[\sin(i)\sin(\theta(t) + w_o)] \quad \text{(Eq. 8.20)}$$

Longitude of SSP:

$$\lambda(t) =$$

$$\lambda_o - (-1)^{n_2+n_4}\arccos\left[\frac{\cos(\theta(t)+w_o)}{\cos(\phi(t))}\right] - t/4 - |t_p|/4 \quad \text{(Eq. 8.21)}$$

$$n_2 = \begin{cases} 0 \text{ when } 90° \le i \le 180° \\ 1 \text{ when } 0° \le i < 90° \end{cases}$$

$$n_4 = \begin{cases} 0 \text{ when } \phi(t) \ge 0 \text{ (Northern Hemisphere)} \\ 1 \text{ when } \phi(t) < 0 \text{ (Southern Hemisphere)} \end{cases}$$

Sign Conventions

Latitude
- North: positive
- South: negative

Longitude
- East: positive
- West: negative

Time
- After perigee: positive
- Before perigee: negative

All angles are in degrees and time is in minutes

t time from perigee (at perigee $t=0$)

t_p satellite travel time between ascending node and perigee (Can be calculated from Kepler's Equation (Eqs. 8.9 and 8.10) when w_o is known)

i inclination of orbit

θ polar angle in satellite plane (see Fig. 8-20 and Kepler's Equation)

w_o argument of perigee (The angle in the orbital plane locating the perigee with respect to the line of nodes, see Fig. 8-20)

λ_o longitude at ascending node (If one knows the longitude at perigee and not at ascending node, see Eq. 8.22)

Longitude of SSP at ascending node:

$$\lambda_o = \lambda_p + (-1)^{n_2+n_3}\arccos\left[\frac{\cos(w_o)}{\cos(\phi_p)}\right] + |t_p|/4 \quad \text{(Eq. 8.22)}$$

$$n_3 = \begin{cases} 0 \text{ when } 0° \le w_o \le 180° \text{ (Perigee in N Hemisphere)} \\ 1 \text{ when } 180° < w_o < 360° \text{ (Perigee in S Hemisphere)} \end{cases}$$

Latitude of SSP at perigee: $\phi_p = \phi(t=0)$ (Use Eq. 8.20)

Special Orbits

The Geostationary Orbit

A satellite launched into an orbit with an inclination of zero degrees will always remain directly above the equator. If such a satellite is in a circular orbit (constant velocity), traveling west to east, at a carefully selected height (35,800 km), its angular velocity will equal that of the earth about its axis (period = 24 hours). As a result, to an observer on the surface of the earth the spacecraft will appear to be hanging motionless in the sky. Satellites in such orbits are called *geostationary* (or stationary for short).

The geostationary orbit has a number of features that make it nearly ideal for a communications satellite. Of prime importance, Doppler shift on the radio links is nonexistent, and ground stations can forget about orbit calendars and tracking. These features have not gone unnoticed — so many commercial spacecraft are spaced along the geostationary arc above the equator that a severe "parking" problem exists.

From an amateur radio point of view, a geostationary satellite is not without problems. The biggest shortcoming is that a single spacecraft can only serve slightly less than half the earth. It's sometimes stated that a geostationary satellite provides poor east-west communications coverage to radio amateurs at medium to high latitudes. This may be true when Molniya-type orbits (see next section) are the standard of comparison, but take a good look at the map shown in Fig. 8-21 before adopting an opinion.

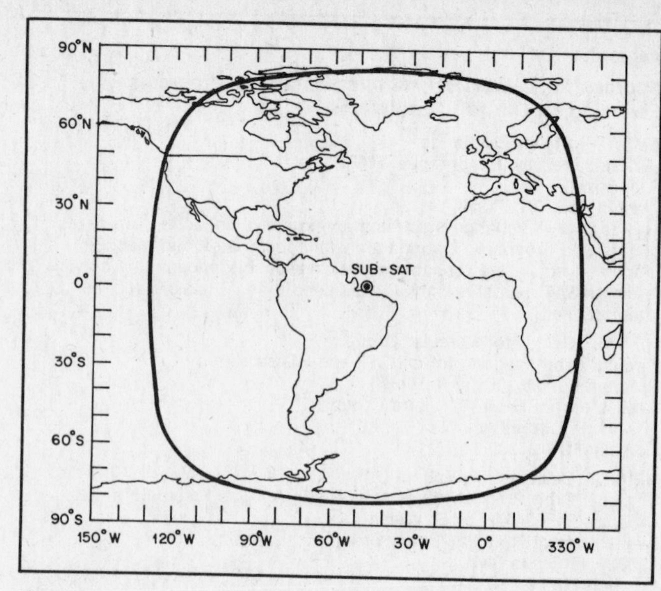

Fig. 8-21 — Coverage provided by a geostationary satellite at 0°N, 47°W. The access region corresponds to 0° elevation angle at ground station.

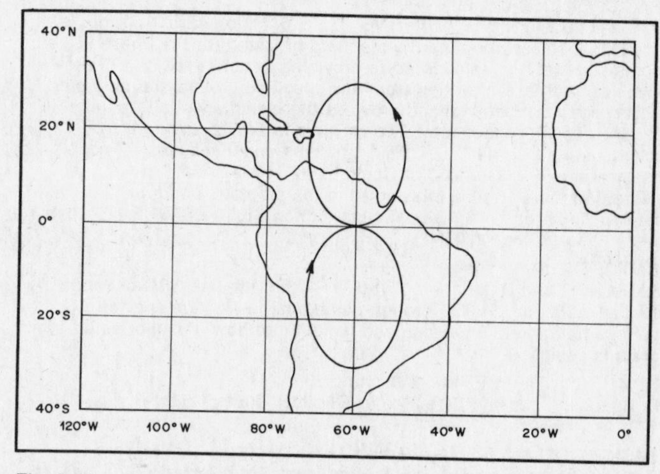

Fig. 8-22 — Ground track for satellite in circular orbit with a period of 24 hours and an inclination of 30°.

Designing an entire spacecraft destined for a geostationary orbit would introduce radio amateurs to new difficulties in attitude stabilization, antenna, and propulsion systems. Such problems can be resolved, but they do make producing a satellite like Phase III-A, -B or -C more attractive and less expensive. Furthermore, a realistic assessment of those launch opportunities that give access to geostationary orbits does not lead to optimism. It may, however, be possible to catch a ride into geostationary orbit aboard a large host satellite designed to accommodate several passengers. This concept was discussed earlier (see the SYNCART project in Chapter 3). Work on a transponder package for such a satellite is well underway.

If the orbital inclination angle of a satellite is not zero the spacecraft cannot appear stationary; stationary satellites can only be located above the equator. A 24-hour-period circular orbit of non-zero inclination will have a ground track like a symmetrical figure eight (see Fig. 8-22). Note that the ascending and descend-

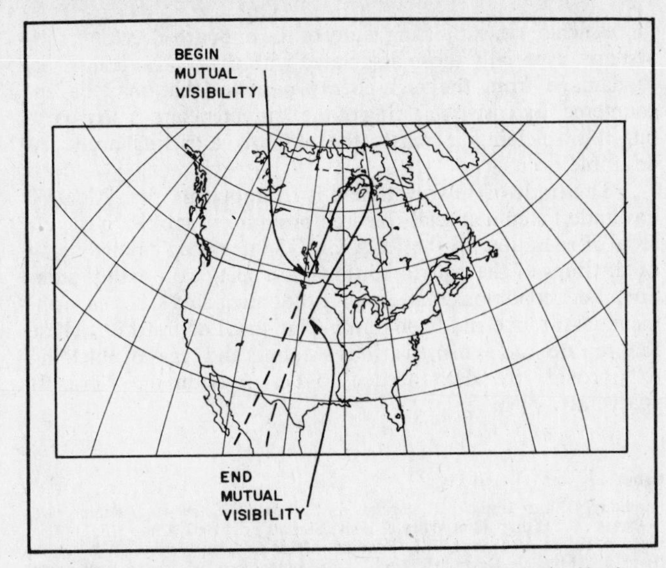

Fig. 8-23 — Typical Molniya II ground track with apogee over North America. The mutual visibility window is for the Washington, DC to Moscow path.

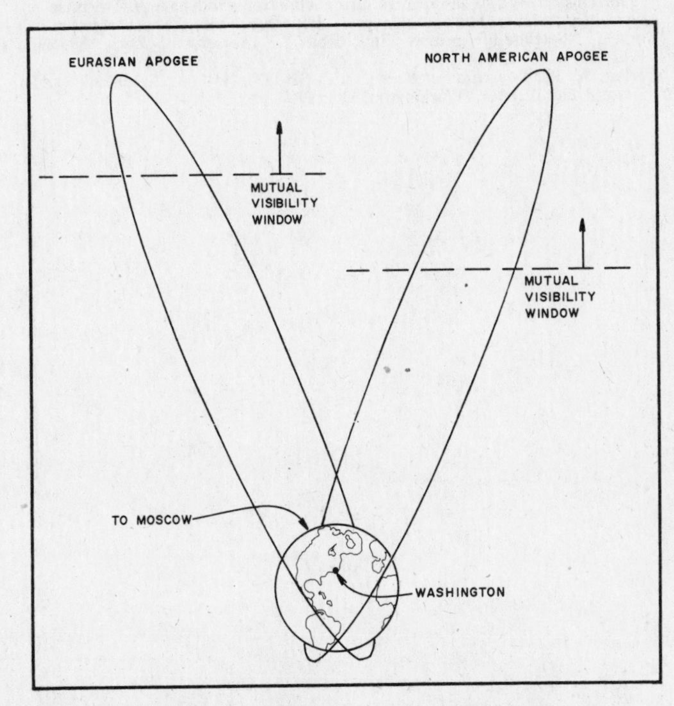

Fig. 8-24 — The relative positions of two successive Molniya II apogees. In actuality, the orbit plane remains fixed in space while the earth rotates. The mutual visibility windows shown are for the Washington, DC to Moscow path.

ing nodes of such an orbit coincide and the longitude of ascending node is constant (the increment is nearly zero). The 24-hour circular path is known as a *synchronous orbit*. The geostationary orbit is a special type of synchronous orbit: one with a zero-degree inclination.

Note that some authors apply the term synchronous (or geosynchronous) to other types of orbits, ones that are circular or elliptical with periods that are an exact divisor of 24 hours, such as 8 hours or 12 hours. Because this might lead to unnecessary confusion, we'll avoid this use of the term synchronous.

Molniya-Type Orbits

Looking at elliptical orbits earlier in this chapter we noted that the position of the perigee in the orbital plane (the argument of perigee) changes from day-to-day at a rate given by Eq. 8.12.

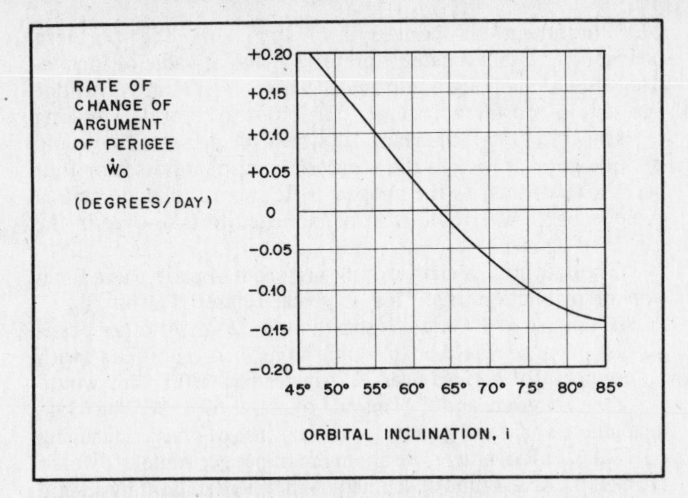

Fig. 8-25 — Rate of change of argument of perigee vs. orbital inclination for Phase III-type elliptical orbit (h_a = 35,800 km, h_p = 1500 km). See Eq. 8.12. When $\dot{w}_o$ is positive the argument of perigee rotates in the same direction as the satellite. When $\dot{w}_o$ is negative the argument of perigee rotates in the opposite direction.

An interesting feature of this equation is that when i = 63.4° the argument of perigee remains constant regardless of the values of the period and eccentricity. As a consequence, the argument of perigee, period and eccentricity can be chosen independently to satisfy other mission requirements.

Orbits with i = 63.4°, eccentricities in the 0.6 to 0.7 range, and periods of 8 to 12 hours have a number of features that make them attractive for communications satellites. Spacecraft in the Russian Molniya series were designed to take advantage of this type of orbit.

Let's take a brief look at a Molniya II series communications satellite of the type used for the Moscow-Washington Hotline. (The Hotline uses redundant Molniya and Intelsat links). The spacecraft is maintained in an orbit with an inclination of 63.4°, an argument of perigee constant at 270°, and a period of 12 hours. Because of the 12-hour period the ground track tends to repeat on a daily basis (there is a slow drift in longitude of ascending node). Apogee, where the satellite moves slowly, always occurs over 63.4° N latitude. At apogee nearly half the earth, most in the northern hemisphere, is in view. A typical Molniya orbit ground track is shown in Fig. 8-23. The primary Washington-Moscow mutual visibility window lasts 8 to 9 hours. Fig. 8-24 represents the orbit geometry for two consecutive orbits with apogees occurring 180° apart in longitude. In reality, the orbit remains fixed as the earth rotates so don't take the illustration too literally. Notice how on the Eurasian circuit there's a second good 4 to 5 hour Washington-Moscow window. A single spacecraft is accessible to the Washington station about 16 hours per day, to the Moscow station about 18 hours per day, and simultaneously to both stations about 12 hours each day. Thus, a three-satellite Molniya system is used to provide a reliable Washington-Moscow link 24 hours a day.

Interestingly, the Russians also use Molniya spacecraft for relaying domestic TV signals in the 4-GHz range and these satellites are active on the North American apogee as well as on the Eurasian apogee. Since relatively wide-beamwidth antennas are used and the format (except for color) is compatible with the U.S. system, anyone with a TVRO (4-GHz satellite *TV receive only* terminal) can monitor these transmissions by pointing his antenna northward (see Chapter 11 for additional information).

From an amateur radio point of view the Molniya-type orbit is very attractive. Amateurs, however, have not reached a consensus as to the best combination of orbital parameters. Many amateurs want an orbit with i = 63.4° and an argument of perigee of 270°. This inclination value simplifies tracking since the orbit overlay on a ϕ3 TRACKER never has to be changed. And, the

270° argument of perigee strongly favors the northern hemisphere. Other amateurs prefer a different value of inclination since a changing argument of perigee eventually gives one access to a considerably larger portion of the world. Thinking in terms of a long-term Phase-III system, such a satellite would provide good service to the southern hemisphere three or four years after launch as the apogee drifted south of the equator. A new spacecraft would then be launched to take over in the northern hemisphere.

The daily rate of change of the argument of perigee as a function of inclination angle for a typical Phase-III orbit (h_a = 35,800 km, h_p = 1500 km) is shown in Fig. 8-25. Another possible compromise would be to couple an inclination of 63.4° with an argument of perigee near 225° instead of 270°. This would place the apogee near 40° N instead of above 63.4° N. The 63.4° inclination would still give us the convenience of a never-changing ϕ3 TRACKER overlay. The argument of perigee trade-off would increase the access time of southern hemisphere stations by several hundred percent while decreasing the access time of stations north of latitude 40° N by only 25 percent (from roughly 16 to 12 hours daily). Stations north of 40° N latitude would no longer have access to apogees occurring on the opposite side of the earth.

Although the Molniya-type orbit clearly has several desirable features, it's not without shortcomings. Most are minor. Greater attention must be given to antenna aiming and Doppler shifts, but to a lesser degree than with low-altitude spacecraft. The major problem is one for the AMSAT spacecraft engineers. A satellite in a Molniya orbit traverses the Van Allen radiation belts twice each orbit, subjecting many of the onboard electronic subsystems, especially those associated with the central computer, to damage from the high-energy particles that may be encountered. Extensive shielding of the computer chips is necessary, but this shielding increases the weight, restricting access to desirable orbits.

The trade-off involved here is so important that AMSAT has undertaken a special research program to look into the effects of radiation on the RCA CMOS integrated circuits being used. Chips of the type to be flown are being exposed to radiation under conditions that simulate the anticipated space environment at Argonne and Brookhaven National Laboratories. These failure-rate studies using various amounts and types of shielding will provide the data needed to design optimum Phase-III spacecraft.

References

[1]Halliday, D. and Resnick, R., *Physics for Students of Science and Engineering Part I.* New York: John Wiley & Sons, Second Ed., 1962, Chap. 16.
[2]Symon, K. R., *Mechanics, 3/E.* Reading, Mass: Addison-Wesley, 1971.
[3]Bate, R., Mueller, D. and White, J., *Fundamentals of Astrodynamics.* New York: Dover Publications, 1971. In addition to being an excellent book, this text is a bargain. If you're interested in additional information on astrodynamics this is the first book to buy. Dover Publications, 180 Varick St., New York, NY 10014.
[4]Escobal, P., *Methods of Orbit Determination.* New York: John Wiley & Sons, 1976. This text is also an excellent introduction to astrodynamics. The price is typical for technical books at this level, about three times that of Reference 3.
[5]Kork, J., "Satellite lifetimes in elliptic orbits," *J. Aerospace Science,* Vol. 29, 1962, pp. 1273-1290.
[6]Corliss, W. R., *Scientific Satellites* (NASA SP-133), National Aeronautics and Space Administration, Washington, DC, 1967, p. 104.

Chapter 9
Tracking Topics

Chapter 9

Tracking Topics

To communicate through an Amateur Radio satellite, you must know where it is, whether (or when) it's accessible from your location and, if you're using directional antennas, what arc it will trace in the sky above. For a geostationary satellite, you'll need to know where to point and fix your antennas; for others, you'll want to know where to move your antennas and when. This task, tracking, is met and solved by every satellite user.

Several popular and useful satellite tracking methods were presented in Chapter 5. In the opening sections of this chapter we'll step back to view tracking in a broader sense. From our perch we'll look at a number of additional tracking approaches that amateurs have developed over the years and compare them with each other, and to the methods discussed earlier. By focusing on the advantages and disadvantages of each approach you'll see why OSCARLOCATOR-type devices are so popular for circular orbits and understand why the closely related ϕ3 TRACKER works so well for high-inclination elliptical orbits. You'll also be able to determine when other methods may be preferable for special applications.

The later sections of this chapter treat several special computational topics essential to tracking: determining satellite coverage, predicting azimuth and elevation angles from your ground station to a satellite at any time, and obtaining data to produce a range-circle "spiderweb" around any point on the earth.

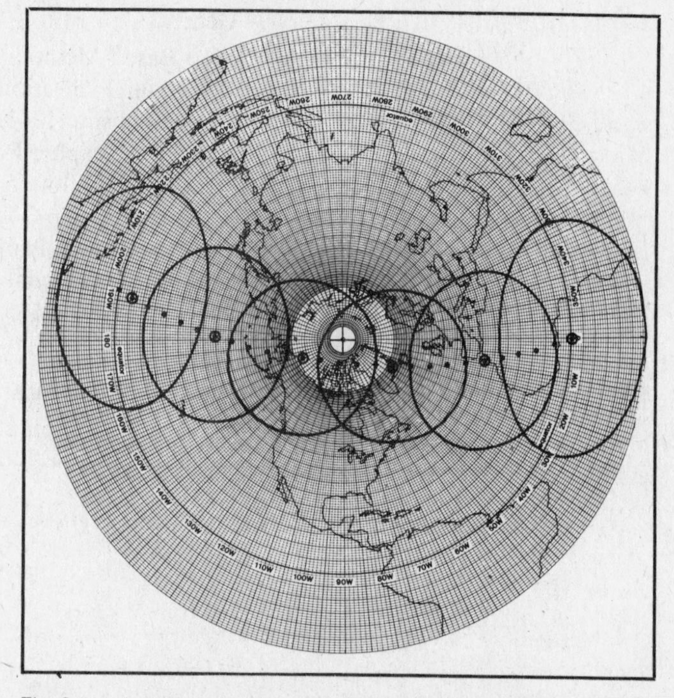

Fig. 9-1 — Access circles drawn about AMSAT-OSCAR 8 every 10 minutes beginning at ascending node at 0°W longitude.

General Description

Every tracking method presents selected pieces of information in a format that's designed to be convenient. Radio amateurs have developed and tested a great many approaches to satellite tracking because each of the several different groups of amateurs interested in the topic needs somewhat different information and has access to different resources (maps, computers, etc.). The user group can be partitioned in several ways: amateurs interested in communicating vs. those interested in satellite system-design and management, amateurs with microcomputers and those without, amateurs at high latitudes vs. those near the equator, and so on. Each partitioning will likely lead to tracking methods that have different features.

To illustrate the contrasting attitudes of different subgroups, consider the tracking approaches for a low-altitude, near-polar, circular-orbit satellite such as OSCAR 8. Both the system design/management and general-user viewpoints will be taken. The general user, following the suggestions in Chapter 5, might choose an OSCARLOCATOR and draw a spiderweb around his location. The spacecraft perspective, however, is much more useful to the design/management group. They might, for example, modify the OSCARLOCATOR (see Fig. 9-1) to determine what kind of coverage OSCAR 8 provides. Note that Fig. 9-1 shows access circles drawn about the spacecraft at fixed time intervals. Alternatively, by drawing a set of access circles at selected latitudes on the OSCARLOCATOR, a system designer could use it to compute the average daily access time as a function of ground station latitude. The averages can be figured quite accurately by setting the ground-track overlay for ascending nodes at longitude increments of 10°, noting the access times and tabulating the results.

Of course, the same information could be obtained using a computer but the map-based approach is often considerably quicker. A couple of morals are worth noting here: A computer should only be used if it's quicker, easier, cheaper, or provides a necessary precision that is unavailable with other methods; you don't need a computer to take an active role in providing impor-

Table 9.1
Tracking Device Requirements from User's Point of View

A satellite tracking method should enable one to predict:
1) when the satellite will be in range (times for AOS and LOS);
2) proper antenna azimuth and elevation at any time;
3) the region of the earth that has access to the spacecraft at any instant.

The tracking aid should, in addition, be simple to construct or program, easy to use, and inexpensive.

tant technical information to the AMSAT satellite user group.

Since readers of this handbook are most likely in the user subgroup, we'll focus on that viewpoint. The requirements outlined in Chapter 5 for a tracking device designed entirely from the users' point of view are restated in Table 9.1.

One common categorization of tracking approaches is whether they are computer-based (including sophisticated programmable calculators) or map-based. Although this appears to be a clear and distinct partition of approaches, it really isn't. Map-based methods have their roots in a set of calculations, performed one time, using the formulas in Chapter 8 (ground track) and Chapter 9 (azimuth, elevation, spiderweb). The results of these calculations are then used to construct ground-track overlays and spiderwebs that are keyed to a particular map. Most computer methods are based on the same set of formulas, but the calculations are repeated for each orbit and the results are usually presented in numerical form or used to drive an antenna automatically. Depending on the power, flexibility and human engineering that has gone into the programming, microcomputer methods may or may not be more convenient than map-based approaches. The real difference (besides cost) between the two methods is in how information is presented.

Each tracking method attempts to present data in a format that will be most useful to a particular set of users. The problem of data presentation is not confined to tracking, Amateur Radio or even science in general. It's one that occurs wherever a large amount of numerical material is handled. As computers make it easier and quicker to do thousands of calculations, developing methods for presenting the results in a meaningful format becomes more and more important. A good picture is often worth considerably more than a long list of numbers. Today, in response to this situation, the design of graphic display systems for microcomputers is receiving a great deal of attention. Indeed, this is currently one of the most exciting areas in microcomputing; applications to satellite tracking are just beginning to appear.

In the following section we discuss and compare map-based and computer-based tracking methods separately, keeping in mind that the distinction rests mainly on how data has been presented historically. In the future we'll likely see a merging of the two approaches as computer tracking programs begin to offer map-based visual displays that will probably bear a strong resemblance to the OSCARLOCATOR.

Map-Based Methods

Map-based methods such as the OSCARLOCATOR have passed the test of time. Their popularity endures because they satisfy user requirements well (Table 9.1).

Every map-based approach shows (1) a ground track and (2) a spiderweb, usually drawn about a ground station but sometimes shown in reference to the satellite position. The popularity of a particular approach depends on how easy it is to construct, to reposition the ground tracks and spiderwebs when necessary, and to use. Each approach usually depends on a particular type or class of maps. As all two-dimensional maps distort the globe, most projections are designed to minimize particular distortions such as area, distance or bearing on at least certain portions of the map. We'll discuss the important characteristics of each map as we look at it.

On certain types of maps (polar and rectangular) the shape of the ground track for circular orbits does *not* change. With these projections it's possible to draw a permanent ground track on a transparent overlay that can then be repositioned for each pass. On a polar map, repositioning means rotating the overlay about the pole; on a rectangular-coordinate map, it involves shifting the overlay horizontally along the equator. Polar maps have proved most popular among amateurs for several reasons: Their ground-track overlays are easier to reposition; mid-latitude ground stations cn approximate spiderwebs with circles and incur only a minor penalty in accuracy; and there's a quick and simple way to sketch ground tracks that we'll outline shortly. Other types of maps, however, may have advantages for certain orbits or ground station applications.

With elliptical orbits the ground track shape on polar and rectangular coordinate maps changes as the argument of perigee shifts. As we saw in the last chapter (Fig. 8-25), the argument of perigee changes slowly for high-inclination orbits ($i > 50°$) of the type planned for early Phase III missions. A single overlay should, therefore, be adequate for a month or longer, though a periodic change will be needed.

Polar Projection Maps

Polar projection maps, centered about either the north or south poles, are readily available (see Table 5.9). On these maps latitude curves are represented by a set of concentric circles centered on the pole, and longitude curves (meridians) by lines that radiate straight out from the pole. The various projections differ primarily in the spacing between latitude circles.

Tracking devices based on polar maps have proven extremely popular for satellites in low-altitude circular orbits and they should do likewise for high-altitude elliptical orbits with inclination angles greater than 50°. Note that the recommended polar projections feature a modest level of geographical shape distortion and extend well beyond the equator into the opposite hemisphere.

The three most common polar map projections are the equidistant, the stereographic and the orthographic. The equidistant is designed to show true distances from the pole, the stereographic is designed so that all circles on the globe will be shown as circles, and the orthographic shows what the earth would actually look like from a particular height above the pole. The first two are excellent for constructing OSCARLOCATORs and ϕ3 TRACKERs. We'll look at them in detail shortly. Because the vantage point of the orthographic projection severely compresses geographic features near the equator and does not, by definition, show any of the opposite hemisphere, the orthographic projection is poorly suited to tracking. Of course, if you were planning an arctic expedition, you might find the characteristics of the orthographic polar projection very useful.

As alluded to earlier, rough ground tracks can be sketched on polar maps using a shortcut that bypasses the calculations of Chapter 8. The shortcut, suitable only for low-altitude circular orbits has been useful several times in situations like getting a quick fix on RS-1 and RS-2 shortly after launch, and taking a look at non-amateur spacecraft such as the Shuttle on its first flight. Assume that a northern hemisphere polar projection map is being used and that you have a rough estimate of the satellite period, T (in minutes), and orbit inclination, i (in degrees). If an *a*scending *n*ode occurs at latitude $\phi_{an} = 0°$ N, longitude $\lambda_{an} = 0°$ W, then a *d*escending *n*ode will occur T/2 minutes later at $\phi_{dn} = 0°$ N, $\lambda_{dn} = 180° + (T/8)°$ W. Midway between these two points the spacecraft will be at its *n*orthernmost *p*oint: $\phi_{np} = i°$ N, $\lambda_{np} = 270°$ W + $(T/16)°$ W when i is between 0° and 90°, or $\phi_{np} = (180-i)$ °N, $\lambda_{np} = 90°$ W + $(T/16)$°W when i is between 90° and 180°. A roughly sketched curve joining just these three points gives a surprisingly good picture of the ground track of this particular orbit. Once this curve is transferred to a rotatable overlay, you're all set to track the satellite when it's over the northern hemisphere.

Equidistant Polar Projection

An equidistant polar projection map is characterized by equal

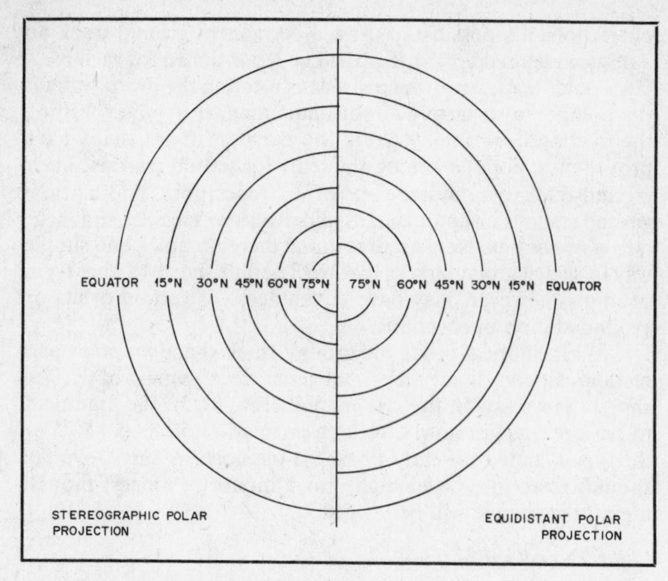

Fig. 9-2 — A comparison of latitude-circle spacing on stereographic and equidistant polar projection maps.

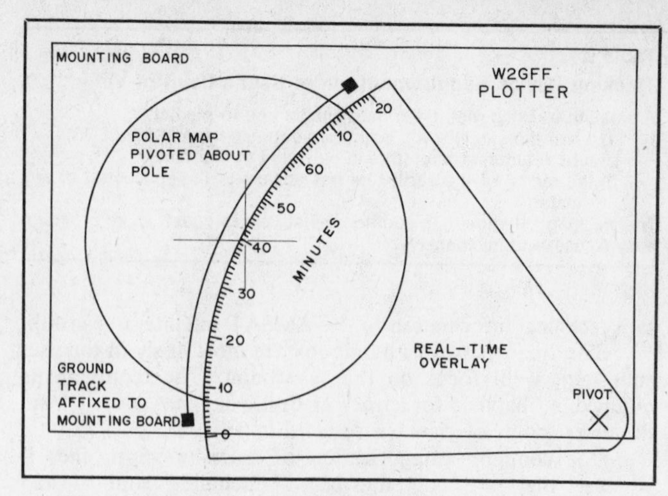

Fig. 9-3 — Key elements of W2GFF Plotter.

spacing between the concentric latitude circles. Fig. 9-2 illustrates the difference between equidistant and stereographic projections. The difficulty in using this type of map is that accurate spiderwebs are tedious to draw since range circles about a specific location are distorted. Despite this drawback equidistant polar maps have been very popular. We've standardized on this projection, extending the geographic coverage out to 30° S latitude, for the master ground-track overlays and spiderwebs in Appendix B.

Stereographic Polar Projection

Stereographic polar projection maps are characterized by increased spacing between latitude lines as one gets further from the pole (Fig. 9-2). The formula for latitude line position is

$$S = S_o \tan[(90° - \phi)/2]$$

where

S = distance between pole and latitude line
ϕ = latitude
S_o = distance between pole and equator, arbitrarily chosen to adjust overall size of map.

The stereographic projection has the interesting characteristic of preserving circles: A circle on the globe is also a circle on the map. Therefore, drawing range circles, acquisition circles or elevation circles about a particular location, or locating a mutual window for two ground stations is relatively easy.

To draw a range circle about a particular ground station (latitude ϕ_g, longitude λ_g), note that the center of the circle (ϕ_o, λ_o) does *not* coincide with the ground station; both do lie along the same meridian ($\lambda_o = \lambda_g$) however.

To find the latitude of the center of the circle (ϕ_o) and the radius of the circle:

1) Transform the range you're interested in into degrees of arc along the surface of the earth using 1.000° of arc = 111.2 km. Call the result $\Delta\phi$.

2) Compute $\phi_g + \Delta\phi$ and $\phi_g - \Delta\phi$ and plot both these points along meridian λ_g.

3) Bisect the line joining the two points found in Step 2. This gives the center of the circle.

4) With the center of the circle (Step 3) and two points on the circumference (Step 2) sketch the circle using a drawing compass.

The main drawback of the stereographic projection is that the mid-northern latitudes become significantly compressed if one draws a map extending out to latitude 30° S.

W2GFF Plotter

One variation of the OSCARLOCATOR, developed by R. Peacock, has gained widespread acceptance. It provides users of circular orbit spacecraft the convenience of a real-time readout. The key elements of the W2GFF Plotter© are shown schematically in Fig. 9-3. An equidistant polar map is attached to a mounting board at the pole so that it's free to rotate. A ground-track overlay, drawn on transparent stock, is permanently affixed to the mounting board above the map. Last, an adjustable real-time scale, also drawn on transparent stock, is placed on top of the other components and pivoted so it can be set at the ascending node. To preview a particular orbit one uses an orbit calendar to set first the polar map for the correct longitude of ascending node and then the real-time overlay for time at ascending node. The position of the spacecraft can then be followed in real time without the burden of mental arithmetic. Though it isn't shown in Fig. 9-3, the polar map of the W2GFF Plotter includes a spiderweb to provide azimuth, elevation, AOS and LOS data. An equidistant projection must be used so that the equally spaced time ticks on the real-time overlay will provide accurate SSP position data.

Rectangular Coordinate Maps

Rectangular coordinate maps (Mercator, Miller Cylindrical, etc.) have also been used for tracking by radio amateurs. A ground-track overlay for AMSAT-OSCAR 7 and a matching spiderweb as they would appear on a Mercator map are shown in Fig. 9-4. Note the severe distortion of the spiderweb. Users who have tried trackers of this type and the polar map models for low-altitude, near-polar, circular-type orbits almost universally prefer the polar map trackers. With high-altitude, high-inclination satellites the preference for polar maps will likely be even more emphatic, since over-the-pole communications paths are certain to be of special interest, and rectangular coordinate maps are poorly suited for analyzing what happens beyond the pole.

All this negative publicity for Mercator-type maps may tempt you to dump your cache into the nearest wastepaper basket. Don't do it. Rectangular coordinate maps may be very useful to mid-latitude ground stations for satellites in low-inclination orbits and to ground stations near the equator for a variety of orbits.

The careful reader of Chapter 5 probably noted that none

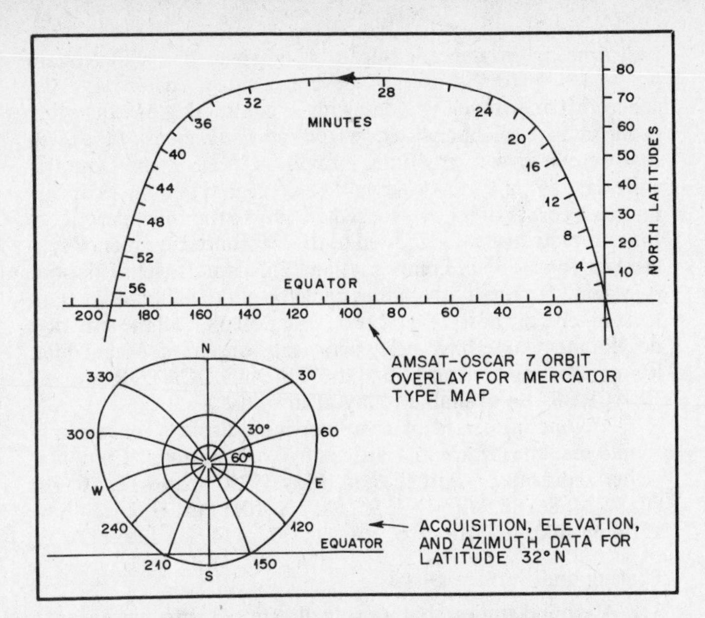

Fig. 9-4 — Ground-track overlay for AMSAT-OSCAR 7 and matching spiderweb for ground station at 32°N, prepared for Mercator type map.

of the tracking devices presented were described as being suitable for high-altitude, elliptical orbits with low inclinations. This wasn't an oversight. Effective map-based techniques for this type of orbit haven't yet been developed, in part because we haven't had the incentive of such an operational amateur spacecraft in recent years. Since the transfer orbit phases of Phase III spacecraft will probably involve such orbits for a considerable period of time during which the transponders may be opened for general communication, it's time to start looking into this problem.

Although a ϕ3 TRACKER could be used for following a high-altitude, elliptical-orbit, low-inclination spacecraft, the rapidly shifting argument of perigee would probably mean changing the ground-track overlay for every few orbits. A file of 36 to 72 ground-track overlays would therefore be required. This is obviously not a convenient approach.

One alternative that appears promising is to retain the color coding and fixed range circles of the ϕ3 TRACKER but switch to a Mercator projection map. As with any map method, we must consider the effects of the map projection on both the spiderweb and ground track.

Focusing on the ground track first, note that its shape is not affected by the longitude of the ascending node. So, for a particular argument of perigee, the ground track can be drawn on a transparent overlay and repositioned by shifting it horizontally along the equator. Interestingly, each ground-track overlay can serve for at least two values of argument of perigee if we flip it over and rotate it 180°. The big question is how close does the overlay have to be to the actual argument of perigee to provide reasonable azimuth and elevation predictions? The answer isn't in yet. If it turns out that an offset of less than 9° in argument of perigee is acceptable then we could use a set of only 10 ground-track overlays prepared with 18° increments for the argument of perigee. Experience with high elliptical-orbit satellites will tell.

Now let's take a close look at spiderwebs on the Mercator map. Because the low-inclination spacecraft orbits we're interested in almost always lie to the south of ground stations at mid-northern latitudes, each ground station need only include the southern half of the spiderwebs on the map. Looking at Fig. 9-4 we see that the southern half of the spiderweb can be approximated relatively accurately with half circles. East and West azimuth lines can be approximated roughly by straight lines connecting the ground station to its antipodal point.

These comments suggest that a tracker based on a Mercator map may be much more convenient than one based on a polar map for low-inclination elliptical orbits. These ideas are, however, largely untested. They're offered in the hope that someone will test, develop and provide how-to-do-it details if the approach is successful.

Table 9.2
Sources for Equidistant Projection Maps

Source: Defense Mapping Agency Hydrographic Center, Washington, DC 20390.
Central city and identification information:

Fairbanks, Alaska	WOXZP5180
Seattle, Washington	WOXZP5181
Honolulu, Oahu, Hawaii	WOXZP5182
San Francisco, California	WOXZP5184
Washington, DC	WOXZP5185
San Diego, California	WOXZP5190
Balboa, Panama	WOXZP5192
Yosami, Japan	WOXZP5193

Source: National Ocean Survey, NOAA, Chart Distribution Division — C44, Rockville, MD 20852. This agency also handles USAF Aeronautical Charts and Publications.
Central city and identification information:
New York City NOS 3042 (This map also contains two small, 6"-diameter, maps centered on London and Tokyo)

Source: Rand McNally & Co., P.O. Box 7600, Chicago, IL
Central city:
Wichita, Kansas

Source: William D. Johnston, N5KR, 1808 Pomona Dr., Las Cruces, NM 88001.
Central city:
Anyplace: Johnston can provide a custom computer-generated map centered about any coordinates you desire.

1983 prices for most are under $5, except for the custom computer-generated map, which is under $15.

Equidistant Projection (Ground Station Centered)

Earlier, we discussed equidistant projections that are centered on the pole. Maps of this type can also be drawn using any point on the earth as the center. On such a map, azimuth curves radiating from the center will be straight lines and range curves about the center will be true circles. Azimuthal equidistant projection maps are available centered on many large cities and a computer-generated map made to order for your particular location can be obtained at modest cost (see Table 9.2).

The following approach is suitable only for low-altitude circular orbits. Although it hasn't received the publicity of the OSCARLOCATOR, a good percentage of the people who have tried both do prefer the method based on the ground-station-centered equidistant projection.

One of the most tedious parts of building a map-based tracker is plotting the spiderweb. When using an equidistant projection centered on your location, spiderweb construction is trivial: Azimuth lines are already in place and range circles corresponding to particular distances or elevation angles are easily added using a drawing compass. The ground track situation is more involved. Unlike all the other map-based methods so far considered, the shape of the ground track depends on the longitude of the ascending node. As a result, we *cannot* draw a single ground-track overlay to reposition for each pass. Instead, we draw representative ground tracks for every 20° of longitude for ascending nodes that enter our acquisition circle (see Fig. 9-5). Latitude lines may be labeled with the time to nearest ascending or descending node. As a last step, the map is covered with a sheet of clear plastic. To preview any particular orbit, one uses "eyeball interpolation" with respect to the representative orbits previously drawn to locate the ground track of the orbit in question. Sketching in the orbit of interest with a felt-tipped marker, and erasing it with a tissue when no longer needed, works well.

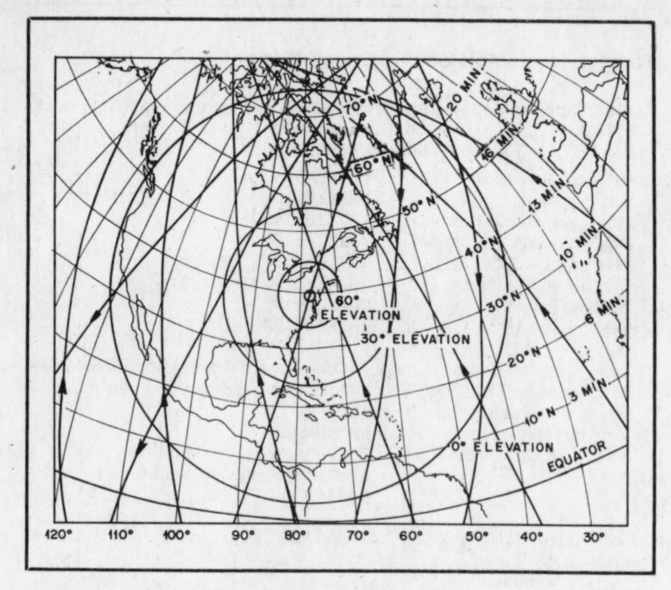

Fig. 9-5 — Azimuthal equidistant projection map centered on Washington, DC used as orbit calculator for AMSAT-OSCAR 7.

For additional information see: K. Nose, "Making Your Own Satellite Tracking Nomogram," *QST,* March 1974, pp. 40-41, 78.

Computer Methods

It's difficult to discuss computer methods in general terms since a great many computer-based approaches, ranging from the very simple to the very complex, have been described over the years. Nonetheless, several representative approaches will be mentioned.

Programming a small hand calculator, such as the HP-25, HP-41, HP-67, TI-58 or TI-59, to solve the circular-orbit ground-track equations of Chapter 8 and the azimuth and elevation equations to be presented later in this chapter is fairly easy. If you have a printer, a three-column listing giving clock time, azimuth and elevation can be printed for any orbit of interest. If a printer isn't available, the calculator can provide updated values every few minutes during a pass on request since the actual calculations take only a few seconds. Some users have included an appropriate time-killing delay loop in the program so that the calculator can be started as the satellite of interest crosses the equator or some other point. Azimuth and elevation are then read out in real-time. Moreover, azimuth and elevation can be displayed alternately or simultaneously. Discussing all these details would take us too far afield into the programming tricks of different calculators.

The powerful microcomputers now available have mainly been applied in an evolutionary manner. Programs (see Table 8.4) now handle elliptical orbits, provide several pieces of data simultaneously — in real time if desired — and automatically control antennas. Although systems of this type have a certain appeal, they generally have serious shortcomings: the time and expertise needed to develop or adapt software, an inability to present the overall picture that is provided by a map, the need to load programs, the effects of short power outages, RFI and so on. Many of these shortcomings are disappearing as prices drop and software becomes available. Computers are likely to take up a significantly larger portion of the tracking load when software gets away from the numbers on a screen approach and starts presenting information with good graphics.

One day soon many of us may use video terminals for tracking. The display might show a ground track for the entire orbit of interest (elliptical or circular orbit satellite of any inclination) superimposed on a world map (polar or Mercator projection at the user's option). A blinking light moving along the ground track would indicate the satellite position in real time, some future or past time, or in a speeded-up orbit preview. The choice again would be up to the user. A coverage circle, centered on the spacecraft and traveling along with it, contracting or expanding as the satellite height changed, would show all regions of the earth that were in view at any time. Azimuth, elevation and Doppler-shift data for any location could be presented in numerical format on a corner of the screen, fed directly to the antenna rotators for automatic tracking and fed to the transmitter frequency synthesizer for Doppler compensation. This is not science fiction. A system with these capabilities could be put together today. Unfortunately, the cost, several thousand dollars, and the software development time have been significant obstacles. Meanwhile, it's comforting to know that the OSCARLOCATOR and ϕ3 TRACKER are available today at low cost.

Anyone interested in computer map displays for tracking should read the informative articles by W. Johnston: "Computer Generated Maps," Part I, *Byte,* May 1979, pp. 10-12, 76, 78, 80, 82-84, 86, 88, 90, 92-94, 96, 98, 100-101; Part II, *Byte,* June 1979, pp. 100, 102, 104, 106, 108, 110, 112, 114, 118-119, 122-123.

Bearing and Surface Distance

A ground station that is using directional antennas needs to know where to point them. This information is usually presented as two angles: azimuth (angle in a plane tangent to the earth at the ground station, measured with respect to true north) and elevation (angle above the tangent plane). Bearing (the azimuth angle) can be computed when four quantities are known — the positions in latitude and longitude of both the ground station and the SSP. If, in addition, one knows the instantaneous height of the satellite, then elevation can also be computed. We'll look at each of these problems separately.

Finding the surface distance between two points on the earth and the bearing from one to the other is the basic problem of navigation. Although it can be solved using the information on right spherical triangles presented in Chapter 8, there's a more direct method based on the law of cosines for oblique spherical triangles. Since the derivation is readily available in introductory books on navigation we'll just include the results for reference. But first, note that given two points on the surface of the earth, only one great circle goes through both. If we were to look at the surface distances along various paths joining the two points, the minimum distance would be along the great circle path.

The formula for distance between two points along a great circle arc is

$$s = R\beta \qquad \text{(Eq. 9.1)}$$

where

- s = surface distance in the same units as R
- R = radius of earth in kilometers, statute miles or nautical miles
- β = central angle at geocenter in *radians* (angle between the line segments joining the geocenter to the two points of interest).

With this equation in hand, we can discuss surface distance in terms of either s or the associated central angle, β. The formula relating surface distance to the coordinates (latitude and longitude) of the two points is

surface distance:

$$\cos\beta = \sin\phi_1 \sin\phi_2 + \cos\phi_1 \cos\phi_2 \cos(\lambda_1 - \lambda_2 \qquad \text{(Eq. 9.2)}$$

where

- ϕ_1, λ_1 = latitude and longitude of point 1 (ground station)
- ϕ_2, λ_2, = latitude and longitude of point 2 (SSP or second ground station)
- β = central angle representing the short path (β between 0° and 180°) distance. Note: 1.000° of arc = 111.2 km = 60.00 nautical miles = 69.05 statute miles.

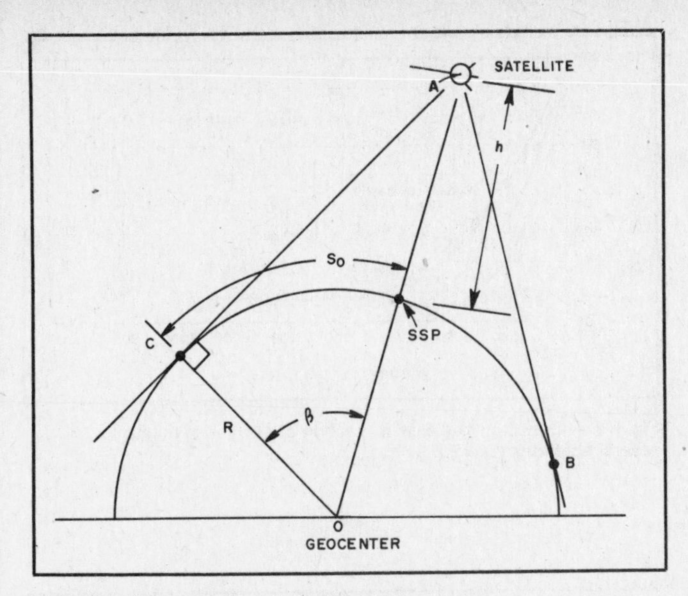

Fig. 9-6 — Cross section of satellite coverage cone.

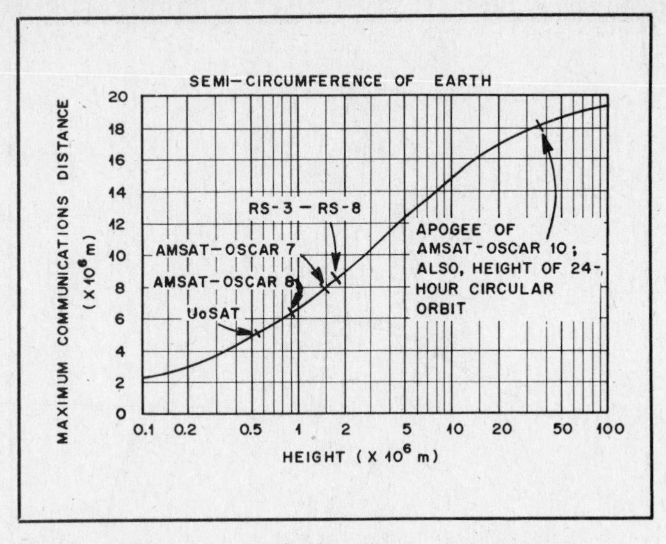

Fig. 9-7 — Maximum communications distance vs. instantaneous satellite altitude.

Sign conventions

Latitude: North (positive), South (negative).
Longitude: East (positive), West (negative).

The azimuth of point 2 as seen from point 1 is given by

$$\text{Azimuth: } \cos A = \frac{\sin\phi_2 - \sin\phi_1\cos\beta}{\cos\phi_1\sin\beta} \qquad \text{(Eq. 9.3)}$$

where

A = azimuth parameter. To obtain true azimuth of point 2 as seen from point 1, measured clockwise from north (0 to 360°) we note that the arccos function on calculators and computers only returns a value of A between 0° and +180°. If $\lambda_1 - \lambda_2$ is not between −180° and +180°, add or subtract 360° to bring it into this range. If the adjusted value of $\lambda_1 - \lambda_2$ is:
1) negative or zero, then true azimuth is given by A
2) positive, then true azimuth is given by 360° − A

Coverage

Because the radio frequencies used in conjunction with most satellites normally propagate over line-of-sight paths only, we will consider a communications satellite to be within range whenever the elevation angle at the ground station is greater than zero degrees. Depending on the actual propagation conditions, however, communication might not be possible until the satellite is well above the local horizon. Commercial satellite users typically use an elevation angle of +5° as their cutoff point for determining when a satellite is in range.

The locus of all lines through the satellite and tangent to the earth at a specific instant of time forms a cone (See Fig. 9-6). The intersection of this cone with the surface of the earth is a circle whose center lies on the line through the satellite, SSP and geocenter. Any ground station inside the circle has access to the satellite. Any two suitably equipped ground stations inside the circle can communicate via the satellite. The maximum terrestrial distance (between ground station and SSP) at which one can hear signals from the satellite is s_o. The maximum surface distance over which communication is possible is $2s_o$; see, for example, stations B and C in Fig. 9-6.

Solving for s_o as a function of satellite height requires only plane trigonometry. Since line AC is tangent to the earth, triangle AOC is a right triangle, $\cos\beta = R/(R+h)$, and s_o is given by $R\beta$. Therefore,

$$2s_o = 2R \arccos [R/(R+h)] \text{ (max. communication distance)} \qquad \text{(Eq. 9.4)}$$

where s_o = maximum access distance. See Sample Problem 9.1.

Sample Problem 9.1

Find the maximum communications distance for AMSAT-OSCARs 7, 8 and 10 at apogee given that

A-O-7, h = 1460 km,
A-O-8, h = 910 km,
A-O-10, h_a = 35,500 km.

Solution

Plugging the given values for h into Eq. 9.4 and using R = 6371 km we obtain for

A-O-7: $2s_o$ = 7907 km;
A-O-8: $2s_o$ = 6439 km; and for
A-O-10 at apogee: $2s_o$ = 18,069 km.

If your answers don't agree you probably forgot to convert the arccos [R/(R+h)] term into radians before multiplying by 2R. For reference, Eq. 9.4 has been plotted in Fig. 9-7.

Elevation and Slant Range

We now consider satellite elevation angle and slant range. The instantaneous elevation angle (ϵ) of a satellite can be obtained if (1) the instantaneous height, h, of the satellite above the surface of the earth and (2) the surface distance, s, between the SSP and one's ground station are known. We just saw (Eq. 9.2) that s can be found if the latitude and longitude of the ground station and SSP are known. Our object is to express ϵ in terms of s (or β) and h. In the course of determining the elevation angle, the slant range (line-of-sight distance between satellite and ground station) will also be found.

Once again, this problem can be solved using only plane

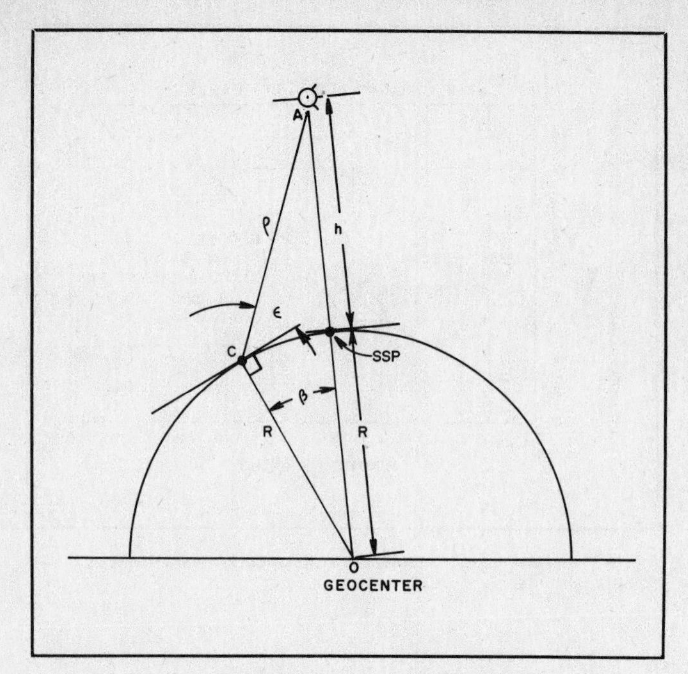

Fig. 9-8 — Diagram for determining satellite elevation angle and slant range as a function of height and distance to subsatellite point.

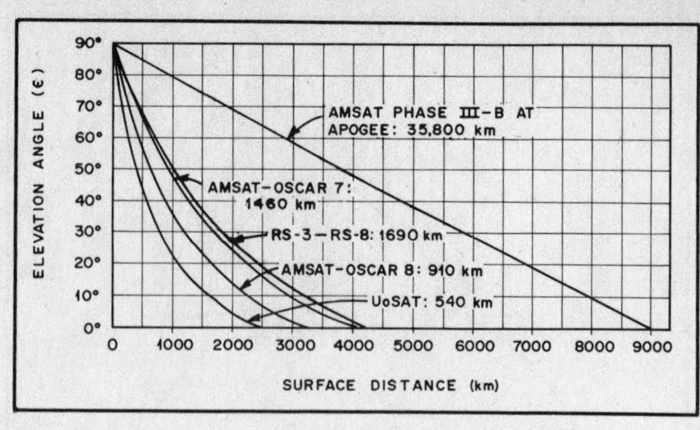

Fig. 9-9 — Elevation angle as a function of surface distance for several satellites (see Eq. 9.5b).

trigonometry. The parameters involved are shown in Fig. 9-8. Note the difference between this figure and Fig. 9-6. The solution is obtained as follows:

Focus attention on triangle AOC formed by the satellite, the geocenter and the ground station. Since the angles in a plane triangle must add up to 180° the included angle at the satellite is:

$$A = 180° - \beta - (\epsilon + 90°) = 90° - \beta - \epsilon.$$

Applying the Law of Sines to sides R and R+h we get

$$\frac{R+h}{\sin(\epsilon+90°)} = \frac{R}{\sin(90°-\beta-\epsilon)}$$

Using the basic trigonometric identity $\sin(90° \pm x) = \cos(x)$, we reduce this to

$$\frac{R+h}{\cos(\epsilon)} = \frac{R}{\cos(\epsilon+\beta)}$$

Next, the addition formula for the cosine function,

$$\cos(x+y) = \cos(x)\cos(y) - \sin(x)\sin(y)$$

is applied. This gives

$$\frac{R+h}{\cos(\epsilon)} = \frac{R}{\cos(\epsilon)\cos(\beta) - \sin(\epsilon)\sin(\beta)}$$

Finally, isolating all terms containing ϵ on the left hand side, we arrive at the desired formula

elevation angle: $$\tan(\epsilon) = \frac{(R+h)\cos(\beta) - R}{(R+h)\sin(\beta)} \qquad \text{(Eq. 9.5a)}$$

Using Eq. 9.1 we can rewrite the elevation angle formula in terms of surface distance instead of central angle:

elevation angle: $$\tan(\epsilon) = \frac{(R+h)\cos(s/R) - R}{(R+h)\sin(s/R)} \qquad \text{(Eq. 9.5b)}$$

Sample Problem 9.2

A ground station using a satellite in a circular orbit may find it convenient to have a graph of elevation angle vs. surface distance between ground station and SSP, since the latter is easily estimated from an OSCARLOCATOR. On a single set of axes, prepare graphs for OSCARs 7 and 8, UoSAT and RS-3 — RS-8. Also include OSCAR 10 at apogee. Since the height of OSCAR 10 changes very slowly near apogee, this curve will serve for several hours during each orbit.

Solution

Use Eq. 9.5b in conjunction with the following values

R = 6371 km
RS-3 — RS-8: h = 1690 km
OSCAR 7: h = 1460 km
OSCAR 8: h = 910 km
UoSAT: h = 540 km
AMSAT-OSCAR 10 at apogee: h = 35,500 km

The results are shown in Fig. 9-9. Note that the curve for OSCAR 10 is nearly a straight line. As a result, the following approximate expression

$$\epsilon[\text{in degrees}] \sim 90° - 0.01\, s[\text{in km}]$$

can be used for determining satellite elevation during the six hours centered on apogee.

Note that the arguments of the angles in Eq. 9.5b are given in radians. If you try to evaluate this expression with hand calculators that don't work in radians, be sure to input your angles in degrees.

An expression for slant range can be obtained by applying the Law of Cosines to the satellite-ground-station-geocenter triangle of Fig. 9-8.

slant range: $$\rho = [(R+h)^2 + R^2 - 2R(R+h)\cos(s/R)]^{1/2} \qquad \text{(Eq. 9.6)}$$

Elevation angle and slant range depend only on (1) the height of the satellite and (2) the surface distance between SSP and ground station. Eqs. 9.5 and 9.6, therefore, are valid for elliptical as well as circular orbits. See Sample Problems 9.2, 9.3 and 9.4.

Table 9.3
Terrestrial Distances for Drawing Elevation Circles

Satellite	*Mean Altitude (km)*	*Surface Distance (km)*					
		ϵ = 0°	ϵ = 15°	ϵ = 30°	ϵ = 45°	ϵ = 60°	ϵ = 75°
OSCAR 7	1460	3953	2580	1691	1099	667	316
OSCAR 8	910	3219	1924	1193	754	451	212
UoSAT	540	2535	1342	781	480	284	133
RS-3-RS-8	1690	4201	2806	1869	1226	748	355

Spiderweb Computation

A set of range circles about a specific location on the earth, and a set of azimuth curves that radiate outward from it are commonly referred to as a *spiderweb*. Most map-based tracking techniques use spiderwebs as an effective way to present information on satellite azimuth and elevation. To draw a spiderweb about a specific location on a map one needs the coor-

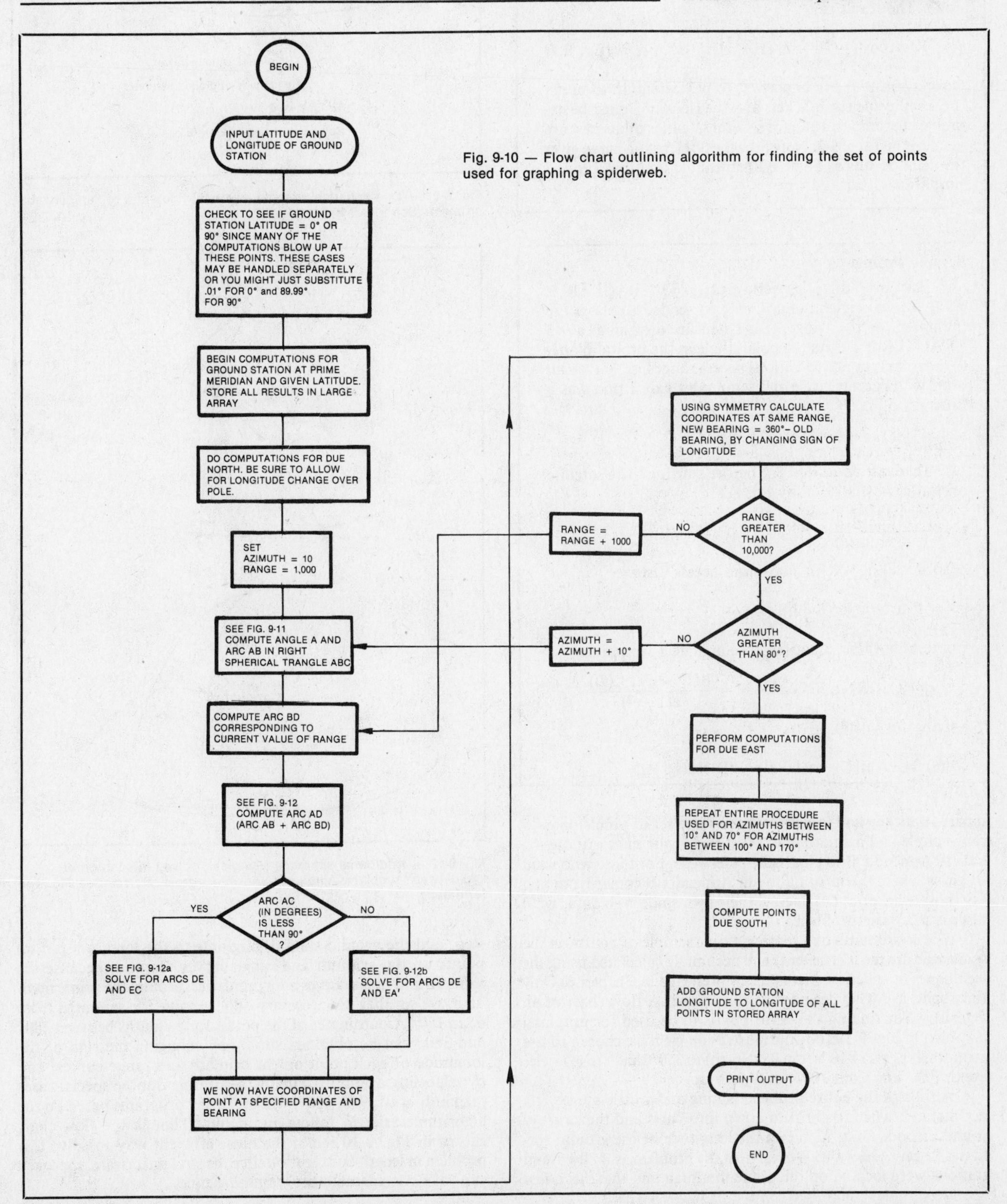

Fig. 9-10 — Flow chart outlining algorithm for finding the set of points used for graphing a spiderweb.

Sample Problem 9.3

The concentric circles that form a spiderweb on an OSCARLOCATOR are iso-elevation curves. Find an expression that gives the surface distance, s, corresponding to a given elevation angle, ϵ, for a satellite at height h.

Solution

To solve this problem we just rearrange Eq. 9.5b to obtain s as a function of ϵ and h. This is easier said than done. The result is

$$s = R[\arccos((\frac{R}{R+h})\cos \epsilon) - \epsilon] \quad \text{(Eq. 9.7)}$$

Once again it's very important to pay attention to units; the term inside the brackets and the isolated ϵ must be expressed in radians. Data for several satellites of interest are included in Table 9-3. Note how the 0° elevation curve gives the same results as Sample Problem 9.1 and that Eq. 9.7 simplifies to Eq. 9.4 when $\epsilon = 0°$.

Sample Problem 9.4

The ground-track overlay on a ϕ3 TRACKER is divided into segments that are color-coded to show the minimum access range. One step in designing a ϕ3 TRACKER is to find the polar angle in the orbital plane, θ, that corresponds to various access ranges, s_o. Given an orbit with eccentricity e and semimajor axis a find θ as a function of s_o.

Solution

Take a look at Fig. 8-2 before starting. The satellite orbital radius is given by Eq. 8.11:

$$r = \frac{a(1-e^2)}{1 - e\cos(\theta)}$$

Also note Eq. 9.4 for maximum access distance

$$s_o = R \arccos[R/(R+h)]$$

Combining these equations to eliminate r we obtain

$$\cos(s_o/R) = R/(R+h) = R/r = \frac{R(1 - e\cos(\theta))}{a(1-e^2)}$$

Solving for $\cos(\theta)$,

$$\cos(\theta) = (1/e)[1 - a(1-e^2)(1/R)\cos(s_o/R)]$$

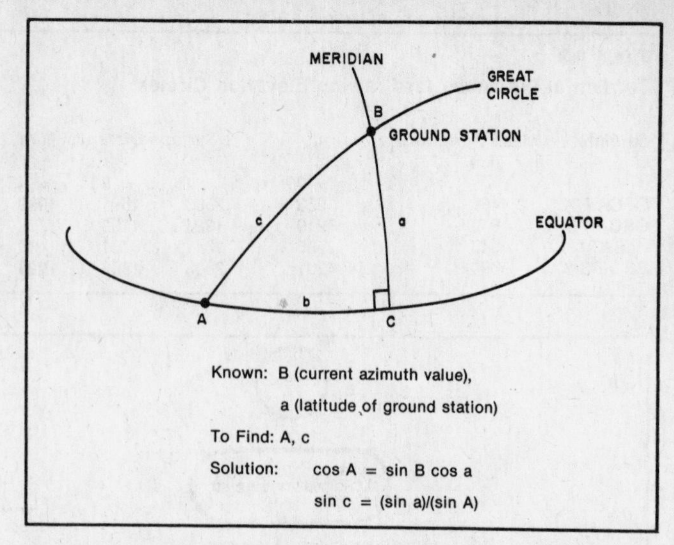

Fig. 9-11 — One spherical triangle involved in solution of "spiderweb" problem. (See Fig. 9-10 for details).

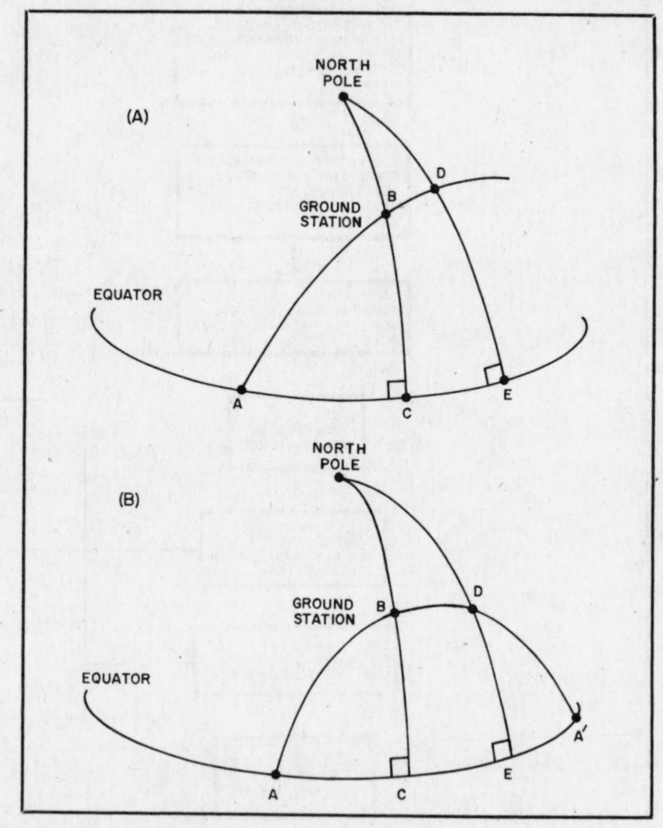

Fig. 9-12 — Additional spherical triangles involved in solution of "spiderweb" problem. All curves shown are great circles. (See Fig. 9-10 for details).

dinates (latitude and longitude) for a large set of points on each range circle and azimuth line. Most radio amateurs will use the data in Appendix B to plot their spiderwebs. For those who want to know where the information in Appendix B comes from and for those who want to generate their own spiderweb data, we'll discuss the basic problem.

The coordinates of a representative sample of points on the spiderweb are most efficiently and accurately calculated using the techniques of spherical trigonometry presented in Chapter 8. Only right spherical triangles need be considered. A flow chart of an algorithm for finding the set of points to be used for graphing is shown in Fig. 9-10. For the illustration we have chosen to use azimuth lines every 10° from 0° (North) to 350°, and range circles every 1000 km from 1000 to 10,000 km.

Although the entire problem is long and a little messy, it's not difficult when it's broken down into parts and the strategy is understood. First, note that the longitude of the ground station is largely irrelevant. We can treat the problem as if all ground stations were located on the prime meridian and then, as a *last* step, add the ground station longitude to the longitudes of all points on the azimuth and range curves. Second, because the azimuth and range curves are symmetrical around the prime meridian, we see that it's necessary only to solve for azimuths from 0° to 180°. Coordinates of the points for azimuths between 180° and 360° can be obtained simply by changing the sign of the longitude of each point on the azimuth and range curves, and by relabeling azimuths. Finally, separating out the special cases (azimuth = 0°, 90°, 180°) will make any programs based on the algorithm easier to follow and debug. Though the flow chart shown in Fig. 9-10 is not the most efficient way to solve this problem in length or execution time, its straightforward approach can save considerable programming time.

Chapter 10
Satellite Radio Links

- The Doppler Effect
  - Doppler Shift
  - Doppler Shift at Closest Approach
  - Doppler Shift Limits
    - Rotation of Earth
    - Satellite Motion: Circular Orbits
    - Satellite Motion: Elliptical Orbits
  - Doppler Shift and Transponders
  - Doppler: General
  - Anomalous Doppler
  - Doppler: Orbit Determination and Navigation
- Faraday Rotation
- Unusual Propagation
  - Sporadic-E
  - FAI
  - Antipodal Reception
  - Auroral Effects
  - General
- Frequency Selection
  - Legal Constraints
  - Technical Factors
    - Predicting Signal Levels: An Example
  - Frequency Management

Chapter 10

Satellite Radio Links

This chapter focuses on the radio signals linking satellites and ground stations. The topics we'll cover include basic physical phenomena such as Doppler shift and Faraday rotation, unusual forms of propagation that may be encountered, and a discussion of the process of selecting transponder frequencies.

The Doppler Effect

Have you ever noticed how the pitch of a whistle on a passing train appears to decrease? A passenger on the train, listening to the same whistle at the same time, wouldn't notice a change in frequency. Who's right? Both of you. The frequency of the sound you hear depends on the relative motion between the source (the whistle) and you, the observer. Since the train passenger moves along with the source while the distance between you and the source is continually changing, each of you observes a different audio frequency. The phenomenon is known as the Doppler effect (after Johann Doppler, 1803-1853).

Though radio waves are very different from the sound waves we've just been discussing, they do exhibit a similar effect: A monitor that is at rest with respect to a transmitter will measure a frequency f_o, while an observer who is moving with respect to the transmitter will measure a different frequency, f^*. The relation is given by

$$f^* = f_o \pm \frac{v_r}{c} f_o \qquad \text{(Eq. 10.1)}$$

where

f_o = frequency as measured by a monitor at rest with respect to the source (source frequency)

f^* = frequency as measured by an observer who is moving with respect to the source (apparent frequency)

v_r = relative velocity of observer with respect to source

c = speed of light = 3.00×10^8 m/s

Sign Convention

When this distance between source and observer is
decreasing, a "+" sign is used ($f^* > f_o$)
increasing, a "−" sign is used ($f^* < f_o$)

Eq. 10.1 is often written

Doppler shift: $$\Delta f = f^* - f_o = \pm \frac{v_r}{c} f_o \qquad \text{(Eq. 10.2)}$$

We'll settle for an intuitive understanding of Eq. 10.1 and leave the details of the derivation to a physics text. The situation is easier to grasp when expressed in terms of period; since period = 1/frequency, we don't lose anything by doing so. Note that the period we use here is *not* the orbital period of the satellite but the time for one complete cycle of the transmitted radio wave to pass. Refer to the moving satellite and fixed observer located at G in Fig. 10-1. For convenience consider that the source is transmitting a linearly polarized wave and that the period is the time interval between two successive crests (occurring at A and B) in the transmitted electric field (E-field). From the diagram it's clear that the slant range $\overline{AG}$ is longer than the slant range $\overline{BG}$. Therefore it takes less time for a signal that is sent from B to reach G than it does for a signal that is sent from A. Our observer at G, recording the time interval between the two successive E-field crests, will therefore record the time as being shorter than that measured by someone who remains equidistant from the satellite or who is sitting on it. So, for an approaching satellite, our observer records a shorter period than that measured at the spacecraft. Consequently, the observed frequency is higher.

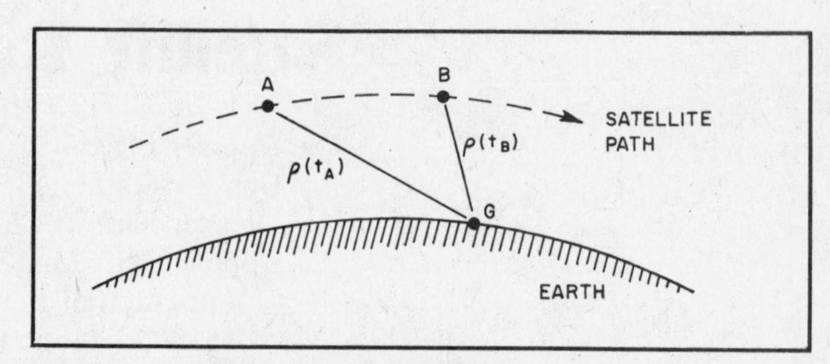

Fig. 10-1 — Doppler shift is observed when the distance between the satellite and ground station is changing.

Two primary questions related to Doppler shift are of interest to satellite users: (1) What will the actual Doppler shift be on a given link at a given time? and (2) What is the maximum Doppler shift that can be expected on a given link? We'll look at each of these questions in turn.

Doppler Shift

To calculate the instantaneous Doppler shift (the shift at any instant) we apply Eq. 10.1 to Fig. 10-1. If we assume that the source frequency is known, then the only unknown quantity in Eq. 10.1 is v_r, the relative velocity. Relative velocity during a short time interval can be approximated by dividing the change in slant range by the change in time, i.e.

$$\overline{v}_r = \frac{\rho(t_B) - \rho(t_A)}{t_B - t_A} \qquad \text{(Eq. 10.3)}$$

where

$\overline{v}_r$ = approx. relative velocity
t_B = time satellite passes point B
t_A = time satellite passes point A
$\rho(t_A)$ = slant range at time t_A
$\rho(t_B)$ = slant range at time t_B

To calculate the slant range at two times we can apply Eq. 9.6 that gives us slant range as a function of satellite height and the ground station to SSP distance. In sum, if you have a computer or calculator program that is written to predict basic tracking information (latitude and longitude of SSP and satellite height), adding Eq. 9.6 for slant range and Eqs. 10.2 and 10.3 for Doppler shift is a simple matter.

Doppler Shift at Closest Approach

A graph of apparent frequency against time for a specific satellite pass and ground station is called a Doppler curve. A typical Doppler curve, plotted from observations made during an AMSAT-OSCAR 7 pass, is shown in Fig. 10-2. For circular orbits the steepest part of the graph occurs at the point of closest approach (position where slant range is a minimum) and the observed frequency at this point is equal to the actual source frequency. Referring to Fig. 10-2 we can determine that closest approach occurred at time 10:31:20 and the source frequency is 145.9727 MHz ± the accuracy of our frequency measurement. From the steepness (slope) of the curve at closest approach we can compute the minimum slant range using the formula

$$\rho_o = -\frac{f_o v^2}{c\, m^*} \qquad \text{(Eq. 10.4)}$$

where

ρ_o = slant range at closest approach (minimum slant range)
f_o = transmitter frequency
v = magnitude of satellite velocity (note: this is *not* the relative velocity discussed above)
c = speed of light = 3.00×10^8 m/s
m* = slope of tangent line at TCA

To obtain m* from an experimental graph like the one in Fig. 10-2, align a transparent ruler over the central part of the curve until you get the steepest match and draw that line. Using any two convenient times complete the right triangle as shown. Slope m* is given by the ratio of the vertical side to the horizontal side of the triangle.

While situations occur in which one may want to predict the actual Doppler shift, most radio amateurs using a satellite transponder are satisfied to monitor the downlink and just twiddle their transmitter frequency control while sending a few dits until the downlink is positioned at a desired spot. One particularly useful piece of information for operators, however, is the value of the maximum Doppler shift on a link that any ground station might see.

Doppler Shift Limits

At any given time there's a maximum and minimum Doppler shift that can be seen by any ground station. During most orbits, however, most stations will observe a shift somewhere between these two extremes. For a circular orbit the two limits remain constant; for an elliptical path the limits vary over the course of the orbit.

We'll consider both cases, but first let's look at the two factors that contribute to the relative velocity term in Eq. 10.1: (1) satellite motion in the orbital plane and (2) rotation of the earth about the N-S axis. In any given situation these two factors can be combined (velocities add as vectors) to produce a relative velocity having a magnitude that can range from the arithmetic difference to the arithmetic sum of the two components. Since our objective is to determine the worst-case limits for a practical situation, we need only calculate each contribution separately and then form the sum and difference. First we look at the angular rotational velocity of the earth.

Rotation of Earth: The earth rotates about its N-S axis at an angular velocity of approximately

$$\dot{w}_E \sim 360°/\text{day} = 15°/\text{hour} = 0.25°/\text{min} = 0.000{,}073 \text{ radians/sec.}$$

The tangential velocity of a point on the surface of the earth at latitude ϕ, is

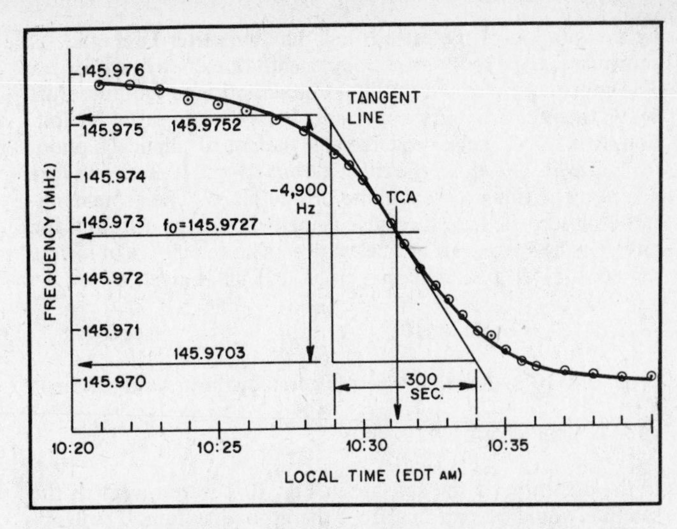

Fig. 10-2 — AMSAT-OSCAR 7 Doppler curve, orbit 7603, 14 July 1976 as observed from Baltimore, MD. Using the triangle shown we evaluate the slope at TCA: m* = (−4900 Hz)/(300 sec) = −16.3 Hz/s. The satellite velocity was determined in sample problem 8.4, v = 7.13 km/s. Applying Eq. 10.4 we obtain the slant range at closest approach, ρ_o = 1520 km.

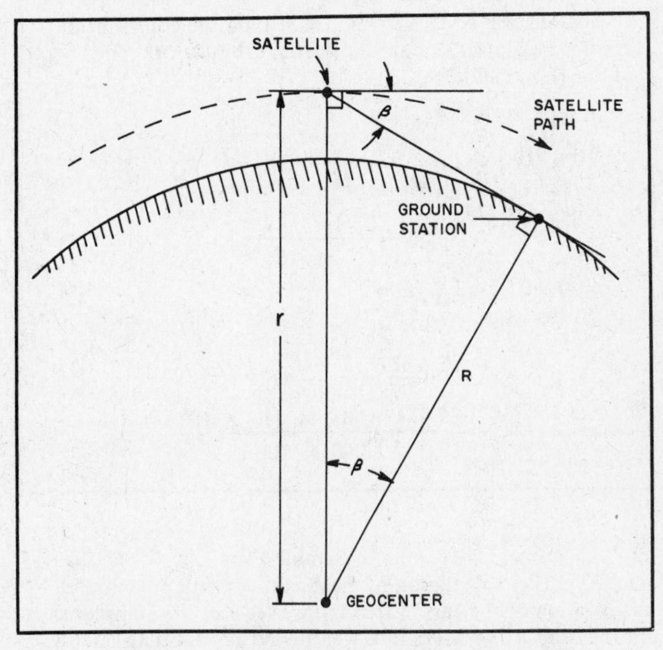

Fig. 10-3 — Geometry for computing contribution to worst-case Doppler shift from satellite motion only (circular orbit).

$$v_E = \dot{w}_E R \cos\phi$$

where

R = radius of earth
ϕ = latitude
$\dot{w}_E$ = angular velocity of earth (expressed in radians)

The maximum value of v_E will occur at the equator, v_E (max.) = 465m/s. To get a handle on the size of the Doppler shift that arises exclusively from the rotation of the earth, assume a link frequency of 146 MHz and a ground station on the equator that sees a satellite due east on the horizon. (In this position the tangential velocity of the earth and the relative ground-station-to-satellite velocity are equal.) Using Eq. 10.2,

$$\Delta f = \frac{465}{3.00 \times 10^8} \times 146 \times 10^6 = 226 \text{ Hz}$$

So, at 2 m, the worst-case contribution to Doppler shift produced by the rotation of the earth is less than a quarter kilohertz. As shown in Fig. 10-2, observed Doppler shifts are often much larger. The contribution of the satellite orbital motion to Doppler shift must, therefore, be very important. Let's look at this contribution, first for a circular orbit, then for the case of elliptical motion.

Satellite Motion: Circular Orbits. Fig. 10-3 shows the geometry of this problem in the orbital plane. The ground station that sees the largest relative velocity lies in the orbital plane and sees the spacecraft at 0° elevation. The velocity (v in meters per second) of a satellite in a circular orbit is given by

$$v^2 = \frac{GM}{r} = 3.986 \times 10^{14} (1/r) \qquad \text{(see Eq. 8.7)}$$

From Fig. 10-3, the relative velocity seen by the ground station is

$$v_r = v \cos\beta = v\frac{R}{r}$$

If the direction of the satellite in Fig. 10-3 were reversed, the satellite would be receding from the ground station. The Doppler shift would be equal in magnitude, but would represent a decrease in frequency. A short sample problem shows how this information can be applied.

Sample Problem 10.1

Consider AMSAT-OSCAR 8. Find the contribution to the maximum Doppler shift from orbital motion on the 146-MHz uplink.

Solution

$h = 910$ km

$R = 6371$ km

$r = 7281$ km

$$v^2 = 3.986 \times 10^{14} \left(\frac{1}{7.281 \times 10^6}\right)$$

$$= 0.5475 \times 10^8 \text{ (m/s)}^2$$

$v = 7399$ m/s

$$v_r = v\frac{R}{r} = 7399 \left(\frac{6371}{7281}\right) = 6474 \text{ m/s}$$

$$\Delta f = \frac{v_r}{c} f_o = \frac{6.474 \times 10^3 \times 146 \times 10^6}{3.00 \times 10^8}$$

$= 3150$ Hz

Combining the results of Sample Problem 10.1 (3150 Hz) and the earlier calculation of the maximum shift contributed by the rotation of the earth (226 Hz), we see that the Doppler shift on the A-O-8 146-MHz link will always be less than 3376 Hz. For our needs it's more appropriate to express the maximum shift as less than ±3.4 kHz. Note that this is the maximum shift from the actual, transmitted frequency, 3.4 kHz above transmit frequency as the satellite approaches and below transmit frequency as the satellite recedes. Consequently, the actual range of observed frequencies would be 6.8 kHz for a 146-MHz satellite link. This conclusion is true for a ground station monitoring a downlink and for an "imaginary observer" aboard A-O-8 listening to an uplink.

Satellite Motion: Elliptical Orbits. We now consider the contribution to Doppler shift provided by satellite motion when the orbit is an ellipse. The geometry, shown in Fig. 10-4, is somewhat more involved. The ground station observing the largest velocity (station A) lies in the orbital plane and sees the spacecraft at a 0° elevation. The ground station observing the smallest relative velocity (station B) also lies in the orbital plane and sees the spacecraft at a 0° elevation. The following series of steps enables one to compute the Doppler for stations located at the special points A and B. Note that the locations of A and B change; we're not considering two fixed stations. We assume that the semimajor axis (a) and eccentricity (e) are known.

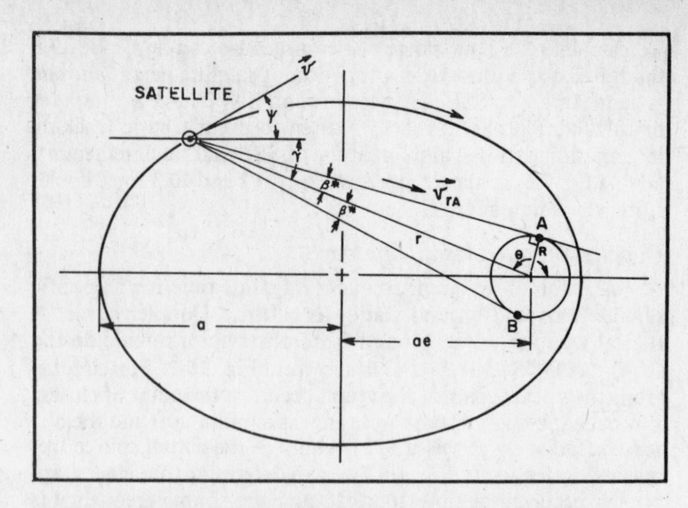

Fig. 10-4 — Geometry for computing contribution to worst-case Doppler shift produced by satellite motion only (elliptical orbit). Ground station A sees maximum shift. Ground station B sees minimum shift.

Step 1. Use Eqs. 8.9 and 8.10 to solve for θ.

Step 2. Use Eq. 8.11 to solve for r.

Step 3. Solve for the angle ψ shown in Fig. 10-4.
$\psi = \arctan[(e^2 - 1)(r\cos\theta + ae)/(r\sin\theta)]$
(Note: This equation was derived using the techniques of elementary calculus to solve for the slope of a line that is tangent to an ellipse.)

Step 4. Solve for β^* using $\sin\beta^* = \frac{R}{r}$

Step 5. Solve for $\gamma = 180° - \theta$

Step 6. Solve Eq. 8.7 for the satellite velocity, v.

Step 7. Solve for the relative velocity seen by
ground station at A: $v_{rA} = v\cos(\psi + \gamma - \beta^*)$ and
ground station at B: $v_{rB} = v\cos(\psi + \gamma + \beta^*)$

Step 8. Solve for the Doppler shift using Eq. 10.2.

In preparation for the AMSAT Phase III-A launch, this procedure was applied to the transfer orbit since Doppler shift limits were needed to design the command station network. The resulting graph that includes the effects of satellite motion and rotation of the earth is shown in Fig. 10-5. Keep in mind that this is *not* the Doppler shift seen by a particular ground station. The graph represents the maximum possible shifts that could be seen from somewhere on the earth at any instant during the orbit. Note that for elliptical orbits Doppler shift limits are generally inversely related to satellite height: greatest near perigee, least near apogee.

Doppler Shift and Transponders

So far, all our Doppler shift calculations have focused on a single link. When communicating via a transponder there are two links involved — the uplink and the downlink. When a non-inverting transponder is used the Doppler shifts on the two links add, so one can use Eq. 10.2 twice or, equivalently, add uplink and downlink frequencies and plug the result into Eq. 10.2 in place of f_o. For example, all Mode-A transponders to date have been non-inverting. Doppler shift calculations for the combined link should be performed with the combined frequency value of 175 MHz (146 + 29) for f_o. When an inverting transponder is used (as on Modes B, J and L) the frequency shifts on the two links are in opposite directions, one link an apparent increase, the other an apparent decrease. To calculate the total shift for the combined link, use the difference between the two link frequencies for the value of f_o in Eq. 10.2. For example, on Mode B the appropriate value for f_o is 289 MHz (435-146). For reference, several values of maximum expected Doppler shifts

Table 10.1
Maximum Doppler Shifts on Various Satellite Radio Links

	Maximum Doppler Shift (kHz) Satellite/height				
Beacon frequency	*UoSAT 530 km*	*OSCAR 8 910 km*	*OSCAR 7 1460 km*	*Phase III perigee 1500 km*	*Phase III apogee 35,800 km*
29 MHz	0.8	0.7	0.7	(0.7)	(0.1)
146 MHz	3.7	(3.4)	3.1	3.9	0.4
435 MHz	10.9	10.1	9.1	11.6	1.1
1260 MHz	31.5	(29.2)	(26.4)	(33.4)	(3.1)
2401 MHz	59.9	(55.6)	50.2	(63.6)	(5.9)
10.47 GHz	262.0	(243.0)	(220.0)	(278.0)	(25.3)
Transponder Frequencies					
146/29 MHz non-inverting	(4.4)	4.1	3.7	(4.7)	(0.5)
435/146 MHZ or 146/435 MHz inverting	(7.3)	6.7	6.1	7.7	0.7
1260/435 MHz inverting	(20.6)	(19.1)	(17.3)	21.9	2.0

Figures shown in parentheses are for links that are not now in use and are given for reference purposes only. Values quoted include contributions of both satellite motion in the orbital plane and the rotation of the earth.

for beacons and transponders are presented in Table 10.1.

Doppler: General

From our earlier analysis of the two factors (satellite motion and earth rotation) that contribute to the relative velocity between a satellite and a ground station, it should be clear that different ground stations will see different relative velocities at the same time. As a result, on a communication link involving two or more ground stations, each will receive the downlink signal at a slightly different frequency. For example, if we're in contact and I adjust my transmitter so I hear both of our downlinks on the same frequency, you're likely to notice a small offset. For an operator this is only a minor inconvenience as long as everyone understands what's going on and doesn't waste time hopelessly trying to adjust transmitter frequencies so that everyone hears everyone else on the same downlink frequency.

The following operational procedure is offered to minimize the interference caused to other stations by unwitting operators "chasing each other across the band." In an ssb contact among any number of individuals, one station with a stable receiver should be designated as the frequency reference. The station so designated should *not* touch his receiver dial. As needed during the pass, the reference station should adjust his transmitter frequency for clear ssb reception of his own signal. All other stations should touch up their transmission frequencies whenever

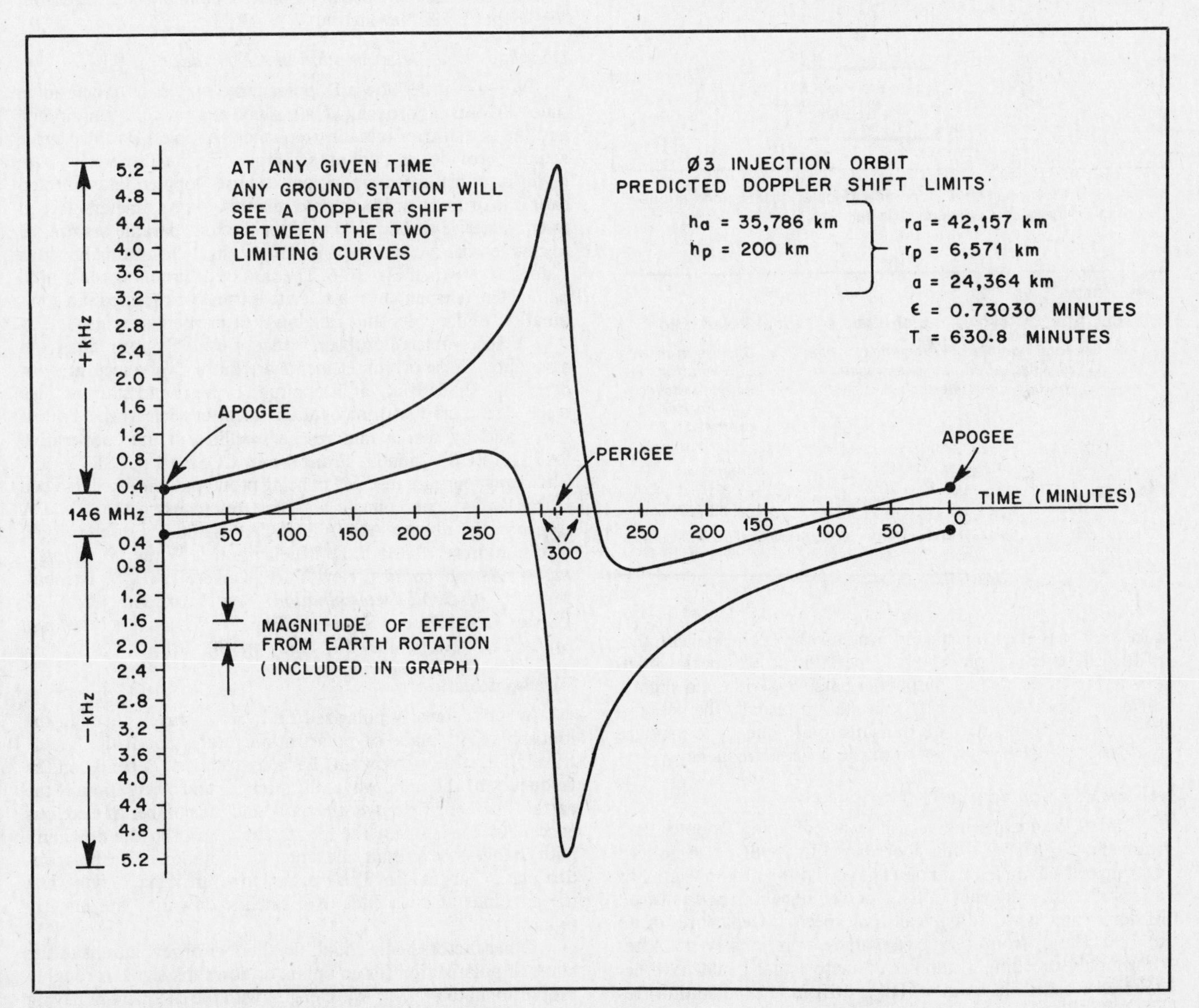

Fig. 10-5 — Doppler shift limits for Phase III injection orbit as seen by ground station anywhere on earth.

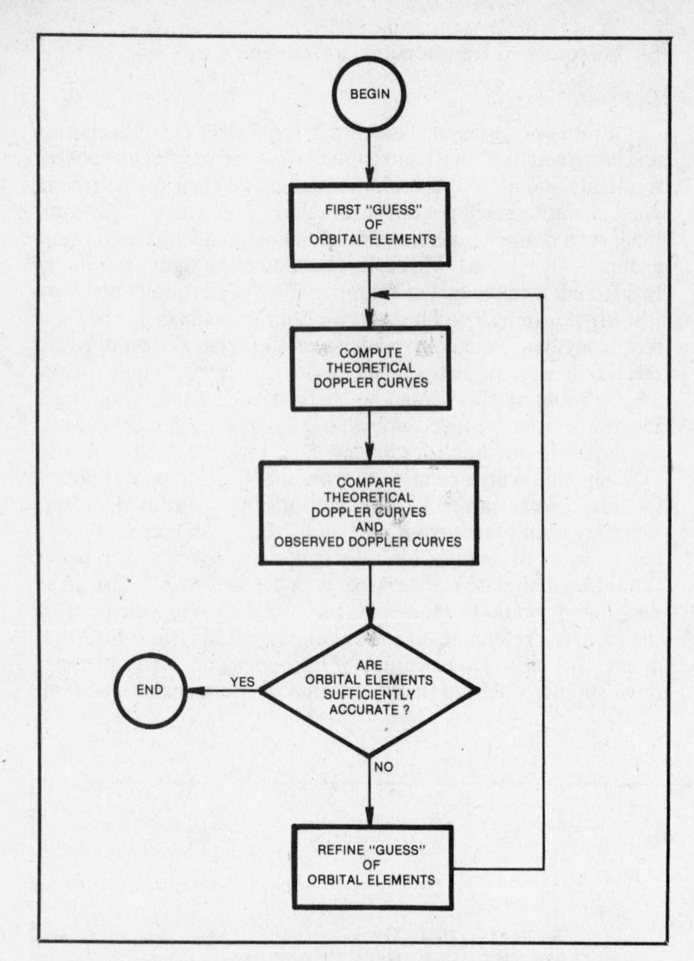

Fig. 10-6 — Flow chart illustrating how Doppler observations are used to compute the orbital elements of a satellite.

Table 10.2

Faraday Effect: Revolutions of Plane of Signal Polarization†

Frequency	Total number of revolutions of plane of polarization as signal passes through ionosphere: 0° elev.	90° elev.	Change in number of revolutions as satellite moves from horizon to overhead
29 MHz	89	28	60
146 MHz	4.4	1.4	3
435 MHz	0.8	0.25	0.5

† As predicted by the simple model outlined in G. N. Krassner and J. V. Michaels, *Introduction to Space Communications Systems*, New York: McGraw-Hill 1964.

necessary so that each hears his own downlink coincide with that of the reference station. Stations operating on the special channels at the edges of the transponder passband will use a slightly different procedure. The reference station (usually the net control operator) will adjust his transmission frequency to produce the desired downlink offset from the beacon frequency.

Anomalous Doppler

In 1972 an experimenter who was collecting Doppler data from the 435-MHz beacon aboard AMSAT-OSCAR 6 noticed a strange effect on a northbound pass. For the first few minutes after AOS, the frequency of the observed signal increased instead of decreasing as would normally be expected. Departures of up to 700 Hertz from predicted values were observed. After thoroughly checking a number of factors that could have accounted for the observations (e.g., drift in ground station frequency measuring equipment, change in satellite temperature that would affect the beacon oscillator frequency, etc.) it was concluded that an interesting physical anomaly was being seen. The effect was later observed on navigational satellites that were operating near 400 MHz.

An exhaustive experimental investigation was undertaken to delineate the spatial and temporal (time of day, season of year, etc.) extent of the anomaly and to determine the frequency range over which it occurred. It was hoped that this data would make it possible to correlate the effect with physical changes in the ionosphere that were suspected of being related. Although a great deal of data was collected, no firm conclusions have ever been reached as to the cause of anomalous Doppler. For additional information see W. Smith, "Doppler Anomaly on OSCAR 6 435-MHz Beacon," *QST*, May 1973, pp. 105–106; J. Fox and R. Dunbar, "Inverted Doppler Effect," *Proceedings of the ARRL Technical Symposium on Space Communications*, Reston, VA, Sept. 1973, Newington, CT: ARRL, pp. 1-30.

Table 10.3

Variables Affecting Strength of the Downlink Received Signal.

1) Satellite (antenna) orientation with respect to ground station.
2) Satellite spin producing a time-dependent antenna pattern.
3) Changing slant range (inverse power law).
4) Signal absorption in the ionosphere.
5) Ground station antenna pattern.
6) Faraday rotation.

Doppler: Orbit Determination and Navigation

We saw earlier how a Doppler curve enables us to determine time of closest approach (TCA), slant range at closest approach and the actual transmission frequency. A single Doppler curve actually provides us a unique signature for a satellite's orbit. Applying a sophisticated model, one can use Doppler data collected over one or more orbits to determine the six parameters needed to characterize an elliptical orbit or the four parameters needed to characterize a circular orbit. A flow chart illustrating how this is done is shown in Fig. 10-6. The task of determining the orbital parameters (elements) for a satellite is usually simplified if a combination of Doppler and ranging measurements are used.

A closely related problem is that of using Doppler data from a satellite whose orbital elements are known very accurately to determine the latitude and longitude of a ground station. This technique is used with navigation satellites such as the Transit series and by search and rescue satellites in the cooperative SARSAT (US, Canada, France) and COSPAS (USSR) series. Radio amateurs are justified in being proud that the satellite-aided search and rescue concept was first tested using the Mode A transponders aboard AMSAT OSCAR 6 and 7. For additional information see R. Bate, D. Mueller and J. White, *Fundamentals of Astrodynamics*, New York: Dover, 1971; P. Escobal, *Methods of Orbit Determination*, New York: Krieger, 1975; "Space-Based Rescue System to Expand," *Aviation Week and Space Technology*, July 18, 1983, pp. 99, 101.

Faraday Rotation

When a linearly polarized radio wave travels through the ionosphere, its plane of polarization rotates about the line of travel. The effect, known as Faraday rotation, depends on the frequency of the radio wave, the strength and orientation of the earth's magnetic field over the path, and the number of electrons encountered. As a satellite moves along its orbit the downlink path changes. As a result, the amount of rotation of the polarization plane changes also. This can lead to severe signal fading when the antennas at both ends of a satellite downlink are linearly polarized.

Other factors being equal, the number of revolutions of the plane of polarization for an uplink or downlink signal is roughly proportional to λ^m (λ = wavelength) with the exponent m having a value between 2 and 3. The predictions of a simple model of

Faraday rotation are summarized in Table 10.2. The absolute number of revolutions that the plane of polarization undergoes is not of much practical interest. The change in this number is, however, because we expect to experience one deep fade in signal strength for each unit change in the total number of revolutions. While the model is very simple, comparing the predictions with observations is interesting. Let's consider the 29-MHz beacon on AMSAT-OSCAR 8. The model (Table 10.2) predicts 60 deep fades in the 9 minutes it takes A-O-8 to go from horizon to overhead. This amounts to one deep fade every 9 seconds on the average. The model actually predicts a shorter time interval between fades when the elevation angle is near 90° and a longer interval at low elevation angles. Observations of the beacon generally show a time interval between fades of 20 to 100 seconds. This is not necessarily in contradiction to the model since the maximum spacecraft elevation angle on most passes is relatively low and the results do depend, to a significant extent, on the actual satellite path and location of the ground station.

From an operational viewpoint, Faraday rotation is important at 29 MHz, of minor concern at 146 MHz and of little effect at higher frequencies. Variations in downlink signal strength may be caused by many factors, including those listed in Table 10.3. From a communications standpoint we're interested mainly in minimizing fading. Using circular polarization at the ground station or the spacecraft does reduce fading from several of the factors listed.

Experimenters who are interested in observing Faraday rotation directly will look at Table 10.3 in a different light. How can the effects of Faraday rotation be separated from the other factors? One strategy would be to concentrate on those links where Faraday rotation is very prominent. A 29-MHz beacon on a low-altitude satellite that uses a linearly polarized antenna is clearly the link of choice. If the ground station uses two linearly polarized antennas that are mounted at right angles to one another and perpendicular to the incoming wave, we can switch back and forth between them to monitor the signal strength alternately at each polarization. With this information we can separate out most of the factors listed in Table 10.3. Faraday rotation will appear as fading on the two antennas, one reaching a peak as the other reaches a minimum, with the period changing slowly in a regular manner. Studies of the Faraday effect are often used to deduce electron concentrations in specific regions of the ionosphere. For additional information on the Faraday effect see G. N. Krassner and J. V. Michaels, *Introduction to Space Communications Systems,* New York: McGraw-Hill, 1964 (this text discusses the model upon which Table 10.2 is based); J. D. Kraus, *Radio Astronomy,* New York: McGraw-Hill, 1964, Chapter 5, section 5; and W. A. S. Murray and J. K. Hargreaves, "Lunar Radio Echoes and the Faraday Effect in the Ionosphere," *Nature,* Vol. 173, no. 4411, May 15, 1954, pp. 944-945.

Unusual Propagation

While ionospheric effects on 29-MHz satellite links were both expected and observed, most discussions of vhf and uhf links treat the ionosphere as if it ceases to have any impact, other than Faraday rotation, above 40 MHz. Contrary to traditional thought, amateur measurements show that satellite links are clearly affected by the ionosphere at both 146 and 435 MHz. Significant signal attenuations of 12 dB or greater that were attributable to the state of the ionosphere were frequently observed on the downlinks of Phase II spacecraft.

When the F2-layer is efficiently reflecting terrestrial 10-m signals back to the earth, it's also reflecting 10-m signals arriving from space back to whence they've come. As a result, an open 10-m band often coincides with an absence of observable 29-MHz Mode-A downlink signals.

Turning to higher frequencies, John Branegan (GM8OXQ/GM4IHJ) collected detailed quantitative information on 70-cm downlink signal strength over a large number of orbits involving OSCARs 7 and 8 and other spacecraft. Statistical procedures were then used to separate the temporal and spatial extent of the attenuating region(s) from antenna orientation and other effects. For details see J. Branegan, "Reception of 70-cm Signals from Satellites, Summary of Results March to Oct. 1978," *AMSAT Newsletter,* Vol. X, no. 4, Dec. 1978, pp. 10-14.

Sporadic-E. In a later study Branegan reported that high attenuation levels on vhf/uhf satellite downlinks were correlated positively with enhanced terrestrial propagation attributable to sporadic-E. Sporadic-E refers to relatively dense clouds or patches of ionized particles that often form at heights approximately the same as the E-layer. To monitor terrestrial sporadic-E he selected vhf TV and fm stations located so that the same general region of the ionosphere was shared by both satellite and terrestrial links. For additional information see J. Branegan, "Sporadic-E Impact on Satellite Signals," *Orbit,* Vol. 1, no. 4, Nov./Dec. 1980, pp. 8-10.

FAI. OSCAR operation was directly responsible for the discovery of a new mode of vhf propagation, called FAI (magnetic-field-aligned irregularities), after the mechanism thought to be responsible. The first observations of signals via this medium were reported by stations in equatorial zones who listened for direct signals from amateurs uplinking to OSCAR spacecraft at 146 MHz. The positive results led to direct terrestrial experiments at 2 m and 70 cm, which helped determine the properties of the FAI mechanism. For details see J. Reisert and G. Pfeffer, "A Newly Discovered Mode of VHF Propagation," *QST,* Oct. 1978, pp. 11-14; and T. F. Kneisel, "Ionospheric Scatter by Field-Aligned Irregularities at 144 MHz," *QST,* Jan. 1982, pp. 30-32.

Antipodal Reception. Soon after Sputnik I was launched observers noticed that the 20-MHz signal from the satellite was often heard for a short period of time when the satellite was located nearly antipodal to the observer. (If you were to dig a hole right through the center of the earth the spot where you'd re-emerge is called the antipodal point). The phenomenon was quickly dubbed the "antipodal effect" and a number of articles appeared in IEEE journals during the late 1950s discussing its

Table 10.4

ASAT Frequency Allocations

Effective 1971 (see Ref. 1)	*Effective upon modification of Part 97 (see Ref. 2 and Ref. 3)*	
7.000-7.100 MHz	7.000-7.100 MHz	
14.000-14.250 MHz	14.000-14.250 MHz	
	18.068-18.168 MHz	
21.000-21.450 MHz	21.000-21.450 MHz	
	24.890-24.990 MHz	
28.000-29.700 MHz	28.000-29.700 MHz	
144.000-146.000 MHz	144.000-146.000 MHz	
435.000-438.000 MHz†1	435.000-438.000 MHz†	3644/320A
	1.26-1.27 GHz†(uplink only)	3644/320A
	2.40-2.45 GHz†	3644/320A
	3.40-3.41 GHz†(in Region 2 and 3 only)	3644/320A
	5.65-5.67 GHz (uplink only)†	3644/320A
	5.83-5.85 GHz (downlink only)	3761C
	10.45-10.50 GHz	3780A
24.000-24.050 GHz	24.00-24.05 GHz	
	47.0-47.2 GHz (Amateur Exclusive)	
	75.5-76.0 GHz (Amateur Exclusive)	
	76-81 GHz	
	142-144 GHz (Amateur Exclusive)	
	144-149 GHz	
	241-248 GHz	
	248-250 GHz (Amateur Exclusive)	

†The ASAT may use these frequencies subject to *not* causing harmful interference to other services operating in accordance with provisions of Allocation Table. This applies to space stations and ground stations.

Ref. 1: *Federal Register,* 47 CFR 97 [Docket No. 19852; FCC 80-419], Amendment of Rules to Provide for the Amateur-Satellite Service; Agency: Federal Communications Commission; Effective Date: Nov. 3, 1980.

Ref. 2: "Extracts From the International Radio Regulations for the Amateur and Amateur-Satellite Services," *QST,* Feb. 1980, pp. 62-71. Paragraph numbers refer to relevant sections.

Ref. 3: "Second Report and Order in FCC General Docket 80-739," released December 8, 1983, implementing into Part 2 of the Commission's Rules and Regulations the Final Acts of the 1979 World Administrative Radio Conference.

causes. Antipodal effects were later observed on the 29.5-MHz beacon of OSCAR 5. See: R. Soifer, "Australis-OSCAR 5 Ionospheric Propagation Results," *QST*, Oct. 1970, pp. 54-57.

The likelihood of antipodal reception is correlated positively with solar activity. During sunspot maxima it's relatively common on 29.5-MHz beacon signals. Although most occurrences are thought to result from normal multihop propagation under the influence of a favorable Maximum Usable Frequency, signal strength is at times exceptionally high. This suggests that a ducting mechanism may sometimes be responsible.

Auroral Effects. Radio signals that pass through zones of aurora activity acquire a characteristic distorted sound, described as raspy, rough, hissy, fluttery, growling and so on. Low-altitude satellites in near-polar orbits are excellent tools for studying auroral effects. One can, for example, use the various beacons on UoSAT to map the extent of the auroral zone experimentally at various frequencies and note changes. The changes at particular frequencies or locations may turn out to be excellent predictors of hf or vhf openings that are caused by various modes. See: K. Doyle, "10 Meter Anomalous Propagation with Australis OSCAR [AMSAT-OSCAR 5]," *CQ*, May, 1970, pp. 60-64, 89.

General. A great deal is still to be learned about through-the-ionosphere propagation of vhf and uhf signals and amateurs are in a unique position to collect important data. UoSAT, with its array of beacons ranging from hf to microwave frequencies, is especially well suited to propagation studies.

This discussion of unusual propagation has touched upon only a few of the interesting phenomena that occur. Other topics of possible interest include rf noise (atmospheric, man-made, cosmic, terrestrial, oxygen and water-vapor, solar), attenuation (electron, condensed water-vapor, oxygen and water-vapor), refraction (ionospheric, tropospheric) and scintillation. For a general overview of all these propagation effects see G. N. Krassner and J. V. Michaels, *Introduction to Space Communications Systems*, New York: McGraw-Hill, 1964.

Frequency Selection

The selection of frequencies for an Amateur Radio spacecraft transponder is a complicated process. Consideration must be given to

(1) legal constraints (national and international laws governing the use of the rf spectrum);
(2) technical factors (propagation and the ability of the amateur community to produce necessary flight and ground station hardware); and
(3) frequency management (cooperative agreements concerning frequency use among the worldwide amateur community)

We'll look at each of these factors in the selection process.

Legal Constraints

The U.S. *A*mateur-*Sat*ellite Service (ASAT) was formally established by the FCC in 1973. The frequencies allocated to ASAT, the subset of the regular amateur bands shown in Table 10.4, had been agreed upon at a World Administrative Radio Conference (WARC) for Space Telecommunications held in 1971. The tremendous gap between 438 MHz and 24.0 GHz placed serious limitations on the future development of amateur space communications.

In 1979 another international conference was held — WARC 79. At this meeting, a concerted lobbying effort by IARU, ARRL, AMSAT and other interested amateur groups succeeded in securing several additional frequency bands for the ASAT Service. The complete list of allocations proposed at the end of this meeting is given in Table 10.4. The new allocations were ratified by the U.S. Senate and a formal announcement was issued by the FCC on December 8, 1983, implementing into Part 2 of the Commission's Rules and Regulations the Final Acts of the 1979 WARC. The ASAT may use these frequencies when the appropriate sections of FCC Part 97 are eventually modified to reflect the changes.

Table 10.5
High-Altitude Satellite Link Performance at Several Frequencies†

band[1]	*Relative free-space path-loss*	*Spacecraft antenna gain*	*Ground-station antenna gain*	*Relative performance*
29 MHz	+14 dB	0 dB	+ 5 dB[2]	+19 dB
146 MHz	0 dB	+ 8 dB	+13 dB	+21 dB
435 MHz	−10 dB	+10 dB	+16 dB[2]	+16 dB
1.26 GHz	−19 dB	+12 dB	+21 dB[3]	+14 dB
2.4 GHz	−24 dB	+12 dB	+26 dB[3]	+14 dB

Notes
[1]50 MHz and 220 MHz are not included because these frequencies are not authorized for ASAT use.
[2]Based on well designed 8-ft boom Yagi.
[3]Based on 4-ft-diameter dish.
†These figures take into account path loss, practical spacecraft antennas and ground station antennas of similar physical size.

Meanwhile, in 1980 the FCC amended the 1973 ASAT rules and regulations. One result was that the procedure for obtaining a license for space operation (i.e., a satellite) was greatly simplified. Frequency allocations were not changed.

Technical Factors

Technical factors for choosing frequencies for satellite links include propagation, satellite design and ground-station complexity. We'll begin with some general considerations and then go on to specifics.

Many of the desirable features of a satellite communication link depend on operating the transponder and ground station in a duplex mode (simultaneous transmission and reception). A cross-band transponder permits duplex operation without inordinately complex equipment. Our objective here, therefore, will be to pick (from the list in Table 10.4) the two optimal bands, based purely on technical considerations. We'll also consider how other bands compare to our optimal choices in case we come across obstacles that prevent their use.

From a system viewpoint, the downlink is the "weaklink" in the communication chain. If necessary, ground station transmitter power levels can exceed the power allocated to a single user at the spacecraft by considerably more than 20 dB. And, even with a sophisticated attitude stabilization system, satellite antenna gain must be limited to provide a sufficiently broad pattern (footprint) for full earth coverage during most of the orbit. Consequently, the "best" band should be used for the downlink unless there are compelling reasons to do otherwise. Frequencies will therefore be evaluated as downlinks. If a band provides good downlink performance it almost certainly will be excellent as an uplink.

We will assume a high-altitude spin-stabilized spacecraft to compare downlink performance at several frequencies. We'll also assume that transponder power and bandwidth are constant and that ground stations provide good, but not necessarily state-of-the-art performance.

Free-space path loss increases with frequency. Comparing link performance at different frequencies taking only path loss into account gives us the information in column 2 of Table 10.5. Since we're interested in relative performance, we can choose a convenient reference level — 2 m in this example. Though beam antennas can be used with spin-stabilized high-altitude satellites, for reasonable earth coverage the satellite antenna gain should be limited to approximately 12 dB_i. The quantitative reasons for this limitation are discussed in the next chapter. This gain can be achieved at uhf and higher frequencies, but at 146 MHz the problems mount. At 29 MHz a gain antenna presents monstrous mechanical problems that so far have made it impossible to place such a device on an AMSAT satellite. Reasonable estimates for achievable antenna gain on a spin-stabilized Phase III satellite have been included in column 3 of Table 10.5.

The ground-station antenna-gain entries are based on a constant boom length of 8 feet at 29 MHz, 146 MHz and 435 MHz. At 1.26 GHz, a 4-ft-diameter dish produces about the same gain as an 8-ft-boom loop Yagi; therefore we've selected a 4-ft dish as a comparable ground station antenna at 1.26 and 2.4 GHz. Finally, the last column in Table 10.5 summarizes relative link performance taking into account all factors mentioned so far: path loss, satellite antenna gain and ground-station antenna gain.

Before we attempt to interpret Table 10.5 look back at Fig. 7-1 that shows sky noise arriving at an antenna. As a result of the steep increase in noise below 1 GHz, we can hear much weaker signals with a good receiver at 146 MHz than at 29 MHz. The relative advantage of 146 MHz over 29 MHz can reach 15 dB. Also, atmospheric absorption at 29 MHz may amount to as much as 20 dB, especially during the peaks of the sunspot cycle. Taking these facts and the data in Table 10.5 into account, we'd conclude that 146 and 435 MHz are the best links from a purely technical standpoint. Our analysis so far, however, has considered only relative performance; we have somehow to make contact with the absolute levels of the real world.

Absolute link performance can be calculated if factors such as receiver noise figure and bandwidth, and cosmic noise are taken into account. We show how to do this in the example that concludes this section. Although calculations of absolute link performance are very useful, there's no substitute for experience. Amateurs have accumulated an extensive data base on 29-, 146- and 435-MHz link performance. Calculations of relative performance, coupled with actual experience, provide our best projections.

A 50-watt PEP 146-MHz downlink with a 200-kHz bandwidth would give users of a high-altitude satellite a signal-to-noise ratio of 20-25 dB. Such a link would provide excellent performance. Other possibilities will be compared to this link.

From a technical viewpoint, the 21- and 28-MHz bands are suitable for low-altitude links during periods of low sunspot activity if one is willing to put up with frequent ionospheric disruptions. Users have expressed considerable interest in seeing Mode A continue, and support for a 21/29-MHz or 29/21-MHz transponder for use during sunspot minima would probably materialize if a group of amateurs decided to undertake such a project. Link calculations show that 21 and 29 MHz could not serve as a down link on a high-altitude spacecraft. Though Table 10.5 shows 29 MHz and 146 MHz as being relatively equal, cosmic noise and atmospheric absorption at 29 MHz takes this band right out of the picture. The allocation at 29 MHz could conceivably be used as a Phase III uplink if amateurs were willing to generate roughly 5000 watts EIRP (500-1000 watts and a 3-element beam). Since this approach is contrary to the fundamental objective of providing reliable long-distance communication using relatively low power and small antennas, it doesn't have much appeal.

Although Table 10.5 suggests a 5-dB advantage for 146 MHz over 435 MHz as a downlink, both bands appear capable of providing good performance. It therefore seems appropriate to emphasize more subtle technical concerns such as urban rf-noise environments and equipment availability, and extremely important nontechnical factors such as frequency usage, discussed in the next section, when comparing these two bands as possible downlinks.

The band of frequencies at 1260 MHz is near our prime choices. Since legal constraints restrict this band to use as an uplink, we look at it from this viewpoint. The 2-dB penalty compared to 435 MHz that is shown in Table 10.5 is a reasonable indication of relative uplink performance. The satellite receiver front end, however, will probably add a few decibels. In any event, 1260 MHz looks promising as an uplink if signal margins are adequate (e.g., a user will hardly notice going from a 22-dB signal-to-noise ratio (SNR) to an 18-dB SNR, but going from a 10-dB SNR to a 6-dB SNR will degrade performance seriously). Calculations show that 5000 watts EIRP (50 watts + 20 dB_i antenna gain) at a ground station should produce a 15- to 20-dB SNR on a 1260-MHz uplink. So, by performance standards, this link is perfectly acceptable. Clearly, however, this band will present a significantly greater challenge so far as building ground station transmitting equipment is concerned. At 1260 MHz, Doppler should pose little problem with high-altitude satellites, except that near perigee ssb will probably be borderline.

A 2.4-GHz uplink coupled with an earth-coverage satellite antenna could probably produce an adequate SNR at the satellite if ground stations used sufficient power; a microwave oven magnetron phase-locked to a crystal oscillator harmonic might serve as a transmitter. It's not clear whether 2.4 GHz could provide adequate performance as a downlink with a full earth-coverage antenna but experience with Mode L should provide the answer. The new allocations at 3.4, 5.6 and 5.8 GHz are also potentially valuable if amateurs decide to equip future spacecraft with multiple spot-beam antennas. Equipment for this part of the radio spectrum should become available as a result of the commercial development of a 12-GHz direct-to-home satellite-TV service (see next chapter).

Predicting Signal Levels: An Example. The following example illustrates how the performance of a 435-MHz Phase III satellite downlink may be calculated.

Spacecraft Characteristics

Transponder
- Total power = 35 watts average
- Bandwidth = 800 kHz

Antenna
- Gain = 10 dB_i
- Apogee height = 35,800 km

Ground Station Characteristics

Antenna
- Gain = 13 dB_i
- Sky temperature as seen by antenna (T_s) = 150 K (often much better — see Fig. 7-1)

Receiver
- Total noise figure (F_T) = 2.2 dB
- Bandwidth (B) = 3 kHz (for ssb)

Ground station-Satellite distance (slant range) = 42,000 km (ground station at edge of coverage cone at apogee)

Objective: To calculate the expected signal-to-noise ratio (SNR) of a typical downlink ssb signal.

$$SNR = \frac{\text{Received Signal power } (W_s)}{\text{Received Noise power } (W_n)}$$

$$SNR[\text{in dB}] = 10 \log W_s - 10 \log W_n = P_s - P_n$$

Our approach will be to (i) compute P_s, (ii) compute P_n and (iii) evaluate SNR.

Step 1: Computation of Received Signal Power

Assume that the transponder is handling 70 equal-power ssb contacts. The average power allocated to each user is, therefore, 0.5 watt. For unprocessed ssb this represents about 3 watts PEP (34.8 dBm). (Note: dBm = dB above 1 milliwatt). Free-space path-loss may be calculated from

$$L = 10 \log \left(\frac{4\pi\rho}{\lambda}\right)^2$$

where

L = free-space path-loss in decibels
ρ = slant range in meters
λ = wavelength in meters

For calculation it's easier to use the equivalent formula

$$L[\text{in dB}] = 32.4 + 20 \log f\ [\text{in MHz}] + 20 \log \rho\ [\text{in km}] \quad \text{(Eq. 10.5)}$$

where

f = frequency.

When applying Eq. 10.5 be sure to express the variables in the units indicated. In our example

$$L = 32.4 + 20 \log (435) + 20 \log (42{,}000) = 177.6 \text{ dB}.$$

We now evaluate P_s.

P_s = transmitted signal power [in dB_m]
+ satellite transmit antenna gain [in dB_i]
+ ground station receive antenna gain [in dB_i]
− free-space path-loss [in dB].

$$P_s = 34.8 \text{ dBm} + 10 \text{ dB}_i + 13 \text{ dB}_i - 177.6 \text{ dB} = -119.8 \text{ dBm}.$$

Step 2: Computation of Received Noise Power

Received noise power is given by

$$W_n \text{ [in milliwatts]} = k\, T_e\, B$$

where

k = Boltzmann's constant = 1.38×10^{-20} [in (milliwatts)/(Hertz) (Kelvin)]

T_e = effective system temperature (discussed below)

B = receiver bandwidth [in Hertz]

Note: All temperatures are in Kelvins. Temperatures in the Kelvin scale are referenced to absolute zero and are given by the Celsius temperature + 273°. Room temperature is defined as 17° C = 290 K. "290 K" is read as "290 Kelvins" — there's no degree sign and the expression "degrees Kelvin" is *not* used.

The effective system temperature (T_e) takes into account (1) noise picked up by the receive antenna (cosmic noise plus radiation from the earth at ~290 K that enters the main or side lobes) and (2) noise generated in the receiver. The temperature of the receive system (T_R) can be computed when the system noise figure (F_T in dB) is known. (See Fig. 7-3 for a discussion of F_T.)

$$T_R\text{[in K]} = 290(10^{F_T/10} - 1) \qquad \text{(Eq. 10.6)}$$

When using Eq. 10.6 any feed-line losses between the antenna and first active receiver stage must be included in the receiver noise-figure computation as illustrated in Fig. 7-3. Applying Eq. 10.6 to our example we obtain

$$T_R = 290(10^{0.22} - 1) = 191 \text{ K}$$

The sky temperature was given as T_s = 150 K so we now have everything needed to evaluate T_e

$$T_e = 191 + 150 = 341 \text{ K}$$

The total received noise power can now be calculated

$$\begin{aligned} P_n\text{[in dBm]} &= 10 \log W_n\text{[in milliwatts]} = 10 \log (k\, T_e\, B) \\ &= 10 \log k + 10 \log T_e + 10 \log B \\ &= 10 \log (10^{-20}) + 10 \log (1.38) + 10 \log (341) \\ &\quad + 10 \log (3000) \\ &= -200 + 1.4 + 25.3 + 34.8 = -138.5 \text{ dBm} \end{aligned}$$

Step 3: Calculation of SNR

$$\begin{aligned} \text{SNR[in dB]} &= P_s\text{[in dBm]} - P_n\text{[in dBm]} \\ &= -119.8 - (-138.5) \\ &= 18.7 \text{ dB} \end{aligned}$$

An 18.7 dB SNR indicates a very good quality signal. Note that the calculations were based on a good (but in no way exotic) 70-cm receive setup situated as far as possible from the spacecraft. For paths where the slant range is shorter, the sky noise temperature behind the satellite is lower, or the preamp in use is better, the link SNR will exceed the value calculated. For situations where a user doesn't have sufficient uplink EIRP to drive the transponder to the indicated output power, or where the number of stations simultaneously using the transponder is greater than the 70 assumed, the link SNR will be less than the calculated value.

For additional information on calculating link performance see the following references:

J. Fisk, "Receiver Noise Figure and Sensitivity and Dynamic Range," *HR*, Oct. 1975, pp. 8-25. This article contains an extensive list of additional references.

B. Atkins, "Estimating Microwave System Performance," *QST*, Dec. 1980, p. 74. Contains a brief but very clear example of calculating link performance at 10 GHz.

J. D. Kraus, *Radio Astronomy*, New York: McGraw-Hill, 1966. Chapters 3 and 7 (by M. E. Tiuri) contain advanced level information on calculating ultimate receiver sensitivity.

Frequency Management

Now that the legal constraints and the technical trade-offs have been considered we get down to the difficult problem: All frequencies allocated to the Amateur Satellite Service are shared with the Amateur Radio Service. Therefore, it's extremely important for satellite users and the general amateur community to establish guidelines for frequency use. Satellite buffs can pursue two paths: (1) Use bands that are sparsely populated; (2) educate the general amateur population as to the goals and constraints of the Amateur Satellite Service so that, even if not personally interested in space communications, they'll understand the importance of space activities to all of Amateur Radio. The significance and difficulty of the educational task should not be underestimated. When a local radio club has pioneered in the development of fast-scan TV repeaters over many years, it will take considerable tact to convince them to invest time and cash in switching to a different segment of the 70-cm band. And, when a country has 100 times as many amateurs equipped for 2-m fm as for satellite operation, a dedicated effort will certainly be needed to explain why they should part with a significant segment of a crowded 2-m band.

The Amateur Satellite Service is taking a balanced approach. For several years worldwide support for establishing exclusive satellite segments at 29.300-29.500 MHz, 145.800-146.000 MHz and 435.000-438.000 MHz has been growing. A decade ago, when the 10-m and 2-m proposals were made, the band segments were almost empty. Today, nearby crowding often leads amateurs who are unaware of the "gentlemen's agreements" on frequency management to move into the "open space," not realizing that they're disrupting satellite links. A continuing, tactful educational program is a necessity. At present, there are no nationally agreed on band plans for 23 cm and above, but consideration is being given to this situation before crowding becomes significant.

Returning to our primary concern, choosing transponder frequencies, it's clear that a Phase III satellite can use the preferred 435 MHz-to-146 MHz frequency combination to support, at most, one 200-kHz-wide transponder. Once this is done we *must* turn to a second choice: 1260/435 MHz. Keep in mind that the phrase "second choice" does not imply inferior performance. Though this combination does require greater effort in setting up ground stations, in return it offers the bandwidth necessary to support a great many users. The 1260/435 MHz transponder for AMSAT OSCAR 10 is 800-kHz wide and capable of supporting four times as many users as the 200-kHz-wide Mode B unit.

In sum, from a purely technical viewpoint, the transponder of choice would use 146 MHz and 435 MHz for both high- and low-altitude satellites, with either band serving as the downlink. Frequency management considerations will mandate the use of other frequency combinations in the near future. A 1260/435-MHz transponder should provide good performance on high-altitude spacecraft but Doppler makes this combination only marginally acceptable for cw and ssb on a low-altitude spacecraft. As several OSCARs have demonstrated, 146/29 MHz is suitable for low-altitude spacecraft if users are willing to accept frequent ionospheric disturbances. A 21/29-MHz transponder on a low-altitude spacecraft represents a possible alternative to the traditional Mode A for newcomers.

An understanding of the frequency selection process should make it clear that AMSAT designers have not rushed up the frequency ladder with callous disregard for users' needs. The complex constraints and trade-offs involved in transponder frequency selection have always been considered carefully and some difficult decisions have had to be made.

For additional information on the frequency selection process see the comprehensive paper by R. Soifer: "Frequency Planning for AMSAT Satellites," *Proceedings of the ARRL Technical Symposium on Space Communications,* Reston, VA, Sept. 1973, Newington, CT: ARRL, pp. 101-127. Additional information may be found in a paper elsewhere in the *Proceedings* by K. Meinzer: "Spacecraft Considerations for Future OSCAR Satellites," pp. 137-143. Despite advances in low-noise microwave transistors, the new emphasis on high-altitude satellites and the availability of new bands, the conclusions of both of these studies are still basically valid.

Chapter 11
Weather, TV, And Other Satellites

Chapter 11

Weather, TV, and Other Satellites

Our focus, so far, has been on OSCAR satellites. Over the years several other satellite systems, primarily weather-picture and TV-broadcast, have attracted the interest of radio amateurs. This chapter will provide an overview of some of these systems and discuss satellite sleuthing: locating unidentified radio signals from space and determining their origin.

Weather Satellites

In the early 1960s, when satellites designed to provide cloud-cover pictures were first being developed, a group of farsighted individuals realized that such spacecraft held tremendous potential for improving the quality of life of people all over our planet. To achieve this potential fully, however, life-saving weather satellite information had to be made as accessible as possible. As these ideas took hold, the U.S. weather satellite program adopted an automatic-picture-transmission (APT) encoding system that required relatively simple receiving and processing equipment at the ground station. Potential beneficiaries around the world were encouraged to use the system without charge; information on the satellite system and ground station construction was widely distributed; and a commitment was made to support the APT format for a long time.

The APT system was first put into service in late 1963 aboard TIROS VIII. Even though satellite-borne imaging equipment has changed drastically over the years since, the APT format for encoding downlink pictorial information has undergone only minor revision. Indeed, receiving equipment built in the mid 1960s can still be used with only slight changes, and plans are to continue supporting the APT format indefinitely.

Today several satellite systems take pictures from space. These images are used for natural resource assessment, crop management and air/sea navigation, as well as for weather prediction. Different types of surface features register most clearly in different parts of the electromagnetic spectrum. Imaging equipment is designed to select special segments of the infrared (IR), near-IR or visual light spectra during the recording process. In this way we can choose to focus on cloud cover, grain crops or forests, or land/water, land/ice or ice/water boundaries. The resolution of current satellite imaging equipment is often much greater than the APT system can accommodate, so two downlink formats are used: a high-resolution mode that requires extensive ground station equipment and a lower-resolution APT mode that is suitable for the majority of applications. Don't let this comparison mislead you; the APT system is capable of producing very high quality images.

So, if you're interested in weather prediction, the myriad other applications of satellite photos or just being an "armchair astronaut," stay with us as we discuss the APT services provided by the U.S. National Environmental Satellite System (NESS) and the ground-station equipment needed to receive them.

The NESS Satellite System

Although for historical reasons it's customary to discuss weather satellites, our real interest is in receiving APT services. Some APT information is not directly related to weather and there is a considerable amount of non-pictorial weather information available from NESS satellites that we completely ignore.

Two NESS satellite systems provide APT data: (1) a series of low-altitude, near-polar, sun-synchronous spacecraft and (2) a series of geosynchronous satellites. Both systems carry imaging equipment and both have APT downlinks.

The low-altitude NESS satellites, such as NOAA-6, NOAA-7 and NOAA-8, provide *direct readout services* 24 hours per day. Each records a continuous picture that is immediately forwarded to the ground in an APT format near 137 MHz and in a high-resolution format near 1700 MHz. The orbit height used by these spacecraft, 800 to 900 km, provides ground stations with two or three morning passes and two or three evening passes for each satellite. Plans call for keeping at least two satellites in orbit at any given time. Their pictures cover an area of roughly 2000 km by 2000 km with a resolution of about 4 km.

The geostationary NESS satellites serve several functions. As noted earlier, they're equipped to record high-resolution images in segments of the IR, near-IR, and visible light spectra. This data is typically downlinked directly to special NESS ground stations in a very high data rate format. At the NESS stations the information is processed by computer, reformatted in the APT mode and sent back up to the satellite for retransmission.

The satellite rebroadcast function is known as the WEFAX (Weather Facsimile) service. WEFAX broadcasts, on a standard frequency of 1691 MHz and using the APT format, include pictures taken from geostationary satellites, pictures taken from low-altitude spacecraft, operational messages such as photo schedules, selected weather charts and the like. WEFAX pictures are often enhanced by computer processing or the superposition of grids that outline land masses and states. European WEFAX broadcasts are transmitted on a second frequency, 1694.5 MHz. Prior to 1979, U.S. WEFAX services from geostationary orbit were also provided in the 137-MHz band but these broadcasts have been discontinued permanently.

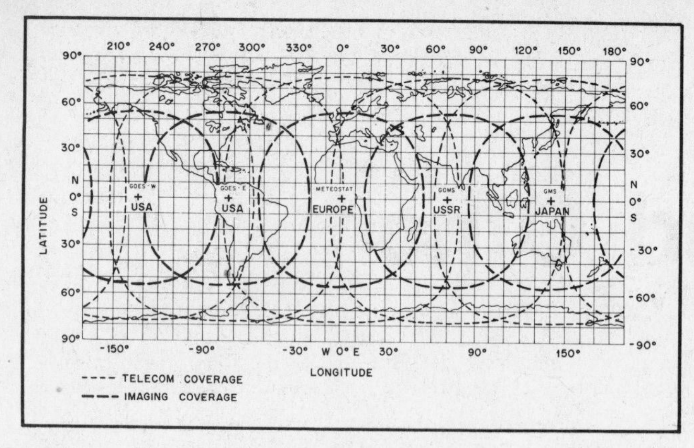

Fig. 11-1 — Five geostationary WEFAX satellites provide weather satellite pictures worldwide.

Table 11.1

International WEFAX (Weather Facsimile) Geostationary Satellite System

Designation[1]	*Operated by*	*Nominal Subpoint Longitude*
GOES-East	USA	75° W
GOES-Central[2]	USA	105° W
GOES-West	USA	135° W
METEOSAT	ESA[3]	0°
GMS	Japan	220° W (140° E)
GOMS	USSR	290° W (70° E)

Notes

[1]*Satellite Names*
GOES: Geostationary Operational Environmental Satellite
GMS: Geostationary Meteorological Satellite
GOMS: Geostationary Operational Meteorological Satellite
SMS: Synchronous Meteorological Satellite
[2]Tentative
[3]European Space Agency

WEFAX broadcasts are a cooperative worldwide venture. Fig. 11-1 shows the communications and picture coverage provided by geostationary weather spacecraft located at the five internationally agreed upon positions. Additional information on the international system is given in Table 11.1. It's clear that most U.S. amateurs will have access to at least two spacecraft, GOES-East and GOES-West. Another satellite, at a position tentatively called GOES-Central, will likely be added to the system and will be accessible to U.S. ground stations. Note that the names designate locations, not satellites. During 1979, for example, the GOES-East position was occupied at various times by the spacecraft GOES-2, SMS-1 and SMS-2.

The APT Ground Station

A ground station that is set up to receive APT direct readout services from low-altitude satellites or WEFAX services from geostationary satellites will have two major, distinct subsystems: (1) an fm receive system and (2) the picture reconstruction system (see Fig. 11-2). An *audio* tape recorder usually separates the two units. We'll look at each subsystem in some detail.

The FM Receive System

Specifications of the fm receive systems needed to tune in APT transmissions from both low-altitude and geosynchronous satellites include frequencies, i-f bandwidths and system sensitivity (including antenna and receiver front end performance). Table 11.2 summarizes the main characteristics of the two rf links of interest.

Table 11.2

Link Characteristics for APT Services from High- and Low-Altitude Spacecraft

	Low-altitude U.S. satellites[1]	*Geostationary satellites*
Satellites of primary interest	NOAA-6 (79-57A)[2] NOAA-7 (81-59A) NOAA-8	GOES-1, 2, 3 SMS-1, 2
Name of APT Service	Direct readout	WEFAX
Frequencies	137.500 MHz 137.620 MHz	1691.000 MHz 1694.500 MHZ (Europe METEOSAT only)
Frequency stability	± 3 kHz	—
Maximum Doppler	± 4 kHz	0.0 kHz
Transmitted Bandwidth	± 17 kHz	± 9 kHz
Transmitter Power	5 watts	—
Satellite EIRP (nominal)	37 dBm	56 dBm (GOES) 52 dBm (METEOSAT)
Satellite Antenna Polarization	RHCP	Linear
Free Space Path Loss	141.3 dB (at 2000 km)	189.5 dB
Ground Station Receiver i-f Bandwidth	50 kHz	30 kHz (300 kHz, Japan GMS only)

[1]Russian Meteor series low-altitude weather satellites have been heard in operation over the U.S. and several amateurs have printed pictures from them. See section on Soviet weather satellites for information.
[2]International designations are included in parentheses. For orbital data see Table 11.5.

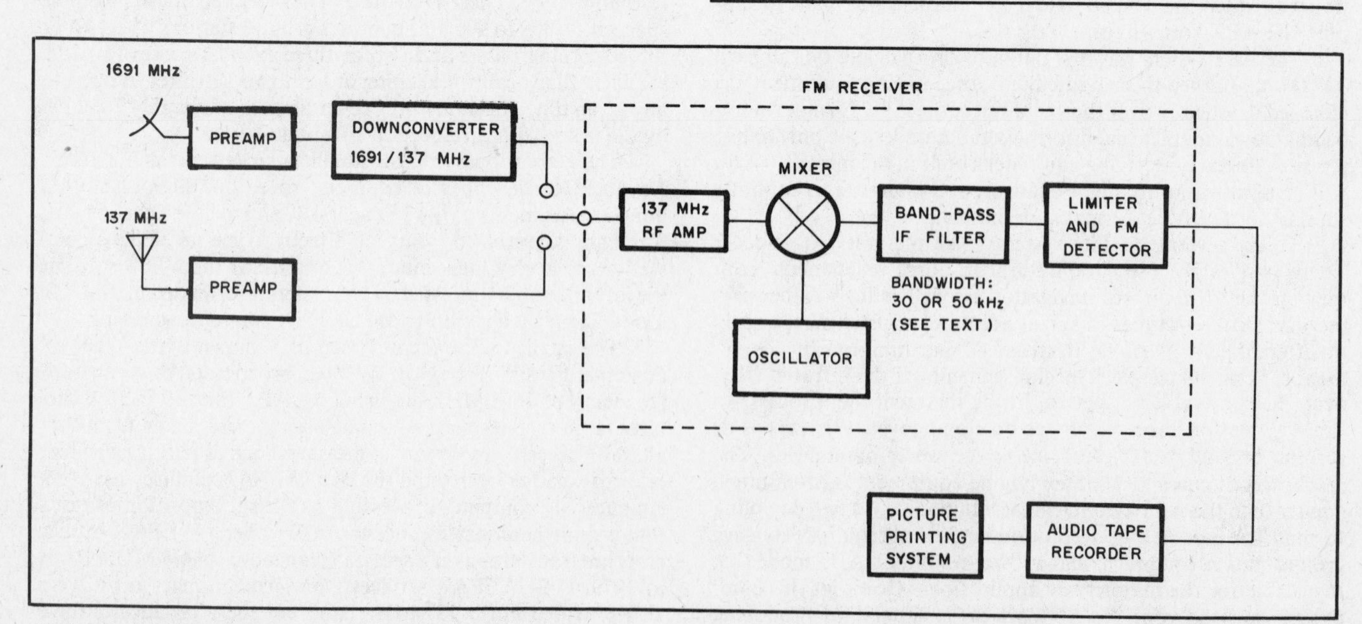

Fig. 11-2 — APT ground station for 137 MHz and 1691 MHz.

Geostationary satellites. The typical WEFAX receive station uses a 1691-MHz downconverter that is fed into an fm receiver in the range 130 to 160 MHz, or 20 to 40 MHz, that is used as an i-f amplifier and detector. Several relatively inexpensive receivers designed expressly for 137-MHz APT reception are made by Vanguard, Hamtronics and others; these work well as the i-f/detector system. Public-service-band monitor receivers are usually suitable — check the i-f bandwidth. Older, wideband commercial fm and amateur 2-m fm equipment also yield good performance while recent narrow-band equipment will not work without i-f modifications. As most of the units mentioned suffer from poor sensitivity, adding a good low-noise amplifier between the converter and i-f system can improve the overall noise figure significantly. Several converter designs for 1691 MHz have been published, and most amateur 1296-MHz or 2304-MHz models will work well if tuned circuit values are properly scaled and the oscillator injection is configured to produce the correct i-f frequency. An extensive list of references to construction articles follows this section.

The typical antenna is a fixed-aim parabolic dish with a 1691-MHz preamp mounted at the feed. Most bipolar preamps using inexpensive devices (MRF 901 or BRF 91) provide only 6 to 8 dB of gain per stage and noise figures of roughly 3 dB. A single GaAs FET preamp will provide a gain of 16 to 18 dB and a noise figure under 2 dB. Suitable GaAs devices (MGF 1200) now sell for under $10 and prices will continue to drop. With a single-stage antenna-mounted GaAs preamp and a 1.5-m diameter dish, your link margin (signal strength in excess of that needed to produce a noise-free picture) should be about 5 dB. Using a 50-kHz i-f filter in place of the 30-kHz unit will reduce the link margin by about 2 dB and make the i-f suitable for receiving the 137-MHz direct broadcast service.

One's perception of how difficult it is to set up a 1691-MHz receive station really depends on one's experience. Operators who are familiar with 137-MHz reception from geostationary spacecraft or the powerful 137-MHz signals of low-altitude satellites naturally consider stepping up to microwaves to be a big challenge. But anyone who's ever tried to receive EME signals or experimented with receiving 4-GHz TV signals directly from geostationary satellites (see next section) thinks of 1691-MHz WEFAX reception as relatively simple.

Low-altitude satellites. If you've worked with OSCAR satellites you'll appreciate the powerful 137-MHz signals provided by low altitude NESS imaging spacecraft. Most of the receiver options just discussed in conjunction with geostationary satellites are also suitable at this frequency, though vhf public-service-band or commercial fm receivers are usually converted to 137.5 or 137.62 MHz operation so no separate down converter is needed. Note that a 50-kHz i-f filter is required to accommodate the wide deviation and Doppler shift. Inexpensive receivers that have 30-kHz-wide filters can be used if (1) the i-f filter roll off is gradual, (2) lots of rf amplification is used, and (3) the operator adjusts receiver tuning during each pass to follow the Doppler shift. Good receivers with steep-sided 30-kHz i-f filters will not work. Though most users over the years have used beam antennas that require tracking, signal levels are sufficient for omnidirectional antennas if a low-noise (under 1.5 dB) preamp is mounted at the antenna. Despite a circularly polarized downlink, users report fading problems when linearly polarized antennas are used. Therefore, it's best to use a right-hand circularly polarized antenna; the omnidirectional Lindenblad is an excellent choice.

Picture Reproduction System

Image display systems currently used by radio amateurs generally employ either a Facsimile (FAX) recorder, or a camera and cathode-ray-tube (CRT) combination. Microcomputer-based image-display systems are just beginning to be developed. The percentage of amateurs using the microcomputer approach will probably increase greatly in the mid and late 1980s. All three approaches are capable of producing excellent pictures. Though other techniques for displaying images have not received widespread acceptance, amateurs who already own SSTV equipment should be sure to note the interface for linking a weather-satellite receiver to an SSTV monitor described by R. Taggart (1974). (References to construction articles are contained in Table 11.3).

The APT signal coming out of the receiver fm detector consists of an audio signal with a 2400-Hz subcarrier. The subcarrier is amplitude modulated with video information. Modulation percentage varies from 5% (black level) to 80% (white level). In addition to the video modulation, square-wave pulses are used to indicate the beginning of picture, the phasing and the end of picture. The phasing pulses keep the starting point of each scan line synchronized with the transmitted video. Satellites generally transmit 120 lines/minute. Video and sync processing equipment depends, to some extent, on the type of readout device being used. An experienced electronics experimenter will find the processing electronics described in the referenced construction articles (Table 11.3) straightforward and easy to reproduce.

Camera/CRT System. In the camera/CRT system the picture is painted, line by line as it was sent, on the face of a standard oscilloscope. Since each picture takes several minutes to complete, and the glow of CRT phosphors dies out after a fraction of a second, this approach doesn't work for direct viewing. To see the entire picture you have to take a time-exposure photograph of the CRT screen. Nonetheless, the camera/CRT method is relatively easy to implement using published circuits. Moreover, it's flexible in its ability to accommodate changes in APT scan rates or number of lines if changes are made in the APT format, for viewing Russian spacecraft and so on. The drawbacks are that if you use instant film the expense quickly mounts; if you use low cost roll film you have to put up with a signficant delay before viewing the results. Since the timeliness of space photos is a major attraction of APT reception, the significance of the delays inherent with film processing shouldn't be underestimated.

FAX System. FAX recorders are a popular alternative to camera/CRT methods. A FAX recorder consists of a rotating, rolling-pin-like drum wrapped with special sensitized paper. The APT signal consists of several hundred lines of information, each sent sequentially. In an operating system the FAX drum rotates once for each incoming line; the video modulation is transferred to the sensitized paper by a special stylus. For the next line the stylus is moved slightly and the whole process is repeated. Three types of sensitized paper are in common use: photo sensitized, electrosensitive and electrolytic. In each case, the incoming video information controls the exposure. With photo-sensitized paper the stylus is a carefully focused beam of light that varies in intensity. Standard photographic techniques must be used to print the picture. With electrosensitive and electrolytic paper the stylus is a pin-like device that directs a variable-intensity arc. Electrolytic paper must be treated chemically to view the image; electrosensitive paper has the image burnt directly into it so no processing is necessary. Electrosensitive paper is relatively inexpensive (less than a dime for a complete picture) and a great convenience since no special chemicals, lighting or extra processing is required. A complete FAX recorder and associated electronics designed to be used with electrosensitive paper was described by R. Taggart in *73* (Nov., Dec. 1980, and Jan. 1981). A modern design for a FAX readout using photosensitive paper was presented by G. Emiliani and M. Righini in *QST* (April, 1981). Though FAX recorders are complex mechanical devices, they currently are the most popular readout units.

Microcomputer Image System. With the rapidly growing availability of microcomputer systems that use video display terminals, amateurs have considered using these devices to display APT images. Several radio amateurs attending the 1983 NOAA/NASA International Direct Broadcast [Satellite] Services Users' Conference have reported success with relatively simple setups. In addition to a satellite receiver and a microcomputer with a video display, one needs an interface unit containing the electronics and appropriate software. The output of the APT receiver is fed to the interface unit which generally consists of a sample-

and-hold stage followed by an analog-to-digital converter. (Note: The remainder of this section assumes the reader is familiar with basic computer terms. You may wish to skip to the next section.)

The most popular amateur approach to digitizing an APT picture is to sample each line 256 times (256 pixels per line) and form a picture from 256 lines. One presenter at the 1983 conference exhibited reasonably good pictures using just one bit (on or off, no gray scale) for each pixel in memory. This extremely modest system required only 8K of memory for storing an image. Matjaz Vidmar (YU3UMV) has described a system (see references) using 256 pixels per line by 256 lines with 64 gray levels that produces quality images with less than 64K of memory per picture. Commercial systems generally use 512 pixels per line by 512 lines with one byte per pixel. With the rapidly dropping prices of memory, the 256K of RAM required for this approach is no longer a major stumbling block.

Software is, of course, an extremely critical part of a computer-based display system. Programs are needed to control the sampling of the analog-to-digital converter, to store the values in RAM or on disk, and later to read out values for display. Pioneers will have to write their own software. In fact, it's the software aspect that provides the greatest opportunity for creative work since the programmer can experiment with image processing. For example, one may be able to combine stored frames of IR and visible images in a manner that clearly shows both land/water transitions and cloud cover in a single picture. This is not normally possible. In addition, one can experiment with non-linear gray scales or false color to bring out aspects of an image that are normally difficult to observe. Although this discussion has centered on producing images on a video display it's possible to obtain hard copies by using a dot-matrix printer that has graphics capabilities.

General Information

Our objective has been to introduce you to weather satellites by discussing the image services available, the characteristics of the NESS satellite systems, and the more popular approaches to setting up a ground station so that you'll have some feel for what's required. Construction details for receiving gear and picture-reproduction equipment will be found in the references at the end of this section.

Tracking weather satellites is simple. Azimuth and elevation angles for aiming your antenna at one of the geostationary WEFAX satellites can be obtained from Fig. 5-10. The OSCARLOCATOR is suitable for low-altitude weather satellites. If you use an omnidirectional antenna, the quick approach to drawing a ground-track overlay (discussed in Chapter 9, "Polar Projection Map" section) is sufficient for calculating AOS and LOS. To use this approach the only information you need about the orbit is either the period or the height. If you start with the period, compute the height using Eq. 8.6c. As these spacecraft are in sun-synchronous orbits, Fig. 8-14 can be used to determine the orbit inclination. With the inclination and period known, an OSCARLOCATOR orbit overlay can be sketched quickly.

Table 11.3

Information Resources: Weather Satellites

U.S. Agencies

NESS: The *N*ational *E*nvironmental *S*atellite *S*ervice is the agency responsible for operating and disseminating information related to U.S. Weather Satellites. NESS is part of NOAA. You may wish to write NESS to request that your name be added to the "APT Information Notes" mailing list, and a list of NOAA Technical Memorandums (TMs) and Reports (TRs) concerning weather satellites, and other information (always be as specific as possible). Address requests to Coordinator, Direct Readout Services, OA/S131, NOAA/NESS, Washington, DC 20233.

NOAA: The *N*ational *O*ceanic and *A*tmospheric *A*dministration is part of the U.S. Department of Commerce. For lists of NOAA publications related to weather satellites contact Environmental Data Service (D822), NOAA, 6009 Executive Blvd., Rockville, MD 20852. NOAA usually has a limited supply of each publication for free distribution. Once these are gone copies may be purchased from NTIS.

NTIS: The *N*ational *T*echnical *I*nformation *S*ervice acts as a central clearinghouse for specialized publications by NOAA and other government agencies. Documents are generally available in hard copy or in microfiche format. Microfiche copies are much cheaper so check your local library to see if they have the special reader required. Ordering is usually a two-step process. First request the price of the documents (specify accession number if known and give complete details as to author, title, agency, date, etc.). After you receive the details, send your check to NTIS, Dept. of Commerce, 5285 Port Royal Rd., Springfield, VA 22151.

Books on Weather Satellites that Users Shouldn't be Without

Summers, R.J. and T. Gotwald, *Teachers' Guide for Building and Operating Weather Satellite Ground Stations,* NASA EP-184, 1981. Educators may obtain free copies from Educational Programs Officer, Code 202, Goddard Space Flight Center, Greenbelt, MD 20771.

Taggart, R.E., *Weather Satellite Handbook,* 2/E, 73 Publications, Peterborough, NH 03458, 1981.

Vermillion, C.H. *Weather Satellite Picture Receiving Stations: Inexpensive Construction of Automatic Picture Transmission Ground Equipment,* NASA, SP-5080, 1969. NTIS Accession no. N69-31985.

System Documentation

Corbell, R., et al., "GOES/SMS User's Guide," NOAA/NESS, 1977.

Nagle, J., "A Method of Converting the SMS/GOES WEFAX Frequency (1691 MHz) to the Existing APT/WEFAX Frequency (137 MHz)," NOAA TM NESS-54, April 1974.

Nelson, M., "Ground Stations to Receive GOES WEFAX — the Engineering Considerations," NOAA/NESS Preliminary Report, Aug. 1978.

Schneider, J., "Guide for Designing RF Ground Stations for TIROS-N," NOAA TR NESS-75.

Schwalb, A., "The TIROS-N / NOAA A-G Satellite Series," NOAA TM NESS-95, March 1978.

WMO No. 411, "World Weather Watch, Global Observing System — Satellite Subsystem, Information on Meteorological Satellite Programmes Operated by Members and Organizations." Available from Secretariat, World Meteorological Organization, Geneva, Switzerland, or UNIPUB, Box 433, New York, NY 10016.

Construction Articles

Christieson, M., "A METEOSAT Earth Station," *Wireless World,* June and July, 1979.

Emilani, G., and M. Righini, "An S-band Receiving System for Weather Satellites," *QST,* Aug. 1980, pp. 28-33.

Emilani, G., and M. Righini, "Printing Pictures for 'Your' Weather Geostationary Satellite," *QST,* April 1981, pp. 20-25.

Petit, N. J., and P. Johnson, "Weather Satellite Pictures and How to Obtain Then," *The Physics Teacher,* Sept. 1982, pp. 381-387, 390-393.

Ruperto, E., "A Satellite Receiver for the Home," The Amateur Scientist, *Scientific American,* Jan. 1974, pp. 114-120.

Ruperto, E., "The Microwave Midget," *73,* Dec. 1980, pp. 106-109. Provides details of a 1691/137 MHz downconverter using an active mixer.

Ruperto, E., "Weather Satellite Pix Printer," *73,* Jan. 1978, pp. 82.

Shuch, H., "Variable Tuning for WEFAX Receivers," *73,* Dec. 1979, pp. 70-75.

Taggart, R., "Weather Satellite Pictures On Your SSTV Monitor," *73,* Sept. 1974, pp. 79-83.

Taggart, R., "Be a Weather Genius," *73,* Nov. 1978, pp. 198. Lots of good information on monitoring GOES.

Taggart, R., "Attention, Satellite Watchers! — A Solid-State Monitor for GOES," *73,* Feb. 1979.

Taggart, R., "New Weather Eye in the Sky," *73,* Nov. 1980, pp. 176-181. A primer on TIROS-N.

Taggart, R., "Direct Printing FAX." Part I, *73,* Nov. 1980, pp. 90-98; Part II, *73,* Dec. 1980, pp. 52-56; Part III, *73,* Jan. 1981, pp. 54-57. Contains complete construction information for a FAX printer (using electrosensitive paper) and all associated electronics.

Vidmar, M., "A Digital Storage and Scan Converter for Weather Satellite Images," *VHF Communications* (English language edition published in Germany), Part I, Winter (4), 1982; Part II, Spring (1), 1983.

Winkler, L., "Producing Weather Satellite Pictures at Lower Cost," *QST,* June 1978, pp. 32-34.

Specialized Communications Techniques for the Radio Amateur, ARRL, 1975. Chapter 4 contains a considerable amount of valuable information on FAX printers for those planning to use these devices (out of print).

Reference orbit data are provided on ARRL phone and cw bulletins (see *QST* for bulletin schedule).

When you decide to set up for weather satellite reception, you'll have two important decisions to make: (1) the type of readout device you'll use and (2) the service you'll receive (WEFAX on 1691 MHz or direct readout at 137 MHz). In the long run either a FAX printer using electrosensitive paper or a microcomputer video system will give the most satisfactory results. If you already own a microcomputer with at least 64K RAM and a disk drive, opt for the latter approach. Most users eventually want to receive the 1691-MHz WEFAX services. Since the rf equipment for receiving the 137-MHz direct readout APT is simpler, you may choose to set up for this service first. If you do, think of it as a first step. Make sure that any major equipment you purchase will be useful as part of your future 1691-MHz station. Table 11-3 lists resources for information on Weather Satellite reception.

Note: The pictures of the earth to be taken by UoSAT will be downlinked in a special format that is not compatible with APT. Information on reconstructing UoSAT image data will be presented in *Orbit* and/or *QST* as soon as it becomes available.

Soviet Weather Satellites

Although several western world radio amateurs regularly copy pictures from Soviet weather satellites, very little information on these spacecraft has appeared in print. Here's a summary of what is known about them.

Operating frequencies. The Soviets appear to have two distinct series of low-altitude, near-polar satellites downlinking pictorial information in an APT compatible format. Meteor 1 series satellites are probably used for experimental and developmental purposes. They can, a times, be heard operating over the U.S., but such operation is limited and unpredictable. Meteor 2 series spacecraft are assumed to be part of an operating system. Several spacecraft in this series have been operational at all times since about 1976 and they're heard over the U.S. on a fairly regular basis.

Meteor satellites reported active after 1981 include

Meteor 1 Series
- Meteor 30 (80 51A) on 137.120 or 137.130 or 137.150 MHz
- Meteor 31 (81 65A) on 137.130 MHz

Meteor 2 Series
- Meteor 2-5 (79 95A) on 137.300 MHz
- Meteor 2-6 (80 73A) on 137.400 MHz
- Meteor 2-7 (81 43A) on 137.300 MHz
- Meteor 2-8 (82 25A) on 137.850 MHz
- Meteor 2-9 (82 116A) on 137.300 MHz

For orbital data on these satellites see Table 11.5. The numbers in parentheses are international designations for the specified spacecraft. The frequencies 137.170 and 137.200 MHz have been used on Meteor satellites which are no longer active.

Technical information. The following brief description of key technical features of the Meteor APT system should enable experienced NOAA spacecraft users to make the transition to Soviet APT reception.

The Meteor satellites use an fm deviation of about ±10 kHz so a 30-kHz-wide receiver i-f filter is optimal.

The video subcarrier disappears during synchronization pulses. As a result, receive systems that derive FAX motor speed control by locking onto the subcarrier may not work properly. FAX motor speed control should be obtained locally from an oscillator designed for this purpose as is done in many modern image recovery systems used with NOAA spacecraft. If may be necessary to change the oscillator frequency, however, since Meteor spacecraft tend to deviate somewhat from the 2400 Hz used by NOAA satellites. A stable variable frequency source capable of tuning 2400 Hz to 2520 Hz plus some overlap should suffice.

Meteor 2 spacecraft generally transmit a single image at a 120 line/minute scan rate. Images are usually registered in the visible spectrum. (NOAA spacecraft transmit two side-by-side images: one visible, one infrared.) The portion of the spectrum sampled by Meteor 2 spacecraft typically provides poor land-water boundary definition. The system produces excellent resolution of snow cover, however, so that when conditions are favorable, good land-water definition can be obtained. Resolution of Meteor 2 pictures appears to exceed that of NOAA APT images. At times, Meteor 2 spacecraft transmit a very low scan rate (20 lines/minute) signal that exhibits characteristics of IR imaging. Although Meteor 2 satellites operate a large percentage of the time when in range of the U.S., they are sometimes switched off. Such actions are consistent with prudent spacecraft operation (protecting vidicons, spacecraft power system or temperature management, etc.) and should not be assumed to be designed to limit use of the spacecraft by other nations.

Marciano Righini (I4MY) has pointed out that the black and white bars along the edge of the Soviet APT image (created by the synchronization pulses) can be used to identify a particular spacecraft: Meteor 2-7 (13 black bars), Meteor 2-8 (14 black bars), Meteor 2-9 (17 black bars). Greg Roberts (ZS1BI) suggests that these bars may also contain data on aperture setting and grey-scale calibration.

Meteor 1 series spacecraft use a scan rate of 240 lines/minute. Experienced users agree that spacecraft in this series provide the best definition of any APT signal currently available. Resolution of land-water boundaries is excellent. Unfortunately, operation in range of the U.S. is infrequent. Meteor 1 images exhibit some panoramic distortion. Meteor 2 and NOAA spacecraft eliminate this type of distortion by processing the images before they're transmitted.

The information presented in this section is from the excellent detective work of Greg Roberts (ZS1BI) and Grant Zehr (WA9TFB). For additional information see: G. Roberts, "Soviet Weather Satellites," *OSCAR NEWS*, no. 34, Summer 1981, pp. 2-4.

Satellite TV

There are two distinct groups of geostationary satellites designed to downlink TV program material to U.S. ground stations. One group, transmitting in the 3.7- to 4.2-GHz band allocated to the Common Carrier Service, is currently providing more than 40 channels of first run movies, sports and other special programming. This is *not* a broadcast service. It's meant to be a private distribution system. The other group consists of direct broadcast satellites (DBS) operating in the 11.7- to 12.2-GHz band. These spacecraft are designed to provide direct-to-home services in conjunction with simple, low-cost receiving equipment. DBS systems are currently in a preoperational testing stage; in the U.S., regular service is tentatively planned for about 1984. The 4-GHz and 12-GHz systems are of immense interest both to technically oriented radio amateurs and, if the numerous popular magazine and newspaper articles are any indication, the general public. Unfortunately, the popularized reports tend to confuse the two systems hopelessly. Our aim here will be to introduce you to both services, to discuss *TV r*eceive *o*nly (TVRO) ground stations and to look at what satellite TV will mean to Amateur Radio.

4-GHz Satellite TV

More than 10 geostationary satellites, parked between 70° W and 135° W, are capable of relaying TV programming in the 4-GHz band (3.7 to 4.2 GHz) down to North American ground stations. The heart of each spacecraft is an fm repeater designed for crossband uplink (5.9- to 6.4-GHz) operation. A typical satellite, SATCOM F1 for example, has 24 channels, each of which is 36 MHz (+ 4-MHz guard band) wide. Each channel, or transponder as it's often called, has its own 5-watt amplifier. If you're following closely you may wonder how we fit all these channels (40 MHz × 24 = 960 MHz) in a band that's only 500

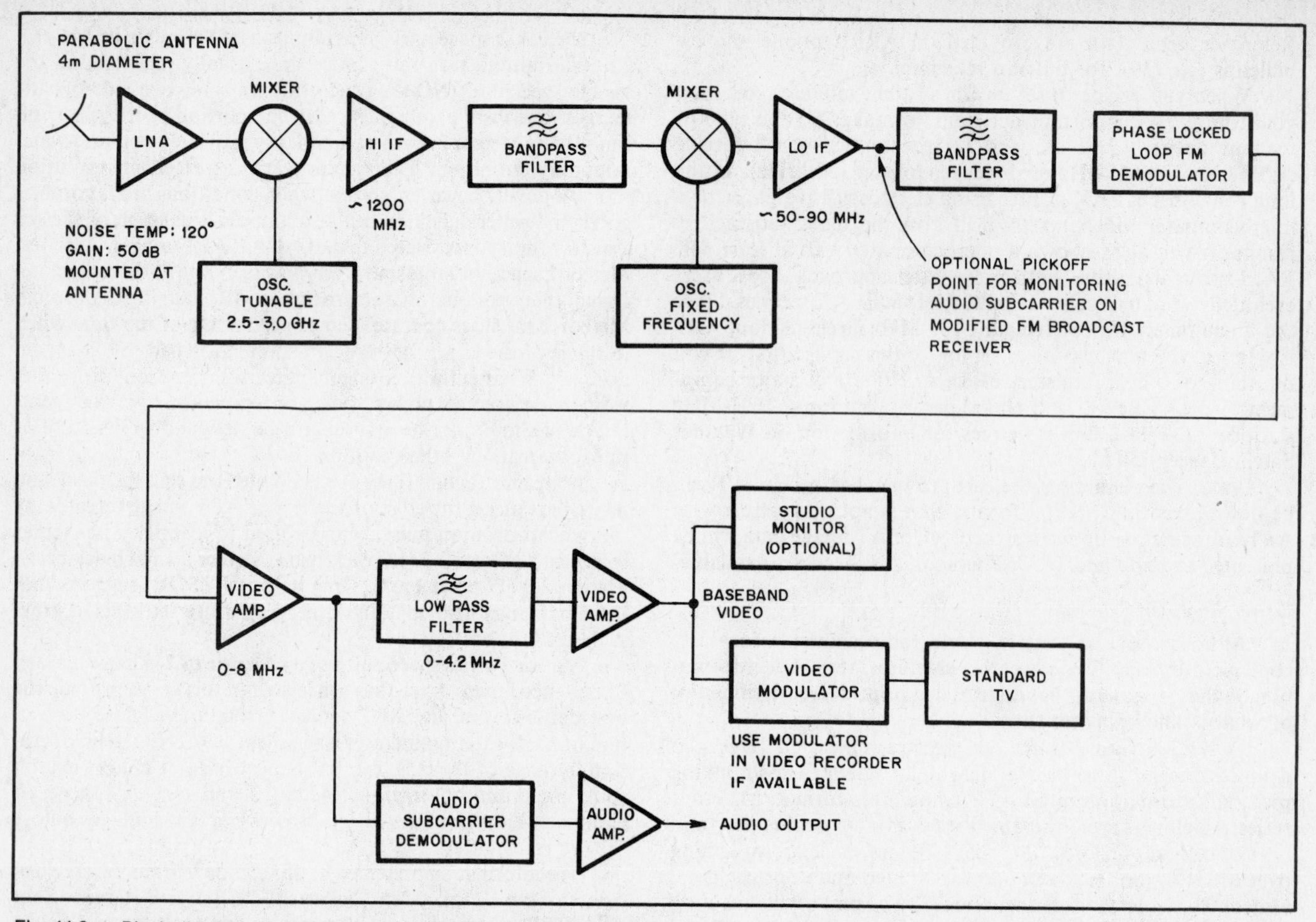

Fig. 11-3 — Block diagram of typical double conversion TVRO with nominal values indicated. Excellent results can be obtained using either single- or double-conversion approaches if good engineering practices are followed.

MHz wide. The trick is to use linear polarization with alternate transponders, spaced 20 MHz center-to-center, polarized at right angles. In effect, we double the available frequency spectrum by using each segment twice, once with each polarization, without any interference between neighboring channels. Anyone familiar with U.S. NTSC color standards based on a 4-MHz-wide amplitude-modulated vestigial sideband video signal, will realize that the 36-MHz-wide fm downlink just described is quite different. As a result, you *can't* just build a downconverter to shift one of the 4-GHz satellite channels to an unoccupied TV channel.

An effective TVRO terminal must capture some downlink rf, filter out the 36-MHz-wide channel of interest, demodulate the video and audio information and reconstruct a standard NTSC TV signal to feed into a regular TV set. If this sounds like a big job, it is. In difficulty, it's comparable to putting a 432-MHz or 1296-MHz EME station on the air. Although the video processing may sound complex to radio amateurs with no practical video experience, it turns out to be relatively easy. The real challenge lies in the antenna and 4-GHz rf equipment (especially the preamp, which is always referred to as a *l*ow *n*oise *a*mplifier or LNA).

A 24-transponder satellite like the one described costs about 50 million dollars by the time it reaches geostationary orbit, the expense split about 50-50 between the spacecraft hardware and the launch. With all this money being invested and the technical expertise available, you may wonder why the system wasn't designed to be easier to receive. The reason is that the system is a commercial venture designed for limited private distribution in accordance with the objectives of the Common Carrier allocations. The satellites are owned by large corporations such as RCA (SATCOM), Western Union (WESTAR) and ATT/GTE (COMSTAR) who aim to recoup their construction and launch costs by renting transponders. TV program distributors may rent transponders for a number of reasons: They're cheaper and more reliable, and may provide more timely programming than the alternatives, such as using a large number of terrestrial, tower-mounted microwave relay stations, or mailing out video tapes a week or two in advance. At present, the corporations owning the satellites have no economic incentive to invest more money so you and I can build cheaper ground stations. Thus, we're not likely to see a significant (more than 3 dB) increase in power density per channel reaching the ground over the next several years. The requirements for a 4-GHz TVRO terminal should, therefore, remain relatively stable throughout the 1980s.

Laws Governing Reception

To say that there's a lot of confusion as to the legality of intercepting 4-GHz satellite TV programming for personal, noncommercial use, would be a gross understatement. The rules governing reception of "private transmissions" were written nearly 50 years ago (Communications Act of 1934). With the rapid pace of technological progress, the real wonder is not the confusion, but that the laws still serve so well. The applicable rules, found mainly in Section 605, are unfamiliar to most radio amateurs. To paraphrase a portion of the relevant law beyond legal recognition: If the originator of common carrier material didn't intend for you to be a specific recipient, you shouldn't tune in; if you happen to tune in by accident, you shouldn't (1) divulge what you've seen or heard, (2) tell anyone you happened to tune in or (3) profit from your transgression.

Let's look at the actual situation. In 1981, program originators ran the gamut. For a small one-time fee, or no charge at all, some would designate you a "specific intended recipient" upon request. Others would do the same for a modest yearly fee.

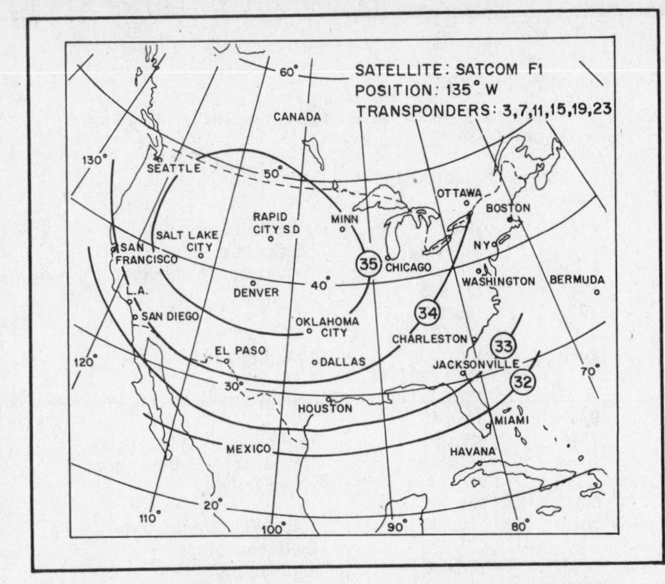

Fig. 11-4 — This figure shows the footprint (EIRP contours in dBw) for one set of transponders on SATCOM F1. The boresight is near Rapid City, South Dakota.

Still others would quote outlandish charges or refuse to respond to letters. With an "intended recipient" letter in hand one can obtain a one-year renewable experimental/developmental class license from the FCC to set up a TVRO terminal for equipment development and testing. The FCC has, on several occasions, indicated that the registration process for individual terminals that are operated by noncommercial entities would be further simplified or eliminated in the near future. Contact your local FCC field office for the latest information. Even with a legal TVRO terminal, however, you must still contract with each program provider before viewing the material.

This brief discussion of the laws governing satellite TV reception is meant to make you aware that you are faced with a number of legal concerns related to setting up a TVRO terminal, even when it's for your personal, noncommercial use.

The 4-GHz TVRO

One possible TVRO terminal configuration is shown in Fig. 11-3. A TV picture with a 48-dB video signal-to-noise ratio (SNR) is defined as being of excellent quality. To obtain a 48-dB SNR at the detector output, we need an 11-dB link margin on the downlink signal; this margin must be maintained up to the detector. It's critically important that the antenna gain and LNA noise temperature be chosen to provide the needed link margin. LNAs are almost always specified in noise temperature, values that are easily converted into noise figure (see Eq. 10.6).

Sample link calculations are shown in Chapter 10 in the section "Frequency Selection: Technical Constraints." One required input was the satellite EIRP in the direction of the ground station. For the 4-GHZ satellites of interest this information is provided in the form of contour charts (in dBw) like the one shown in Fig. 11-4 for SATCOM F1. The central region (where the strongest signal is received) is referred to as the *boresight.* The entire pattern is called the *footprint.* From Fig. 11-4 we see that signal levels across the continental U.S. vary by more than 4 dB, making it considerably easier for stations in Chicago to put together a ground station than those in Miami.

As an example, let's consider a station in Chicago that is working with a 35-dB signal. A link calculation assuming a 30-MHz receiver i-f bandwidth would show that this station could obtain the desired 11-dB link margin using a 4-m diameter parabolic dish antenna and a 200-K LNA. Note that there's a direct trade-off between antenna size and LNA temperature: The Chicago station could obtain the same 11-dB link margin with a 5-m dish and a 300-K LNA, or with a 3-m dish and a 120-K LNA. A station in Miami would need at least a 5-m dish and a 120-K LNA to produce the desired 11-dB margin.

On most Amateur Radio communications circuits we don't worry about a 1- or 2-dB change in signal level. But, on a TVRO downlink, a difference of this magnitude can be extremely significant. Since (1) the signal level on a particular downlink path rarely varies by more than +/−0.5 dB from propagation, (2) the picture does not perceptibly improve once we exceed the "magic" 11-dB link margin threshold and (3) each decibel of extra link margin generally costs at least several hundred dollars, the incentive is to design the system with just enough sensitivity. While system performance degrades rapidly below the 11-dB margin level, most viewers of noncommercial installations find an 8- or 9-dB margin acceptable.

In 1983 components for a complete commercial TVRO terminal could be purchased for under $2000. This includes $400 for an LNA with 120-K noise temperature and 50-dB gain, $800 for a tunable receiver and $800 for a 3.5-m diameter dish and mounting kit. Building parts of the system from scratch can, of course, save money. Sources of construction information are listed at the end of this section.

Although our discussion has focused on U.S. domestic satellites that are parked between 70° W and 135° W, other satellites, including INTELSAT and Soviet birds parked along the geostationary arc and in Molniya orbits (see Fig. 8-24 and accompanying text) also distribute video information. The

Table 11.4
Information Resources: TV Satellites

4-GHz TV Satellites

Introductory Material

Cooper, R., "The Satellite TV Primer," *73,* Nov. 1979, pp. 120-133.

Shuch, H., "Low-Cost Receiver for Satellite TV," *73,* Dec. 1979, pp. 38-43.

Cooper, R., "Television Home Reception via Satellite," *Radio Electronics,* (in 7 parts): Part I, Vol. 50, no. 8, Aug. 1979, pp. 47-49; Part II, Vol. 50, no. 9, Sept. 1979, pp. 47-50; Part III, Vol. 50, no. 10, Oct. 1979, pp. 81-85; Part IV, Vol. 51, no. 1, Jan. 1980, pp. 55-59, 65; Part V, Vol. 51, no. 2, Feb. 1980, pp. 47-52, 83; Part VI, Vol. 51, no. 3, March 1980, pp. 38-42; Part VII, Vol. 51, no. 4, April 1980, pp. 47-52. A reprint book containing these seven articles is available from Radio-Electronics, 45 East 17 St., New York, NY 10003.

Hopengarten, F., "Backyard Satellite-TV Reception, Fact or Fantasy?" *Radio Electronics,* Vol. 51, no. 6, June 1980, p. 68.

Specialized Material

CATJ (Community Antenna Television Journal, ISSN-0194-5963), published monthly by TPI, 4209 NW 23rd, Suite 106, Oklahoma City, OK 73107. A number of excellent construction articles focusing on specific pieces of TVRO terminals were described in this publication in 1978 and 1979. The articles by Steve Birkell are especially valuable. In recent years this magazine has severely decreased coverage of TVRO topics.

Coop's Satellite Digest, published monthly by Satellite Television Technology, P.O. Box G, Arcadia, OK 73007. This monthly magazine is an outstanding source of technical and operational information on satellite TV topics. The subscription price, geared to commercial TVRO operators, is formidable. The information is invaluable, however, so get a few friends to share a subscription.

12-GHz Direct Broadcast Satellites

Coop's Satellite Digest.

Harrop, P., P. Lesartre, and C. Tsironis, "Low-cost 12 GHz Receiver Heralds Satellite-to-home TV," *Electronics,* No. 17, 1981, pp. 125-127.

Harrop, P., J. Margarshack, R. Dessert, J. Forrest, "Satellite Communications II: Television for Everyone," *IEEE Spectrum,* March 1980, pp. 54-56.

Gosch, J., "Germans, French Plan TV Satellites," *Electronics,* Sept. 27, 1979, pp. 98, 100.

Pritchard, W., and C. Kase, "Getting Set for Direct-Broadcast Satellites," *IEEE Spectrum,* August 1981, pp. 22-28.

Table 11.5

Some Radio Transmissions Observed in the 136- to 138-MHz Band Between 1978 and 1982.

All identifications are tentative. The table is based on articles by G. Roberts in *Orbit* and *OSCAR NEWS* (see Table 11.6 for full references), "Satellite Situation Report" (NASA) and several other sources.

International designation	*Satellite name*	*Period (minutes)*	*Inclination*	*Apogee km*	*Perigee km*	*Frequencies*	*Comments*
62A Alpha 1	Tiros 5	100.1	58.1°	939	588	136.230, 136.920	Continuous tone but periodic pulsing, occasional hiccoughs
62B Alpha 1	Alouette 1	105.3	80.5°	1026	993	136.980	
62 Upsilon 1	Relay 1	185.1	47.5°	7436	1323	136.140, 136.620	
64 03A	Relay 2	194.7	46.4°	7476	2025	136.140, 136.620	Continuous tone
64 83D	Transit 5B-5	106.2	89.8°	1083	1018	136.650	Musical sequence, fm
65 32A	Explorer 27	107.7	41.1°	1313	933	136.740	
65 51A	Tiros 10	100.5	98.3°	824	735	136.230, 136.920	Continuous tone but slow frequency variations apart from Doppler
65 98A	Alouette 2	120.3	79.8°	2888	502	136.980	
66 77B	EGRS 7 (Secor 7)	167.5	89.9°	3698	3673	136.800	Typical EGRS multi-tone sequence, fm
66 77C	ERS 15	167.6	89.9°	3698	3681	136.440	Modulated fm
66 89B	EGRS 8 (Secor 8)	167.6	90.3°	3694	3688	136.830	
66 110A	ATS-1 [1]	1436.2	9.7°	35,792	35,786	136.470, 137.350	See note 1.
67 40D	ERS 20 (OV5 3)	2840	32.9°	111,529	8619	136.260	Modulated signal having period of 4.56 seconds
67 65A	EGRS 9 (Secor 9)	172.1	89.8°	3937	3801	136.840	
67 100A	OSO 4	94.3	33.0°	493	472	136.710	
67 111A	ATS-3 [1]	1436.1	8.3°	35,856	35,719	136.470, 137.350	See note 1.
69 09A	ISIS 1	128.2	88.4°	3514	577	136.410, 136.080, 136.590	Continuous tone
69 37B	EGRS 13 (Secor 13)	107.2	99.5°	1128	1068	136.800	Standard EGRS fm signal
69 46B	OV5-6	3114.8	32.9°	113,084	15,460	136.380	
69 82B	Timation 2	103.3	70.0°	931	900	137.380	Musical tones, fm
69 82E	—	103.4	70.0°	934	902	137.410	Continuous tone
70 09A	Sert 2	106.1	99.1°	1047	1039	136.920, 136.230, 136.928	Rapid periodic pecking, fm
70 25A	Nimbus 4	107.1	99.6°	1102	1091	136.500, 136.797	Operating illuminated passes only?
70 25B	TOPO 1	106.9	99.7°	1086	1082	136.840	
71 24A	ISIS 2	113.6	88.1°	1426	1360	136.410, 136.080, 136.590	.410-continuous tone .590 occasionally fm
71 30A	Tournesol	96.2	46.4°	697	457	136.630	
71 71A	Eole 1	100.5	50.2°	891	672	136.350	
71 80A	Shinsei	113.2	32.1°	1869	873	136.694	Continuous tone
71 96A	Explorer 45	326.8	3.5°	18,315	362	136.830	
71 110A	—	104.8	69.9°	989	984	136.800	
71 110C	—	104.8	70.0°	992	982	137.080	
71 110D	—	104.8	70.0°	992	982	136.320	
71 110E	—	104.8	70.0°	991	982	137.050	
72 65A	Copernicus	99.5	35.0°	742	731	136.260, 136.440	Modulated fm carrier
72 97A	Nimbus 5	107.2	99.8°	1105	1092	136.500	Operating illuminated passes only?
73 78A	Explorer 50	17,462	51.1°	230,086	203,072	137.980, 136.800	
74 33A	SMS 1	1437.2	4.4°	35,822	35,795	136.380	
74 39A	ATS-6	1435.8	2.2°	35,796	35,767	136.230, 136.112	
74 101A	Symphonie-1	1436.1	1.1°	35,801	35,775	137.020	Modulated fm carrier

Russian Molniya satellites are of special interest since they're generally operated during both Eurasian and North American apogees. They usually use earth coverage antennas and typically provide about 30-dBw EIRP in your direction. TVRO terminal operators in most of the U.S., with a few extra decibels in margin, can catch these transmissions by searching for spacecraft using nominal values for height (35,800 km) and subsatellite latitude (62° N). Azimuth and elevation settings for a search from your location can be computed using the techniques outlined in Chapters 8 and 9. Downlinks are primarily in the range 3.75 to 3.95 GHz, with 3.895 GHz being most common. The video format is compatible with U.S. systems so you won't have any trouble obtaining a picture. Decoding the color information is complex, however, so most experimenters settle for black-and-white viewing.

Let's look more closely at how the downlink signal from a domestic 4-GHz TV satellite is decoded. The output of the TVRO terminal detector (Fig. 11-3) consists of (1) a video waveform, (2) a frequency modulated audio subcarrier and (3) a triangular energy dispersal waveform. The *baseband* video waveform, which contains components from dc to 4.2 MHz, is similar to the signal provided by video cameras and video tape recorders. It can be fed directly into a studio monitor or into an rf modulator for viewing on standard TVs. Since video monitors are relatively expensive, most people use the rf-modulator-to-TV approach. Video recorder owners generally patch the excellent rf modulators in these units into the TVRO setup. The fm audio subcarrier is usually at 6.8 MHz or 6.2 MHz. With a peak deviation of 75 kHz it's so similar to standard fm broadcast-band (88-108 MHz) signals that some homemade TVRO systems use modified fm broadcast receivers to tune across the low i-f looking for subcarrier signals.

The 30-Hz triangular energy dispersal waveform, which has a peak deviation of 750 kHz, needs some explanation. Its sole

International designation	Satellite name	Period (minutes)	Inclination	Apogee (km)	Perigee (km)	Frequencies	Comments
75 04A	Landsat 2	103.2	99.1°	919	904	137.860	Modulated fm carrier
75 11A	SMS 2	1436.0	0.4°	35,810	35,763	136.380	
75 27A	GEOS 3	101.7	114.9°	863	821	136.320	Strong modulated fm carrier
75 33A	Aryabhata	96.1	50.7°	591	553	137.440	Strong modulated rasping fm carrier
75 49B	SRET 2	736.4	64.0°	40,504	763	137.530	Broad modulated fm carrier (see note 2)
75 52A	Nimbus 6	107.4	99.9°	1116	1105	136.500	
75 72A	COS B	2202.5	96.5°	89,407	9985	136.950	Modulated fm carrier
75 77A	Symphonie-2	1436.1	1.6°	35,840	35,734	136.800	Modulated fm carrier
75 100A	GOES 1	1425.5	0.0°	35,591	35,566	136.380	
75 107A	Explorer 55	93.6	19.6°	449	447	137.230	
76 23D	Solrad 11B	7333.8	28.0°	119,817	117,505	136.530	
77 48A	GOES 2	1436.2	0.7°	35,809	35,770	136.380	
77 80A	Sirio	1437.6	1.6°	37,049	34,582	136.140	Strong, fm modulation
77 108A	Meteosat 1	1436.2	0.2°	35,803	35,774	137.080	Strong, fm modulation
77 117A	Meteor 2-3	102.3	81.2°	887	850	137.300	Soviet APT
78 12A	IUE	1435.4	28.3°	45,691	25,856	136.860	Modulated carrier, fm
78 26A	Landsat 3	103.1	99.0°	917	898	137.860	Modulated carrier, fm
78 41A	HCMM	—	97.7°	—	—	137.170	Continuous tone, cw
78 44A	OTS 2	1436.1	0.0°	35,796	35,779	137.050	Strong fm modulation
78 62A	GOES 3	1436.0	0.0°	35,795	35,776	136.380	
78 71A	ESA GOES	1436.0	0.4°	35,814	35,757	137.200	Modulated carrier, fm
78 87A	Jiki'ken	473.4	31.1°	27,215	268	136.695	
78 96A	Tiros-N	102.0	99.0°	876	839	137.620 (APT), 137.770, 136.770	
78 99A	Intercosmos 18	94.6	82.9°	618	375	137.850	Strong wide fm, slow tone sequence, about 1 min/frame
78 99C	Magion	94.9	82.9°	648	382	137.150	Pulses (about 1 sec)
79 14A	Corsa-B	95.5	29.9	554	527	136.725	Strong carrier
79 21A	Meteor 2-4	102.2	81.2	891	833	137.300	Soviet APT
79 47A	UK 6	97.0	55.0°	651	585	136.560, 137.560	Strong, fm modulation
79 51A	Bhaskar	95.0	50.7°	529	509	137.230	Strong, fm modulation
79 57A	NOAA-6	101.2	98.7°	824	807	137.500 (APT), 136.770	
79 95A	Meteor 2-5	102.5	81.2°	894	874	137.300	Soviet APT
80 15A	Tansei 4	95.9	38.7°	606	520	137.725	Continuous carrier
80 51A	Meteor 30	97.5	97°	640	—	137.150, 137.130	Soviet APT (Experimental?)
80 73A	Meteor 2-6	102.3	81.2°	899	851	137.400	Soviet APT
81 43A	Meteor 2-7	102.4	81.3°	899	859	137.400	Soviet APT
81 57C	Ariane LO3	627.	10°	35,838	202	136.610	(see note 3)
81 59A	NOAA-7	102.0	98.9°	869	851	137.620 (APT), 136.770	
81 65A	Meteor 31	97.8	97.8°	670	630	137.130	Soviet APT
82 25A	Meteor 2-8	104.0	81.2°	960	940	137.850	Soviet APT
82 116A	Meteor 2-9	102.3	81.2°	910	850	137.300	Soviet APT
83 xx	NOAA-8	101.4	98.7°	834	834	137.500	

Notes

[1]ATS-1 (149° W) and ATS-3 (105° W) have operational transponders on 149.195/135.575 MHz, 149.220/135.600 MHz and 149.245/135.625 MHz. NASA allocates transponder time slots for imaginative experimental proposals deemed worthwhile. For information request "ATS VHF Experiments Guide," ATS Experiments Manager, Office of Applications, Code ECS, NASA, Washington, DC 20546.

[2]For additional information see G. Roberts "Radio Tracking of SRET 2," *OSCAR NEWS*, no. 28, Winter 1979, pp. 27-30.

[3]For about three days after launch a beacon (a two-tone carrier switching every 8 to 9 seconds) was in operation on the indicated frequency. Speculation is that it was on the rocket and used for tracking. Whether future Ariane launches will also carry tracking beacons at this frequency is not known, but it's worth checking.

purpose is to move the carrier around when no modulation is present. This reduces the rf energy density at any single frequency and helps prevent interference to terrestrial microwave links that share the 4-GHz band. At the ground station, our main interest is in removing the energy dispersal waveform, an easily accomplished task.

12-GHz Direct Broadcast Satellites

Geostationary Direct Broadcast Satellites will use the 12-GHz band (11.7 to 12.5 GHz in U.S.). Experiments with this service began in the late 1970s with CTS in North America (also known as Hermes), BSE in Japan and OTS in Europe. Projections call for activating operational systems in Japan (about 1983), in Europe (about 1984) and in the U.S. (about 1986). Since the 12-GHz TV downlinks are meant to be a broadcast service to ground stations numbering in the tens of millions, the economics dictate designing the spacecraft to minimize ground-terminal cost. The projected price of a typical ground station, including antenna (roof-mounted 75-cm diameter dish) and all electronics, is under $500. These cost estimates are based on a satellite EIRP per channel of 60 dBw, a figure 25 dB higher than that of the 4-GHz service.

Since these satellites will be similar in size and power budget to the 4-GHz models currently in operation, where will this extra power come from? By restricting the number of channels to four, the power available per channel could be raised to 30 watts, a 7.8-dB increase. The additional 17-dB EIRP needed will be attained by high-gain spot-beam antennas. A 17-dB change roughly equals a 50-fold reduction in coverage: from the entire U.S. to a circular area approximately 300 km in diameter.

How these channels will be used is not clear. Some proposals call for a service similar to that currently being operated at 4 GHz. Others suggest adopting a new high-resolution TV system that would produce better pictures than the three major systems

(NTSC, SECAM, PAL) currently in use around the world. To take advantage of a high-resolution channel, users would need to purchase an entirely new TV receiver. Whatever the outcome, by 1990 50 to 100 million households worldwide will likely be equipped for direct reception of 12-GHz satellite TV. As a result of this new mass market for microwave electronics, the availability of equipment should increase exponentially while prices nose dive. Government and commercial research laboratories are now working on the GaAs FET and MESFET (metal-semiconductor field-effect transistors) devices of tomorrow. MESFETs, consisting of metal source, gate and drain contacts on a chip of gallium-arsenide semiconductor substrate, are a recent development. In fact, 12-GHz LNAs that operate at room temperature have been constructed using MESFETs. Noise figures in the 2-dB range were reported. Spinoff from the 12-GHz satellite development effort is bound to benefit all amateurs interested in microwave communications as new, inexpensive solid-state microwave equipment for the 1 to 10 GHz spectrum becomes available. See Table 11.4 for a listing of information sources related to TV satellites.

Satellite Sleuthing

Ever since the early days of the space program, radio amateurs have been captivated by the challenge of identifying "unknown" transmissions from space, an activity sometimes referred to as satellite sleuthing. In Chapter 2 we mentioned a group of U.S. amateurs who were asked in the late '50s and early '60s to apply their hf-radio expertise to help locate unannounced low-altitude satellites. They were to do so by noting propagation anomalies produced by the ionized trails such spacecraft left behind. A school group in Kettering, England, led by Geoff Perry, has become famous for often providing details of Soviet space launches long before official announcements. By studying orbital and launch data carefully, and by correlating this information with known astronomical facts and available details of the Soviet space program, the Kettering group has been able to predict mission objectives with uncanny accuracy. Another well known satellite detective is Greg Roberts (ZS1BI), a professional astronomer who first became involved in Amateur Radio and the OSCAR program as a result of his satellite monitoring activities. Roberts shared nearly 20 years of accumulated sleuthing experience in a series of articles in *Orbit*. The articles present the best available overview of satellite transmissions in various parts of the rf spectrum, from hf to microwave. (References are listed at the end of this section in Table 11.6.)

Although eavesdropping is usually approached as a fascinating game in which the objectives are to (1) locate an unidentified transmission, (2) determine the orbital elements and (3) identify the transmitting spacecraft, its real value is as an educational tool. Those who pursue this activity gain a practical understanding and knowledge of space communications that few professionals ever acquire. The Kettering group has on occasion, for example, left the professionals dumbfounded by decoding downlink telemetry without prior knowledge of the content or format. (G. Perry and C. Wood, "The Russian Satellite Navigation System," *Phil. Trans. Royal Society,* Vol. A 294 [1980], pp. 307-315.)

Monitoring interest always perks up when astronauts or cosmonauts are involved in space missions. As mentioned in Chapter 3, a few radio amateurs successsfully monitored lunar missions in the late 1960s and early 1970s. Undoubtedly, they'll be eavesdroppers when the first visitors from earth set foot on Mars.

136- to 138-MHz Satellite Band

The best band in which to begin searching for unidentified space transmissions is 136 to 138 MHz. It's best because many spacecraft transmit at these frequencies and most amateur 2-m converters can be modified easily to cover this range. Some of the signals you may hear are listed in Table 11.5, based primarily on data provided by G. Roberts. Although NASA plans to phase out vhf downlinks, this segment of the spectrum should continue to be well populated with space signals until at least 1990.

I personally use an old vacuum-tube 2-m converter picked up at a hamfest for $12 to monitor the 136- to 138-MHz range. My approach to modifying the unit, which was originally configured for 144 to 146 MHz in, 28 to 30 MHz out, was determined by my desire to start listening as soon as possible. As a result, I tuned the rf circuits down to the desired frequency and adjusted the output coupling network to 20 to 22 MHz without so much as touching the oscillator chain. The entire procedure took about 15 minutes and, since the Drake R4-C being used as a tunable i-f had crystals for 20.0 to 20.5 MHz and 21.0 to 21.5 MHz, half the band could immediately be checked using a 2-m ground-plane antenna. Two extra crystals for the R4-C gave full

Table 11.6

Information Resources: Satellite Sleuthing

Introductory Material

Roberts, G. R., "Radio Transmissions from Outer Space," *Orbit,* Part I, Vol. 1, no. 1, March 1980, pp. 15-18; Part II, Vol. 1, no. 3, Sept./Oct. 1980, pp. 28-30.

Roberts, G. R., "Transmitting Satellites, Sept. 1980," *OSCAR NEWS,* no. 31, Autumn 1980.

Orbital Data: Comprehensive

Satellite Situation Report. Published by NASA. Contains orbital data on more than 2500 space objects. Also lists transmitting frequencies for a limited number of satellites being monitored by the NASA spaceflight Tracking and Data Network. A copy may be obtained from NASA, Office of Public Affairs, Code 502, Goddard Space Flight Center, Greenbelt, MD 20771.

TRW Space Log. Extensive unofficial compilation of orbital data dating back to 1957. Published annually by the Public Relations staff of TRW Systems group. Copies are available to professional personnel in the aerospace industry, military and other government agencies. Requests must be on organization letterhead. Write Editor, Space Log, Public Relations Dept., TRW Systems Group, One Space Park, Redondo Beach, CA 90278.

Orbital Data: Recent Launches

Satellite News by G. Falworth. This inexpensive newsletter provides the most comprehensive, up-to-date information available on satellites and space activity. For subscription information write G. Falsworth, 12 Barn Croft, Penworthan, Preston PR1 ØSX, England. Requests should include a small donation to cover mailing costs.

Satellite Log in *Orbit* and *Satellite Activity* in *OSCAR NEWS* are columns by G. Falworth that include much of the information from *Satellite News* which is of interest to radio amateurs. Generally includes orbital parameters of recent launches and transmission frequencies if announced.

Aviation Week and Space Technology. Published weekly by McGraw-Hill. Contains orbital data for all announced launches. Often carries good descriptions of new satellite systems, launch vehicles and facilities. This magazine is very widely distributed; check your local library.

General Information

Spaceflight. Published monthly by the British Interplanetary Society. This journal contains by far the best available in-depth descriptions of U.S., USSR, and other satellite systems. Their articles are frequently more informative than official system documentation. For subscription information contact: British Interplanetary Society, 27/29 South Lambeth Rd., London SW8 1SZ, England.

OSCAR NEWS. Published quarterly by AMSAT-UK, 94 Herongate Road, Wanstead Park, London, E12 5EQ. Often contains hard to find information on non-amateur-satellite rf transmissions.

Aviation Week and Space Technology. (See previous description).

Interferometer Design

Swenson, G. W., "An Amateur Radio Telescope," *Sky & Telescope,* Part I, Vol. 55, no. 5, May 1978, pp. 385-390; Part II, Vol. 55, no. 6, June 1978, pp. 475-479; Part III, Vol. 56, no. 1, July 1978, pp. 28-33; Part IV, Vol. 56, no. 2, Aug. 1978, pp. 114-120; Part V, Vol. 56, no. 3, Sept. 1978, pp. 201-205; Part VI, Vol. 56, no. 4, Oct. 1978, pp. 290-293. This series of articles provides the best practical introduction to Amateur Radio-Astronomy instrumentation available. Reprinted in booklet form by Pachart Publishing House, Tucson, Arizona (1980). Unfortunately, the reprint booklet omits the valuable photographs accompanying the original articles. Parts I and II and the April 1979 article by Swenson contain very useful information on interferometers.

Swenson, G. W., "Antennas for Amateur Radio Interferometers," *Sky & Telescope,* Vol. 57, no. 4, April 1979, pp. 338-341.

coverage and a JFET preamp produced a big improvement in sensitivity.

As a first step in the monitoring game, try to associate some of the signals you hear with those listed in Table 11.5. You'll soon be recognizing various signals by their sounds. The APT signal, for example, is very distinctive. Now, suppose one day you hear an APT signal on 137.170 MHz that's not in Table 11.5. Good record keeping (an extremely important aspect of this activity) and several days of monitoring will provide you with an estimate of the orbital period. AOS, LOS and Doppler observations will help you verify and refine the period measurements and give you an estimate of the orbital inclination. Finally, a search of your launch announcements file (which you've carefully been keeping up-to-date) may show a likely source, a spacecraft with no announced downlink that was recently placed in a closely matching orbit.

Of course, identifying the source of a signal is not always so easy; sometimes it takes years. An expert like Roberts may eventually be able to identify 95% of signals heard. But newcomers can take pride in a 25% identification record. The listener has basically two important keys to work with: (1) recognizing the sound of modulation schemes used by different countries for various series of satellites and (2) accurate orbit determination. The ability to recognize modulation schemes develops with eavesdropping on clearly identified spacecraft. Orbit determination, using passive techniques, involves applying ideas found in Chapters 5, 8 and 9 and applying the more advanced concepts found in the references in these chapters. Valuable information can be collected by using highly directional antennas with accurate azimuth and elevation readouts or by using an array of antennas in the form of an interferometer (see the reference in Table 11.6 by Swenson).

You'll quickly learn that being a good detective requires meticulous record keeping, an up-to-date collection of information on the thousands of spacecraft that have been launched in the past 25 years and data on the new launches that occur almost daily. Much of this information can be obtained from the sources listed in Table 11.6. Good hunting!

Chapter 12
Satellite Systems

Chapter 12

Satellite Systems

Even a simple satellite is a complex collection of hardware. To manage the design and construction of communications and scientific satellites, it's convenient to think of a spacecraft as comprising a standard set of subsystems, each with a specific function.[1] See Table 12.1. This makes it possible to parcel out the tasks of analyzing and optimizing each subsystem. Several design objectives almost always apply: minimizing weight and cost, maximizing reliability and performance, and insuring compatibility. These aims often result in conflicts. On Phase III spacecraft, for example, extra radiation shielding of the central computer increases its reliability while producing a severe weight penalty. Even when designers focus on a single subsystem they must keep in mind how it impacts on other subsystems. As an illustration, a small reduction in transponder power-amplifier efficiency may have little effect on signal strength of the downlink, but it might completely upset spacecraft thermal design.

In this chapter we'll look at each of the systems listed in Table 12.1. We'll discuss various methods of accomplishing system objectives, emphasizing the approaches that were used on past OSCAR satellites or that might be appropriate for future OSCAR missions.

Communications, Engineering and Mission Subsystems

The communications subsystem provides a direct link with the satellite, enabling us to observe what's happening inside the spacecraft as it happens and to modify the operation of the spacecraft. Three communications links are of interest: (1) downlink beacons providing telemetry (TTY), (2) uplink telecommand and (3) communications links supported by a transponder.

Beacons: Function

The beacons aboard the OSCAR satellites serve a number of functions. In the *telemetry mode* (TLM mode) they convey

Table 12.1
Satellite Subsystems: Emphasis OSCAR

Subsystem	*Function*	*OSCAR Series Equipment*
Attitude-control	To modify and stabilize satellite orientation	Phase II: magnet, gravity boom Phase III: solid-propellant or gas spin up motors, torquing coils
Central-computer	To coordinate and control other subsystems; provides memory, computation capability	Digital logic, microprocessor, d/a converter, command decoder
Communication	To receive uplink commands and transmit downlink telemetry	Command receiver, transmitters (beacons), antennas
Energy-supply	To provide power for all onboard subsystems	Batteries, solar cells, conditioning electronics
Engineering-telemetry	To measure operating status of onboard subsystems	Electronic sensors, telemetry encoders
Environment-control	To regulate temperature levels, provide electromagnetic shielding, provide high energy particle shielding	Mechanical design, thermal coatings
Guidance and control	To interface computer with sensors, attitude-control and propulsion subsystems	Hard-wired electronics, sun and earth sensors
Mission-unique-equipment	To accomplish mission objectives	Transponders, scientific and educational instruments
Propulsion	To provide thrust for orbit changes	Phase II: none Phase III: solid-fuel kick motor, liquid-fuel rocket, ignition system
Structure	To provide support and packaging function, thermal control, protect modules from stress of launch, mate to launch vehicle.	Mechanical structure, aluminum sheet wherever possible to minimize machining

Table 12.2

Beacon Functions

1) Telemetry
   a) Morse code
   b) Radioteletype (RTTY)
   c) Advanced encoding techniques
   d) Digitized (digitally synthesized) Speech
2) Communications
   a) Store-and-forward
3) Miscellaneous
   a) Tracking
   b) Propagation measurements
   c) Reference signal

Each OSCAR includes some, but not necessarily all, of the telemetry options listed.

Table 12.3

Telemetry Encoding Methods, Ground Station Complexity and Telemetry Data Transmission Rates

Telemetry Encoding Method	*Relative Ground Station Complexity*	*Telemetry Data Rate*	*First OSCAR Utilization*
Digitized speech	Very low	Very low	UoSAT
Morse code	Low	Low	OSCAR 6
Radioteletype	Moderate	Moderate	OSCAR 7
Advanced encoding techniques	High	High	UoSAT, Phase III

information about onboard satellite systems (solar cell panel currents, temperatures at various points, storage battery condition, etc.); in the *Codestore mode* they can be used for delayed time (store-and-forward) communication; in either the telemetry or Codestore modes they can be used for tracking and for propagation measurements, and as a reference signal of known characteristics. Beacon functions are summarized in Table 12.2.

Beacon telemetry. From the user's point of view, each telemetry mode can be characterized by the capacity (the amount of information that can be conveyed in a given time interval) and the complexity of the decoding equipment at the ground station. To a certain extent, there is a trade-off between these two factors (see Table 12.3). Let's look at each of the beacon telemetry modes.

Morse code telemetry. The Morse code telemetry system is one of the ingenious features that have made the OSCAR series of satellites so valuable to educators and amateur scientists.[2] In the cw telemetry mode, information on satellite systems is transmitted in Morse code using a numbers-only format, usually at either 25 or 50 numbers/minute (about 10 or 20 words/minute). Because AMSAT put the telemetry information-processing equipment aboard the satellite, ground stations do not need specialized decoding electronics. The information capacity of this mode is inherently limited in that any attempt to speed up the Morse code transmission would interfere with the ability of untrained users to decode it without special equipment.

Radioteletype telemetry. During the early and mid 1970s the most cost-effective way to provide moderate-speed telemetry was with standard radioteletype equipment. Radioteletype telemetry was, therefore, included on AMSAT-OSCAR 7. This feature was especially valuable to the advanced experimenter, to stations that were responsible for managing the satellite and to the scientists who would design and build future AMSAT spacecraft.

Advanced encoding techniques. Recent advances in digital electronics and widespread experimentation with microcomputers among OSCAR satellite users and designers have displaced RTTY equipment. Microcomputers offer advantages in price, power and flexibility. Once a ground-station-to-computer interface is constructed the computer can be used not only to decode TLM, but to perform calculations on the data as it's received, store it as part of a large data base, automatically check for values that indicate developing problems and alert the operators, graph the data over time and so on. Phase III satellites will have integral computers that can be instructed from the ground to use ASCII, Morse or other codes. Present plans are to use Morse code telemetry on one or more *general beacons* and ASCII or some other code on an *engineering beacon* simultaneously. The performance of these telemetry formats has been tested operationally by generating a signal on the ground and relaying it through the transponders aboard operating spacecraft.

Digitized speech mode. The digitally synthesized speech mode for telemetry produces the ultimate simplicity in ground station decoding requirements. This mode is excellent for general demonstration and for educational applications at lower grade levels but the extremely low data rate makes it unsuitable for most engineering studies.

Beacon communication mode (Codestore). The beacon Codestore mode relies on an onboard digital memory system that can be loaded by suitably licensed and equipped ground stations for later rebroadcast in the form of Morse or other codes. The system has proved very useful with OSCARs 6 and 7 in disseminating information to the worldwide network of OSCAR users and command stations.

Beacon: Miscellaneous functions. In either the telemetry or Codestore modes, a beacon with a well-known intensity and frequency can serve a number of useful functions. For example, it can be used for Doppler shift studies, propagation measurements and testing ground-based receiving equipment. In addition, stations communicating via a satellite transponder can optimize their uplink power levels so the strength of their downlink signals, compared with the beacon, is at the desired level.

Beacons: Design

Engineering beacon power levels are chosen to provide adequate signal-to-noise ratios at well-equipped ground stations. Overkill (too much power) serves only to decrease the power available for other satellite subsystems (especially the transponder), reduce reliability and cause potential compatibility problems with other spacecraft electronics systems. *General beacons* may run at a higher power level so that they will be effective with relatively simple ground-station equipment. Moreover, since beacons are often used for Doppler studies, frequency stability over a wide range of temperatures and battery conditions is important. Telemetry systems should include provisions for monitoring beacon power output, a valuable piece of the performance puzzle for monitoring stations. As beacons are our primary diagnostic tool, redundant systems, often at different frequencies, are flown to enhance reliability. Beacon frequencies are usually set just outside the transponder-downlink passband, a location convenient both to users (the same ground-station receiving system can be used for both transponder and beacon downlinks) and to the spacecraft designers (the same satellite antenna can serve both systems). As with all spacecraft subsystems, high power-efficiency is essential. Typical power levels at 146 and 435 MHz are 50 to 100 mW on low-altitude spacecraft and 0.5 to 1.0 W on high-altitude spacecraft.

Command Links

The OSCARs are designed so that authorized volunteer ground stations with the necessary equipment can command them to switch from one operating mode to another. The ability to command the satellites is both a necessity and a convenience. Legally, telecommand capability is necessary because AMSAT must be able to turn off a malfunctioning transmitter that might conceivably cause harmful interference to important radio services worldwide. The ability to command satellites can also extend their operational lifetimes and usefulness: Subsystems that are not working properly can be turned off to conserve energy; operating schedules can be adjusted to suit the changing needs of the user community; telemetry modes can be switched when

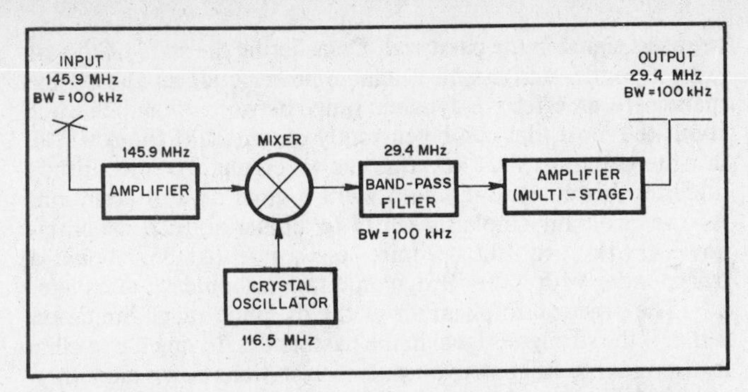

Fig. 12-1 — Block diagram of a simple 2-m/10-m linear transponder: input passband 145.850-145.950 MHz, output passband 29.350-29.450 MHz.

needed and so on. Intensive commanding of the AMSAT-OSCAR 6 spacecraft contributed significantly to its serving for about 4.5 years, well beyond its design lifetime of only one year. Command stations are built and manned by dedicated volunteers. Though command frequencies, access codes and formats are considered confidential, they are available to responsible stations for projects approved by AMSAT. To date, command stations have operated from more than 8 countries.

Transponders: Function

A *transponder* (sometimes called a translator) is a device that receives signals in a narrow slice (passband) of the radio-frequency spectrum, amplifies the signals, shifts the frequency of the entire passband, and then transmits it. On most AMSAT spacecraft the transponder is the primary mission subsystem. (To transmit signals to an OSCAR, one needs a government-issued license. In the United States these are Amateur Radio Service licenses issued by the Federal Communications Commission to individuals who pass appropriate exams. See page 202.) The translators currently in orbit are linear: Any type of signal put in (single-sideband, fm, cw, a-m, facsimile, slow-scan television, etc.) comes out the same, except for the shift in frequency and the great amplification — on the order of 10^{13} (130 dB). Because all stations using the transponder must share the limited power available, continuous-carrier modes such as fm and a-m, inefficient with respect to power consumption, are discouraged.

Transponders: Design

Transponder design is, in many respects, similar to receiver design. Input signals are typically on the order of 10^{-13} watts and the output level is on the order of 1 watt. A major difference, of course, is that the transponder output is at radio frequency while the receiver output is at audio frequency. The convention is to specify a transponder by first giving the approximate input frequency followed by the output frequency. For example, a 146/29-MHz transponder would have an input passband centered near 146 MHz and an output passband centered near 29 MHz. The same transponder could be specified in wavelength, as a 2-m/10-m unit.

A block diagram of a simple transponder is shown in Fig. 12-1. For several reasons, flight-model transponders are more complex than the one shown. As with receiver design, such considerations as band-pass filter availability, image response, wide variations in input signal level and required overall gain often cause designers to use multiple-frequency conversions. A block diagram of the basic Mode-A transponder used on OSCARs 6, 7 and 8 is shown in Fig. 12-2.

Linear vs. Nonlinear Transponders. A transponder that *only* shifts the frequency and amplifies the power of an incoming signal is called a *linear* transponder. Input signals of any mode will be retransmitted in the same mode but on a different frequency and at higher power. Of course, no transponder is perfectly linear, but with a well designed unit, undesired sum and difference frequencies that are caused by input signals beating against each other should be down by 30 dB or more.

One way to build linear transponders is to use linear ampli-

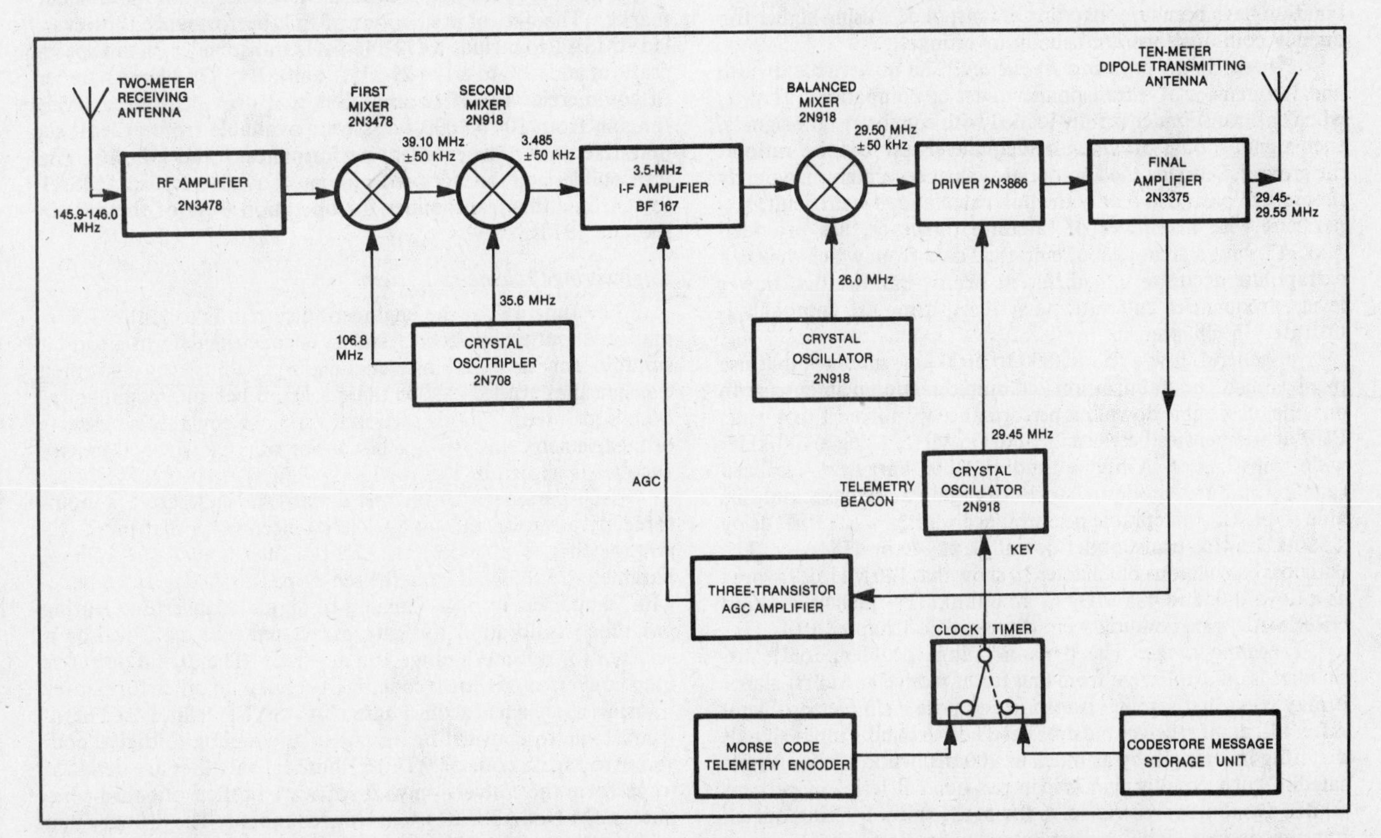

Fig. 12-2 — Block diagram of Mode-A transponder used on OSCARs 6, 7, and 8. For additional details on this transponder see: J. A. King, "The Sixth Amateur Satellite," Part I, *QST*, Vol. LVII, no. 7, July 1973, pp. 66-71; Part II, *QST*, Vol. LVII, no. 8, Aug. 1973, pp. 69-74, 106. This article is highly recommended for anyone interested in satellite design.

fiers wherever possible. Unfortunately, such devices are generally inefficient, a characteristic that cannot be tolerated aboard a spacecraft for transponder stages running above 1 watt. Several special techniques for constructing high-efficiency linear transponders have been developed by Dr. Karl Meinzer, DJ4ZC, and his coworkers at the University of Marburg in the Federal Republic of Germany. One method, known as envelope elimination and restoration (EER), operates somewhat like a class D amplifier.[3,4] Although individual stages are not linear, the overall transponder is a linear device. This technique proved very successful on the Mode-B transponder on OSCAR 7 but is not suitable for the wide-bandwidth, high-power transponders planned for Phase III. The problem results because EER transponder operation depends on switching large currents at a high frequency, with current proportional to transponder power and switching-frequency proportional to bandwidth. Currently available solid-state switching devices simply don't provide the long-term reliability needed for the power and bandwidth combinations planned for Phase III. To solve the problem, Dr. Meinzer developed an approach using EER and Doherty amplifiers for Phase III-A. Development work on the Mode-L transponder has led to the EER/Doherty technique being superseded by HELAPS (*H*igh *E*fficiency *L*inear *A*mplification by *P*arametric *S*ynthesis). Technical details of HELAPS will appear in a forthcoming issue of *Orbit*.

Inverting vs. non-inverting transponders. In any multiple-stage conversion mixing scheme, local-oscillator frequencies can be above or below the incoming frequency. If the local-oscillator frequencies are chosen so that signals entering a linear transponder are inverted before being retransmitted, we have an *inverting transponder*. Such a transponder will change upper-sideband signals into lower-sideband (and vice versa), transpose relative mark-space placement in RTTY, and so on. An important advantage of an inverting transponder is that Doppler shifts on the uplink and downlink are in opposite directions and will, to a limited extent, cancel. With the 146/29-MHz link combination, Doppler is not serious; transponders using this frequency combination have been non-inverting. Transponders using higher frequency combinations are usually inverting.

Power, bandwidth and frequency. The power, bandwidth and frequencies of a transponder must be compatible. That is, when the transponder is fully loaded with equal-strength signals, each signal should provide an adequate signal-to-noise ratio at the ground. Selecting the appropriate values accurately on a purely theoretical basis using only the link calculations is difficult. Experience with a number of satellites, however, has provided AMSAT with a great deal of empirical data from which they can extrapolate accurately to different orbits, bandwidths, power levels, frequencies and antenna systems, using the approach illustrated in Chapter 10.

In general, low-altitude (800 to 1600 km) satellites that use passive magnetic stabilization and omnidirectional antennas can provide reasonable downlink performance with from 1 to 4 watts PEP at frequencies between 29 and 435 MHz, using a 100-kHz-wide transponder. A high-altitude (35,000 km) spin-stabilized satellite that uses modest (7 to 10 dB_i) gain antennas should be able to provide acceptable performance with 35 watts PEP using a 500-kHz-wide transponder downlink at 146 or 435 MHz. The path loss calculations of Chapter 10 show that 146 MHz is favored as a downlink and 435 MHz as an uplink. (Frequency selection criteria for transponding were discussed in Chapter 10.)

Dynamic range. The dynamic range problem for transponders is quite different from that for hf receivers. At first glance it may seem that satellite transponders pose a simpler problem. After all, an hf receiver must be designed to handle input signals differing in strength by as much as 100 dB, while a low-altitude satellite will encounter signals in its passband differing by perhaps 40 dB. Good hf receivers solve the problem by filtering out all but the desired signal before introducing significant gain. A satellite, however, has to accommodate all users simultaneously. The maximum overall gain can, therefore, be limited by the strongest signal in the passband. Considering the state-of-the-art in transponder design and available power budgets aboard the spacecraft, an effective dynamic range between 20 and 25 dB is about the most that can be currently obtained. If the AMSAT satellite program were to continue to emphasize low-altitude satellites this constraint would merit a great deal of attention. As the program emphasis shifts to higher-altitude satellites, however, the problem becomes less severe. At 35,000 km a transponder with a 20-dB dynamic range should be adequate.

The practical implication of the dynamic-range limitation is that if the strongest signal in the passband is driving the satellite transponder to full power output, then stations down more than about 22 dB from this level will not be heard on the downlink, even though the weaker stations might be perfectly readable if the strong signal were not present. To remedy this problem by changing the transfer characteristic of the transponder from linear to, say, logarithmic would produce intermodulation between signals, decreasing the dynamic range.

Transponder design is an interesting area for innovative developmental work. In particular, increasing effective dynamic range by using channelized transponders (linear and non-linear) in which each channel has its own automatic gain control, and evaluating the potential utility of limiting-repeaters that are suitable for fm voice or digital signals deserve attention.

Redundancy. Since the transponder is the primary mission subsystem, reliability is extremely important. One way to improve system reliability is to include two transponders on each spacecraft; if one fails, the other would be available full time. Rather than use identical units, AMSAT has chosen to work with different frequency combinations. This has provided AMSAT with practical data on the performance of different link frequencies. Furthermore, during the period between satellite conception and the later part of the satellite's useful life (a period on the order of 10 years for AMSAT-OSCAR 7) drastic changes in the availability of equipment may occur. For example, in 1972 when planning for AMSAT-OSCAR 7 began, 432-MHz power amplifiers were not being produced commercially for the amateur market. Theoretical predictions of link performance, however, led AMSAT to include a 432-146-MHz transponder on this spacecraft (in addition to a 146-29-MHz unit). By 1978 a large number of commercial 432-MHz amplifiers that provided power levels ranging from 10 to 1000 watts were available from at least six manufacturers. The excellent performance of the 432-146 MHz transponder and the increased equipment availability led AMSAT to schedule this transponder for operation 67% of the time in the late 1970s.

Engineering/Telemetry System

The function of the engineering system is to gather information about all onboard systems, encode the data in a format suitable for downlinking (engineering subsystem) and then transmit the encoded data on the spacecraft beacons (communications subsystem). In this section we look at engineering aspects of the telemetry subsystem; a block diagram of a typical telemetry encoder is shown in Fig. 12-3.

Each parameter of interest aboard the spacecraft is monitored by a sensor and its associated electronics that put out a voltage that is proportionate to the measured value of that parameter. The signal from the sensor passes through a variable-gain amplifier into an analog-to-digital converter. During prelaunch calibration the gain associated with each sensor is selected for optimum range and accuracy. The digital output is then converted to Morse code, RTTY or some other format for transmission via a beacon. Phase II AMSAT satellites used hard-wired logic to convert the output of the analog-to-digital converter to Morse code or RTTY. Phase III satellites are designed to perform the conversion with software in their onboard computers. On UoSAT the system was redundant: The primary unit was hard wired but this was backed up by software in the on-board computer. Telemetry control logic (either hard wired or in software) selects the proper input sensor, chooses the ap-

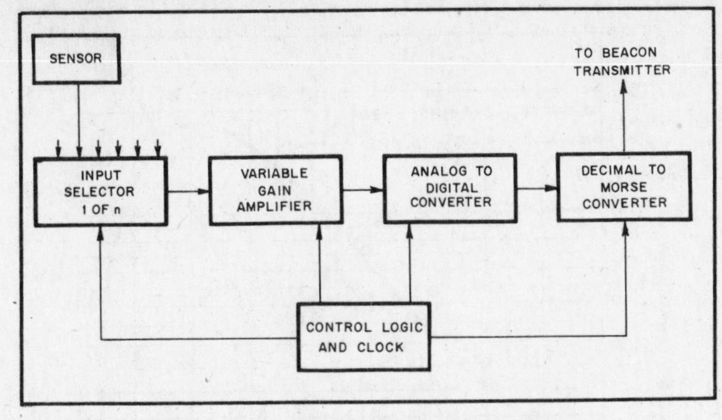

Fig. 12-3 — Typical AMSAT-OSCAR Morse code telemetry encoder.

propriate amplifier gain and conducts other bookkeeping chores.

Sensors are usually sampled sequentially *(serial mode)* and the measurements are transmitted as they are made. Under flexible computer control, the sampling strategy can be modified by instructions on the command link. If the situation warrants, we can dwell on a particular sensor *(dwell mode)* or sample it frequently so short-term changes can be studied. The TLM capabilities of each satellite are described in Appendix A: Spacecraft Profiles. In general, the control logic on future satellites will be handled by software that can be programmed from the ground, an approach that provides AMSAT with a great deal of flexibility in encoding the telemetry of engineering and general beacons.

Morse code format. The Morse code telemetry systems aboard OSCARs 6, 7 and 8 had several features in common. The parameters being measured were sampled in a fixed sequence. One complete series of measurements was called a *frame.* The beginning and end of each frame were marked by a distinctive signal; the Morse code letters HI were used on OSCAR 7 and OSCAR 8. Each transmitted value consisted of three integers called a *channel.* To interpret a channel we needed to identify which parameter was being monitored and obtain a raw data measurement that could be converted into a meaningful value. AMSAT uses the first integer in a channel for parameter identification. When the number of channels is small (OSCAR 8 had six channels), a single digit can uniquely identify the parameter being measured. When the number of channels is large, the user must also note the order in which the channels are being sent to identify the particular parameter being sampled.

As an example, a telemetry frame for an imaginary satellite might consist of nine channels as shown in the top row of Table 12.4. Channel identity information, shown in the bottom row of the table, uses the order in which the channels are transmitted to label them uniquely. The last two digits in each channel represent the encoded information of interest. To decode a telemetry channel one refers to published information about the specific satellite to determine which parameter is measured by the channel and to obtain the simple algebraic equation that should be used to decode the last two digits in the channel into the quantity being measured. For example, the spacecraft description for the imaginary satellite that transmitted the data in Table 12.4 might tell us that channel 1A is total solar panel current, and that multiplying the significant digits (42) by 30 will yield the value of total solar panel current in milliamperes (1260 mA) at the time the measurement was taken.

Advanced telemetry formats. The advanced telemetry systems aboard Phase III spacecraft are more flexible than those aboard earlier OSCARs. The selection of the particular sensors to be scanned is under the control of the central computer, which is, in turn, under the control of command stations. Because of the relatively high speed of TLM it's possible to include as part of each channel information that uniquely identifies the parameter being measured. Thus, knowledge of the location of a channel within a frame is not needed by users. The relatively high speed also makes it possible to use three digits to encode measured levels for greater accuracy. Decoding information for each operational satellite is included in Appendix A: Spacecraft Profiles.

Table 12.4

A Morse Code Telemetry Frame with Nine Channels

Raw data begin)	*HI*	*142*	*116*	*178*	*239*	*202*	*216*	*392*	*352*	*365*	*HI (end)*
Channel ID		1A	1B	1C	2A	2B	2C	3A	3B	3C	

The top row is the actual data as received. The bottom row assigns a unique label to each channel. Channel 1A is the first one received, channel 1B is the second, 1C is the third, 2A is the fourth, etc. Such data is sometimes written in the form of a 3 × 3 matrix in which case the ID integer is a line number and the ID letter is a column label.

Operating Schedules: Frequencies, Telemetry Modes, Etc.

Every OSCAR has a number of beacons and transponders that use various frequencies and modes. Since all units don't operate at the same time, knowing the latest operating schedule is important. To conform to the needs of users or spacecraft condition, schedules change from time to time. To obtain the latest information check periodicals such as *QST, Orbit, World Radio, 73, OSCAR NEWS,* and so on, or tune into the weekly AMSAT nets.

Antennas

In selecting the antennas, design engineers must consider other spacecraft systems, mission objectives, type of orbit, transponder and beacon frequencies, attitude stabilization, spacecraft structure, launch-vehicle constraints and seemingly countless other issues. With low-altitude satellites (OSCARs 5, 6, 7 and 8) spacecraft designers were able to choose relatively omnidirectional antennas and passive magnetic stabilization schemes, a combination that greatly simplified other aspects of satellite design. When mechanical considerations permitted, circularly polarized antennas such as the canted turnstile (a turnstile with drooping elements) were used. The advantages of circular polarization were, of course, only partially realized since radiation from the canted turnstile was not fully circular from a perspective off the main axis. At 10 m, mechanical constraints made circular polarization on the spacecraft impossible; therefore, simple dipole antennas were used. Since the relatively large 10-m antenna must be folded out of the way during launch, several schemes for antenna deployment have been tried. With OSCARs 5 and 6, springy flexible elements made from material similar to that used in a carpenter's rule was used. The explosive bolts that released the satellite from the launch vehicle also released the folded antenna, allowing it to extend to its full, precut length.

On OSCAR 7 a different technique, producing a much stiffer antenna, was tried successfully. Imagine a sheet of newspaper rolled into a two-inch diameter tube. Grab an inside corner and pull it till the tube reaches about three times its original length. You've just modeled the OSCAR-7 antenna elements. The design and fabrication of antenna elements of this type, using springy metals that self-deploy, is a very difficult and specialized procedure. Commercial products are prohibitively expensive for most OSCAR applications.

On OSCAR 8 yet another method was tried. The 10-m antenna consisted of motor-deployed concentric tubes much like the car radio antennas that automatically extend when the radio is turned on. Producing motors that work reliably in the vacuum and temperature extremes of space is also a tricky and expensive business. A lot of nail biting went on during the hours between OSCAR 8's launch and the successful commanding of the 10-m antenna deployment mechanism. In sum, while all the approaches to 10-m dipole design for OSCAR spacecraft have worked, none of the methods are completely satisfactory in cost, operation and reliability.

At 146 and 435 MHz, rod-like antenna elements are small enough to be formed from springy, whiplike material that is freed

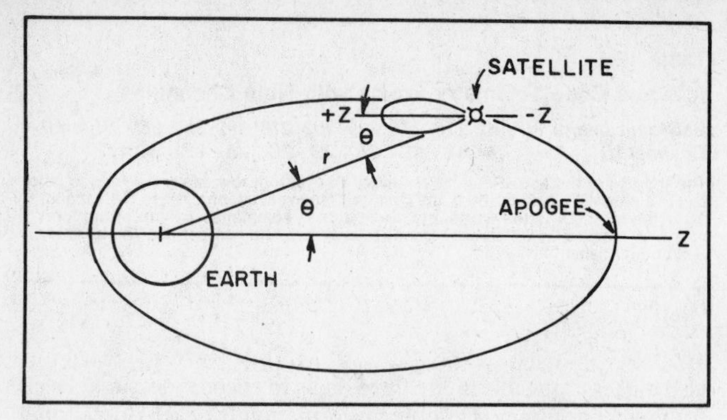

Fig. 12-4 — Orbit geometry for comparing possible Phase III satellite antennas.

when the satellite is released from its launch vehicle. Mechanical constraints make using arrays such as the helix difficult, so most antennas consist of whip sections that can be used to form monopole, dipole, canted turnstile, driven-element-and-reflector array, or driven-element-and-director array antennas. At 146 MHz and higher frequencies, spacecraft dimensions (in wavelengths) are significant and the structure itself can affect the antenna pattern. As a result, antenna design involves sophisticated theoretical models and considerable empirical testing for optimal results. As antenna dimensions decrease (at 1.26 GHz and higher frequencies) design options increase. The helix and quadrifilar helix are of special interest, but many other possibilities are well suited to different orbits and attitude control systems.

One aspect of antenna selection is choosing a radiation pattern that, in conjunction with the satellite orbit and attitude control system, strikes a good balance between coverage and signal level at the ground. For example, consider the AMSAT Phase III-B orbit shown in Fig. 12-4. Because of the large slant range at apogee, we want a beam antenna on the spacecraft. But clearly the narrow beamwidth of a high-gain antenna can lead to poor results when the satellite is away from apogee if ground stations are too far off to the side of the satellite antenna pattern. Let's look at one simple approach to modeling the situation.

J. Kraus has shown[6] that the radiation patterns of a great many common beam antennas can be approximated by the expression.

$$2(n+1)\cos^{n}(\theta) \qquad \text{(Eq. 12.1)}$$

In other words, the gain in a given direction can be calculated approximately using only θ and n, where n is a derived parameter that is related to the maximum gain of antenna gain patterns and θ is the polar angle of the satellite antenna pattern measured from apogee in the orbital plane (see Table 12.5). With this formula and our knowledge of satellite orbits we can calculate the signal power at the subsatellite point (SSP) as the satellite travels around its orbit (see Fig. 12-4). The results for beams of 6 dB_i and 10 dB_i gain, and for an isotropic antenna, are shown in Fig. 12-5.[7]

For AMSAT-OSCAR 10, a beam will be used during the apogee portion of the orbit and a 1/4-wavelength whip will be switched in at the point away from apogee where it provides better signals. Radiation from a whip along the +Z axis will spill over into the hemisphere centered about the −Z axis. As a crude approximation let's assume that the signal level from the whip is similar to that of the isotropic antenna near the edge of the spacecraft +Z hemisphere. Referring to Fig. 12-5, we see that the switch from beam to whip should be made when θ is approximately equal to 56° for the 10-dB_i beam and 76° for the 6 dB_i beam. Also, the 6 dB_i antenna begins to outperform the 10 dB_i beam when θ increases past 43°. Each angle corresponds to a specific time from apogee. From Fig. 12-5, the 10-dB_i beam will provide superior performance (by 0 to 4 dB) during a 7.5-hour

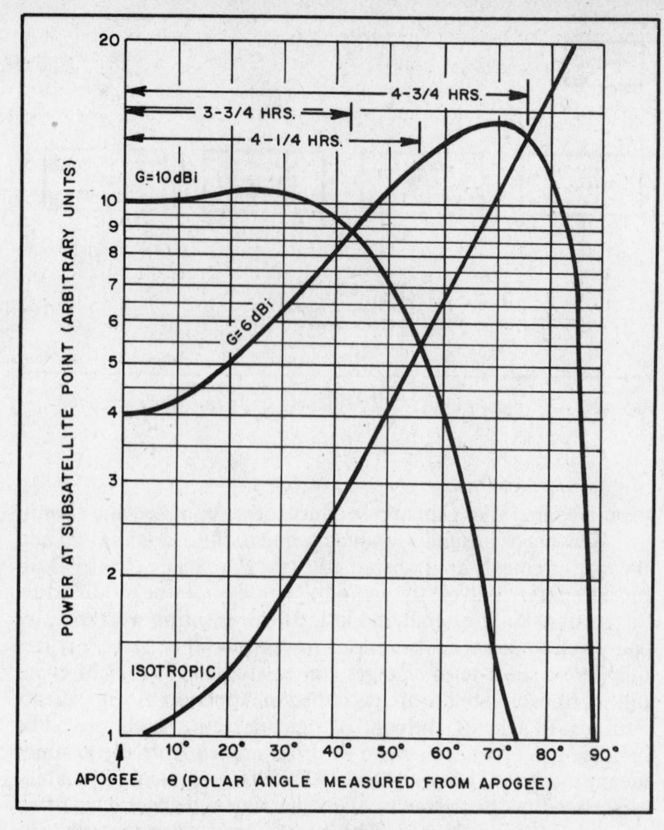

Fig. 12-5 — Relative power at subsatellite point as a function of Phase III satellite position in orbital plane (measured from apogee) for three possible antennas. Based on a period of 11.0 hours and an eccentricity of 0.688.

Table 12.5

n	*Gain* *2(n+1)*	*Gain* *dB_i*	*Half-power beamwidth*
isotropic	—	0	—
0	2	3.0	180°
1/2	3	4.8	151°
1	4	6.0	120°
2	6	7.8	90°
3	8	9.0	74.9°
4	10	10.0	65.5°

The radiation pattern of a great many common beam antennas can be approximated by $2(n+1)\cos^{n}(\theta)$, where n is a derived parameter that is related to maximum gain as specified in the way shown in the table.

segment of each orbit (3.75 hours on either side of apogee). The 6-dB_i antenna will provide the best performance (by 0 to 3.5 dB) during 2.0 hours of each orbit (a one-hour segment centered at $\theta = +60°$ and a one-hour segment centered at $\theta = -60°$).

From the viewpoint of stations at the SSP, the 10-dB_i antenna appears preferable. In fact, you might even wonder why we don't consider higher-gain antennas. A more careful analysis would take into account (1) signal levels at ground stations located away from the SSP and (2) the possible necessity to align the spacecraft Z-axis in a slightly different orientation to account for poor sun angle on the solar cells or to modify the spacecraft temperature. When this is done it makes higher gains less appealing. There is no clear-cut "best" choice, but gains between 6 dBi and 12 dBi appear to be a good compromise for Phase III elliptical orbits.

Structural, Environmental-Control, And Energy-Supply Subsystems

Structural Subsystem

The spacecraft structural subsystem, the frame that holds

it all together, serves a number of functions including physical support of antennas, solar cells and internal electronics; protection of onboard subsystems from the environment during launch and while in space; conduction of heat into and out of the satellite interior; mating to the launch vehicle and so on. Structural design (size, shape and materials) is influenced by launch vehicle constraints and by the spacecraft's function. AMSAT low-altitude satellites have fallen within the 20 to 30 kg range. Phase III high-altitude spacecraft, with their own kick motors and fuel, will probably weigh in at close to 70 kg at launch. Insofar as possible, AMSAT satellite structures are fabricated from sheet aluminum to minimize the machining operations.

The prominent features one observes when looking at a satellite are the attach fitting used to mount the satellite on the launch vehicle, antennas for the various radio links, solar cells, the heat-radiative coating designed to achieve the desired thermal equilibrium aboard the spacecraft and, for Phase III, the nozzle of the apogee kick motor. Satellite shape has a significant effect on the equilibrium temperature and overall solar-cell efficiency.

Environmental Control

The function of a spacecraft environmental control subsystem is to regulate temperatures at various points, shield against high-energy particles and protect the onboard electronics from rf interference. We'll focus on thermal control.

The temperature of a satellite is determined by the inflow and outflow of energy. More specifically, the satellite temperature will adjust itself so heat inflow equals heat outflow. This is known as the energy balance concept. Although we sometimes talk about the "temperature" of the satellite, different parts of a spacecraft are at different temperatures, which vary over time. The objective of the spacecraft designer is to create a model of the spacecraft and its environment that will accurately predict the average and extreme temperatures that each unit of the spacecraft will exhibit during all phases of satellite operation. This includes prelaunch, where the satellite may sit atop a rocket more than a week, baking under the hot tropical sun; the launch and orbit-insertion sequence of events; and the final orbit where, during certain seasons of the year, the spacecraft may go for months without any eclipse time (an eclipse occurs when the earth passes between the spacecraft and the sun), while during other seasons it may be eclipsed for several hours each day. Since excessive temperature extremes, either too hot or too cold, may damage the electronic subsystems or battery permanently, the thermal design must keep the temperatures of susceptible components within bounds at all times.

Once the satellite is in the vacuum of space, heat is transferred only by radiation and conduction; convection need not be considered. The complete energy balance model of Phase III-A depicted the satellite as comprising 121 subunits, each connected via conduction and radiation links to several other subunits. To solve the resulting energy-balance equations mathematically, the designers had to manipulate 121 nonlinear simultaneous equations, each consisting of about three or four terms. This was not a job for pad and pencil. A fairly large computer was needed. Even with sophisticated computer models, achieving a precision of ± 10 K is difficult; commercial satellite builders usually resort to testing the thermal balance of full scale models in space simulators. The Phase III thermal design problem was handled by Richard Jansson, WD4FAB, using computer time donated by the Martin Marietta Corp. Earlier OSCARs used a far simpler and less accurate approach that nevertheless provided reasonable results. See, for example, Fig. 2-2 that shows the thermal behavior of OSCARs I and II. Since the details of the simple approach provide a good introduction to the science (art?) of thermal design, we'll go through an example.

Thermal Design: A Simple Example

The sun is the sole source of energy input to the satellite. Quantitatively we can write:

$$P_{in} = P_o A^* \propto \beta \qquad \text{(Eq. 12.2)}$$

where

P_{in} = energy input to the satellite
P_o = Solar constant = incident energy per unit time on a surface of unit area (perpendicular to direction of radiation) at 1.49×10^{11}m (earth-sun distance) from the sun.
P_o = 1380 watts/m^2
A^* = effective capture area of the satellite for solar radiation
$\propto$ = absorptivity (fraction of incoming radiation absorbed by the satellite)
β = eclipse factor (fraction of time satellite is exposed to the sun during each complete orbit)

Power output from the satellite consists of blackbody radiation at temperature T, and the radio emissions. Since blackbody radiation is very much greater than the radio emissions, we can ignore the latter.

$$P_{out} = A\, \sigma e T^4 \qquad \text{(Eq. 12.3)}$$

where

P_{out} = energy radiated by satellite
A = surface area of satellite
σ = Stefan-Boltzmann constant = $\dfrac{5.67 \times 10^{-8} \text{ joules}}{K^4 m^2 s}$
e = average emissivity factor for satellite surface
T = temperature (K)

For equilibrium, incoming and outgoing radiation must balance,

$$P_{in} = P_{out} \quad \text{or}$$

$$P_o A^* \propto \beta = A \sigma e T^4 \qquad \text{(energy balance equation) (Eq. 12.4)}$$

Solving for temperature, we obtain

$$T = \left[\frac{P_o \propto \beta A^*}{\sigma e A}\right]^{0.25} \qquad \text{(Eq. 12.5)}$$

Reasonable average values for the various constants are $\propto$ = 0.8, β = 0.8 and e = 0.5. For AMSAT-OSCAR 7, A = 7770 cm^2 and A* = 1870 cm^2.[8] Inserting these values in Eq. 12-5 we obtain T = 294 K = 21° C, which is close to the observed equilibrium temperature of OSCAR 7. Over the course of a year, as β varied from 0.8 to 1.0, the temperature of OSCAR 7 varied between 275 K and 290 K. The transponder final amplifier, of course, ran considerably hotter.

One of the techniques used to achieve the desired spacecraft temperature is to adjust the absorptivity ($\propto$) and emissivity (e) by a variety of techniques. OSCAR spacecraft have used passive techniques such as roughening or painting the surface. The RS spacecraft designers reported that they used an active technique (no details available) that proved very successful. Active techniques include shutters, or louvers, that are controlled by bimetallic strips; conducting pipes that can be filled with Helium gas or evacuated; selective activation of different subsystems that generate heat at various points in the satellite, and so on.

Future Phase III satellites may use some of these approaches. To a certain extent these factors have already been exploited for temperature control as, for example, with OSCAR 7. Its Mode-A and -B transponder operating schedules were juggled for this purpose. On OSCAR 8, simultaneous A/J operation was at times used to help control temperature.

Energy-Supply Subsystem

Communications satellites can be classified as active or passive. An example of a passive satellite would be a big balloon (Echo I, launched August 12, 1960, was 30 meters in diameter when fully inflated) that is coated with conductive material that reflects radio signals. When used as passive reflectors, such satellites do not need any electronic components or any power source. While such a satellite is appealingly simple, the radio power it reflects back to earth is less, by a factor of 10 million (70 dB), than the signal transmitted by a transponder aboard an active satellite (assuming equal uplink signal strength and a comparison based on equal satellite masses in the 50-kg range).[9] Ground station antenna and power requirements for use with passive satellite systems are, therefore, prohibitively large and expensive.

An active satellite (one with a transponder) needs power. The energy source supplying the power should be reliable, efficient, low-cost and long-lived. By efficient we mean that the ratio of available electrical power to weight and the ratio of available electrical power to waste heat should be large. We examine three energy sources that have been studied extensively: chemical, nuclear and solar.[10]

Chemical power sources. Chemical power sources include primary cells, secondary cells and fuel cells. Early satellites such as Sputnik I, Explorer I and the first few OSCARs were flown with primary cells. When the cells ran down, the satellite "died." Spacecraft of this type usually had lifetimes of a few weeks, although Explorer I with low-power transmitters (about 70 mW total), ran almost four months on mercury (Hg) batteries. These early experimental spacecraft demonstrated the feasibility of using satellites for communications and scientific exploration and thereby provided the impetus for the development of longer-lived power systems. Today, batteries (secondary cells in this case) are used mainly to store energy aboard satellites; they are no longer used as the primary source. Nickel-Cadmium batteries are used almost universally. Scientists at COMSAT Laboratories have recently developed sealed nickel-hydrogen batteries that operate at energy densities of up to 75 watt-hours/kg, about five times the value for Ni-Cd cells. If the lifetime of these cells meets expectations, they will probably be used on future spacecraft. Another chemical power system, the fuel cell, has been used as a source of energy on manned space missions, such as Apollo and Gemini, that required large amounts of power over a relatively short time. Though the development of fuel cells is continuing, they don't seem appropriate for OSCAR missions.

Nuclear power sources. One nuclear power source to be flight-tested is the radioisotopic-thermoelectric power plant. In devices of this type, heat from decaying radioisotopes is converted directly to electricity by thermoelectric couples. Some U.S. transit navigation satellites, the Snap 3B and Snap 9A (25 watts), have flown generators of this type. These generators have a high available-power-to-weight ratio but generate large amounts of waste heat and have a high cost per watt because of the fuel. They are most useful when solar cells are unsuitable, in orbits inside the Van Allen Belts, on deep-space missions where solar intensity is greatly reduced or when very large amounts of power are required.

Nuclear power sources unfortunately present serious safety hazards in cases of launch accidents or reentries such as occurred in early 1978 when COSMOS 954 spread radioactive debris over a 1000-km-long track in Canada. Under the international laws governing such occurrences, the USSR paid Canada 3 million dollars in compensation for the cleanup costs. Will AMSAT ever consider a nuclear power system? Because of the initial cost, the need for launch accident contingency capability and associated documentation and insurance costs, it is very unlikely.

Solar power sources. The third power source we consider is solar. The first solar cells were built in 1954.[11] Since then cell technology has evolved to the point where today, solar cells power the great majority of satellites in orbit. Solar cells do have a number of undesirable features, however. They compete for mounting space on the outer surface of the satellite with antennas and heat-radiating coatings. They are subject to degradation, particularly when the satellite orbit passes through the Van Allen radiation belts (roughly at altitudes between 1600 and 8000 km). They work most efficiently below 0° C though the electronics systems aboard satellites are usually designed for a 10° C environment. They call for attitude control (spacecraft orientation) that often conflicts with other mission objectives. Their power output decreases with distance from the sun, rendering them unsuitable for missions to the outer planets and beyond. Finally, they produce no output when eclipsed from the sun. Nonetheless, power sources that use solar cells to produce electrical energy and secondary cells to store energy are by far the simplest for long-lifetime satellites. They are also comparatively low in cost for the power produced, generate little waste heat and produce acceptable ratios of available-electric-power to weight. When satellite power requirements are greater than about 50 watts, the satellite structure must include paddles or panels to mount the solar cells on (paddles that can be oriented toward the sun), since the body of the satellite may not have sufficient surface area to support all the cells needed. A 1-meter-square solar panel, oriented perpendicularly to the sun-panel line, will intercept about 1380 watts of solar energy (panel-sun distance assumed equal to earth-sun distance). New solar cells are typically 12.5 percent efficient, but their efficiency decreases with time; the exact degradation rate is a function of the cells' environment. Although new cells cost roughly $30,000 per square meter at the present time, a great deal of development work is underway to increase their efficiency and reduce their cost. Scientists at COMSAT Laboratories, for example, have developed solar cells with efficiencies of 13% (violet cell) and 15% (nonreflective cell). Solar cells that are mounted on a satellite are usually protected by glass cover slides to reduce the rate of degradation from radiation damage. The glass cover slides also reduce the efficiency of the cells (the thicker the slide, the greater the reduction) and increase the weight of the spacecraft.

Practical energy subsystems. The typical AMSAT satellite energy system consists of a source, a storage device and conditioning equipment (shown in Fig. 12-6). The source consists of silicon solar cells. A storage unit is needed because of eclipses (satellite in earth's shadow) and the varying load; Nickel-Cadmium secondary cells are currently being used on AMSAT satellites. These are quite different from the NiCds available at the local electronics supply store. Even when using the best available cells, achieving lifetimes of more than four years is difficult. The ultimate failures of OSCARs 6, 7 and 8 were attributed to battery problems, though with improvements in technology we hope the cells aboard more recent AMSAT spacecraft will last at least seven years.

Power conditioning equipment typically flown on AMSAT spacecraft includes a battery charge regulator (BCR) and at least one instrument switching regulator (ISR) to provide dc-to-dc conversion with changes of voltage, regulation and protection. Because failures in the energy subsystem could be catastrophic to the mission, special attention is paid to insuring continuity of operation. Battery charge regulators and instrument switching regulators usually are built as redundant twin units with switchover between redundant units controlled automatically, in case of internal failure, and by ground command. Solar cell strings are isolated by diodes, so a failure in one string will lower total capacity but will not otherwise affect spacecraft operation. These diodes also prevent current from flowing in the reverse direction through the strings of cells on the dark side of the satellite.

When the energy supply subsystem provides sufficient energy to operate the satellite continuously, we say it has a *positive power budget.* If some satellite subsystems must be turned off periodically for the storage batteries to be recharged, we say the spacecraft has a *negative power budget.* We leave the detailed calculations of how spacecraft geometry can be taken into account when estimating the amount of power a solar cell array can provide to other reference sources.[12] Ideally, a satellite is designed with initial power sufficiently high to ensure that a positive power budget will exist at the end of the design life.

Attitude-Control, Propulsion, Computer, and Guidance- and-Control Subsystems

Attitude-Control Subsystem

The orientation of a satellite (its attitude) with respect to the

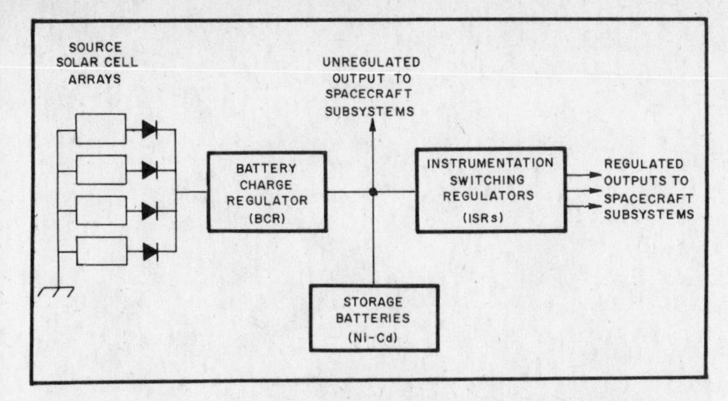

Fig. 12-6 — Long-lifetime communications satellite energy subsystem.

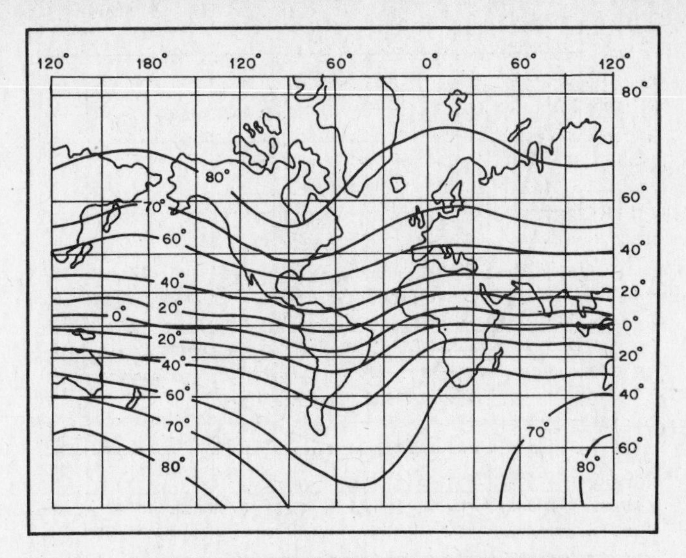

Fig. 12-7 — Inclination (dip) angle of earth's magnetic field.

earth and sun greatly affects the effective antenna gain, solar cell efficiency, thermal equilibrium and scientific instrument operation. Attitude-control subsystems vary widely in complexity. A simple system might consist of a frame-mounted bar magnet that tends to align itself parallel to the earth's magnetic field; a complex system might use cold gas jets, solid rockets and inertia wheels, all operating under computer control in conjunction with a sophisticated system of sensors. Attitude-control systems can be used to provide three-axis stabilization, or to point a particular satellite axis toward the earth, in any fixed direction in inertial space (with respect to the fixed stars), or along the earth's magnetic field. Single-axis orientation is often achieved by spinning the spacecraft about its major axis (spin-stabilized). Attitude-control systems are classified as *active* or *passive.* Passive systems do not require power or sensor signals for their operation. Consequently, they are simpler and more reliable, but also less flexible and accurate. Some of the attitude-control systems in general use are described below.

Mass expulsors. Devices of this type are based on the rocket principle and are classified as active and relatively complex. Examples are cold gas jets, solid-propellent rockets and ion-thrust engines. Mass expulsors are often used to spin a satellite around its principal axis. The resulting angular momentum of the satellite is then parallel to the spin axis that tends to maintain a fixed direction in inertial space (the principle of angular momentum conservation).

Angular momentum reservoirs. This category includes devices based on the inertia (fly) wheel principle. Assume that a spacecraft contains a flywheel as part of a dc motor that can be powered up on ground command. If the angular momentum of the flywheel is changed, then the angular momentum of the rest of the satellite must change in an equal and opposite direction (the principle of conservation of an angular momentum). Systems of this type are classified as active.

Moment-of-inertia changers. The spin rate of a satellite can also be changed by deploying booms. The booms change the satellite's moment of inertia causing the spin rate to change, again in accordance with the principle of conservation of angular momentum. These systems are classified as active.

Environmental-force couplers. The satellite is coupled to (affected by) its environment in a number of ways. In the two-body central force model (outlined in Chapter 8) we discussed how the satellite and earth were first treated as point masses at their respective centers of mass. We went on to see that the departure of the earth from spherical symmetry caused readily observable perturbations of the satellite's path. The departure of the satellite's mass distribution from spherical symmetry likewise causes readily observable effects. An analysis of the mass distribution in the satellite defines a specific axis that tends to line up pointing towards the geocenter as a result of the earth's gravity gradient. Gravity-gradient devices exploit this tendency. Anyone who's been on a sailboat, however, knows that gravity can produce two stable states. This gravity gradient effect is greatly accentuated if one of the satellite dimensions is much longer than the others. To simulate this condition, a long boom with a weight at one end may be attached to the spacecraft.

Another environmental factor that can be tapped for attitude control is the earth's magnetic field. A strong bar magnet carried by the satellite will tend to align itself parallel to the direction of this field. One important characteristic of the earth's magnetic field, the dip angle, is shown in Fig. 12-7. Instead of using permanent magnets it's possible to use electromagnets consisting of coils of wire. By passing current through these coils, one forms a magnet temporarily. With proper timing, the coils can produce torques in any desired direction. Devices of this type are often called torquing coils.

Note that even if a satellite designer does not exploit magnetic or gravity-based environmental couplers for attitude control, these forces are always present and their effect on the satellite must be taken into account.

Energy absorbers. Energy absorbers or dampers convert undesired motional energy into heat. They are needed in conjunction with many of the previously mentioned attitude control schemes. For example, if dissipative forces did not exist, gravity gradient forces would cause the satellite's principal axis to swing pendulum-like about the local vertical (the line from the satellite to the geocenter) instead of pointing toward the geocenter. Similarly, a bar magnet carried on a satellite would oscillate about the local magnetic field direction instead of lining up parallel to it. Dampers may consist of passive devices such as springs, viscous fluids or hysteresis rods (eddy-current brakes). At times, torquing coils are used to obtain similar results.

Practical attitude control. Let's look at some of the tradeoffs involved in choosing an attitude-control system for low-altitude satellites such as AMSAT-OSCARs 6, 7 and 8. If stabilization systems were not designed into these satellites, isotropic satellite antennas would be desirable. Consequently, the mechanical complexity of the antenna system would increase and power levels on all uplinks would have to be raised to provide the desired signal levels at the satellite and at earth. The need for higher power aboard the satellite would, in turn, mean a bigger transmitter, larger power supplies, more solar cells and batteries or less operating time, and generally greater weight and complexity. For efficient illumination of the solar cells the physical structure of the satellites must be matched to the attitude-control system. If no attitude control is used, then a spherical distribution of cells would be most efficient. An attitude-control system is clearly desirable if its cost in complexity, weight and so on is small compared to the benefits it provides.

The system chosen for AMSAT-OSCARs 6, 7 and 8 was passive; bar magnets mounted in the satellite aligned a specific axis along the earth's magnetic field. Permalloy hysteresis damping

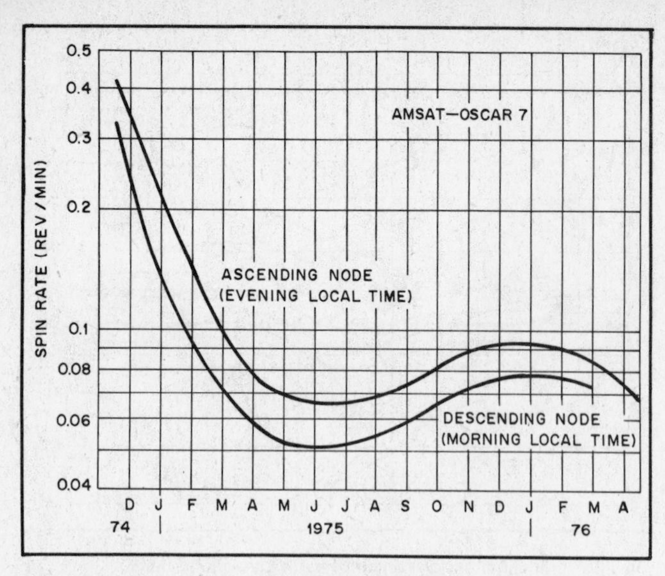

Fig. 12-8 — The spin rate of AMSAT-OSCAR 7 during the year following launch. (Data provided by John Fox, WØLER)

The Thiokol TE-45 motor that was installed along the spin axis inside the AMSAT-OSCAR Phase III-A tri-star structure. Such "kick motors" are fired once Phase III satellites are separated from their launch vehicles, lifting the satellites to their final orbits. (A liquid-fuel kick motor donated by Messerschmitt-Bolkow-Blohm (MBB) was used to lift AMSAT-OSCAR 10 from its initial transfer orbit to its final operational orbit.)

rods, mounted perpendicular to the primary magnet, were used to reduce spin about the bar-magnet axis and small oscillations. Unless this spin about the bar-magnet axis had been reduced, it would have caused radio link fades when the relative orientations of the spacecraft and ground station antennas changed. Using a passive magnetic stabilization technique of this type causes the principal axis of the satellite to rotate 720° in inertial space during each orbit. By referring to Fig. 12-7 and using an OSCARLOCATOR to track OSCAR 8, you should be able to picture how the spacecraft antennas are oriented with respect to your location.

Magnetic stabilization was first tested by radio amateurs on OSCAR 5 and has proved satisfactory for subsequent low-altitude communications spacecraft. Fig. 12-8 shows the residual spin of AMSAT-OSCAR 7 for 15 months following launch. Part of this spin was introduced purposely to help regulate temperature. The technique was a novel one wherein the elements of the canted turnstile antenna were painted with reflective paint on one side and absorbent paint on the other. Solar radiation pressure then produced a radiometer-like rotation dubbed by users at the time as the "barbecue rotisserie" technique.

Because the camera on UoSAT must be pointed directly at the earth, a magnetic stabilization system wasn't appropriate. To accomplish their objective, UoSAT engineers chose a relatively complex gravity-gradient stabilization system. UoSAT also contains torquing coils for stabilization before the gravity-gradient boom is deployed and for damping unwanted oscillations later on.

High-altitude Phase III satellites will require more sophisticated approaches to attitude control. Phase III-A, and -C and AMSAT-OSCAR 10 were designed with active stabilization systems. Each spacecraft is spun about its principal axis that lies in the orbital plane and is aligned so it points toward the geocenter at apogee. If sun angles on the solar cells are very poor or if temperature control warrants, a slightly different orientation will have to be used. Initial spin up may be accomplished with small solid-propellant rockets or the torquing coils. The torquing coils will then be used to maintain and, when necessary, adjust the spin rate or orientation. Near perigee, where the magnitude of the earth's magnetic field is greatest, is the region where the torquing coils will be pulsed with current at the proper rate and time. Pulsing is controlled by the onboard computer using orientation data from the sun and earth sensors and is ultimately under the control of ground command stations. Phase III-A and AMSAT-OSCAR 10 were also built using viscous-fluid nutation-dampers consisting of roughly a 50/50 mixture of glycerine and water sloshing around in thin tubes (about 0.2 cm in diameter and 40 cm long) that run along the outer edge of each arm of the spacecraft.

Propulsion Subsystem

The simplest type of space propulsion system consists of a small solid-propellent rocket which, once ignited, burns until the fuel is exhausted. Rockets of this type are often used to boost a satellite from a near-earth orbit into an elliptical orbit with an apogee close to synchronous altitude (35,800 km) or to shift a satellite from this type of elliptical orbit into a circular orbit near synchronous altitude. Such rockets are known as "apogee kick motors" or "kick motors." The first AMSAT satellite to use a kick motor was Phase III-A; unfortunately, AMSAT never had the chance to fire it. A kick motor used in conjunction with a sophisticated system of sun and earth sensors will enable AMSAT to raise the perigee, alter the orbital inclination or make both changes on a satellite orbit. Since the amount of energy available is limited, its expenditure will be carefully evaluated to select the most desirable final orbit that can be attained from a given injection orbit. Kick motors are very dangerous devices and their use, handling, shipping and storage must conform to rigid safety procedures.

The kick motor used on AMSAT Phase III-A was a solid-propellant Thiokol TEM 345-12 containing approximately 35 kg of a mixture of powdered aluminum and organic chemicals in a spherical shell with a single exit nozzle. The unit was capable of producing a velocity change of about 1600 m/s during its single

20-second burn. AMSAT has a similar unit being held in storage for Phase III-C.

AMSAT-OSCAR 10 used a liquid-fuel rocket. Produced by the West German company Messerschmitt-Bolkow-Blohm (MBB), the unit produces a thrust of 400 Newtons and is similar to one used for the Symphonie satellite. As this motor can be ignited several times, the strategic options available to AMSAT for shifting from transfer orbits to operational orbits are greatly expanded.

Computer and Guidance-and-Control Subsystems

On OSCAR 5 through OSCAR 8, hard-wired logic was used to interface the various spacecraft components to both the telemetry system and the command system. As overall spacecraft complexity grows, at some point it becomes simpler and more reliable to use a central computing facility in place of hard-wired logic. Once the decision to incorporate a computer is made, the design of the spacecraft must be reevaluated totally to take advantage of the incredible flexibility provided by this approach.[13] Ground telecommand stations need no longer send immediate commands; they uplink pretested computer programs. After correct reception is confirmed, these programs take control of the spacecraft. Using positional data from the sun and earth sensors, the computer triggers the firing of the apogee kick motor and pulses the torquing coils at the appropriate times to maintain the correct spacecraft attitude.

Computer programs also control telemetry content and format. If we want to change the scale used to monitor a particular telemetry channel or to sample it more frequently, we simply add a couple of bits to the computer program and it's done. Want to send out a daily Codestore message at 0000 GMT? No problem; uplink the message and control program whenever it's convenient and the message will be broadcast on schedule. There's no need to have a control station that is in range and manned at the specified time.

Phase III-A, -C, AMSAT-OSCAR 10 and UoSAT contain central computers using 8-bit microprocessors. Two very important criteria for a spacecraft microprocessor that will operate in the hostile environment of space are low power consumption and high resistance to radiation damage. These requirements led to the selection of a relatively low-speed CMOS device, the RCA COSMAC 1802 for Phase III missions and for the primary computer on UoSAT. Depending on the mission and the available power, anywhere from 8K bytes to 64K bytes of dynamic random-access memory (RAM) may be included. The 1802 processor chips used on Phase III-A and AMSAT-OSCAR 10 underwent special radiation hardening. Additional radiation protection was provided by bonding thin sheets of tantalum to critical ICs. Localized shielding saves considerable weight compared with shielding the entire spacecraft. To enhance reliability, almost no read-only memory (ROM) is used on the spacecraft; only the bare minimum is included so that the computer can be bootstrapped up from step zero if necessary.

The language used on the spacecraft, known as IPS (Interpreter for Process Structures), was developed by Dr. Karl Meinzer, DJ4ZC, for multi-task industrial control type operations. It has several features that make it especially suitable for a spacecraft control system.[14] Command stations need IPS compilers to test their software, but knowledge of IPS is *not* required to use the downlink engineering or general beacons.

The satellite guidance and control subsystem is responsible for measuring position and attitude using onboard sensors, activating active attitude control components and controlling the status of all onboard systems in response to orders issued by telecommand or the spacecraft computer. The dividing line between this subsystem, the computer and the attitude-control system is typically hazy. In truth, each satellite is unique and it's far more logical to adjust our subsystem definitions to fit specific spacecraft than to make a given satellite fit a preconceived mold.

The Satellite Profiles in Appendix A are organized under the standard subsystem breakdown outlined in this chapter. The block diagram of AMSAT-OSCAR 10 illustrates the utility of this subsystem format. Referring to the block diagram, you'll see that the attitude-control system comprises sun and earth sensors and their associated electronics, and the torquing coils. The attitude-control system interfaces primarily with the central computer. The computer subsystem (also known as the integrated housekeeping unit or IHU) consists of a microprocessor, dynamic RAM, a command decoder, an analog-to-digital converter and a 64-channel analog multiplexer.

Table 12.6
Major Launch Sites of the World

Site	Location
Major U.S. Launch Sites	
Eastern Test Range (ETR) (Cape Kennedy)	28°22' N, 80°36' W
Western Test Range (WTR) (Vandenberg Air Force Base, Lompoc, CA)	34°38' N, 120°27'W
Major European Space Agency Launch Site	
Kourou, French Guiana, S. America	5°08' N, 52°37' W
Major USSR Launch Sites	
Kapustin Yar	48°31' N, 45°48' E
Plesetsk	62°42' N, 40°21' E
Tyuratam (Baikonur)	45°38' N, 63°16' E

Locations specified are of cities named, not of actual launch complex. Reference: *NY Times Index-Gazetteer of the World,* Boston: Houghton Mifflin Co., 1966.

Launch Considerations

Launch Sites

Launching a satellite takes a lot of energy. The amount depends on the final orbit, the location of the launch site and the relation between the two. Since energy constraints related to launch and orbit transfer affect AMSAT's selection of orbits, you might be interested in some of the major ones.

To place the largest possible payload in orbit using a specific rocket and launch site, the launch azimuth should be due east, taking full advantage of the relative "boost" given by the earth's rotational velocity. When this is done the orbital inclination of the satellite will equal the latitude of the launch site. The coordinates of several U.S., USSR and European Space Agency (ESA) launch sites are listed in Table 12.6. Fig. 12-9 shows the locations of these sites. These locations were chosen for several reasons including safety (it's best to launch over water or very sparsely populated regions) and energy considerations. Looking at the launch site lists it's clear that Plesetsk is nearly the ideal location for placing a payload into a Molniya (63° inclination) orbit, while Kourou is excellent for launch to geostationary orbit.

Changing the inclination of the orbital plane of a spacecraft takes a great deal of energy. When a rocket has extra energy available, inclination angle changes can sometimes be programmed into the launch. If the initial azimuth is other than due east or due west, the orbital plane inclination will be *greater* than the launch site latitude. Note that it's impossible to place a payload directly into an orbit having an inclination lower than the launch site latitude unless the upper stages of the launch vehicle expend considerable energy to modify the initial trajectory.

When AMSAT secures a ride into space, it must either accept the orbit provided or include a propulsion system on the spacecraft so a new orbit can be attained. Obviously, the decision to include a propulsion system on a spacecraft is a major one since it requires not only a rocket but a complex support system of sensors, computer and physical structure, and a much higher level of coordination with the launch agency. A satellite that contains a propulsion system involves roughly three to five times more work and expense than one without this facility. With current resources such an investment can be justified only by high-altitude Phase III spacecraft.

The simplest type of propulsion system is a kick motor that

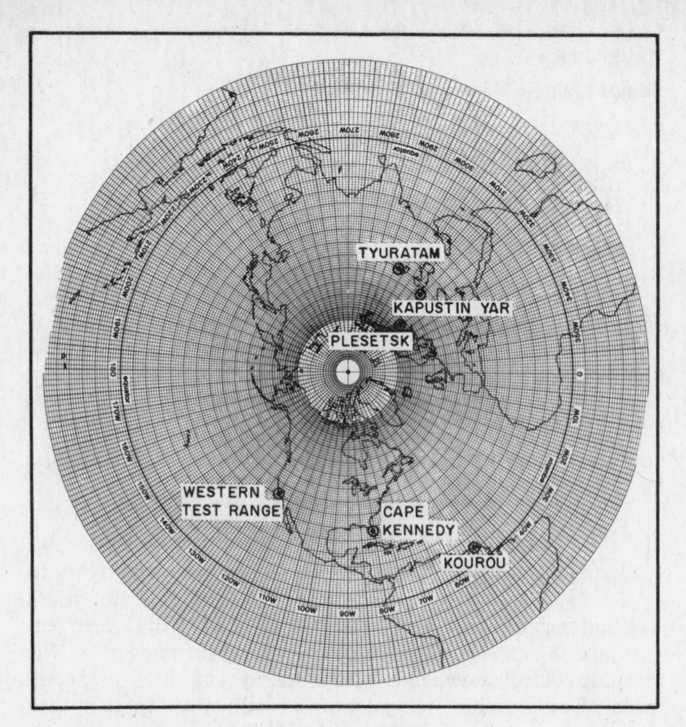

Fig. 12-9 — Locations of major launch sites.

fires only *once*. With such a motor the perigee height of the final orbit can never be greater than the apogee height of the initial orbit. We'll see why this is an important constraint when we look at the Space Shuttle.

The Space Shuttle

The U.S. Space Shuttle, the key element of the Space Transportation System (STS), will continue to produce major changes in space activity. But, how the Amateur Radio satellite program can take advantage of the coming events is not yet clear. As there are several misconceptions about the capabilities of the STS, let's take a closer look at it. The Shuttle is designed for relatively low earth orbits. Most early flights from Cape Kennedy are scheduled for circular orbits with altitudes of 200 to 300 km and inclinations of 30 to 35°. This orbit is too low for our purposes since it would provide most ground stations with only one or two 8- to 10-minute passes each morning and evening. Of even greater consequence is that the Shuttle orbits are *not* desirable as a starting point for transferring to a Phase III elliptical orbit. If AMSAT were to use a solid-propellant kick motor capable of being fired only once, the resulting low perigee height would jeopardize the satellite lifetime. Moreover, a very large kick motor would be required to raise both the perigee height and inclination. Future plans call for Shuttle launches from the Western Test Range into orbits with near polar inclinations and heights of 800 to 900 km. Such orbits would be suitable for low-altitude AMSAT spacecraft of the OSCAR 8 variety or for use as a transfer orbit for high-altitude spacecraft.

Other government and commercial satellite users are, of course, interested in reaching high orbits, so this problem has attracted considerable attention. A special "upper stage" rocket is being designed that will be carried aloft by the Shuttle. This rocket would ferry satellites to higher orbits. Because manned missions, like the Shuttle, have *very* stringent safety requirements, it's unclear whether AMSAT will be able to fly satellites with internal propulsion systems aboard the STS. Consequently, the only way to reach high-altitude orbit via the Shuttle may be to catch a lift on the upper stage to whatever final orbit is available. Since it may be very difficult to procure such a ride and the orbits available may not be desirable, the STS should certainly not be thought of as the answer to all of AMSAT's launch requirements.

Satellite retrieval is another area where misconceptions abound. The Shuttle will be able to retrieve faulty spacecraft in certain low-altitude orbits. The procedure can be relatively inexpensive ($50,000) if the satellite orbit were specially chosen to simplify retrieval, or very expensive if considerable spacecraft maneuvering is required. We must, at least temporarily, assume that retrieval of AMSAT satellites from space, even if technologically possible, will not be economically feasible.

At present, the effects of the STS on the radio amateur satellite program are uncertain. Experience, however, teaches us to expect the unexpected and the Shuttle program will very likely provide AMSAT with unexpected opportunities. If the radio amateur space program remains both flexible and vital, we'll be in an excellent position to take advantage of the opportunities that are sure to come.

Notes

[1]W. R. Corliss, *Scientific Satellites* (NASA SP-133), National Aeronautics and Space Administration, Washington, DC (1967), p. 78.

[2]P. Klein, J. Goode, P. Hammer and D. Bellair, "Spacecraft Telemetry Systems for the Developing Nations," 1971 IEEE *National Telemetering Conference Record,* April 1971, pp. 118-129.

[3]K. Meinzer, "Lineare Nachrichtensatellitentransponder durch nichtlineare Signalzerlegung" (Linear Communications Satellite Transponder Using Non-linear Signal Splitting), Doctoral Dissertation, Marburg University, Germany, 1974.

[4]K. Meinzer, "A Frequency Multiplication Technique for VHF and UHF SSB," *QST,* Oct. 1970, pp. 32-35.

[5]J. King, "The Third Generation," *Orbit,* Vol. 1, no. 4, Nov./Dec. 1980, pp. 12-18.

[6]J. Kraus, *Antennas,* New York: McGraw-Hill, 1950, chapter 2.

[7]M. Davidoff, *Using Satellites in the Classroom: A Guide for Science Educators,* Catonsville Community College, 1978, pp. 6.52-6.56. Micro Fiche copies of this 234-page book are available at a cost of 97¢ plus 20¢ postage (1983 price) from: ERIC Document Reproduction Service, Box 190, Arlington, VA 22210. Specify Document #ED 162 635.

[8]See note 7, pp. 6.24-6.31.

[9]G. Mueller and E. Spangler, *Communications Satellites,* New York: John Wiley & Sons, 1964, p. 12.

[10]See note 1, section 9.5.

[11]D. Chapin, C. Fuller, and G. Pearson, "A New Silicon P-N Junction Photocell for converting Solar Radiation into Electrical Power," *J. Applied Physics,* Vol. 25, May 1954, p. 676.

[12]See note 7, pp. 6.24-6.31, 6.40-6.41.

[13]P. Stakem "One Step Forward — Three Steps Backup, Computing in the US Space Program," *Byte,* Vol. 6, no. 9, Sept. 1981, pp. 112, 114, 116, 118, 122, 124, 126, 128, 130, 132-134, 138, 140, 142, 144.

[14]K. Meinzer "IPS, An Unorthodox High Level Language," *Byte,* Vol. 4, no. 1, Jan. 1979, pp. 146, 148-152, 154, 156, 158-159.

Chapter 13
So You Want To Build A Satellite

Building Satellites
Spacecraft Hardware

Chapter 13

So You Want To Build a Satellite

Many people view satellite construction as meticulously assembling a huge pile of mechanical and electrical components into an OSCAR. They're about 2% right. The visible part of the satellite program, the flight unit, is only the tip of a massive iceburg. Without an effective support structure there wouldn't be any amateur spacecraft: no iceburg, no tip. A partial list of the countless necessary support activities that lead to a finished OSCAR is given in Table 13.1.

Building Satellites

The radio amateur satellite program has been, and will always be, understaffed. We make this statement without qualification. AMSAT attracts people who are both doers and dreamers, doing the nearly impossible while dreaming about what they could accomplish if they only had access to a few more resources. If you share in the dream you'll probably want to help out in some way. Of the many avenues open to you, the first is to become an active AMSAT volunteer. If this appeals to you, the following steps are in order:

1) Learn all you can about the radio amateur space program;

2) Consider seriously how much time and effort you're willing to commit to satellite activities;

3) Pick an area where your personal skills and interests mesh with the needs of the program, identify an unmet need where you feel you can make a special contribution and then present your ideas to AMSAT.

The importance of Step 2 cannot be overemphasized! Space activities have a certain aura of excitement that attracts many of us initially. But the kind of personal involvement AMSAT needs often leads to long hours of tedious work with hardly even

Table 13.1
Support Activities Involved in the Production of an OSCAR

1) Design of flight hardware
2) Construction of flight hardware
   mechanical: machining; sheet metal work, potting; construction of handling fixtures, shipping crates
   electrical: wiring and cabling
3) Testing of flight hardware
   includes arranging for test facilities and people to oversee tests: vibration, environmental, burn-in, performance
4) Finished drafting
   mechanical and electrical subsystems
5) Interfacing with launch agency
   providing documentation related to satellite/rocket interface, safety and protection of primary payload (outgassing tests, etc.);
   attending coordination meetings as required
6) Identifying and procuring future launches
7) Construction Management
   parts procurement (includes locating special components and ensuring timely arrival of long-lead-time items);
   arranging overall timetable and deadlines and monitoring progress of all subgroups;
   allocating available resources (financial and human);
   locating volunteers with special expertise
8) Launch information nets
9) Providing user information and membership services:
   Orbit magazine production
   Amateur Satellite Report newsletter production
   weekly AMSAT Nets and information broadcasts via satellite
   responding to requests for information or services
   producing information programs (slide shows, video tapes)
   maintaining Satellite QSL Bureau
   operating awards
   Area Coordinator Program
   facilitating magazine article and book production
   supporting educational programs
   orbit calendar production
10) Fund raising
   sales of QSL cards, T-shirts, patches, etc.
   handling special contribution campaigns
   artwork production for magazine ads, T-shirts, QSL cards, etc.
11) Coordination with international AMSAT affiliates
12) Technical Studies focusing on future spacecraft design
   thermal
   orbit selection
   orbit determination
   attitude control
   subsystem design
13) Launch operations
   travel to launch site
   shipping satellite and rocket engine
   interface satellite to launch vehicle
   checkout, etc.
14) Command station network
   arranging for construction and operation of worldwide network
   design of special hardware and computer programs
15) Miscellaneous needs
   language translation: German, French, Russian, Japanese, Spanish, other
   legal: procurement contracts, trademark concerns, corporation papers
   insurance of various types
16) Financial
   overseeing record keeping
   auditing as required
   filing of corporation reports as required
   estimating future needs and cash-flow situation
   international cash transfers
17) Maintaining historical records
   general, spacecraft telemetry
18) Construction of test equipment and special test facilities

a "thank you." For their efforts, most volunteers receive little more than indigestion, a continual drain on their petty cash, and an ever-growing sleep deficit! Seriously, you have to be the kind of person who can be satisfied simply with seeing that an important job gets done well and on schedule. If you're after glamour and personal recognition, you've chosen the wrong field. Bringing a new volunteer onboard involves a big investment of effort by current workers who are probably already up to their apogees in work. The decision to volunteer should be given very serious consideration.

Step 3 also needs further explanation. In truth, many volunteers are attracted initially by the idea of building flight hardware. After learning as much as possible about the program, however, they may realize that their special skills in other areas would be an even more significant contribution to the amateur space program. While a few immediate needs are usually announced on the AMSAT nets, many, many other important tasks aren't mentioned. Why? Because long-term efforts to locate the right person to undertake them have been unsuccessful, or because the idea hasn't yet occurred to the AMSAT directors.

Other potential volunteers hesitate to step forward because they fear that their lack of specialized spacecraft-construction skills means there aren't any important jobs for them. Nothing could be further from the truth. From a glance at Table 13.1 it should be obvious that people with any one of a surprisingly large variety of skills or areas of expertise, from graphic arts, writing and editing, language translation and video-tape production, to accounting and the law, can contribute significantly to the success of the satellite program. In fact, many tasks don't take specialized skills but are nonetheless important to AMSAT's success. These are often the most difficult jobs to find volunteers for since a person must be very committed to undertake them.

If, after due consideration you still want to become part of the team, it takes only an informal proposal to AMSAT to get started. Indeed, volunteers are usually amazed at how quickly they can take on major responsibilities.

When I asked several long-term workers if they'd like to pass along any hints to new volunteers, two closely related themes were repeated: Don't be afraid to say "no," and never agree to a schedule you feel isn't possible (and do everything possible to live up to any schedule you've committed to). It's often difficult, especially for a newcomer, to say "no" when asked to take on some extra assignment. Saying "no," however, is best for both the long-term satellite program and everyone else who is involved, when saying "yes" would lead to unmet schedules or severe personal sacrifices that destroy the satisfaction of involvement. Since the best workers are often like magnets that attract additional tasks, one either learns to say "no" or suffers early burnout. Only the individual knows where the critical overload point is.

The need to meet schedules is absolutely essential. Satellite construction is a team effort to meet deadlines imposed by a launch agency, by a laboratory providing special test facilities on certain dates or by another volunteer who has scheduled personal vacation time so it could be devoted to a specific AMSAT task. Under these circumstances, one person's late project can be disastrous.

One outstanding characteristic of almost all long-term AMSAT volunteers is the seriousness with which they accept commitments. Once they agree to a task or schedule they do everything possible to deliver as promised, often at great personal cost. Over the years this sense of commitment has led to a very special camaraderie, trust and respect among AMSAT workers. It's a spirit that I've never seen anywhere else in the academic, scientific, industrial or sports communities.

You don't have to have a formal title or be willing to invest a big chunk of time to be an AMSAT supporter. Everyone who helps a newcomer get started in satellite communications, provides information on the satellite program to other segments of the Amateur Radio community, or makes a modest financial contribution to AMSAT is filling an important need. A great many people helping in a lot of small ways will keep the satellite program vital, so let's all try to be conscious of the little things. For example, if you need something from AMSAT headquarters, an s.a.s.e. will save a few minutes as will a request that is phrased to be answered yes or no, or with an article reprint. Similarly, providing information by telephone requires only a fraction of the time that preparing a written answer does. The point is that the time of key AMSAT volunteers is a very valuable commodity; small efforts by all of us to lighten their work load will pay off cumulatively.

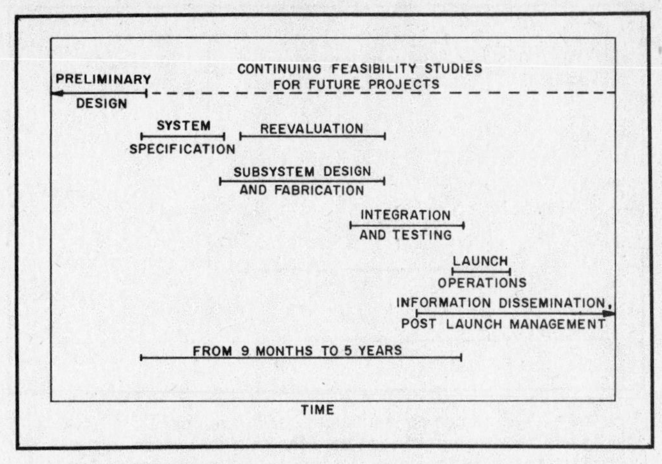

Fig. 13-1 — Time frame for satellite construction from project management perspective.

By now you probably have some idea as to what level of support you'd be comfortable with. For those who are determined to become involved in hardware construction (flight, flight-related or ground-command) we now look at the steps involved in the spacecraft-construction aspect of the OSCAR program.

Spacecraft Hardware

As with most human endeavors, if you want to become directly involved in satellite construction one of your first steps should be to learn everything possible about satellite system design. Chapter 12 is a good starting point, but it's little more than an introduction. You'll want to dig into many of the references.

Before getting down to specifics let's take a brief look at satellite construction from a project-management perspective. The complex process can be reduced to six stages: (1) preliminary design, (2) system specification, (3) subsystem design and fabrication, (4) integration and testing, (5) launch operations and (6) information dissemination and post-launch management. The time frame for these activities is outlined roughly in Fig. 13-1.

The preliminary-design stage involves feasibility studies of new approaches to satellite design. State-of-the-art advances in electronics, cost reductions in components, launch access to unusual orbits, new sources of financial support and the like continually open up new design options. In a long-term multi-satellite program, feasibility studies are continual.

At some point, usually in response to a specific launch opportunity, the decision is made to construct a spacecraft. A set of system specifications must then be agreed upon and the subsystem requirements defined.

Subsystems are then designed, built, tested and refined. With AMSAT satellites, subsystems are designed and built by small core groups scattered around the world. Electronic subsystems usually are built in a number of versions: an *engineering development model,* a *flight prototype* and a *flight unit,* the latter using the most reliable components available. Each subsystem must be tested thoroughly under extreme conditions of temperature and over- and under-voltage so that potential weak spots can be identified and corrected.

Next, the subsystems are integrated into a spacecraft so ad-

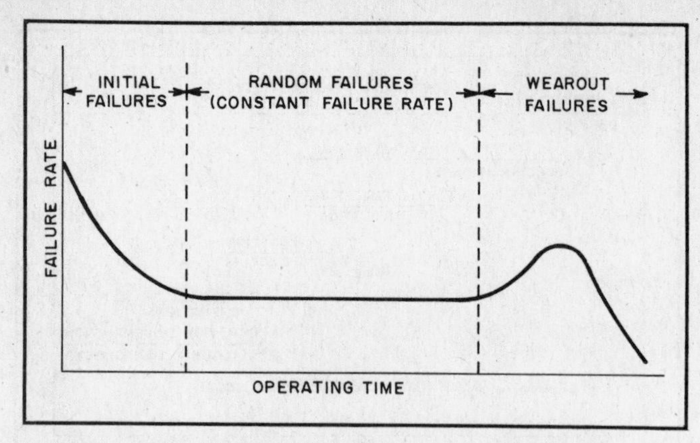

Fig. 13-2 — Typical component failure curve. The object of AMSAT's testing and stress program is to have the satellite operate in the central flat region of the curve. For information on designing for reliability see W. C. Williams, "Reliability: Lessons from NASA," *IEEE Spectrum*, Vol. 18, no. 10, Oct. 1981, pp. 79-84.

Table 13.2

Possible Topics to be Addressed in a Construction Project Proposal†

1) *General Conceptual Plan.* A general description of the proposed subsystem or project.
2) *Trade-off Discussion.* A more detailed discussion of the proposed subsystem including a candid evaluation of (1) its advantages and disadvantages with respect to prior subsystems having a similar function and (2) its impact on other subsystems.
3) *Interface Considerations.* A detailed specification of how the proposed subsystem will interface with other subsystems.
4) *Environmental Design.* A discussion of how the experimenter will attempt to guarantee that the completed unit will perform satisfactorily under anticipated extremes of temperature, vacuum, power variation, radiation and the rf environment, including analysis of waste heat, potential RFI and steps taken to prevent RFI problems.
5) *Component Selection and Construction Techniques.* If the experimenter is planning to construct a flight unit, describe steps to be taken to ensure component and construction quality.
6) *Testing Program.* Detailed description of all tests to be performed on the completed unit.
7) *Required Support.* If the experimenter anticipates calling on AMSAT to provide assistance in financing, design or parts procurement, the required type and level of support should be specified.
8) *Experience of Project Personnel.* List the people expected to work on the project, and the expertise and the expected time commitment of each project member.
9) *Schedule and Delivery.* Set up a timetable for the project, indicating dates for milestones and specifying a realistic delivery date that includes allowances for unanticipated delays.

†These are among the topics that can be addressed by a prospective experimenter who wishes to undertake a major satellite subsystem construction project coordinated with AMSAT.

ditional stress tests, operational checks and rf-compatibility tests can be performed. The stresses include a *burn-in* period for electronics systems during which electrical parameters and temperatures are similar to those expected in space but with the system at atmospheric pressure; *environmental tests* that involve operating the spacecraft in a vacuum chamber under temperatures considerably more severe than those expected in space (for example, −20° C and +60° C); and a vibration test to ensure that the satellite will survive the launch. The objectives of the vacuum test include (1) checking for material sublimation that could contaminate spacecraft systems, (2) testing for corona discharge and (3) verifying the predicted thermal behavior in the absence of convective heat flow. Vacuum and vibration tests are usually performed at large government or commercial laboratories that have the special facilities required.

The electrical testing strategy is based on the fact that high temperatures and overvoltages tend to compress the time scale of the failure curves for most electrical components (a typical curve is shown in Fig. 13-2). Temperature cycling and vibration tests have a similar effect on mechanical components. Consequently, one month of actual testing might be equivalent to two years of testing under normal operating conditions. The aim of the testing program is to discover and correct all weak spots and then bring the spacecraft past the initial hump in the failure curve while it's still on the ground. NASA's experience has clearly proved the validity of this approach to ensuring reliability.

When the satellite has passed all tests it's transported to the launch site, mated to the launch vehicle and checked out one last time. The project, however, doesn't end with the launch. Command stations must be available when and where they are needed; information must be disseminated to users; and data on spacecraft operation must be collected to assist in current operation and for use in the design of future spacecraft. The entire procedure, from system specification to launch, can take anywhere from nine months to five years depending on the complexity of the spacecraft and the available personnel resources.

Once you know something about satellite systems and the stages of satellite construction it's time to pick a particular project, subsystem, or aspect of construction to focus on. This may involve refining a specific subsystem you're particularly knowledgeable about (e.g., analog-digital converter design), applying your skills to various subsystems (e.g., optimizing PCB layouts) or looking into spacecraft subsystems that appear to need improvement even though you have no prior knowledge in the area (when no "expert" is available, as with ion-drive engines, someone has to start from scratch). Or, perhaps you see a technique for accomplishing a spacecraft function that's simpler or more reliable than the approach currently being used.

As a first project pick something modest with clearly defined interfaces. For example, you might elect to work on a high- efficiency transponder for a low-altitude spacecraft with a 146-MHz input and dual outputs at 29.5 MHz and 435 MHz. Initial objectives would probably be limited to producing and testing an engineering development model. One important area that is frequently overlooked is the need for special test equipment and procedures for satellite evaluation and checkout. If you're interested in this area, contact AMSAT to find out what's needed.

Once you identify a specific aspect of construction that you'd like to work on, send a brief memo to the person who is coordinating that core group, to the spacecraft project manager or to the AMSAT directors. It's important to realize that AMSAT workers and directors don't see each other at the office each morning: They're spread around the world and most have full-time jobs. Circulating a letter can take months and the chances of its getting lost are, unfortunately, high. Therefore, make several copies and send one to each person you think may be interested. The object is to establish direct contact with the person or persons responsible for the work you're interested in.

Because of AMSAT's geographical scatter and problems in internal communication, it's always best to establish a single contact, preferably with one person at one of the established AMSAT core groups. If you're working on a Phase III subsystem this probably means the Washington or Marburg teams. The choice can depend on your geographical proximity, your ability in the group's native language, the similarity of your interests and the ease of communication via WATS lines or through business travel. Before undertaking a major effort that will require significant assistance from AMSAT you should be willing to take on some simpler tasks to demonstrate your competence and willingness to adhere to schedules. Undertaking a larger project generally would involve writing a proposal focusing on the concerns listed in Table 13.2. Note that these aren't formal guidelines; they're merely suggestions. AMSAT is most definitely not a huge, faceless organization. Every proposal is treated individually. It's the content that counts.

Many practical aspects of satellite construction aren't obvious to the newcomer. Communicating with other project workers is a major one. You must coordinate with groups working on subsystems that interface with yours. Such communication is much easier if you have access to a WATS line or have a job that takes you (or someone else in your group) to Washington, DC, on business several times each year.

Keep in mind that communication is a two-way street! You must be willing to provide clearly written documentation to other groups that have an immediate need to know, and eventually to users. Another often-overlooked consideration is that passing components across national boundaries can involve a great deal of paperwork. Try to anticipate all of these little loose ends when estimating the total resources needed to accomplish a particular job. Frankly, no matter how thorough you think your estimates are, you'll probably grossly underestimate the real effort needed. Psychologically this might be a good thing: If we knew what we were really committing ourselves to, far fewer might volunteer.

Over the years several individuals in the academic and educational communities have been able to fuse their vocations with their satellite interests. Students in the undergraduate electrical engineering program at Trenton State College, for example, have designed 70-cm to 23-cm transponders for their senior project. Students doing postgraduate work have received Masters and PhD degrees for projects that relate to OSCAR satellite design. Science and engineering educators have even received government and commercial grants to work on particular aspects of the AMSAT program. In addition, AMSAT has recently begun an internship program. The object of the program is to train young radio amateurs who have strong backgrounds in science and engineering in all aspects of satellite construction. This is the wisest investment we can make to ensure the continuation of the radio amateur space program. (As far as we know, the skills necessary to be a spacecraft project manager are not coded in one's DNA at birth.) Interns may be paid employees, but the wages are terrible and the hours ridiculous. Interns nonetheless receive invaluable experience in all phases of satellite design and construction, and a measure of responsibility usually achieved only by senior engineers. In return, AMSAT receives the services of very bright, very committed scientists and engineers at modest cost.

Internships are flexible: Appointments can last from a few months to several years. Don't send for a formal application form — there aren't any. If you'd like to apply, just submit a letter that details your background and explains why you feel you could contribute to the amateur space program. Better yet, volunteer to spend a few weeks as an unpaid intern at the AMSAT OSCAR Spacecraft Laboratory located at Goddard Space Flight Center in Greenbelt, Maryland, so that both you and AMSAT can better evaluate the desirability of making a longer-term commitment.

Though most of the comments in this section have focused on building flight-related hardware, the construction of the telecommand equipment offers many similar satisfactions. In fact, many people working directly on today's spacecraft got their start by building telecommand stations. Most notable are Larry Keyser, VE3QB, who almost singlehandedly kept OSCAR 6 operating during its early days; A. Gschwindt, HA5WH, who set up a command station for OSCAR 6 and later built switching regulators for AMSAT-OSCARs Phase III-A and -B; and Martin Sweeting, G3YJO, who went from commanding OSCARs 6 and 7 to directing the construction of UoSAT-OSCAR 9.

AMSAT is a flexible, vital organization. Its very informality — the lack of straightforward procedures for submitting proposals or volunteering — can make it difficult for a newcomer to become involved. The same lack of formal structure, however, makes it possible for the competent, committed individual to assume important responsibilities quickly. We'd like to hear from you.

Appendix A
Satellite Profiles

AMSAT-OSCAR 8
UoSAT-OSCAR 9
Radio Sputnik
AMSAT-OSCAR 10

Each section of this appendix contains a succinct description of a radio amateur satellite or group of satellites either currently in orbit and operational or soon to be launched. This information has been organized, insofar as possible, using the standard format outlined in Table A.1. For your convenience, the name of the satellite under discussion heads the top of each page. As a result, you should have no trouble finding particular technical data. Though detailed, the profiles are by no means complete. Additional sources of information have been referenced when available.

Table A.1
Standard Format Used to Describe Satellites Listed in Appendix A

SPACECRAFT NAME
GENERAL
- 1.1 Identification: international designation, pre-launch designation
- 1.2 Launch: date, vehicle, agency, site
- 1.3 Orbital Parameters: general designation, period, apogee and perigee altitude (specified over mean radius of earth — 6371 km), inclination, eccentricity, longitude increment, maximum access distance, maximum access time, expected lifetime in orbit if less than 10 years. Parameter values quoted only to those significant digits expected to remain stable over the useful lifetime of the spacecraft (s/c).
- 1.4 Ground Track Data
- 1.5 Operations: Group(s) responsible for coordination and scheduling
- 1.6 Design/Construction Credits: project management, s/c subsystems
- 1.7 Primary References

SPACECRAFT DESCRIPTION
- 2.1 Physical Structure: shape, mass
- 2.2 Subsystem Organization: block diagram

SUBSYSTEM DESCRIPTION
- 3.1 Beacons: frequency, power level, telemetry format, maximum Doppler shift, data sources
- 3.2 Telemetry: formats available, description of each format (including decoding information, sample data, etc.)
- 3.3 Telecommand System
- 3.4 Transponders: for each transponder the type, uplink passband downlink passband, translation equation, output power, uplink EIRP (recommended and maximum values), delay time, etc.
- 3.5 Attitude Stabilization and Control: primary control, secondary control, damping, sensors
- 3.6 Antennas: description, polarization table
- 3.7 Energy-Supply and Power Conditioning: solar cell characteristics and configuration, storage battery, switching regulators, etc.
- 3.8 Propulsion System
- 3.9 Integrated Housekeeping Unit (IHU)
- 3.10 Experimental Systems

AMSAT-OSCAR 8

SPACECRAFT NAME: AMSAT-OSCAR 8

(Though AMSAT-OSCAR 8 ceased operating in mid 1983, it is the most recent in the Phase-II series of active communications satellites built by AMSAT. As much of the design and telemetry approach typifies AMSAT's Phase-II work and may indeed be used in future spacecraft, we have included this reference section on A-O-8.)

GENERAL

1.1 Identification
International designation: 78-026B
Pre-launch designation: AMSAT-OSCAR D

1.2 Launch
Date: 5 March 1978
Vehicle: Two-stage Delta 2910
Agency: U.S. National Aeronautics and Space Administration
Site: NASA Western Test Range, Lompoc, California (Vandenberg Air Force Base)

1.3 Orbital Parameters
General designation: low-altitude, sun-synchronous
Period: 103.2 minutes
Apogee altitude: 916 km
Perigee altitude: 904 km
Eccentricity: 0.0008 (nominally circular)
Inclination: 98.9° (near polar)
Longitude increment: 25.8° West/orbit
Maximum access distance: 3250 km

1.4 Ground Track Data: See Appendix B

1.5 Operations
Coordinating Group: American Radio Relay League
Schedule (subject to change): see *QST* and *Orbit*
Sunday-Monday-Tuesday (UTC): Mode A and 29.402 MHz Beacon
Thursday-Friday-Saturday: Mode J and 435.095 MHz Beacon
Wednesday (UTC): Transponder reserved for special experiments arranged in advance with ARRL.
Monday (UTC): Transponder users requested to observe 10-watt EIRP limit (QRP day).

1.6 Design/Construction Credits
Project Management: Jan King, W3GEY, AMSAT-USA
Spacecraft subsystems: Contributed by groups in Canada, Japan, United States, West Germany

1.7 Primary Reference: P. Klein and J. Kasser, "The AMSAT-OSCAR D Spacecraft," *AMSAT NEWSLETTER*, Vol. IX, no. 4, Dec. 1977, pp. 4-10.

SPACECRAFT DESCRIPTION

2.1 Physical Structure
Shape: Rectangular solid as shown in Fig. 1(A-O-8), approximately 33 cm (height) by 38 cm by 38 cm.
Mass: 25.8 kg

2.2 Subsystem Organization
Block diagram: See Fig. 2(A-O-8)

SUBSYSTEM DESCRIPTION

3.1 Beacons

	Frequency	*Power Output*	*Max. Doppler*
Mode-A Beacon (hf Beacon)	29.402 MHz	110 mW	0.7 kHz
Mode-J Beacon (uhf Beacon)	435.095 MHz	100 mW	10.1 kHz

3.2 Telemetry
Formats available: Morse code, special features
Morse code telemetry
frame: A frame contains six channels (six lines by one column). Parameters are sent in a fixed serial format.
channel: A channel consists of a three digit number. The first digit is a line identifier. Because of the single column format, the first digit uniquely identifies the parameter being measured. The last two digits in a channel are the value of "N" and encode the data as per Table 1(A-O-8).
speed: The telemetry is sent at 20 words per minute. A complete frame requires about 20 seconds.
sample data: See Table 2(A-O-8).
Special features
command enable: When the command system has been enabled and is ready to accept a command, the Morse code telemetry is interrupted and an unmodulated carrier is transmitted on the beacon frequency.

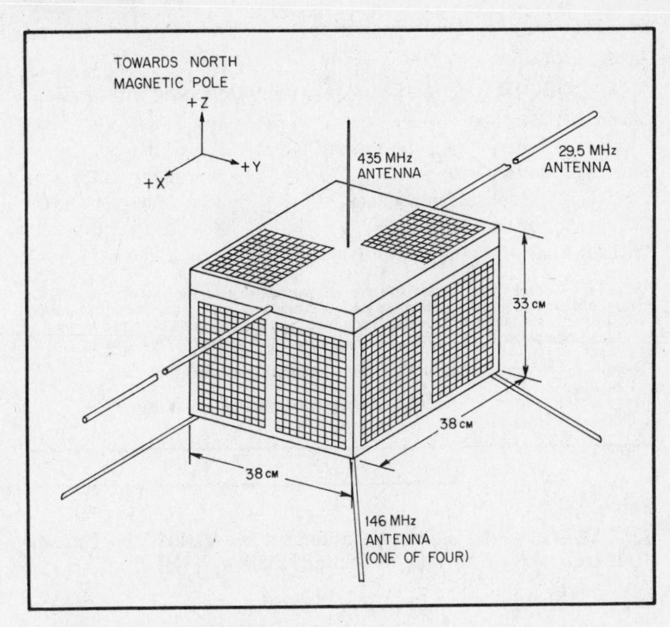

Fig. 1(A-O-8) — AMSAT-OSCAR 8.

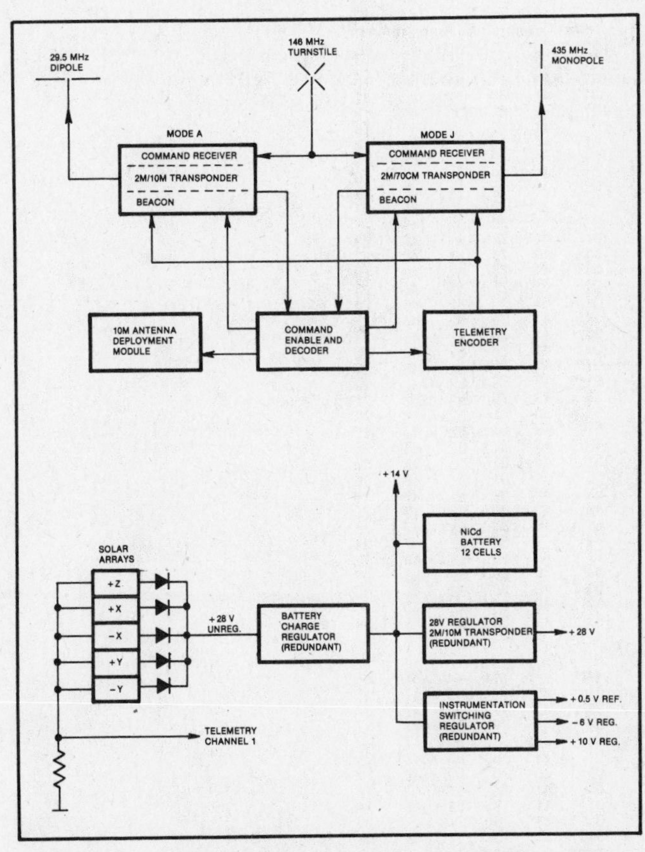

Fig. 2(A-O-8) — AMSAT-OSCAR 8 satellite functional block diagram.

10-m antenna status: When the 10-m antenna deployment command is received at the satellite the beacon transmits a series of pulses. The pulse rate is a function of tip-to-tip antenna length. See 3.6: 29.5-MHz antenna.

AMSAT-OSCAR 8

Table 1(A-O-8)

AMSAT-OSCAR 8 Morse Code Telemetry-Decoding Information

Channel 1: Total Solar Array Current	I = 7.15 (101 − N) mA[1]
Channel 2: Battery Charge-Discharge Current	I = 57 (N − 50) mA[2]
Channel 3: Battery Voltage	V = (0.1N + 8.25) volts
Channel 4: Baseplate Temperature	T = (95.8 − 1.48N) °C
Channel 5: Battery Temperature	T = (95.8 − 1.48N) °C
Channel 6: 435-MHz Transmitter Power Output	P = 23 N mW[3]

[1]Whenever N is less than 10 assume that an overrange condition has occurred. For example, as the satellite enters the earth's shadow a reading of 101 is transmitted. This refers to channel 1, N = 01. Since N is less than 10 we assume that overranging has occurred and the actual N is 101, which corresponds to zero current.

[2]There is a 2-second integration time associated with the current telemetered on this channel.

[3]There is a 2.5-second integration time associated with the power telemetered on this channel.

Table 2(A-O-8)

AMSAT-OSCAR 8 Telemetry Copied on the 29.402-MHz Beacon 10 March 1978. Courtesy of Richard Zwirko, K1HTV.

ORBIT #61

	1N	*2N*	*3N*	*4N*	*5N*	*6N*
[1]	01	47	82	50	48	01
	01	47	82	51	48	01
	01	46	82	—	—	—
	—	—	—	—	—	—
	—	—	—	—	—	—
	01	46	—	—	—	—
	01	46	81	51	47	01
	01	46	81	51	47	01
[2]	mmm	continuous tone				mmmm
[3]	01	41	81	51	47	12
	01	42	81	—	46	18
	01	41	81	51	—	15
	01	41	81	52	47	17
	01	—	82	51	46	17
	01	41	79	51	47	18
	01	44	79	51	47	18
	01	41	79	51	47	11
	01	41	79	51	48	19
	01	41	79	51	46	22
	01	41	79	51	46	17
	01	41	79	51	46	17
	01	41	79	51	46	23
	01	41	78	51	46	23
	01	41	78	51	46	19
	01	41	79	51	46	23
	01	41	78	52	46	21
	01	41	78	52	46	25
	01	41	78	52	46	24
	01	41	78	52	46	26
	01	41	78	52	46	23
	01	43	78	52	47	22
[4]	98	43	79	52	48	22
	61	48	81	52	47	26
	51	48	81	52	47	17
	48	47	81	52	48	23
	51	48	81	52	48	23
	66	51	81	52	48	26
	53	49	81	52	47	19
	46	49	80	52	47	20
	58	49	80	52	47	18
	—	49	80	52	47	21
	64	48	80	52	48	22
	63	49	80	52	48	07
	63	49	80	52	48	13
	52	49	80	52	47	16
[5]	51	49	80	52	48	—

ORBIT #62

	1N	*2N*	*3N*	*4N*	*5N*	*6N*
[6]	01	42	76	50	47	17
	01	41	77	50	47	17
	01	41	77	50	47	17
	01	41	76	50	47	—
	01	41	76	50	47	21
	01	40	76	50	46	21
	01	41	76	51	47	17
	01	41	76	51	47	14
	01	41	76	51	47	20
	01	41	76	51	46	mm
[2]	mmm	continuous tone				mmmm
[7]	01	46	77	51	47	01
	01	46	77	51	47	01
	01	45	77	51	47	01

[1]Acquisition of orbit #61 at 02:12:28 UTC, 10 Mar. 1978 (ascending node 02:09:20 UTC, 69.9° W).

[2]Command station accessing satellite.

[3]Mode J turned on (see channel 6); Mode A remains on (telemetry being copied on 29.402 MHz).

[4]Satellite crossing terminator into daylight (see channels 1 and 2).

[5]Loss of orbit #61 at 02:28:25 UTC.

[6]Acquisition of orbit #62 at 03:59:25 UTC, 10 Mar. 1978 (ascending node 03:52:32 UTC, 95.7° W).

[7]Mode J turned off; Mode A remains on.

Table 3(A-O-8)

AMSAT-OSCAR 8 Commands

Command	*Spacecraft Status*
Mode-A Select	2m/10m transponder and 29.402-MHz beacon ON
Mode-J Select	2m/70cm transponder and 435.095-MHz beacon ON
Mode-D Select	Recharge mode. Both transponders and beacons OFF
10-m Antenna Deployment	Activates 10-m antenna deployment mechanism and switches telemetry to pulse format encoding tip-to-tip length of antenna
10-m Antenna Reset	Stops deployment of 10-m antenna (deployment cannot be reversed). Switches telemetry back to Morse code.

3.3 Telecommand System

The command system recognizes five commands as per Table 3(A-O-8).

3.4 Transponders

Transponder I: Mode A (2m/10m)

type: linear, noninverting
uplink passband: 145.850-145.950 MHz
downlink passband: 29.400-29.500 MHz
translation equation:
downlink freq. (MHz) =
uplink freq. (MHz) − 116.458 MHz ± Doppler
output power: 1-2 watts PEP
uplink eirp: a maximum of 80 watts is recommended
bandwidth: 100 kHz
maximum Doppler: 4.1 kHz
comments: The same basic Mode-A transponder has been used on AMSAT-OSCARs 6, 7 and 8. A block diagram is shown in Fig. 3(A-O-8).

Transponder II: Mode J (2m/70cm)

type: linear, inverting
uplink passband: 145.900-146.000 MHz
downlink passband: 435.100-435.200 MHz
translation equation:
downlink freq. (MHz) =
581.100 − uplink freq. (MHz) ± Doppler
output power: 1 to 2 watts PEP. Telemetry channel six measures the output power using a 2.5-second integration time.
uplink eirp: a maximum of 10 watts is recommended. Under certain conditions of spacecraft temperature and battery voltage, the transponder sensitivity may decrease and 80 watts may be needed.
bandwidth: 100 kHz
maximum Doppler: 6.7 kHz
comments: This transponder was constructed by the Japan AMSAT Association of Tokyo to test the effectiveness of this link for low-altitude spacecraft.

3.5 Attitude Stabilization

Primary control: Four Alnico-5 bar magnets, each approximately 15 cm long and with a square cross-section of about 0.6 cm by 0.6 cm are mounted parallel to the Z-axis of the spacecraft. The resultant far field is similar to that produced by a single 30,000 pole-cm magnet. As the satellite moves along its orbit the Z-axis of the spacecraft constantly changes its direction in inertial space to remain aligned parallel to the local direction of the earth's magnetic field. The +Z-axis (top) of the satellite points in the direction of the earth's north magnetic pole.

Damping: Allegheny Ludlum type 4750 permalloy hysteresis damping rods (0.32-cm diameter) are mounted behind, and parallel to, the +X, −X, +Y and −Y solar panels (perpendicular to the Z-axis) to damp out rotational motion about the Z-axis.

3.6 Antennas (See Fig. 1(A-O-8))

29.5 MHz: The 29.5-MHz transmitting antenna is a half wavelength dipole (about 4.9 m) mounted perpendicular to the Z axis. It is composed of tubular extendable members which are deployed by small motors activated by ground command after launch when the satellite spin rate has decreased below 2 rpm. The non-reversible deployment process takes about 15 seconds. When the satellite receives the 10-m-Antenna Deployment command, the telemetry system transmits a series of pulses, the rate of which is a function of tip-to-tip antenna length. In the fully retracted

AMSAT-OSCAR 8

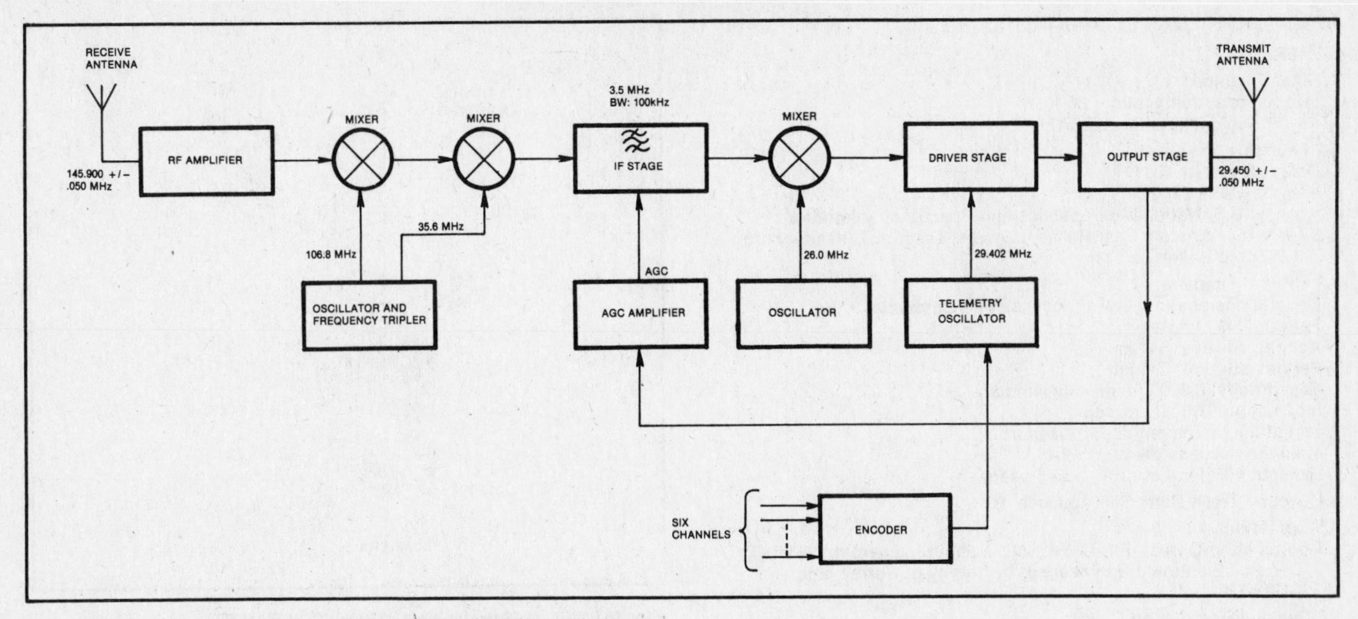

Fig. 3(A-O-8) — Block diagram of AMSAT-OSCAR 8 Mode A transponder.

position (launch state) the rate is about 15 pulses/sec. When the antenna is fully deployed the rate is 1.8 pulses/sec.

146 MHz: The 146-MHz receiving antenna for both transponders is a canted turnstile. It consists of two "inverted V" shaped dipoles mounted at right angles on the base (–Z face) of the spacecraft. Each dipole consists of two 48-cm spokes (1/4 wave-length) constructed from a material similar to 1-cm wide carpenter's rule. The turnstile is fed by a hybrid ring and matching network. It produces an elliptically polarized radiation field (circularly polarized along –Z axis) over a large solid angle. The gain approaches 5 dB along the –Z axis; there's some shadowing along the +Z axis.

435 MHz: The 435-MHz transmit antenna is a 1/4-wavelength monopole mounted on the top (+Z face) of the spacecraft.

Note: Signal polarizations of the spacecraft antennas are summarized in Table 4(A-O-8).

Table 4(A-O-8)
AMSAT-OSCAR 8 Antenna Polarizations

System	*Spacecraft Polarization*
2m/10m transponder uplink (146 MHz)	left-hand circular[1]
2m/10m transponder downlink and 29-MHz beacon	linear
2m/70cm transponder uplink (146 MHz)	right-hand circular[1]
2m/70cm transponder downlink and 435-MHz beacon	linear

[1]Polarization sense referenced to +Z-axis of spacecraft. Ground stations off the +Z-axis will observe elliptical polarization. Stations north of the magnetic equator (see Fig. 3-6) will generally find that the circular component is as indicated in the table. Stations in the southern hemisphere will generally find the circular component reversed.

3.7 Energy-Supply and Power Conditioning

The main components of the AMSAT-OSCAR 8 energy-supply and power conditioning subsystem are shown in Fig. 2(A-O-8).

Solar Cell Characteristics

type: n on p silicon
size: 1 cm × 2 cm
total number: 1920
total surface area: 4005 cm^2
protective cover: 0.015-cm glass cover slide
efficiency: 8% (before launch)
peak array output: 15 w (optimal sun orientation)

Solar Cell Configuration

basic module: 80 cells in series
total number of modules: 24
location: + X, – X, + Y, – Y facets have 5 modules each; + Z facet has 4 modules.

Storage Battery

type of cell: Nickel-Cadmium
voltage/cell: 1.45 V (fully charged)
capacity/cell: 6 Ampere-hours (Ah)
configuration: 12 cells in series
battery (100% charged): 17.4 V, 6 Ah
battery (50% charged): 14.5 V, 3 Ah

Switching regulators

battery charge regulator: Converts 28- to 30-volt solar array bus to 14- to 16-volt main spacecraft power bus. Tapers charge rate to prevent overcharging at a battery voltage of 17.4. Fully redundant and autoswitching if regulator senses open or short.

instrumentation switching regulator: Provides well regulated +10 V, –6 V and precision reference of +0.5 V for all spacecraft systems. Fully redundant.

transponder regulator: Converts 14-16-volt unregulated space-craft bus to 24-28 volts for use by the 2m/10m transponder power amplifier and driver. Fully redundant.

UoSAT-OSCAR 9

SPACECRAFT NAME: UoSAT–OSCAR 9

GENERAL

1.1 Identification
International designation: 81-100B
Pre-launch designation: UoSAT

1.2 Launch
Date: 6 October 1981
Vehicle: Delta 2310
Agency: U.S. National Aeronautics and Space Administration
Site: NASA Western Test Range, Lompoc, California (Vandenberg Air Force Base)

1.3 Orbital Parameters
General designation: low-altitude, sun-synchronous
Period: 95.3 minutes
Apogee altitude: 544 km
Perigee altitude: 536 km
Eccentricity: 0.0006 (nominally circular)
Inclination: 97.5° (near polar)
Longitude increment: 23.8° West/orbit
Maximum access distance: 2545 km
Expected lifetime in orbit: 3 to 5 years

1.4 Ground Track Data: See Appendix B

1.5 Operations
Coordinating Group: UoSAT Project; Dr. Martin Sweeting, G3YJO, Dept. of Electronic Engineering; University of Surrey; England GU2 5XH

1.6 Design/Construction Credits
Project management: Dr. Martin Sweeting, G3YJO, of University of Surrey (UoS) and AMSAT-UK
Spacecraft subsystems: See specific subsystem

1.7 Primary References
M. Sweeting, *UoSAT-OSCAR-9 Technical Handbook,* Published by AMSAT-UK, Oct. 1981. Reprinted in part in *ASR,* Vol. 1, no. 16, 17, 18, 19, 21, 22; 1981.

Several large segments of this profile have been taken from the UoSAT-OSCAR-9 Technical Handbook.

Also see: M. Acuna, UoSAT Magnetometer, *Orbit,* Vol. 2, no. 4, Aug./Sept. 1981, pp. 6-10; B. Ruedisueli, UoSAT Propagation Experiment, *Orbit,* Vol. 3, no. 1, Jan./Feb. 1982, pp. 5-13; *The Radio and Electronic Engineer,* Journal of the Institute of Electronic and Radio Engineers (England), Aug./Sept. 1982, Vol. 52, no. 8/9, Special Issue on: "UoSAT — The University of Surrey's Satellite."

SPACECRAFT DESCRIPTION

2.1 Physical Structure
Shape: rectangular solid as shown in Fig. 1(U-O-9); approximately 67 cm (height) by 42 cm by 42 cm. The separation ring extends ≈ 6.5 cm below the −Z facet of the s/c. The navigation magnetometer, mounted along the Z axis, extends ≈10 cm above the +Z facet of the s/c.
Mass: 59.95 kg.

2.2 Subsystem Description
Block diagram: See Fig. 2(U-O-9)

SUBSYSTEM DESCRIPTION

3.1 Beacons
Data Beacons
Design/Construction Credits: Bob Haining, UoS/AMSAT-UK
Available Data Sources:
Telemetry: ASCII, Baudot, Morse code
Primary s/c computer: serial output port no. 1
serial output port no. 2
speech synthesizer
Video Display Experiment: camera image data, text, graphs, line drawings, other image data.
Modulation: nbfm ±5 kHz deviation
Undesired signal levels: greater than 65 dB below reference carrier

General Data Beacon (VHF Beacon)
Frequency: 145.825 MHz
Power output: 350 mW
Total dc/rf efficiency: 45%
Maximum Doppler: ±3.7 kHz

Engineering Data Beacon (UHF Beacon)
Frequency: 435.025 MHz
Power output: 650 mW
Total dc/rf efficiency: 40%
Maximum Doppler: ±10.9 kHz

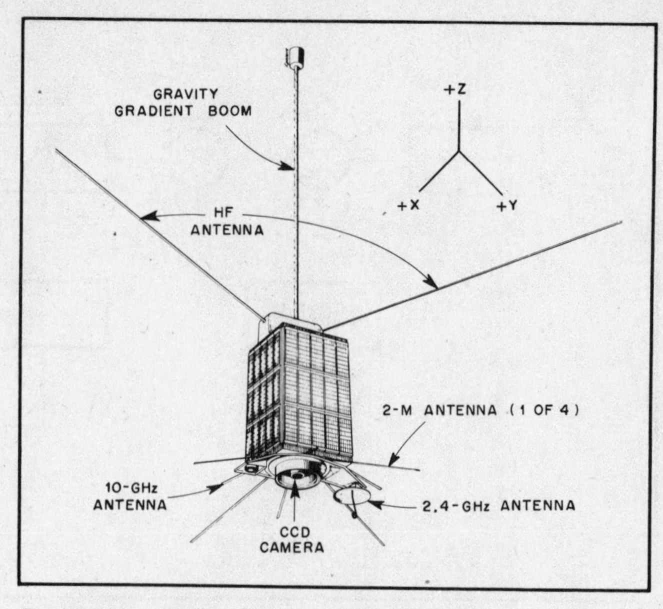

Fig. 1(U-O-9) — Pictorial view of UoSAT-OSCAR 9.

HF Beacons Experiment
Design/Construction Credits: Colin Smithers, G4CWH, UoS/AMSAT-UK
Frequencies: 7.050 MHz
14.002 MHz
21.002 MHz
29.510 MHz
Power output: 100 mW each
Total dc/rf efficiency: 30%
Modulation: Morse code or continuous carrier. Ground control will permit beacons to be either (1) phase locked to each other (phase-coherent) for trans-ionospheric path analysis or (2) operated independently (free-running crystal oscillator)
Reference: For information on applications see *ASR,* "The First Science OSCAR," Vol. 1, no. 13, 10 Aug. 1981, pp. 1-2.

SHF Beacon Experiment
Design/Construction Credits: Richard Porter, Microwave Modules, UK
Frequency: 2.401 GHz
Power output: 125 mW
Total dc/rf efficiency: 4%
Modulation: nbfm ±10 kHz deviation
Maximum Doppler: ±59.9 kHz
Available data sources: same as 146-MHz General Data Beacon
Path loss: −152 dB for 90° elevation angle
−169 dB for 0° elevation angle

Microwave Beacon Experiment
Design/Construction Credits: Jim Arnold, Plessey Research, UK
Frequency: 10.47 GHz
Power output: 125 mW
Total dc/rf efficiency: 6%
Modulation: continuous carrier
Maximum Doppler: ±262 kHz
Path Loss: −169 dB for 90° elevation angle
−181 dB for 0° elevation angle

3.2 Telemetry
Design/Construction Credits: Dr. Lui Mansi, UoS/AMSAT-UK
General: The telemetry system has been designed to provide a high degree of flexibility. It has provisions for monitoring 60 analog sensor channels and 45 digital status points. Decoding information for the sensor channels is given in Table 1(U-O-9). Status points are identified in Table 2(U-O-9).

Formats:
ASCII: 1200, 600, 300 and 75 baud (only one of these speeds is available at any given time).
ASCII: 110 baud
RTTY: 45.5 baud
Morse code: 10 or 20 words per minute (Channels 00 to 09 only)
Synthesized voice: In conjunction with primary s/c computer
Note 1: Any combination of above formats is available to 146-MHz and 435-MHz Data Beacons. Data format on 2.4-GHz beacon will match 146-MHz beacon.

UoSAT-OSCAR 9

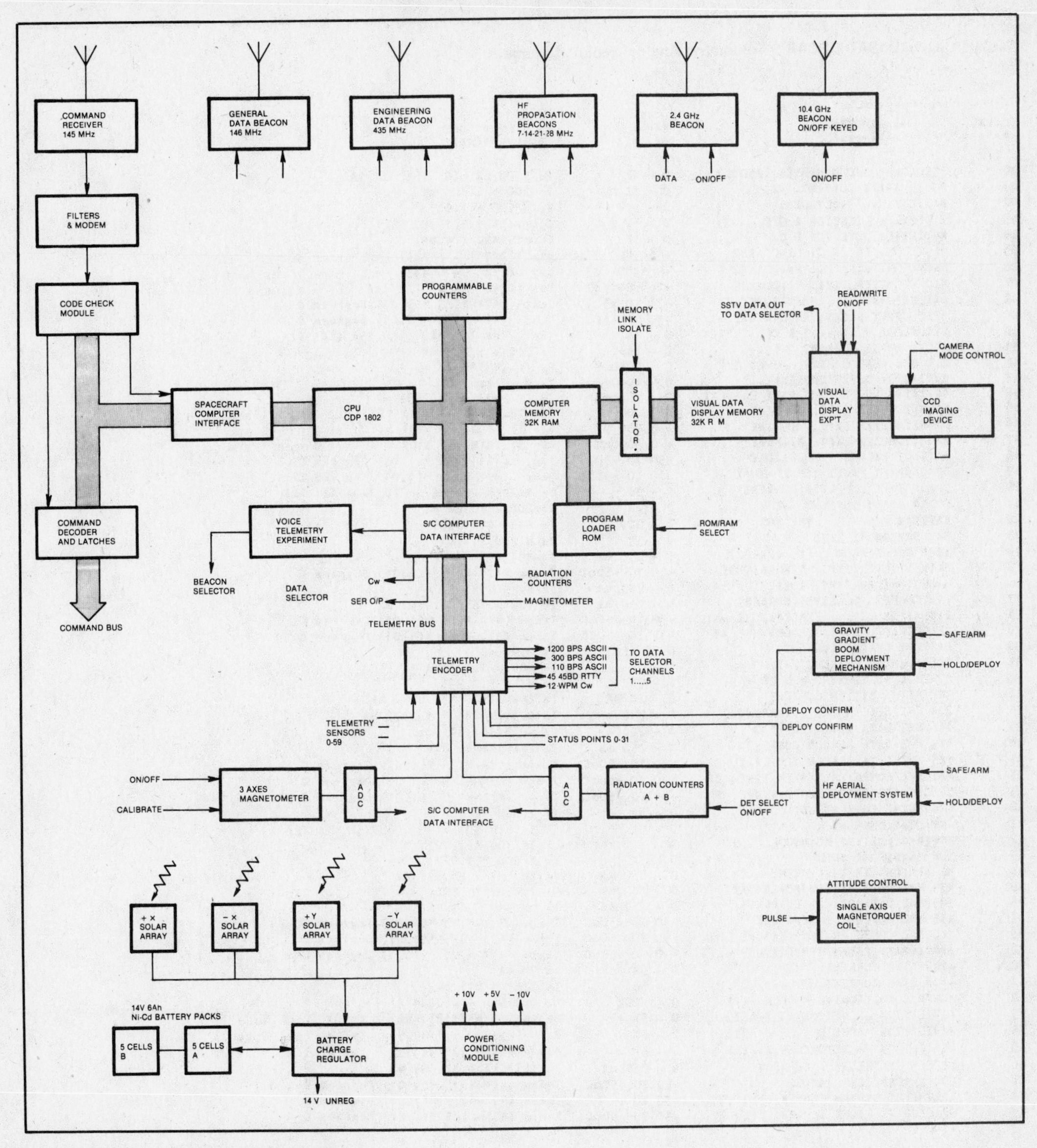

Fig. 2(U-O-9) — UoSAT-OSCAR 9 block diagram.

Note 2: The analog telemetry channels have an accuracy of 2%; the high current-measuring channels, however, suppress the least significant digit. It's possible to dwell on any selected analog channel when using 1200-baud ASCII.

Sample data: Because of U-O-9's flexibility, there is no "typical" frame. A sample frame, received on an early U-O-9 pass, is shown in Figure 3(U-O-9).

Data Encoding: Technical Considerations: The ASCII telemetry format is 1 start bit, 7 data bits, even parity, 3 stop bits. Data at 1200 baud (from telemetry, computer or video display experiment) is transmitted as a series of "ones" and "zeros" using phase-synchronous afsk. Phase-synchronous means that data transitions — both "1" to "0" and "0" to "1" — occur at zero crossings of the tone waveforms. This reduces the dc component of the data modulation spectrum.

exactly *one* complete cycle of 1200 Hz tone = logic "0"
exactly *two* complete cycles of 2400 Hz tone = logic "1"

Data at speeds other than 1200 baud are transmitted asynchronously using a 1200-Hz tone for a logic "1" and a 2400-Hz tone for a logic "0". For Morse code only, a 1200-Hz tone is employed.

UoSAT-OSCAR 9

Table 1(U-O-9) (UoSAT OSCAR 9 Telemetry: Analog Sensor Channels

CHANNEL	PARAMETER	RANGE	Equation
00	SECONDARY S/C COMPUTER (F100L)	0 - 1A	I= 1.2N mA (0.125A <I< 1A)
01	SOLAR ARRAY CURRENT +X	0 - 2A	I= 200 + 1.12N mA
02	BATTERY HALF VOLTAGE	0 - 10V	V= N/100 *(1.01)
03	RADIATION DETECTOR A O/P	0 - 5V	Count= 40N *(1.04)
04	RADIATION DETECTOR B O/P	0 - 5V	Count= 40N *(1.04)
05	MAGNETOMETER EXPT. HX-COARSE	+/-64,000 nT	Bx= (129N - 64824) nT (see Notes 1 and 2)
06	MAGNETOMETER EXPT. HY-COARSE	+/-64,000 nT	By= (129N - 64433) nT (see Notes 1 and 2)
07	MAGNETOMETER EXPT. HZ-COARSE	+/-64,000 nT	Bz= (129N - 64072) nT (see Notes 1 and 2)
08	BATTERY PACK-A TEMPERATURE	-30 TO +50oC	Temp= (474-N)/5 *(1.01) Degrees C
09	SPACECRAFT FACET TEMPERATURE +X	-30 TO +50oC	Temp= (474-N)/5 *(1.01) Degrees C
10	VISUAL DISPLAY EXPT & CCD CURRENT	0 - 1A	I= 1.2*(N-30) mA (0.15A <I< 1A)
11	SOLAR ARRAY CURRENT +Y	0 - 2A	I= 200 + 1.12N mA
12	2.4 GHz BEACON EXPT. POWER O/P	0 - 2000mW	P= (N-145)*0.45 mW
13	RADIATION DETECTORS EXPT. EHT VOLTS	0 - 1000V	V= N volts
14	RADIATION DETECTORS EXPT CURRENT	0 - 250 mA	I= (N+20)/8 *(0.983) mA
15	MAGNETOMETER EXPT. HX-FINE	+/-8,000 nT	Bx= 18.05(N - 511) nT (see Notes 1 and 2)
16	MAGNETOMETER EXPT. HY-FINE	+/-8,000 nT	By= 17.97(N - 510) nT (see Notes 1 and 2)
17	MAGNETOMETER EXPT. HZ-FINE	+/-8,000 nT	Bz= 17.76(N - 510) nT (see Notes 1 and 2)
18	BATTERY PACK-B TEMPERATURE	-30 TO +50oC	Temp= (474-N)/5 *(1.01) Degrees C
19	SPACECRAFT FACET TEMPERATURE -X	-30 TO +50oC	Temp= (474-N)/5 *(1.01) Degrees C
20	SPACECRAFT COMPUTER CURRENT	0 - 1A	I= 1.2*(N-25) mA (0.125A <I< 1A)
21	SOLAR ARRAY CURRENT -X	0 - 2A	I= 200 + 1.12N mA
22	BATTERY / BCR +14V BUS	0 - 20V	V= N/50 *(1.056)
23	SUN SENSOR +Z AXIS	0 - 5V	V= N/200 *(1.01)
24	10.4 GHZ BEACON EXPT. CURRENT	0 - 250 mA	(N-40)/4 * 0.97
25	MAGNETOMETER EXPT. TEMPERATURE	-30 TO +50oC	Temp= (474-N)/5 *(1.01) Degrees C
26	MAGNETOMETER EXPT. CURRENT	0 - 250 mA	(N/8)*0.9945
27	TELECOMMAND RECEIVER CURRENT	0 - 250 mA	I= (N-16)/8 *(0.952) mA
28	RADIATION EXPT. TEMPERATURE +X1	-30 TO +50oC	Temp= (474-N)/5 *(1.01) Degrees C
29	SPACECRAFT FACET TEMPERATURE +Y	-30 TO +50oC	Temp= (474-N)/5 *(1.01) Degrees C
30	BATTERY CHARGE CURRENT	0 TO +5A	I= 2.9N mA
31	SOLAR ARRAY CURRENT -Y	0 - 2A	I= 200 + 1.12N mA
32	POWER CONDITIONING MODULE +10V	0 - 20V	V= N/60 *(0.93)
33	TELEMETRY SYSTEM CURRENT	0 - 20 mA	I= (N-16)/30 *(1.084) mA
34	2.4 GHZ BEACON EXPT. CURRENT	0 - 250 mA	I= 0.4*(N-11) *(1.072) mA
35	145 MHZ DATA BEACO POWER O/P	0 - 2000mW	P= (N-82)*1.67
36	145 MHZ DATA BEACON CURRENT	0 - 250 mA	I= (N-7)/4 *1.014
37	145 MHZ DATA BEACON TEMPERATURE	-30 TO +50oC	Temp= (474-N)/5 *(1.01) Degrees C
38	PRI.S/C COMPUTER TEMPERATURE -X1	-30 TO +50oC	Temp= (474-N)/5 *(1.01) Degrees C
39	SPACECRAFT FACET TEMPERATURE -Y	-30 TO +50oC	Temp= (474-N)/5 *(1.01) Degrees C
40	+14V LINE CURRENT	0 - 5A	I= 2.86N mA
41	+5V LINE CURRENT	0 - 5A	I= 1.28(N-50) mA (0.075A <I< 1A)
42	POWER CONDITIONING MODULE +5V	0 - 10v	V= 2N/300 *(1.12)
43	SUN SENSOR -Z AXIS	0 - 5V	V= N/200 *(1.01)
44	HF BEACONS EXPT. CURRENT	0 - 250 mA	I= (N-36)/3 *1.038 mA
45	435 MHZ DATA BEACON POWER O/P	0 - 2000mW	P= (N-102)*1.792
46	435 MHZ DATA BEACON CURRENT	0 - 250 mA	I= (N-34)/3 *1.053 mA
47	435 MHZ BEACON TEMPERATURE	-30 TO +50oC	Temp= (474-N)/5 *(1.01) Degrees C
48	SEC.S/C COMPUTER TEMPERATURE +Y1	-30 TO +50oC	Temp= (474-N)/5 *(1.01) Degrees C
49	SPACECRAFT FACET TEMPERATURE +Z	-30 TO +50oC	Temp= (474-N)/5 *(1.01) Degrees C
50	+10V LINE CURRENT	0 - 5A	I= 3N mA
51	-10V LINE CURRENT	0 - 5A	I= 1.3*(N-60) mA
52	POWER CONDITIONING MODULE -10V	0 - -20V	V= 0.0158N - 0.0224'N' ('N' of +10v line)
53	NAVIGATION MAGNETOMETER Y-AXIS	0 - 5V	By=(N-663.44)*183.486 (nT) (see Note 2)
54	NAVIGATION MAGNETOMETER Z-AXIS	0 - 5V	Bz= -(N-336.55)*189.54 (nT) (see Note 2)
55	NAVIGATION MAGNETOMETER X-AXIS	0 - 5V	Bx= -(N-496.5)*194.55 (nT) (see Note 2)
56	SPEECH SYNTHESISER CURRENT	0 - 250 mA	I= (N-16)/10 *1.009 mA
57	CCD IMAGER TEMPERATURE	-30 TO +50oC	Temp= (474-N)/5 *(1.01) Degrees C
58	TELEMETRY SYSTEM TEMPERATURE -Y1	-30 TO +50oC	Temp= (474-N)/5 *(1.01) Degrees C
59	SPACECRAFT FACET TEMPERATURE -Z	-30 TO +50oC	Temp= (474-N)/5 *(1.01) Degrees C

1. $B_x = B_x(\text{coarse}) - B_x(\text{fine})$ = ch. 5 - ch. 15
   $B_y = B_y(\text{coarse}) - B_y(\text{fine})$ = ch. 6 - ch. 16
   $B_z = B_z(\text{coarse}) - B_z(\text{fine})$ = ch. 7 - ch. 17

2. $B(\text{total}) = \sqrt{B_x^2 + B_y^2 + B_2^2}$

Data Decoding: Technical Considerations: The most common method for decoding the ASCII telemetry is to feed the ground-station receiver audio output into a device called a *modem* (*mod*ulator/*dem*odulator) which is in turn connected to a micro-computer or video display terminal. A Bell-202 type modem must be used to decode the telemetry. Note that most computer hobbyist modems are not of this variety. Details for constructing suitable modems are contained in the *UoSAT-OSCAR 9 Technical Handbook.* Most "dumb terminals" or popular personal micro-computers can be used for the video display.

Table 2(U-O-9)
UoSAT OSCAR 9 Telemetry: Digital Status Channels

STATUS POINTS		'0' '1'
01	145 MHZ GENERAL DATA BEACON	OFF/ON
02	435 MHZ ENGINEERING DATA BEACON	OFF/ON
03	PRIMARY SPACECRAFT COMPUTER	OFF/ON
04	CCD CAMERA MODULE	OFF/ON
.05	RADIATION DETECTOR — A	OFF/ON
06	MAGNETOMETER EXPT.	OFF/ON
07	7 MHZ BEACON EXPT.	OFF/ON
08	14 MHZ BEACON EXPT.	OFF/ON
09	28 MHZ BEACON EXPT.	OFF/ON
10	21 MHZ BEACON EXPT.	OFF/ON
11	2.4 GHZ BEACON EXPT.	OFF/ON
12	10.47 GHZ BEACON EXPT.	OFF/ON
13	145 MHZ COMMAND RX	SQUELCH 0 = signal present
14	435 MHZ COMMAND RX	SQUELCH 0 = signal present
15	STATUS CALIBRATE	
16	BCR STATUS	A/B
17	HF BEACONS EXPT. SYNTHESIZERS	OFF/ON
18	TELECOMMAND DECODER STATUS	GROUND/PRIMARY COMPUTER
19	MAGNETORQUER	OFF/ON
20	PRIMARY S/C COMPUTER BLOCK LOAD PORT	DISABLE/ENABLE
21	SECONDARY S/C COMPUTER DATA O/P	ACTIVE/INACTIVE
22	SECONDARY S/C COMPUTER CLOCK	ACTIVE/INTERRUPT FAILURE
23	SECONDARY S/C COMPUTER PROCESSOR	OFF/RUNNING
24	SECONDARY S/C COMPUTER POWER-DOWN	ON/OFF
25	14 MHZ HF BEACON SYNTHESIZER LOCK	OUT/IN
26	28 MHZ HF BEACON SYNTHESIZER LOCK	OUT/IN
27	21 MHZ HF BEACON SYNTHESIZER LOCK	OUT/IN
28	RADIATION DETECTOR — B	OFF/ON
29	TIP MASS UNCAGING CONFIRMATION	NO/YES
30	SPEECH SYNTHESIZER POWER	OFF/ON
31	VISUAL DATA DISPLAY MEMORY	OFF/ON
32	GRAVITY GRADIENT. BOOM MOTOR POWER	OFF/ON
33	SECONDARY S/C COMPUTER POWER	OFF/ON
34	HF BEACONS EXPT. POWER	OFF/ON
35	NAVIGATION MAGNETOMETER POWER	OFF/ON
36	S/C COMPUTER MEMORY ERROR BIT – 1	
37	S/C COMPUTER MEMORY ERROR BIT – 2	
38	S/C COMPUTER MEMORY ERROR BIT – 3	
39	STATUS CALIBRATE	
40	PRIMARY S/C COMPUTER DATA UART O/P	INACTIVE/ACTIVE
41	GRAVITY GRADIENT BOOM MOTOR	FORWARD/REVERSE
42	MAGNETORQUER POWER	FORWARD/REVERSE
43	MAGNETOMETER EXPT.	MEASURE/CALIBRATE
44	NAVIGATION MAGNETORQUER	SAFE/ARM
45	GRAVITY GRADIENT BOOM MOTOR	SAFE/ARM

Fig. 3(U-O-9)

Sample UoSAT-OSCAR 9 Telemetry Frame.
date: 8 Oct. 1981
time: 0810 UTC
source: Gordon Hardman, ZS1FE/KE3D

AMSAT	10000	00000	00100	00000	01110	00000	00001	00000	00000	row 1 (status)
AMSAT	10000	00000	00110	00000	01110	00000	00001	00000	00000	row 2 (status)
00110	01040	02689	03000	04000	05000	06000	07000	08509	09506	row 0 (analog)
10100	11090	12000	13000	14025	15000	16000	17000	18508	19505	row 1 (analog)
20110	21090	22659	23000	24004	25000	26004	27287	28512	29508	row 2 (analog)
30000	31060	32665	33344	34012	35268	36363	37440	38513	39504	row 3 (analog)
40000	41090	42742	43300	44043	45000	46002	47496	48513	49521	row 4 (analog)
50000	51090	52272	53210	54901	55318	56012	57499	58510	59484	row 5 (analog)
col. 0	col. 1	col. 2	col. 3	col. 4	col. 5	col. 6	col. 7	col. 8	col. 9	

The first two rows are status data. Each row reports sequentially, from left to right, on the 45 status points. The change in bit 14 indicates that command receiver squelch has been broken (see Table 2(U-O-9)).

The remaining entries in the frame contain analog data. Each analog channel consists of a five-digit number. The two left-hand digits, 00 to 59, uniquely identify the parameter being measured. The remaining three digits encode the measured value (see Table 1(U-O-9)).

3.3 Telecommand System
Design/Construction Credits: Dr. Martin Sweeting, G3YJO, UoS/AMSAT-UK

Two modes of control over the spacecraft are available, with a repertoire of 66 latched, two-state commands:

1) Direct, real-time control of the spacecraft's functions by ground command stations using one of two redundant vhf/uhf command receivers.

2) Indirect, stored-program control executed by one of the two onboard microcomputers according to a "diary" loaded in advance from a ground command station via the telecommand uplink.

Any valid command data received from the ground stations will override any command data simultaneously issued by the on-board microcomputers. The primary computer (RCA 1802) has precedence over the secondary computer (F100L), unless otherwise instructed from the ground.

The Telecommand uplinks also carry high speed data to enable program software and data to be loaded into the on-board microcomputers.

3.4 Transponders

UoSAT-OSCAR 9 does *not* carry any real-time transponders. Nonetheless, a store-and-forward type transponder can be implemented on the telecommand uplink and the spacecraft beacons by using the spacecraft computer or image storage capabilities.

3.5 Attitude Stabilization and Control

Initial orientation of the satellite will be achieved using two magnetorquer coils. The coils, mounted on the +Y and −Y spacecraft (s/c) axes, can produce a magnetic field of approximately 50 A-turns/m^2 (equivalent to 50,000 pole-cm). The interaction of the spacecraft-produced magnetic field with the earth's magnetic field permits a maximum acceleration of the s/c of 0.01°/sec^2.

After preliminary orientation is completed, the passive gravity-gradient stabilization system takes over. The gravity-gradient system employs an 80-m long boom that can be deployed only after initial stabilization has taken place, and a 2.5-kg tip mass. When the boom has been extended the s/c −Z facet (bottom) will constantly point towards the geocenter, important for the camera and the shf and microwave beacon experiments. Small, undesired oscillations (nutation and libration) of the s/c Z axis about the local vertical will be damped by intermittent use of the magnetorquer coils. The s/c will be spun around the Z axis at a very slow rate, about 0.01 rpm, to prevent localized heating.

Attitude sensors include the Navigation Magnetometer (described later), solar cells mounted on the s/c +Z and −Z facets, and the solar panels.

3.6 Antennas
Design/Construction Credits: Tony Brown, UoS/AMSAT-UK; Dr. Mike Underhill, P.R.L., UK

7-, 14-, 21- and 28-MHz Beacons: Center-fed 120° apex angle "V" dipole having 2.5-m arms. Fed via a narrow-band matching network and inductively coupled to the 16-m-long stabilization boom. 145-MHz Beacon: Canted turnstile using quarter wavelength elements. Fed via coaxial hybrid. Produces approximately 3 dBi gain along −Z axis.

435-MHz Beacon: Uses same antenna and matching system employed for 146-MHz beacon, operating in overtone mode. Produces about 5 dBi gain along −Z axis.

2.4-GHz Beacon: 3.5-turn helix. Produces about 6.5 dBi gain along −Z axis.

10.47-GHz Beacon: 4-turn slot helix. Produces about 8 dBi gain along −Z axis.

Antenna Polarization: See Table 3(U-O-9)

3.7 Energy Supply and Power Conditioning
Design/Construction Credits: Dr. Karl Meinzer, DJ4ZC, University of Marburg, AMSAT-DL; Jerzy Slowikowski, UoS/AMSAT-UK

Solar Cell Characteristics
- type: n on p silicon
- size: 2 cm by 2 cm
- total number: 1632
- total surface area: 6528 cm^2
- efficiency: 12.5%

Solar Cell Configuration: The s/c uses four solar arrays, each consisting of 408 cells in a 17 × 24 arrangement. The arrays are mounted on the +X, −X, +Y, and −Y facets of the spacecraft. (Most photographs of the spacecraft were taken during prelaunch testing. They misleadingly show a solar cell array having varying characteristics.)

Solar arrays
Fabricated by: Solarex Corp.
Power: Output per array (408 cells) is 28 watts at a nominal 32 volts when fully illuminated. The total average power available from the arrays, allowing for sun angle and eclipse periods, is about 17 watts.

Table 3(U-O-9)

UoSAT-OSCAR 9 Antenna Polarizations

System	*Antenna Polarization*
7-, 14-, 21-, and 28-MHz beacons	linear
145-MHz beacon	left-hand circular
435-MHz beacon	left-hand circular
2.4-GHz beacon	left-hand circular
10.47-GHz beacon	left-hand circular

Storage Battery: 14-V, 6-Ah, 10-cell NiCd

Power Conditioning: See Fig. 2(U-O-9) for configuration.

Battery Charge Regulator (BCR): two redundant units; each regulates the solar-array power to the NiCd battery with an operating efficiency of approximately 90%.

Power Conditioning Module (PCM): delivers regulated outputs of +10V (1%), −10V (5%), +5V (5%). Total capacity of 10 watts. Average efficiency is about 87%.

Power Budget: The average continuous power budget available to the s/c electronics from battery bus and PCM is about 11.5 watts.

General: The spacecraft consumes around 9.8 watts from the PCM when all experiments are operational with a further 10.5 watts from the unregulated battery bus. Power is distributed around the spacecraft through a central Power Distribution Module that, under the control of the Command System, provides switched power supplies to the various experimental and service modules whilst also allowing central telemetry monitoring facilities. The power switches exhibit resettable current fold-back in the event of malfunction.

3.8 Propulsion System: The s/c does *not* contain an onboard propulsion system.

3.9 Integrated Housekeeping Unit (Spacecraft Microcomputer)
Design/Construction Credits:
Hardware: Tony Jeans, G8ONO, Chris Haynes, UoS/AMSAT-UK
High Level Software: Dr. Karl Meinzer, DJ4ZC, Univ. of Marburg/AMSAT-DL; Robin Gape, Chris Trayner, AMSAT-UK

General: There are two powerful on-board microcomputers that have access to the s/c experiments telemetry and command systems, enabling
1) telemetry surveillance and command & status management;
2) experiment data storage & processing;
3) dissemination of orbital data, operating schedules & spacecraft "news";
4) closed-loop attitude control using the magnetorquers.

The primary s/c computer is based around the RCA 1802 microprocessor and supports 8 parallel ports, 2 serial ports and 16k bytes of dynamic RAM memory with access to a further 32k bytes of dynamic RAM memory in the Video Display Experiment. The parallel ports interface directly to the Telemetry & Command systems and to the Radiation, Magnetometer and Speech Synthesizer experiments allowing high-speed sampling of data. The two serial ports provide redundant data paths and can also generate a wide range of data formats & rates available to the Data Beacons. This computer supports the multi-tasking software system IPS (developed by K. Meinzer) and provided a useful opportunity to evaluate IPS before the launch of AMSAT-OSCAR 10.

The secondary s/c computer is based around the Ferranti F100L microprocessor and is configured as a minimal system with serial interfaces to the s/c telemetry and command systems. This does, however, allow the computer less direct but complete access to the s/c systems. The computer has 2 serial input/output ports and is supported with 32k bytes of CMOS static RAM. The F100L is a 16-bit machine.

The software and accompanying data for both computers are loaded from the ground via the telecommand link and can be modified or replaced during flight by a ground command station to accommodate changes in the mission profile and to allow for rectifying possible inflight software or hardware failures.

3.10A Navigation Magnetometer Experiment
Design/Construction Credits: Dr. Mario Acuna, LU9HBG, AMSAT-USA; Christine Sweeting, G6APF, UoS/AMSAT-UK

General: A three-axis, flux-gate magnetometer mounted on the upper (+Z, +X) facet of the s/c wing is designed to provide information on the orientation of the s/c in orbit by the comparison of measured earth magnetic-field vectors with existing models. It is anticipated that the navigation magnetometer will be able to determine the orientation of the s/c to within 2 degrees. Solar cells mounted on the top and bottom (+Z & −Z) facets of the s/c resolve the up/down ambiguity. The data from the magnetometer is available in real time through the telemetry system.

3.10B Magnetometer Experiment

Design/Construction Credits: Dr. Mario Acuna, LU9HBG, AMSAT-USA

General: A three-axis, multi-range, flux-gate magnetometer on the s/c allows the detection and monitoring of geomagnetic storms and their possible affect on radio propagation. The magnetometer will also be used to study and map the earth's main magnetic field, thus providing amateurs with advanced diagnostic and study capabilities.

Special emphasis has been placed on the acquisition of real-time and stored data over the polar regions.

The basic dynamic range of the magnetometer instrument is ±8000 nT and the output is digitized by a 12-bit A/D converter. Since the strength of the geomagnetic field is approximately 30,000 nT at the equator and 60,000 nT at the poles, the basic range of the magnetometer is increased to 64,000 nT by biasing the zero level in 16 steps.

Dynamic range: ±8000 nT
Resolution: ±2 nT
Zero level stability
sensors (−60°C to +60°C): ±5 nT
electronics (−20°C to +50°C): ±2 nT
Linearity errors: less than 2 × 10^5
Bias Field Generator
dynamic range: ±64,000 nT
quantization step: 8000 nT
temperature coefficient: 2 ppm/°C
Power consumption: 500 mW

Two outputs ("coarse" and "fine") are presented for each axis and the full resolution data are available to the primary s/c computer. Quick-look data are reduced to 10-bit resolution and presented to the analog telemetry system with a resulting maximum resolution of ±8 nT.

Reference: For a more detailed description of the magnetometer experiment see: M. Acuna, LU9HBG, "UoSAT Magnetometer," *Orbit,* Vol. 2, no. 4, Aug./Sept. 1981, pp. 6-10

3.10C CCD Camera Imaging Experiment

Design/Construction Credits: Dr. Paul Traynar, UoS/AMSAT-UK

General: A two-dimensional, charge-coupled device (CCD) imaging array (GEC MA357) is mounted in the bottom (−Z) of the s/c central column that, using the gravity-gradient stabilization mechanism, should point towards the center of the earth and provide images of land, sea and cloud cover over a 500- × 500-km area of the earth's surface. The image is formed by integrating the amount of light falling on the 65,536 light sensitive 'buckets' of the array over a set period of time and then transferring the resulting accumulated charge into a similar, masked storage area alongside. The 'buckets' are organized in a 256 × 256 matrix. The integration time of the CCD is under ground control via the command system and can be set to any of 16 preset periods between 4 ms and 16 ms. The spectral response of the CCD is in the visible/red range and should give good haze penetration.

The charge 'image' in the CCD storage area is then digitized into 4-bit words (each word representing a pixel), and transferred once more to a long-term memory in the Video Display Experiment (VDE) module. The data now resident in the VDE memory can be transmitted to ground stations at 1200 bps (phase-synchronous afsk) through the General or Engineering Data Beacons. The image data is transmitted in a line synchronous manner, that is, 256 × 4 bits are sent (representing one line of image) in one continuous stream preceded by a 'line sync' bit pattern comprising a 32-bit code sequence. The 32-bit code itself comprises an 8-bit word and its one's complement that are repeated twice.

The complete image dump takes approximately 3.5 minutes from the s/c and includes:
1) a frame header comprising one line of 16-line sync codes;
2) 256 lines of 1024 bits (organized as 256 × 4 bits) each preceded by the line sync code:
01011011 10100100 01011011 10100100

The primary s/c computer has direct access to the VDE memory and it may be possible to carry out on-board image processing and annotation. It is also possible to load grahic data into the VDE via the ground telecommand link for later rebroadcast.

Lens characteristics:
focal length: 6.5 mm
speed: 1:18
aperture: 1.5 mm f/4
neutral density filter: 1/32
Field of view: 60 degrees
CCD intensity dynamic range: 35 dB
CCD vertical transfer clock rate: 6.6 MHz, 3-phase

Table 4(U-O-9)
Radiation Detector Characteristics

	LND705 (Tube 1)	*LND710* (Tube 2)
Approximate energy threshold	20keV	60keV
Window thickness	0.35 ± 0.05 mg/cm²	1.75 ± 0.25 mg/cm²
Geometric factor	0.08 mm²ster	0.35 mm²ster
Collimator diameter	8.0 mm	8.0 mm
Angle to s/c Z-axis	13°	18°
Filling gas	neon + halogen	neon + halogen
Operating temperature	−50°C to +150°C	−50°C to +150°C
Sampling rates:		
stored data	10 per second	10 per second
real time data	1 per 8 seconds	1 per 8 seconds

Reference: For additional information on the UoSAT Radiation Detectors Experiment see "Tech. Brief: UoSAT Radiation Counters," *ASR,* Vol. 1, no. 17, Oct. 5, 1981, pp. 1-2.

CCD frame transfer clock rate: 2 MHz
Power consumption: Imaging — 1.5 watts (for one second)
Store/readout — 2.8 watts continuous

Image Display: The *UoSAT-OSCAR 9 Technical Handbook* contains plans for an image display unit. See: "A Decoder & Display for UoSAT-OSCAR-9 Camera Pictures," by T. R. Stockill, G4GPQ, pp. 19-37. Printed circuit boards for a modified version of this unit will be made available by AMSAT-UK. Check recent issues of *Orbit* for details.

3.10D Radiation Detector Experiment

Design/Construction Credits: D. R. Lepine, Appleton Laboratories, UK; Ian Ferebee, G6BTU, UoS/AMSAT-UK

General:

The radiation monitoring experiment employs two LND type Geiger-Mueller tubes to measure integrated fluxes of electrons above threshold energies of approximately 20 and 40 keV. The tubes have thin mica end-windows of thickness 0.35±0.05 mg/cm² and 1.75±0.25 mg/cm² respectively, and are filled with neon and a small quantity of halogen to provide quenching. In addition to detecting electrons the tubes also detect protons of approximately twenty-times-higher energy.

Each tube is contained in a separate housing that also contains a thick-film pulse-amplifier/pulse-shaper to provide 10-V, 50-μs pulses to the on-board data handling system. A single high-voltage converter, generating 560 V and stabilized to ±20 V (−40°C to +60°C), provides the anode supply for the tubes. A collimator consisting of two circular apertures separated by an 8-mm spacer, is located in front of the tube to define the geometry-factor. (See Table 4(U-O-9.)

Data from the experiment will be telemetered to ground using two separate formats. High time-resolution data, where each detector is sampled for ten 0.05-s periods every 0.1s, will be stored by the on-board primary microcomputer and then transmitted to the ground 'on command' using one of the general or engineering data beacons. It is proposed to schedule the experiment so that the data accumulation phase corresponds to the satellite passing over the more interesting charged particle precipitation regions, i.e. the auroral oval and polar caps. The exact quantity of data that can be stored during a pass has still to be determined. In addition to high time-resolution data for the computer, the experiment also averages the count in every 5-s period and makes the result available to the telemetry system where it is transmitted in real time.

Instruments of this type have been used by the RAL Magnetospheric-Plasmas Group on several sounding-rocket flights to measure the intensities of electrons producing auroral displays.

3.10E Speech Synthesizer Experiment

Design/Construction Credits: UoS/AMSAT-UK

General: A 120-word speech synthesizer based on the National Semiconductor "Digitalker" integrated circuit is under the control of the primary on-board computer via a high-speed 14-bit parallel port. Thus, telemetry s/c status and programs, orbit ephemeris data and general s/c news can be encoded in 'English' and relayed via the General, Engineering or 2.4-GHz beacons using nbfm

Power consumption: 2.5 Watts (speaking), 0.25 watts (standby)

Radio Sputnik

SPACECRAFT NAMES:

Radio 3	Radio 5	Radio 7
Radio 4	Radio 6	Radio 8

NOTE: These six Soviet Amateur Radio satellites were launched together on a single launch vehicle. All six are profiled in this section.

GENERAL

1.1 Identification

Satellite	International designation	Telemetry identification
Radio 3	81-120A	RS-3
Radio 4	81-120D	RS-4
Radio 5	81-120C	RS-5
Radio 6	81-120F	RS-6
Radio 7	81-120E	RS-7
Radio 8	81-120B	RS-8

1.2 Launch
Date: 17 December 1981
Site: Pletsetsk, USSR

1.3 Orbital Parameters
General designation: low-altitude
Inclination: 82.95° ± 0.05° (near polar)
Maximum access distance: 4200 km

Satellite	Period (minutes)	Apogee Altitude (km)	Perigee Altitude (km)	Eccen-tricity	Longitude Increment (°W/orbit)
Radio 3	118.46	1688	1577	0.007	29.76
Radio 4	119.34	1692	1641	0.003	29.98
Radio 5	119.50	1690	1653	0.002	30.02
Radio 6	118.66	1691	1593	0.006	29.81
Radio 7	119.14	1689	1634	0.003	29.93
Radio 8	119.71	1693	1657	0.002	30.07

1.4 Ground Track Data: See Appendix B

1.5 Operations
Coordinating Group: Radio Amateur Satellite Committee
Radio Sport Federation
Box 88
Moscow, USSR
Schedule: Wednesday (UTC): Transponder and autotransponder (ROBOT) reserved for special experiments arranged in advance with USSR Radio Amateur Satellite Committee.

1.7 Primary References
Radio 1 and Radio 2 received extensive coverage in *Radio*, a Soviet magazine for radio-electronics experimenters. It's expected that similar coverage will be provided for Radio 3 through Radio 8. Because of the time lag involved in the initial publication, the translation from Russian to English and republication, however, formal information on these spacecraft was not available as this is written. As a result, most of the following technical information must be regarded as tentative. For background information on the Russian amateur satellite program see:

L. Labutin (UA3CR), "The USSR *Radio* Satellites," *Telecommunication Journal*, Vol. 46, no. X, Oct. 1979, pp. 638-639. This report (in English), based on an article published in *Radio*, May 1979, pp. 7-8, summarizes the results of the USSR's first two amateur s/c, Radio 1 and Radio 2.

SPACECRAFT DESCRIPTION

2.1 Physical Structure: Information not available at presstime.

2.2 System Description
General: Each s/c contains two general purpose beacons plus additional equipment.
Radio 3 & Radio 4: These spacecraft are experimental in nature. They do *not* contain either transponders or autotransponders. Details of the experiments will be published in *Radio* at some future date.
Radio 5 & Radio 7: Each spacecraft contains one transponder and one autotransponder.
Radio 6 & Radio 8: Each spacecraft contains one transponder. These s/c do *not* contain autotransponders.

SUBSYSTEM DESCRIPTION

3.1 Beacons
General: Each spacecraft contains two beacons. Usually, only one is operated at any given time but, at least with Radio 5 and Radio 7, both can operate concurrently. Frequencies are as follows:

Satellite	Frequency (MHz) Beacon #1 (0.5-1.5 W)	Frequency (MHz) Beacon #2 (0.1-0.3 W)
Radio 3	29.321	29.401
Radio 4	29.360	29.403
Radio 5	29.331	29.452
Radio 6	29.411	29.453
Radio 7	29.341	29.501
Radio 8	29.461	29.502

Notes
1) Either beacon may be used for Morse code telemetry
2) Either beacon may be used for autotransponder if one is contained on the spacecraft.
3) When a transponder is active the upper frequency (lower power) beacon is generally in operation. The presence of the upper frequency beacon does *not* necessarily imply that the transponder is on.

Table 1(RS)
Sample RS Morse Code Telemetry

RS6	K33	D78	O84	G00	U21	S18	W16
RS6	IK30	ID00	IO38	IG09	IU15	IS00	IW00
RS6	AK24	AD89	AO75	AG90	AU75	AS90	AW75
RS6	MK36	MD00	MO00	MG80	MU08	MS00	MW09

Satellite: RS6
date: 12 Jan. 1982
time: 03:16 UTC
frequency: 29.453 MHz
transponder: on
speed: about 25 wpm
source: K2UBC
ground station QTH: Baltimore, MD

Table 2(RS)
RS Telemetry

Channel	Parameter	Equation
K or EK	transponder output power	$0.2 \times N^2$ (mW)
D or ED	battery voltage	$0.2 \times N$ (Volts)
O or EO	battery charge current	$20 \times (100 - N)$ (mA)
G or EG	telemetry calibration marker	
U or EU	?	
S or ES	temperature of main power regulator	$T = N$ (°C)
W or EW	temperature of 10m tx cooling fins	$T = N$ (°C)
IK or SK	transponder output power	$0.2 \times N^2$ (mW)
ID or SD	telemetry zero level	
IO or SO	beacon output power	$0.2 \times N^2$ (mW)
IG or SG	transponder sensitivity	
IU or SU	transponder 'S' meter	$0.1 \times (N - 10)$ ('S' units)
IS or SS	Robot 'S' meter	$0.1 \times (N - 10)$ ('S' units)
IW or SW	command receiver 'S' meter	$0.1 \times (N - 10)$ ('S' units)
NK or RK	transponder output power	$0.2 \times N^2$ (mW)
ND or RD	solar panel current	$50 \times N$ (mA)
NO or RO	temperature of solar panel 1	$2.7 \times (N - 26)$ (°C)
NG or RG	temperature of solar panel 2	$2.7 \times (N - 26)$ (°C)
NU or RU	temperature of solar panel 3	$2.7 \times (N - 26)$ (°C)
NS or RS	temperature of structure	$0.8 \times (N - 5)$ (°C)
NW or RW	temperature of hermetically sealed casing	$0.8 \times (N - 5)$ (°C)
AK or UK	transponder output power	$0.2 \times N^2$ (mW)
AD or UD	9 V transponder line	$0.1 \times N$ (Volts)
AO or UO	7.5 V transponder line	$0.1 \times N$ (Volts)
AG or UG	9 V regulator #1	$0.1 \times N$ (Volts)
AU or UU	7.5 V regulator #1	$0.1 \times N$ (Volts)
AS or US	9 V regulator #2	$0.1 \times N$ (Volts)
AW or UW	7.5 V regulator #2	$0.1 \times N$ (Volts)
MK or WK	transponder output power	$0.2 \times N^2$ (mW)
MD or WD	autotransponder log	N = number of QSOs ±1
MO or WO	thermal control system heater	$0.1 \times N$ (watts)
MG or WG	input power (Robot or transponder?)	$20 \times N$ (mW)
MU or WU	command unit power	$20 \times N$ (mW)
MS or WS	input attenuator (Robot)	N (dB)
MW or WW	input attenuator (transponder rx)	N (dB)

This table is based on unofficial prelaunch information and must be regarded as highly tentative. It's likely that there will be minor differences between the six spacecraft. N is the two-digit number contained in channel.

Serious experimenters may wish to refer to an article by UA3CR on the Radio 1 and Radio 2 telemetry systems: L. Labutin, *Radio*, March 1979, pp. 18-19. The presence of the extra "dit" in the prefix (changing I to S, A to U, etc.) indicates that the command channel is active.

Radio Sputnik

3.2 Telemetry

General: The following information on the RS telemetry system must be regarded as highly tentative. The spacecraft telemetry systems usually employ Morse code.

frame: A frame usually consists of 28 channels (4 lines by 7 columns) sent in a fixed serial format. Each line begins with RSn where 'n' identifies the spacecraft. Sometimes the s/c dwells on a single line.

sample data: See Table 1(RS)

channel: Each channel consists of a two-letter prefix followed by two digits. The first letter (which is sometimes a blank) indicates channel row (see note 1); the second letter indicates channel column; and the two digits encode the measured parameter.

decoding information: See Table 2(RS)

Notes:

1) Consider the first prefix letter. Two alternate sets are used: [blank, I, A, M] or [E, S, U, W]. The first set can be transformed into the second set by inserting a "dit" in front of the Morse code character representing each letter. The significance of the two sets is not known at this time.

2) Consider the Morse code character representing the second prefix letter. All characters contain three units. Substituting 1 and 0, or True and False, for the "dahs" and "dits" yields a binary sequence that's probably familiar to most amateurs who've worked with digital logic.

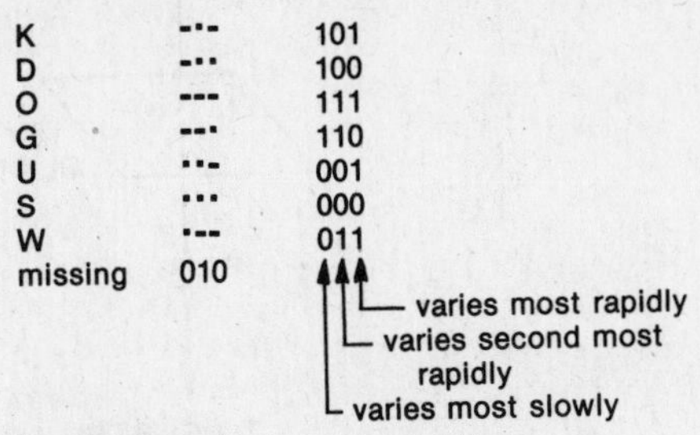

3.3 Telecommand System

Through observations of the RS satellites' operation and published information about their telemetry systems we infer that these spacecraft have a flexible telecommand system whose capabilities include

1) switching major systems (transponder, autotransponder, other) on/off;
2) inserting attenuators in the inputs of the transponders and autotransponders;
3) interconnecting each beacon with various internal systems including transponder, autotransponder, autotransponder memory, codestore, memory dump and so on.

3.4 Transponders

Communications Transponders

Satellite	*uplink passband (MHz)*	*downlink passband (MHz)*
Radio 5	145.910-145.950	29.410-29.450
Radio 6	145.910-145.950	29.410-29.450
Radio 7	145.960-146.000	29.460-29.500
Radio 8	145.960-146.000	29.460-29.500

type: Mode A (2m/10m), linear, non-inverting

translation equation:

downlink freq. (MHz) = uplink freq. (MHz) − 116.495 ± Doppler

output power: 1-2 watts PEP

uplink eirp: 20 watts suggested (do *not* exceed 80 watts)

bandwidth: 40 kHz

maximum Doppler: 3.6 kHz

Autotransponders

The autotransponders (also known as Robots) aboard Radio 5 and Radio 7 are devices which enable you to "contact" the satellite. If you call the spacecraft using the correct protocol, an onboard computer will (1) acknowledge your call, (2) assign you a serial contact number, and (3) store your call letters and contact number for later downlinking when queried by a command station.

Link frequencies follow:

Satellite	*uplink (MHz) ± Doppler*	*primary downlink (MHz)*	*secondary downlink (MHz)*
Radio 5	145.826	29.331	29.452
Radio 7	145.835	29.341	29.501

The uplink window is only 2 to 3 kHz wide centered on the frequency indicated. Be sure to take Doppler into account by transmitting ≈2 kHz low when s/c is rapidly approaching you and ≈ 2 kHz high when s/c is rapidly receding.

The following procedure should be used for contacting the Robot. When it is active (calling CQ) send a few dits on the uplink frequency (*only* a few!). If you hear your dits regenerated on the downlink you're in the capture window. Call the satellite (10 to 30 wpm) as follows:

RS5 DE KA1GD $\overline{\text{AR}}$

If you're successful Radio 5 will respond

KA1GD DE RS5 QSO NR IJK OP ROBOT TU FR QSO 73 $\overline{\text{SK}}$

The letters IJK represent a 3-digit QSO number that is incremented after each contact.

Please do *not* hold your key down on the Robot input frequency as this will simply cause the downlink to generate a continuous tone. Each Robot calls CQ about once per minute when active. If only a partial message is received by the Robot you may hear a response of QRZ, QRM or RPT. In this case just try again. If the Robot wants you to send faster or slower it will respond QRQ or QRS. Clean, high-speed cw usually works best, probably because interference is less likely to be a problem.

A memory dump of Radio 7 listed the first 10 autotransponder QSOs:

00 UK3ACM	05 G3IOR
01 UV3FL	06 G4HUV
02 RS3A	07 G3IOR
03 UA3XBU	08 UK1BI
04 UI8BF	09 KA1GD

AMSAT-OSCAR 10

SPACECRAFT NAME: AMSAT-OSCAR 10

GENERAL

1.1 Identification
International designation: 1983 058 B
Pre-launch designation: AMSAT Phase IIIB, φ3B

1.2 Launch
date: 16 June 1983
vehicle: Ariane
agency: European Space Agency (ESA)
site: Kourou, French Guiana

1.3 Orbital Parameters
General designations: high-altitude, elliptical, synchronous-transfer, Phase III, Molniya
Period: 699.4 minutes
Apogee altitude: 35,500 km
Perigee altitude: 3955 km
Eccentricity: 0.6043
Inclination: 26°
Longitude increment: 175°West/orbit
Argument of Perigee: changing
17 July 1983 at 00:00:00 UTC: 187.4°
Rate of change: 0.27047°/day, 8.22°/month, 98.8°/year
Maximum access distance: 9034 km
Note: AMSAT-OSCAR 10 used its onboard propulsion system to reach the operational orbit after being dropped off in a transfer orbit by the launch vehicle. The transfer orbit was:
Apogee altitude: 35,529 km
Perigee altitude: 199 km
Inclination: 8.5°

1.4 Ground Track Data: See Appendix B
The following information is useful for rough tracking. Given the time, latitude and longitude of a reference apogee, note that apogee *two* orbits later will occur:
time: 40.9 minutes earlier (next day)
longitude: 9.4° further east
latitude: change is small, may be ignored

1.5 Operations
Coordinating Group: AMSAT
Schedule (subject to change)
During AMSAT-OSCAR 10's early years plans are to use Mode B almost exclusively. Scheduling will be periodically reassessed in terms of Mode-B transponder crowding and Mode-L occupancy. A single transponder will generally be in operation 24 hours/day. Exceptions may be made if (1) magnetic torquing is required to re-orient the spacecraft, in which case the transponders will be off near perigee, (2) poor sun angles or excessive eclipse time require curtailment. One alternative under study is to have the Mode-L transponder operate from one hour before apogee to one hour after on two days each week.

1.6 Design/Construction credits:
Project Management: AMSAT-U.S. (Jan King, W3GEY) and AMSAT-DL (Karl Meinzer, DJ4ZC)
Spacecraft subsystems: Contributed by groups in Canada, Hungary, Japan, United States, West Germany

1.7 Primary References
Note: Most of the references cited refer to φ3A, which never reached orbit. AMSAT-OSCAR 10, profiled in this section, is very similar except for (1) changes in the propulsion and antenna systems, (2) the addition of a wide-band Mode-L transponder, (3) increased radiation shielding of the computer system and (4) the selection of a slightly different orbit. These modifications should be kept in mind when checking the references.

1) J. A. King, "Phase III: Toward the Ultimate Amateur Satellite," Part 1, *QST*, Vol. LXI, no. 6, June 1977, pp. 11-14; Part 2, *QST*, Vol. LXI, no. 7, July 1977, pp. 52-55; Part 3, *QST*, Vol. LXI, no. 8, Aug. 1977, pp. 11-13.
2) J. A. King, "The Third Generation," Part 1, *Orbit*, Vol. 1, no. 3, Sept./Oct. 1980, pp. 12-18; Part 2, *Orbit*, Vol. 1, no. 4, Nov./Dec. 1980, pp. 12-18.

SPACECRAFT DESCRIPTION

2.1 Physical Structure
Shape: Tri-star as shown in Fig. 1 (A-O-10)
Mass: Approximately 90 kg at launch

2.2 Subsystem Description
Block diagram: See Fig. 2 (A-O-10)

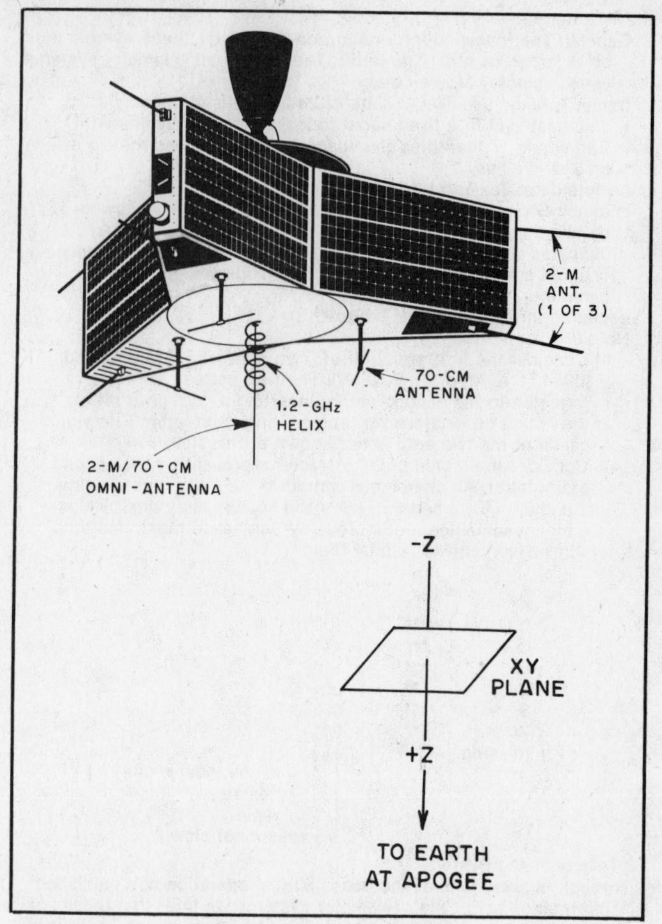

Fig. 1(A-O-10) — Pictorial view of AMSAT-OSCAR 10.

SUBSYSTEM DESCRIPTION

3.1 Beacons

	Frequency	*Power Output*	*Max. Doppler (at perigee)*
Mode B General Beacon	145.810 MHz	~2W†	3.9 kHz
Mode B Engineering Beacon	145.987 MHz	~4W†	3.9 kHz
Mode L General Beacon	436.04 MHz	~2W	11.6 kHz
Mode L Engineering Beacon	436.02 MHz	~2W	11.6 kHz

†Powers specified are with transponder on. With transponder off, power is 200-250% higher.

3.2 Telemetry
Formats Available: Any digital code (Morse code, radioteletype, ASCII, etc.) can be employed since the format is controlled by software residing in the spacecraft computer which can be altered via the command links.
General: Plans are to have the General Beacons carry Morse code to provide updated messages and selected telemetry channels. Engineering Beacons will use 400-baud PSK/PCM. Decoding the Engineering Beacon will require a microcomputer and an AFDEM (*A*udio *F*requency *DEM*odulator) that is connected between the ground station receiver audio output and the computer.

3.3 Command System
See block diagram of spacecraft in Fig. 2 (A-O-10). Uplink will load new program directly into computer memory via microprocessor-interrupt feature.

3.4 Transponders
General
Design/Construction credits: Dr. Karl Meinzer, DJ4ZC, Ulrich Mueller, DK4VW, and Werner Haas, DJ5KQ, University of Marburg, West Germany.

AMSAT-OSCAR 10

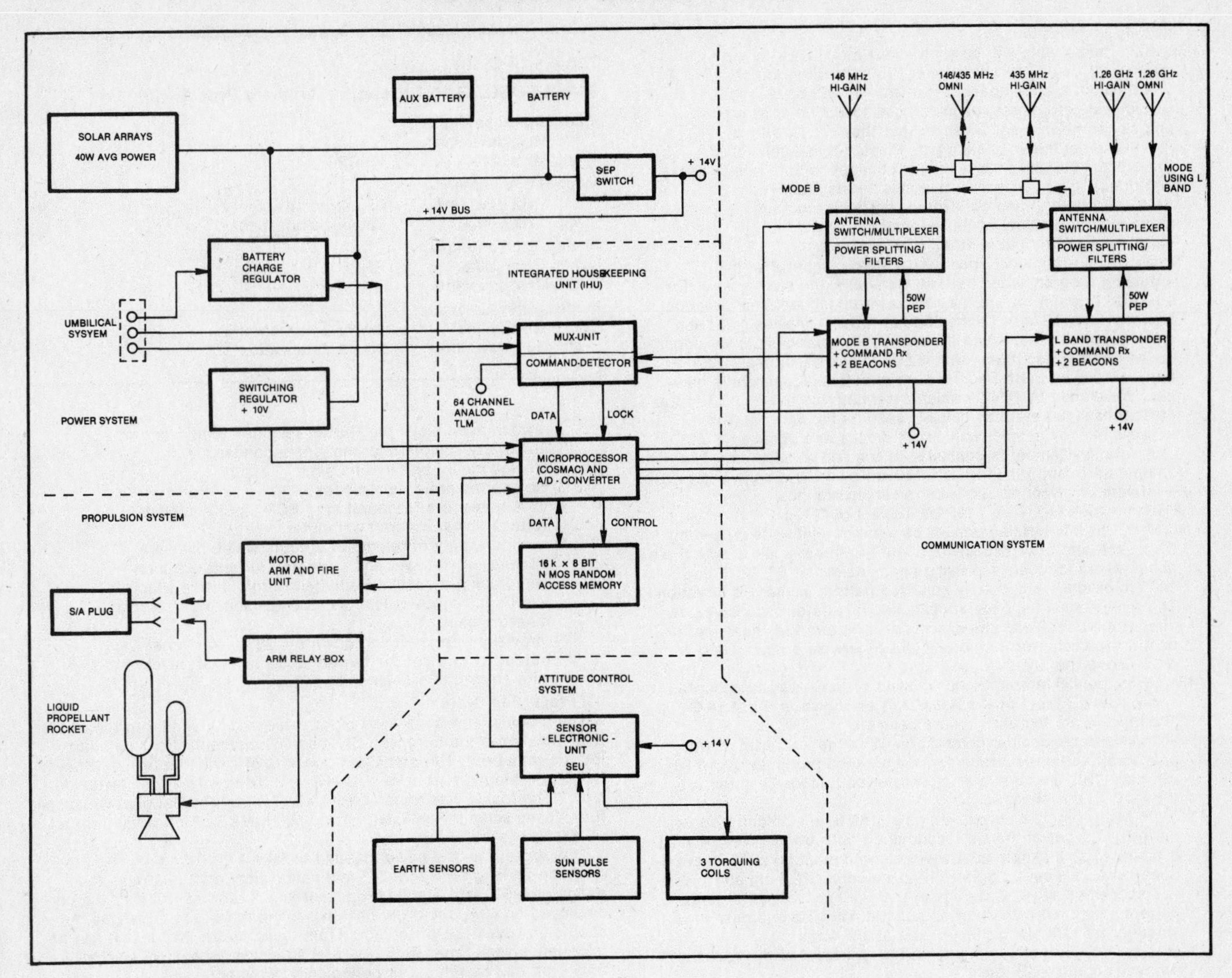

Fig. 2(A-O-10) — AMSAT-OSCAR 10 satellite functional block diagram.

Technical: Both transponders employ a new technique called HELAPS (*H*igh *E*fficiency *L*inear *A*mplification by *P*arametric *S*ynthesis). It's the latest in a series of innovative transponder designs — including EER on AMSAT-OSCAR 7 Mode B and EER-Doherty on AMSAT ϕ3A — developed by the University of Marburg group.

Suggested uplink eirp levels assume (1) circular polarization, (2) nominal path loss (quiet ionosphere, no tree foliage loss, etc.), (3) transponder automatic gain control *not* activated. Any station that sounds louder than the General Beacon is probably activating the transponder agc and thereby reducing the power available to ground stations operating properly.

Transponder I: Mode B (70 cm/2 m)
type: linear, inverting
uplink passband: 435.027-435.179 MHz
downlink passband: 145.825-145.977 MHz
translation equation:
downlink freq. (MHz) =
581.004 − uplink freq. (MHz) ± Doppler
output power: about 50 watts PEP
uplink eirp: about 500 watts
bandwidth: 152 kHz at 3 dB points (steep sides)
maximum Doppler (at perigee): 7.7 kHz (see Fig. 10-5 for additional information)
General: Plans are to use this transponder almost exclusively during regularly scheduled operating days in the early years of the life of the spacecraft. These plans will be reevaluated when crowding becomes a problem and Mode-L activity increases or if the downlink frequencies at 2m must be shared with other spacecraft.
Technical: The heart of this transponder is a 152-kHz-wide band-pass quartz filter having a shape factor of 1.16 and an insertion loss of less than 0.5 dB. The filter was procured by JAMSAT.

Transponder II: Mode L (23 cm/70 cm)
type: linear, inverting
uplink passband: 1269.05-1269.85 MHz
downlink passband: 436.15-436.95 MHz
translation equation:
downlink freq. (MHz) =
1706.000 − uplink freq. (MHz) ± Doppler
output power: about 50 watts PEP
uplink eirp: about 2000 watts (tentative)
bandwidth: 800 kHz
maximum Doppler (at perigee): 21.9 kHz (See Fig. 10-5 for additional information)
General: Plans are initially to use this transponder for experimental purposes. In future years, when other Phase III spacecraft are using the 2-m downlink or when the 150-kHz Mode-B transponder becomes overcrowded, the 800-kHz wide Mode-L unit may be scheduled a larger percentage of the time.
Technical: Original plans for this transponder called for using an 800-kHz-wide crystal filter centered at 53 MHz but all crystal manufacturers contacted said such a filter couldn't be built. As a result, W. Schafer (AMSAT-DL) designed a 10.7 MHz LC band-pass filter. The characteristics of the filter produced — with respect to stability, flatness, shape factor, phase delay, etc. — turned out to be nearly ideal so it was used on the spacecraft.

3.5 Attitude Stabilization

General: The satellite will be spun about its Z-axis at roughly 20 rpm. This will serve to "fix" the direction of the Z-axis in inertial space. Initial spin-up will be accomplished by three magnetorquer coils. The coils are actually somewhat squashed so that they can be fit just inside the perimeter of each arm. Pulsing these coils at the proper rate and time produces changes in spin rate and direction of spacecraft spin axis as the magnetic fields of the earth and coils interact. The interaction magnitude is greatest at those locations where the earth's magnetic field is strongest: near perigee.

A ground station will load the satellite computer with a pulsing program when needed. Generally, the spin axis will be adjusted to point toward the geocenter at apogee. The direction, however, may have to be modified at times if the resultant sun orientation with respect to the spacecraft solar cells is very poor.

The Sensor Electronics Unit (SEU) employs redundant Sun Sensors and an Earth Sensor. Special software algorithms have been developed by DJ4ZC which determine the true center of the crescent-shaped earth as actually seen by the sensors (the sensors employ simple visible-light diodes in a dual-beam unit).

A nutation damper is contained at the end of each arm. The dampers are long tubes (roughly 40 cm by 0.2 cm) containing a mixture of glycerine and water sloshing around.

3.6 Antennas (See Fig. 1 (A-O-10) and Table 1 (A-O-10))

146 MHz: The 146-MHz antenna uses a pair of elements extending from each arm of the spacecraft. The three rear spokes, which are about 48 cm long, are fed using phase delays of 0°, 120°, and 240° to produce a circularly polarized pattern in the +Z direction. The three forward spokes are parasitic directors. Because these directors interact with the spacecraft structure they have had to be cut much shorter than one might otherwise predict. Gain is 8 dBi along the +Z axis.

435 MHz: The 435-MHz antenna consists of three dipoles mounted above the +Z facet of the spacecraft as shown in Fig. 1 (A-O-10). The dipoles are fed using phase delays of 0°, 120° and 240° to provide circular polarization. With the +Z facet of the spacecraft acting as a reflector, the antenna produces about 9.5 dBi gain. This antenna is employed for both Mode-B uplink and the Mode-L downlink.

146/435 MHz "omni" antenna: A single structural element mounted on the +Z facet of the s/c functions on both frequencies. At 2 m it operates as a simple quarter-wavelength monopole. At 70 cm it operates as a sleeve dipole with phase center 28.5 cm above the +Z spacecraft facet. Coverage of the −Z hemisphere is good; pattern measurements show no nulls at 435 MHz and only a single sharp null along the −Z axis at 146 MHz.

1.26 GHz: A 5-turn helix mounted on the +Z facet of the spacecraft produces about 9-dBi gain.

1.26 GHz "omni" antenna: A quarter-wavelength monopole mounted on the −Z facet of the spacecraft is employed.

Note: Signal polarizations associated with each spacecraft link are summarized in Table 1 (A-O-10).

3.7 Energy-Supply and Power Conditioning (See Fig. 2 (A-O-10)

Solar Cell Characteristics
- type: n or p silicon (violet cell technology)
- size: 2 cm × 2 cm (Solarex), 2 cm × 4 cm (AEG-Telefunken)
- efficiency: 12.5% (BOL)
- protective cover: 0.5 mm fused silica cover slides (to minimize radiation degradation, cover slides are thicker than those used on low-altitude AMSAT OSCARs)

Solar Panels
- number: 6 (3 Solarex, 3 AEG-Telefunken)
- maximum output per panel: 27.5 watts
- total surface area (6 panels): 1.1 m²
- peak BOL output of array: more than 50 watts (optimal sun orientation) is available to spacecraft

Storage Battery
- Primary: Nickel-Cadmium, 6 Ah
- Secondary: Nickel-Cadmium, 4 Ah. The secondary (auxiliary or backup) battery is stored discharged. If the primary battery fails, the secondary battery is trickle charged and then switched to the main bus.

Battery Charge Regulator (BCR)
- Design/Construction Credits: The BCR was built by the Amateur Radio Club at the Technical University of Budapest (HG5BME) under the direction of Dr. A. Gschwindt, HA5WH.
- Technical: The fully redundant design (two identical units) includes:
  1) +10 V regulators for all spacecraft logic;
  2) switch-over relays for (i) BCR selection (unit 1 or unit 2), (ii) battery selection (primary or secondary), and (iii) secondary battery charging;
  3) solar-array protection diodes;
  4) D-A converters for monitoring BCR input and output voltages by spacecraft computer
  5) a wide variety of telemetry outputs which measure all relevant voltages and currents. Currents are measured using a special Hall-effect toroid device which results in no power loss as a consequence of the measurement;
  6) provisions for stepping down the 28-V solar panel output to 14 V for the battery and spacecraft loads with an overall efficiency of 87%.

Table 1 (A-O-10)

AMSAT-OSCAR 10 Spacecraft Antenna Polarization

Mode-B gain antennas	
146 MHz downlink	RHCP
435 MHz uplink	RHCP
Mode-B omni antennas	
146 MHz downlink	Linear polarization
435 MHz uplink	Linear polarization
Mode-L gain antennas	
1.26 GHz uplink	RHCP
435 MHz downlink	RHCP
Mode-L omni antennas	
1.26 GHz uplink	Linear polarization
435 MHz downlink	Linear polarization

3.8 Propulsion System

The propulsion system uses a liquid-fuel rocket similar to the one that transferred the European Symphonie communications satellite into its final orbit. The propellant consists of a mixture of fuel (Unsymmetrical DiMethyl Hydrazine [UDMH]) and oxidizer (nitrogen tetroxide, N_2O_4). The rocket was donated to AMSAT-DL by the German aerospace firm Messerschmitt-Boelkow-Blohm GMBH (MBB). It provides a thrust of 400 Newtons.

This rocket has the power needed to take A-O-10 from the 8.5° inclination parking orbit to a 57° inclination final orbit and raise the perigee to 1500 km. A solid-propellant kick motor, identical to the one employed on ϕ3A, could not have produced the desired changes. An additional advantage of the liquid-fuel rocket is the fact that it can be reignited several times (until the fuel supply is exhausted) so that the final orbit can be refined or periodically adjusted.

The liquid fuel motor increases spacecraft complexity and cost, however, because of the need for fuel tanks and associated plumbing that must be provided by AMSAT. And, data on liquid-fuel motor reliability is not as extensive as that for solid-propellant rockets.

Post-launch note: Because of a loss of helium pressure, which is needed to force the propellant into the combustion chamber and to open hydraulic valves, it was only possible to fire the motor once. It's believed that the problem is related to a collision with the launch vehicle shortly after A-O-10 was placed in space.

3.9 Integrated Housekeeping Unit (IHU)

As shown in Fig. 2 (A-O-10) the IHU consists of a CMOS microprocessor (RCA COSMAC 1802), at least 16K of dynamic random access memory (RAM), a command decoder, an analog-to-digital converter, and a 64-channel analog multiplexer (MUX). The IHU is responsible for controlling virtually every function onboard the spacecraft including:
1) execution of all telemetry and command requirements;
2) monitoring conditions of the power and communications systems and taking corrective actions as necessary;
3) establishing clocks needed for various spacecraft timing functions;
4) interacting with the attitude sensors and magnetorquer coils;
5) sending commands to the propulsion system.

Possible radiation damage to IHU components was a major concern of the A-O-10 designers. An experimental program that involved exposing key components to radiation at Argonne National Laboratories to predict failure rates and modes and establish shielding requirements was undertaken. The 1802 processor chip utilizes a special Sandia CMOS radiation hardening process. Selective radiation shielding, consisting of thin sheets of Tantalum bonded to top and bottom of each IC, was used to obtain the maximum protection for a given weight of shielding material. Hadimar error correcting code was used to guard against "soft" memory failures caused by alpha particle radiation.

Appendix B

Map Board (blank polar projection, ARRL standard scale; see also back cover foldout)

AMSAT-OSCAR 8
- ground track table
- ground track tracing master
- 30°N spiderweb table
- 46°N spiderweb table
- 30°N & 46°N spiderweb tracing masters

UoSAT-OSCAR 9
- ground track table
- ground track tracing master
- 30°N spiderweb table
- 46°N spiderweb table
- 30°N & 46°N spiderweb tracing masters

RS-7 (for RS-3 through RS-8)
- ground track table
- ground track tracing master
- 30°N spiderweb table
- 46°N spiderweb table
- 36°N and 46°N spiderweb tracing masters

AMSAT-OSCAR 10
- ground track table
- ground track tracing master
- 30°N spiderweb table
- 40°N spiderweb table
- 50°N spiderweb table
- 40°N spiderweb tracing master

ISKRA 2
- ground track table
- ground track tracing master
- 30°N spiderweb table
- 46°N spiderweb table
- 30°and 46°N spiderweb tracing masters

SPACE SHUTTLE (STS-9)
- ground track table
- ground track tracing master
- 30°N spiderweb table
- 46°N spiderweb table
- 30°N and 46°N spiderweb tracing masters

Geostationary satellite: ground-station antenna aiming

Appendix B

Assembling a tracking aid, such as the OSCARLOCATOR or ϕ3 TRACKER, requires data for drawing ground tracks and spiderwebs. This appendix contains the necessary material in two formats: (1) tables that can be used in conjunction with any map and (2) tracing masters for use with the map board included here or as part of the ARRL OSCARLOCATOR package. See Chapter 5 for a complete description of how the information presented in this appendix is used.

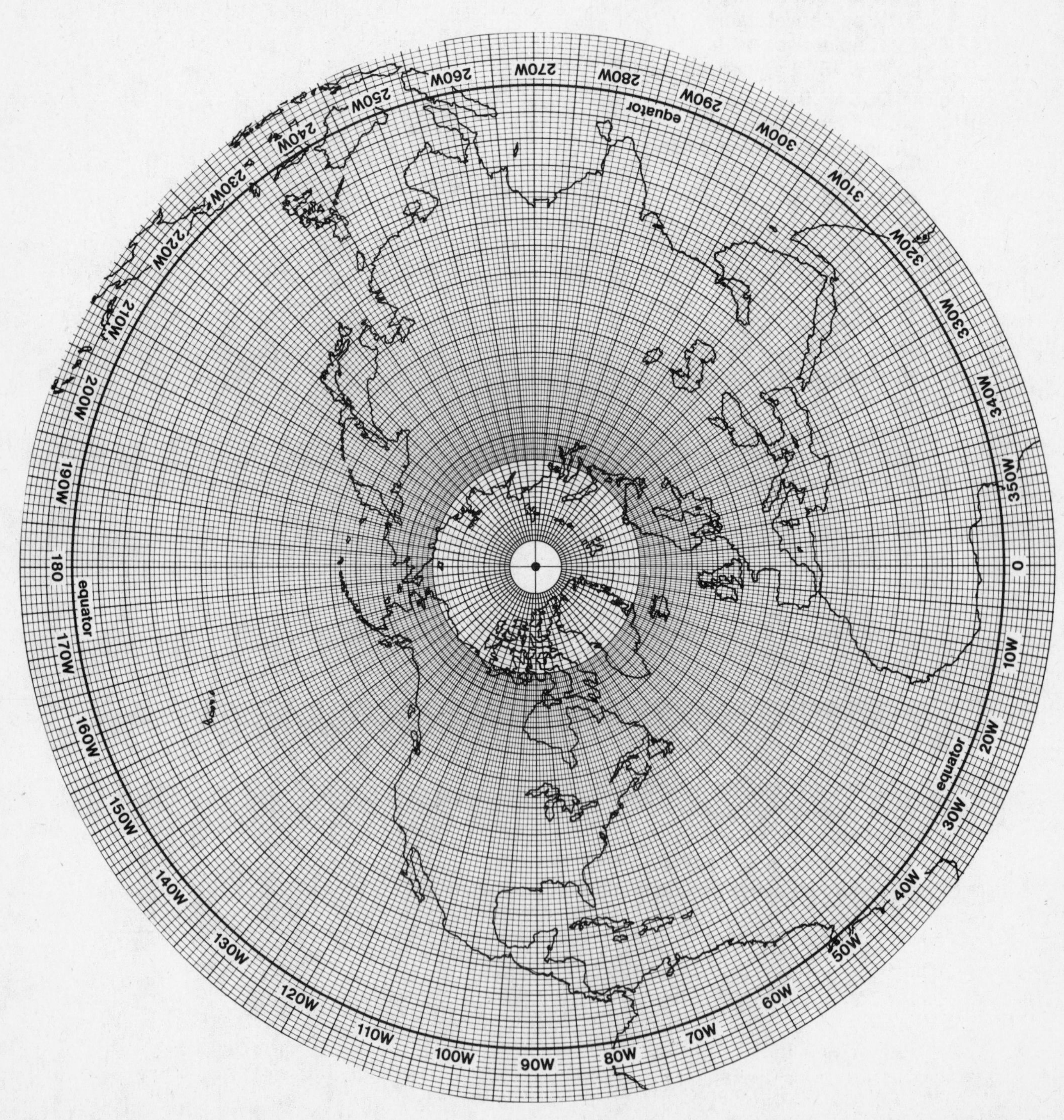

AMSAT-OSCAR 8

OSCAR 8

50 48 46 44 42 40 38 36 34 33 30 28 26 24 22 20 18 16 14 12 10 8 6 4 2 0

MINUTES AFTER EQX

NEXT EQX

Ground Track Data
Satellite: AMSAT-OSCAR 8
Mean altitude: 910 km   Period: 103.2 min.   Inclination: 98.9°

Time after ascending node (minutes)	*Subsatellite point Lat. (°N)*	*Long. (°W)*	*Time after ascending node (minutes)*	*Subsatellite point Lat. (°N)*	*Long. (°W)*
0	0.0	0.0	26	81.1	101.0
2	6.9	1.6	28	78.3	138.1
4	13.8	3.2	30	72.9	156.9
6	20.7	4.9	32	66.7	166.7
8	27.5	6.7	34	60.2	172.7
10	34.4	8.7	36	53.5	176.8
12	41.2	10.9	38	46.7	179.9
14	48.1	13.5	40	39.9	182.5
16	54.8	16.8	42	33.0	184.7
18	61.5	21.2	44	26.2	186.6
20	68.0	27.8	46	19.3	188.4
22	74.1	38.8	48	12.4	190.0
24	79.1	60.6	50	5.5	191.6
25.8	81.1	96.5	51.6	0.0	192.9

Note: Ground track should be drawn directly on map using data presented. It may then be traced on clear plastic. The plastic overlay is then repositioned for each orbit.

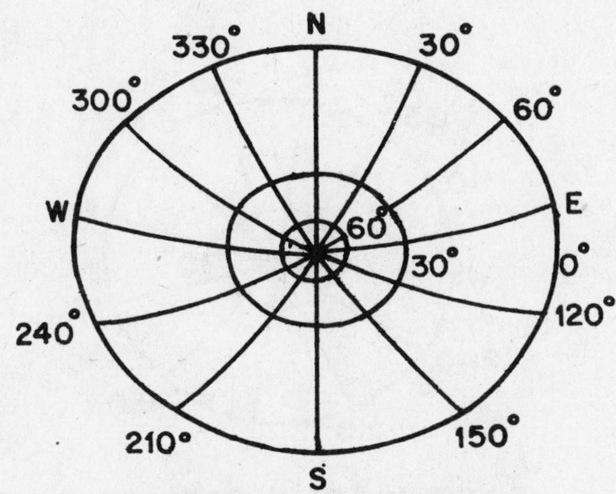

SATELLITE: OSCAR 8

GROUND STATION: 30° N

N 330° 30° 300° 60° W E 60° 30° 0° 240° 120° 210° 150° S

SATELLITE: OSCAR 8

GROUND STATION: 46° N

Spiderweb Data
Satellite: AMSAT-OSCAR 8
Mean altitude: 910 km
Ground station: 30.0° N, 90.0° W

	Latitude (°N)/Longitude (°W)		
Azimuth	*0° elevation 3219 km*	*30° elevation 1192 km*	*60° elevation 451 km*
0° (north)	58.9/90.0	40.7/90.0	34.1/90.0
30°	53.2/66.2	39.1/83.1	33.5/87.6
60°	40.3/56.6	34.9/78.7	32.0/85.9
90° (east)	25.9/57.4	29.4/77.7	29.9/85.3
120°	13.2/64.5	24.3/79.8	27.9/86.0
150°	4.3/76.0	20.6/84.3	26.5/87.7
180° (south)	1.1/90.0	19.3/90.0	25.9/90.0
210°	4.3/104.0	20.6/95.7	26.5/92.3
240°	13.2/115.5	24.3/100.2	27.9/94.0
270° (west)	25.9/122.6	29.4/102.3	29.9/94.7
300°	40.3/123.4	34.9/101.3	32.0/94.1
330°	53.2/113.8	39.1/96.9	33.5/92.4

Note: Spiderweb should be drawn directly on the map using data presented. It may then be traced on clear plastic. The plastic overlay is then repositioned at the user's location.

Spiderweb Data
Satellite: AMSAT-OSCAR 8
Mean altitude: 910 km
Ground Station: 46.0° N, 90.0° W

	Latitude (°N)/Longitude (°W)		
Azimuth	*0° elevation 3219 km*	*30° elevation 1192 km*	*60° elevation 451 km*
0° (north)	74.9/90.0	56.7/90.0	50.1/90.0
30°	67.0/51.7	55.0/80.7	49.5/86.9
60°	52.9/46.0	50.5/75.3	47.9/84.8
90° (east)	39.0/51.5	45.0/74.8	45.9/84.2
120°	27.5/61.8	40.0/77.9	43.9/85.1
150°	19.8/75.1	36.5/83.4	42.5/87.3
180° (south)	17.1/90.0	35.3/90.0	41.9/90.0
210°	19.8/104.9	36.5/96.6	42.5/92.7
240°	27.5/118.2	40.0/102.1	43.9/94.9
270° (west)	39.0/128.5	45.0/105.2	45.9/95.8
300°	52.9/134.0	50.5/104.7	47.9/95.2
330°	67.0/128.3	55.0/99.3	49.5/93.1

Note: Spiderweb should be drawn directly on the map using data presented. It may then be traced on clear plastic. The plastic overlay is then repositioned at the user's location.

UoSAT-OSCAR 9

46 44 42 40 38 36 34 32 30 28 26 24 22 20 18 16 14 12 10 8 6 4 2 0

MINUTES AFTER EQX

NEXT EQX

Ground Track Data
Satellite: UoSAT-OSCAR 9
Mean altitude: 540 km Period: 95.3 min. Inclination: 97.5°

Time after ascending node (minutes)	*Subsatellite point Lat. (°N)*	*Long. (°W)*	*Time after ascending node (minutes)*	*Subsatellite point Lat. (°N)*	*Long. (°W)*
0	0.0	0.0	24	82.5	101.1
2	7.5	1.5	26	78.9	144.4
4	15.0	3.0	28	72.6	162.2
6	22.5	4.6	30	65.6	170.7
8	29.9	6.3	32	58.3	175.7
10	37.4	8.3	34	50.9	179.2
12	44.8	10.5	36	43.5	181.8
14	52.2	13.3	38	36.1	184.0
16	59.6	17.0	40	28.6	185.9
18	66.8	22.4	42	21.2	187.6
20	73.8	31.9	44	13.7	189.2
22	79.8	52.7	46	6.2	190.7
23.8	82.5	96.0	47.7	0.0	191.9

Note: Ground track should be drawn directly on map using data presented. It may then be traced on clear plastic. The plastic overlay is then repositioned for each orbit.

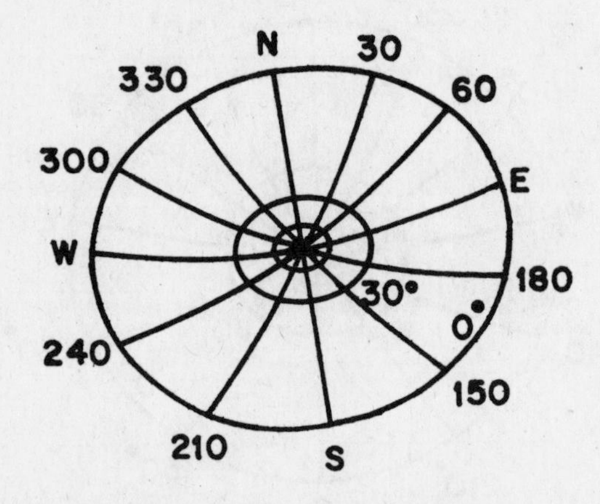

SATELLITE: UoSAT OSCAR 9

GROUND STATION: 30° N

Spiderweb Data
Satellite: UoSAT-OSCAR 9
Mean altitude: 540 km
Ground Station: 30.0° N, 90.0° W

	Latitude (°N)/Longitude (°W)		
	0° elevation	*30° elevation*	*60° elevation*
Azimuth	*2535 km*	*781 km*	*284 km*
0° (north)	52.8/90.0	37.0/90.0	32.6/90.0
30°	48.7/72.9	36.0/85.7	32.2/88.5
60°	39.0/64.4	33.3/82.7	31.3/87.4
90° (east)	27.4/64.1	29.8/81.9	30.0/87.1
120°	17.0/69.5	26.3/83.2	28.7/87.5
150°	9.8/78.7	23.9/86.2	27.8/88.6
180° (south)	7.2/90.0	23.0/90.0	27.4/90.0
210°	9.8/101.3	23.9/93.8	27.8/91.4
240°	17.0/110.5	26.3/96.8	28.7/92.5
270° (west)	27.4/115.9	29.8/98.1	30.0/92.9
300°	39.0/115.6	33.3/97.3	31.3/92.6
330°	48.7/107.1	36.0/94.3	32.2/91.5

Note: Spiderweb should be drawn directly on the map using data presented. It may then be traced on clear plastic. The plastic overlay is then repositioned at the user's location.

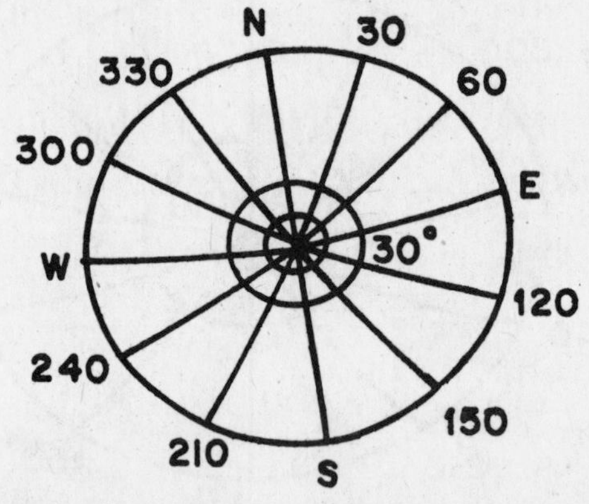

SATELLITE: UoSAT OSCAR 9

GROUND STATION: 46° N

Spiderweb Data
Satellite: UoSAT-OSCAR 9
Mean altitude: 540 km
Ground Station: 46.0° N, 90.0° W

	Latitude (°N)/Longitude (°W)		
	0° elevation	*30° elevation*	*60° elevation*
Azimuth	*2535 km*	*781 km*	*284 km*
0° (north)	68.8/90.0	53.0/90.0	48.6/90.0
30°	63.7/64.1	52.0/84.3	48.2/88.1
60°	52.9/56.2	49.1/80.7	47.2/86.7
90° (east)	41.5/58.8	45.6/79.9	45.9/86.3
120°	31.9/66.7	42.2/81.8	44.7/86.9
150°	25.5/77.6	39.8/85.4	43.8/88.2
180° (south)	23.2/90.0	39.0/90.0	43.4/90.0
210°	25.5/102.4	39.8/94.6	43.8/91.8
240°	31.9/113.3	42.2/98.2	44.7/93.1
270° (west)	41.5/121.2	45.6/100.1	45.9/93.7
300°	52.9/123.8	49.1/99.3	47.2/93.3
330°	63.7/115.9	52.0/95.7	48.2/91.9

Note: Spiderweb should be drawn directly on the map using data presented. It may then be traced on clear plastic. The plastic overlay is then repositioned at the user's location.

Radio Sputnik

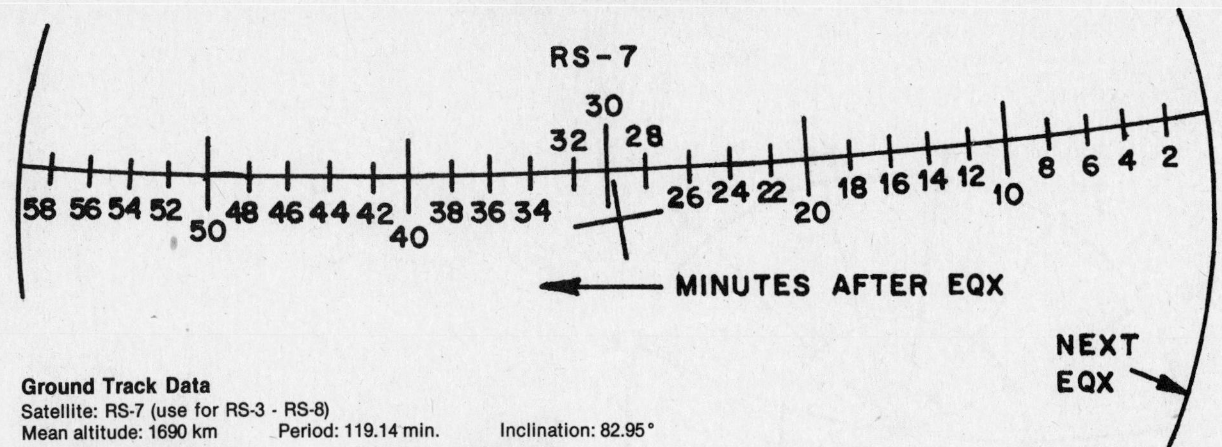

Ground Track Data
Satellite: RS-7 (use for RS-3 - RS-8)
Mean altitude: 1690 km Period: 119.14 min. Inclination: 82.95°

Time after ascending node (minutes)	*Subsatellite point Lat. (°N)*	*Long. (°W)*	*Time after ascending node (minutes)*	*Subsatellite point Lat. (°N)*	*Long. (°W)*
0	0.0	0.0	30	82.9	272.2
2	6.0	359.8	32	80.3	234.3
4	12.0	359.5	34	75.5	217.0
6	18.0	359.2	36	70.0	208.8
8	24.0	358.8	38	64.3	204.4
10	30.0	358.4	40	58.4	201.6
12	35.9	357.9	42	52.5	199.8
14	41.9	357.1	44	46.6	198.5
16	47.9	356.1	46	40.6	197.6
18	53.8	354.8	48	34.7	196.9
20	59.7	352.8	50	28.7	196.4
22	65.5	349.8	52	22.7	196.0
24	71.2	344.7	54	16.7	195.6
26	76.6	335.3	56	10.7	195.3
28	81.1	314.6	58	4.7	195.1
29.8	83.0	277.4	59.6	0.0	194.9

Note: Ground track should be drawn directly on map using data presented. It may then be traced on clear plastic. The plastic overlay is then repositioned for each orbit.

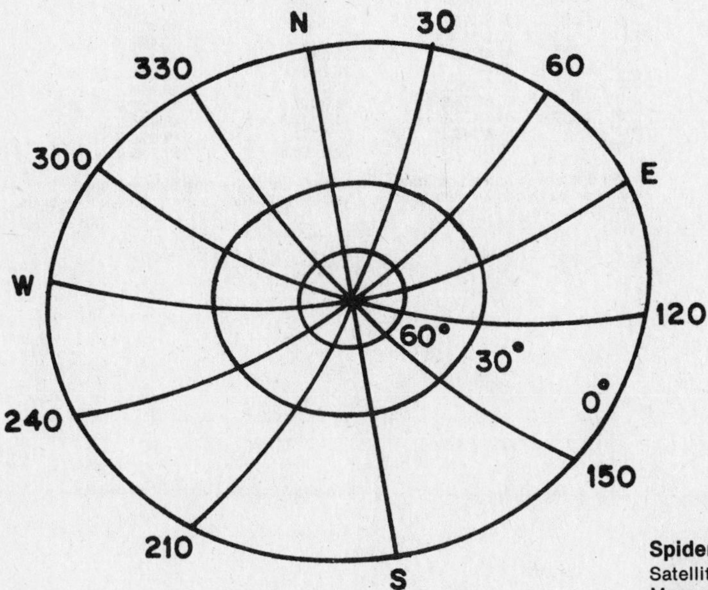

SATELLITE: RS-7
GROUND STATION: 30° N

Spiderweb Data
Satellite: RS-7 (use for RS-3 - RS-8)
Mean altitude: 1690 km
Ground Station: 30.0° N, 90.0° W

	Latitude (°N)/Longitude (°W)		
Azimuth	*0° elevation 4201 km*	*30° elevation 1869 km*	*60° elevation 748 km*
0° (north)	67.8/90.0	46.8/90.0	36.7/90.0
30°	58.7/53.8	44.1/78.4	35.8/85.9
60°	41.3/45.0	37.1/71.7	33.2/83.0
90° (east)	23.3/48.2	28.6/70.8	29.8/82.2
120°	7.5/57.6	20.7/74.5	26.5/83.5
150°	– 3.7/72.1	15.2/81.4	24.1/86.3
180° (south)	– 7.8/90.0	13.2/90.0	23.3/90.0
210°	– 3.7/107.9	15.2/98.6	24.1/93.7
240°	7.5/122.4	20.7/105.5	26.5/96.5
270° (west)	23.3/131.8	28.6/109.2	29.8/97.8
300°	41.3/135.0	37.1/108.3	33.2/97.0
330°	58.7/126.2	44.1/101.6	35.8/94.1

Note: Spiderweb should be drawn directly on the map using data presented. It may then be traced on clear plastic. The plastic overlay is then repositioned at the user's location.

Radio Sputnik

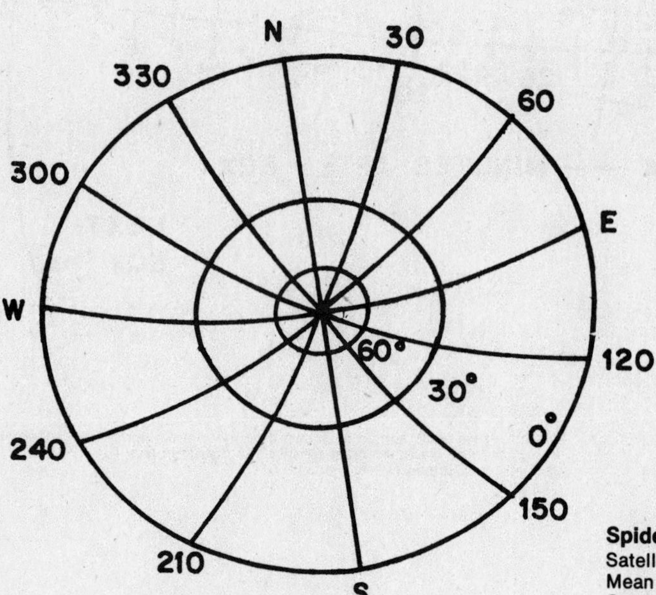

SATELLITE: RS-7
GROUND STATION: 46° N

Spiderweb Data
Satellite: RS-7 (use for RS-3 — RS-8)
Mean altitude: 1690 km
Ground Station: 46.0° N, 90.0° W

	Latitude (°N)/Longitude (°W)		
	0° elevation	*30° elevation*	*60° elevation*
Azimuth	*4201 km*	*1869 km*	*748 km*
0° (north)	83.8/90.0	62.8/90.0	52.7/90.0
30°	69.6/28.7	59.6/73.4	51.7/84.6
60°	51.4/31.8	52.1/65.9	49.0/81.1
90° (east)	34.6/41.9	43.5/66.5	45.6/80.4
120°	20.8/55.4	36.0/72.0	42.4/82.1
150°	11.5/71.8	31.0/80.3	40.1/85.6
180° (south)	8.2/90.0	29.2/90.0	39.3/90.0
210°	11.5/108.2	31.0/99.7	40.1/94.4
240°	20.8/124.6	36.0/108.0	42.4/97.9
270° (west)	34.6/138.1	43.5/113.5	45.6/99.6
300°	51.4/148.2	52.1/114.1	49.0/98.9
330°	69.6/151.3	59.6/106.6	51.7/95.4

Note: Spiderweb should be drawn directly on the map using data presented. It may then be traced on clear plastic. The plastic overlay is then repositioned at the user's location.

AMSAT-OSCAR 10

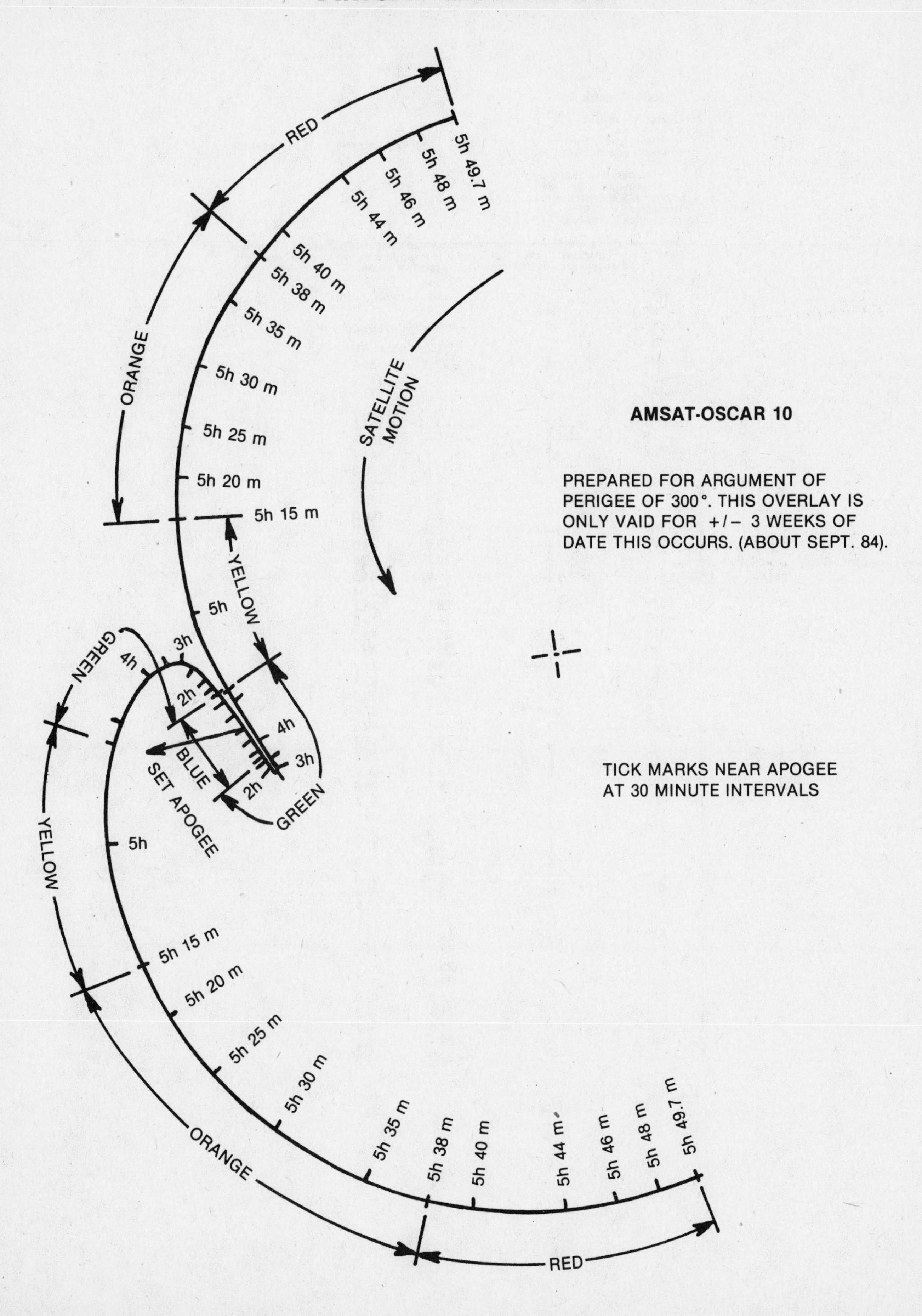

AMSAT-OSCAR 10

Ground Track Data

Satellite: AMSAT-OSCAR 10

Prepared for argument of perigee of 300°. This overlay is only valid for +/– 3 weeks of date this occurs. (About Sept. 1984)

Apogee height = 35,500 km
Perigee height = 3955 km
Period = 699.3 minutes
Inclination = 26.0°
eccentricity = 0.60435

Note: Ground track should be drawn directly on map using data presented. It may then be traced on clear plastic. The plastic overlay is then repositioned for each orbit. With this orbit the overlay must be redrawn about every five weeks.

Time from apogee (minutes)	*Color code*	*Subsatellite point* *Latitude* (°N)	*Longitude* (°West)
– 349.7		– 22.3	180.0
– 348		– 21.2	175.9
– 346	red	– 19.7	171.1
– 344		– 18.2	166.3
– 340		– 14.7	157.5
– 338		– 12.9	153.6
– 335		– 10.1	147.7
– 330		– 5.6	139.2
– 325	orange	– 1.4	131.8
– 320		2.3	125.5
– 315		5.5	120.0
– 300	yellow	12.9	107.0
– 278		19.2	95.0
– 270		20.7	92.0
– 240		24.0	84.3
– 210		25.4	80.4
– 180	green	26.0	78.7
– 150		25.9	78.6
– 120		25.6	79.3
– 90		25.1	80.9
– 75		24.6	81.7
– 60		24.3	82.7
– 30		23.3	85.0
0	blue	22.3	87.4
30		21.1	89.9
60		19.7	92.4
75		18.9	93.6
90		18.2	94.7
120		16.3	96.8
150		14.1	98.4
180	green	11.2	99.5
210		8.0	99.3
240		3.5	97.4
270		– 2.8	91.8
278		– 4.9	89.4
300	yellow	– 12.4	78.4
315		– 18.8	64.7
320		– 21.0	58.2
325	orange	– 23.0	50.7
330		– 24.8	41.6
335		– 25.8	31.1
338		– 26.0	24.1
340		– 25.9	19.2
344		– 25.0	9.3
346	red	– 24.3	4.1
348		– 23.3	359.0
349.7		– 22.3	354.8

Satellite Moving →

AMSAT-OSCAR 10

A spiderweb produced from the data of this table employs range circles in place of elevation angle circles as explained in the discussion of the ϕ3 TRACKER in Chapter 5. Although presented here for use with AMSAT-OSCAR 10 this spiderweb is not keyed to a specific orbit. It can therefore be used with any satellite. The spiderweb should be drawn directly on the map using data presented. It may then be traced on clear plastic. The plastic overlay is then repositioned at the user's location.

LOCATION OF CENTER — **30.0 DEGREES NORTH LATITUDE** — **90.0 DEGREES WEST LONGITUDE**

DISTANCE	NORTH	30	60	EAST	120	150	SOUTH	210	240	WEST	300	330
					AZIMUTH							
1000 KM LATITUDE	39.0	37.7	34.2	29.6	25.2	22.1	21.0	22.1	25.2	29.6	34.2	37.7
LONGITUDE	90.0	84.3	80.6	79.6	81.4	85.2	90.0	94.8	98.6	100.4	99.4	95.7
2000 KM LATITUDE	48.0	45.0	37.5	28.4	20.0	14.1	12.0	14.1	20.0	28.4	37.5	45.0
LONGITUDE	90.0	77.4	70.3	69.4	73.5	80.8	90.0	99.2	106.5	110.6	109.7	102.6
3000 KM LATITUDE	57.0	51.8	39.9	26.5	14.4	6.0	3.0	6.0	14.4	26.5	39.9	51.8
LONGITUDE	90.0	68.5	59.2	59.6	66.1	76.8	90.0	103.2	113.9	120.4	120.8	111.5
4000 KM LATITUDE	66.0	57.7	41.2	23.9	8.6	-2.1	-6.0	-2.1	8.6	23.9	41.2	57.7
LONGITUDE	90.0	56.7	47.4	50.0	59.0	72.9	90.0	107.1	121.0	130.0	132.6	123.3
4500 KM LATITUDE	70.5	60.1	41.4	22.4	5.7	-6.1	-10.5	-6.1	5.7	22.4	41.4	60.1
LONGITUDE	90.0	49.3	41.5	45.4	55.6	71.0	90.0	109.0	124.4	134.6	138.5	130.7
5000 KM LATITUDE	75.0	62.1	41.3	20.7	2.7	-10.2	-15.0	-10.2	2.7	20.7	41.3	62.1
LONGITUDE	90.0	41.0	35.5	40.9	52.2	69.0	90.0	111.0	127.8	139.1	144.5	139.0
6000 KM LATITUDE	84.0	64.2	40.1	17.1	-3.2	-18.2	-24.0	-18.2	'3.2	17.1	40.1	64.2
LONGITUDE	90.0	21.5	23.7	32.2	45.5	64.8	90.0	115.2	134.5	147.8	156.3	158.5
7000 KM LATITUDE	87.0	63.6	37.8	13.1	-9.1	-26.1	-33.0	-26.1	'9.1	13.1	37.8	63.6
LONGITUDE	270.0	1.0	12.5	23.9	38.6	60.3	90.0	119.7	141.4	156.1	167.5	179.0
8000 KM LATITUDE	78.1	60.2	34.5	8.9	-14.9	-33.9	-41.9	-33.9	'14.9	8.9	34.5	60.2
LONGITUDE	270.0	-16.8	2.1	15.8	31.6	55.0	90.0	125.0	148.4	164.2	177.9	196.8
9000 KM LATITUDE	69.1	55.0	30.4	4.5	-20.4	-41.4	-50.9	-41.4	'20.4	4.5	30.4	55.0
LONGITUDE	270.0	-30.5	-7.4	7.9	24.1	48.8	90.0	131.2	155.9	172.1	187.4	210.5

LOCATION OF CENTER — **40.0 DEGREES NORTH LATITUDE** — **90.0 DEGREES WEST LONGITUDE**

DISTANCE	NORTH	30	60	EAST	120	150	SOUTH	210	240	WEST	300	330
					AZIMUTH							
1000 KM LATITUDE	49.0	47.6	44.0	39.4	35.1	32.1	31.0	32.1	35.1	39.4	44.0	47.6
LONGITUDE	90.0	83.3	79.2	78.3	80.5	84.7	90.0	95.3	99.5	101.7	100.8	96.7
2000 KM LATITUDE	58.0	54.7	46.9	37.7	29.5	24.0	22.0	24.0	29.5	37.7	46.9	54.7
LONGITUDE	90.0	74.5	67.0	67.0	72.1	80.3	90.0	99.7	107.9	113.0	113.0	105.5
3000 KM LATITUDE	67.0	60.9	48.3	34.9	23.5	15.8	13.0	15.8	23.5	34.9	48.3	60.9
LONGITUDE	90.0	62.2	53.8	56.4	64.6	76.4	90.0	103.6	115.4	123.6	126.2	117.8
4000 KM LATITUDE	76.0	65.5	48.2	31.3	17.2	7.5	4.0	7.5	17.2	31.3	48.2	65.5
LONGITUDE	90.0	44.9	40.3	46.5	57.8	72.8	90.0	107.2	122.2	133.5	139.7	135.1
4500 KM LATITUDE	80.5	66.9	47.5	29.3	13.9	3.3	-0.5	3.3	13.9	29.3	47.5	66.9
LONGITUDE	90.0	34.3	33.7	41.9	54.6	71.0	90.0	109.0	125.4	138.1	146.3	145.7
5000 KM LATITUDE	85.0	67.5	46.5	27.1	10.6	-0.8	-5.0	-0.8	10.6	27.1	46.5	67.5
LONGITUDE	90.0	22.8	27.2	37.5	51.5	69.3	90.0	110.7	128.5	142.5	152.8	157.2
6000 KM LATITUDE	86.0	66.2	43.5	22.2	3.9	-9.1	-14.0	-9.1	3.9	22.2	43.5	66.2
LONGITUDE	270.0	0.1	15.2	29.1	45.4	65.8	90.0	114.2	134.6	150.9	164.8	179.9
7000 KM LATITUDE	77.0	62.0	39.3	17.0	-2.8	-17.4	-23.0	-17.4	-2.8	17.0	39.3	62.0
LONGITUDE	270.0	'18.3	4.6	21.4	39.4	62.2	90.0	117.8	140.6	158.6	175.4	198.3
8000 KM LATITUDE	68.1	56.1	34.3	11.5	-9.5	-25.6	-31.9	-25.6	-9.5	11.5	34.3	56.1
LONGITUDE	270.0	-31.5	-4.7	14.0	33.4	58.2	90.0	121.8	146.6	166.0	184.7	211.5
9000 KM LATITUDE	59.1	49.1	28.7	5.8	-16.1	-33.6	-40.9	-33.6	-16.1	5.8	28.7	49.1
LONGITUDE	270.0	-41.0	-13.0	7.0	27.1	53.6	90.0	126.4	152.9	173.0	193.0	221.0

LOCATION OF CENTER — **50.0 DEGREES NORTH LATITUDE** — **90.0 DEGREES WEST LONGITUDE**

DISTANCE	NORTH	30	60	EAST	120	150	SOUTH	210	240	WEST	300	330
					AZIMUTH							
1000 KM LATITUDE	59.0	57.5	53.8	49.2	44.9	42.0	41.0	42.0	44.9	49.2	53.8	57.5
LONGITUDE	90.0	81.6	76.8	76.2	79.0	84.0	90.0	96.0	101.0	103.8	103.2	98.4
2000 KM LATITUDE	68.0	64.2	55.9	46.8	39.0	33.8	32.0	33.8	39.0	46.8	55.9	64.2
LONGITUDE	90.0	69.2	61.5	63.2	69.9	79.3	90.0	100.7	110.1	116.8	118.5	110.8
3000 KM LATITUDE	77.0	69.3	55.9	43.1	32.5	25.5	23.0	25.5	32.5	43.1	55.9	69.3
LONGITUDE	90.0	50.2	45.4	51.6	62.2	75.4	90.0	104.6	117.8	128.4	134.6	129.8
4000 KM LATITUDE	86.0	71.3	54.0	38.3	25.5	17.0	14.0	17.0	25.5	38.3	54.0	71.3
LONGITUDE	90.0	24.0	30.1	41.5	55.7	72.1	90.0	107.9	124.3	138.5	149.9	156.0
4500 KM LATITUDE	89.5	70.7	52.3	35.6	22.0	12.8	9.5	12.8	22.0	35.6	52.3	70.7
LONGITUDE	270.0	10.2	23.2	37.0	52.7	70.6	90.0	109.4	127.3	143.0	156.8	169.8
5000 KM LATITUDE	85.0	69.3	50.3	32.8	18.4	8.5	5.0	8.5	18.4	32.8	50.3	69.3
LONGITUDE	270.0	-2.3	16.7	32.8	49.8	69.1	90.0	110.9	130.2	147.2	163.3	182.3
6000 KM LATITUDE	76.0	64.3	45.3	26.8	11.0	0.0	-4.0	0.0	11.0	26.8	45.3	64.3
LONGITUDE	270.0	-21.4	5.6	25.1	44.5	66.2	90.0	113.8	135.5	154.9	174.4	201.4
7000 KM LATITUDE	67.0	57.6	39.4	20.4	3.6	-8.5	-13.0	-8.5	3.6	20.4	39.4	57.6
LONGITUDE	270.00	-33.8	-3.6	18.2	39.4	63.2	90.0	116.8	140.6	161.8	183.6	213.8
8000 KM LATITUDE	58.1	50.1	32.9	13.7	-3.9	-17.0	-21.9	-17.0	-3.9	13.7	32.9	50.1
LONGITUDE	270.0	-42.2	-11.3	11.8	34.4	60.2	90.0	119.8	145.6	168.2	191.3	222.2
9000 KM LATITUDE	49.1	42.1	26.0	6.9	-11.3	-25.4	-30.9	-25.4	-11.3	6.9	26.0	42.1
LONGITUDE	270.0	-48.3	-17.9	5.9	29.3	56.9	90.0	123.1	150.7	174.1	197.9	228.3

AMSAT-OSCAR 10

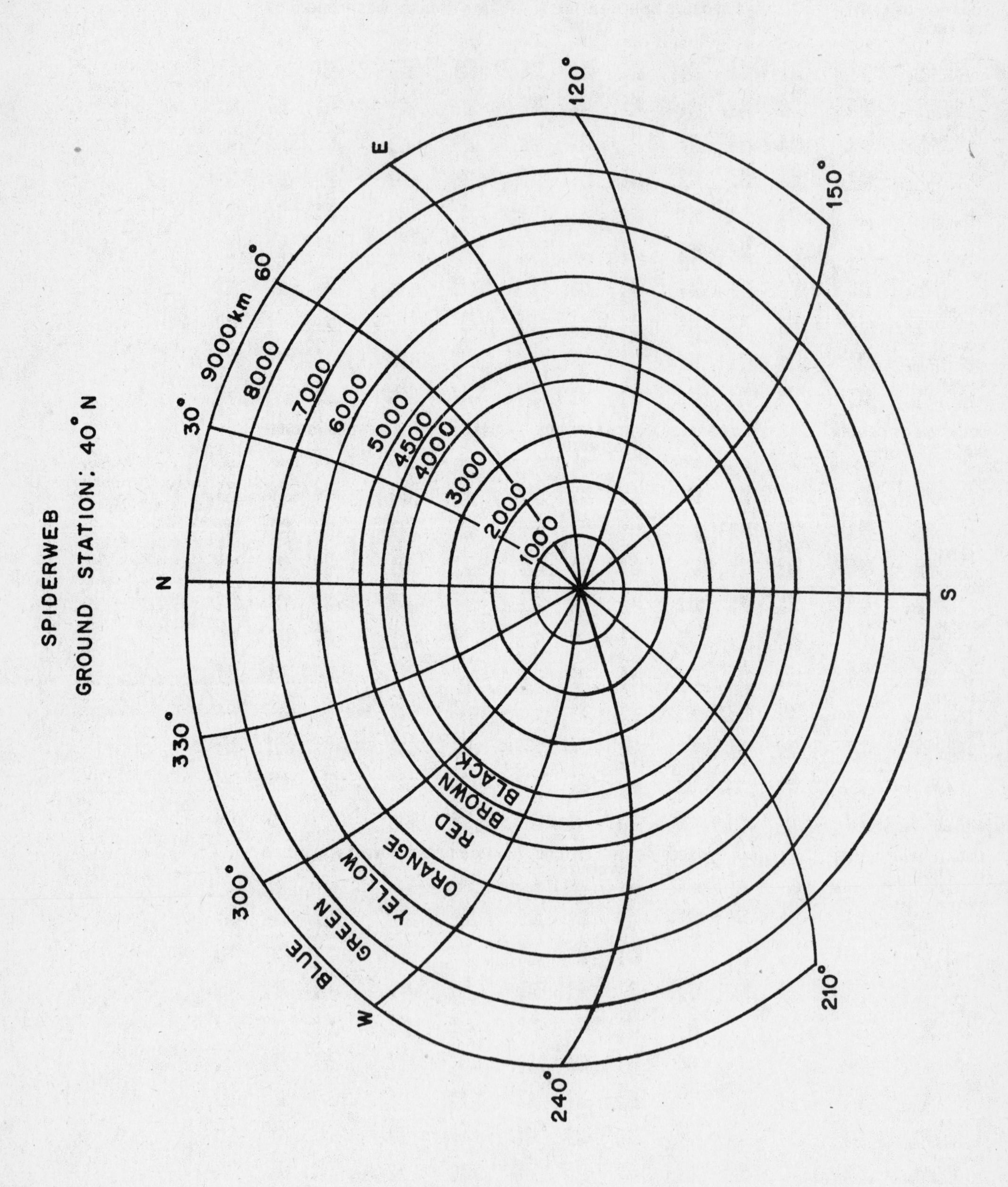

ISKRA 2

44 42 40 38 36 34 32 30 28 26 24 22 20 18 16 14 12 10 8 6 4 2 0

← MINUTES AFTER EQX

(RKØ2) ISKRA

NEXT EQX

Ground Track Data
Satellite: ISKRA 2
Mean altitude: 344 km Period: 91.346 min/orbit Inclination: 51.594°

Time after ascending node (minutes)	*Subsatellite point Lat. (°N)*	*Long. (°W)*
0	0	0
2	6.2	355.6
4	12.3	351.1
6	18.3	346.3
8	24.2	341.1
10	29.8	335.5
12	35.2	329.1
14	40.0	321.7
16	44.3	313.3
18	47.8	303.6
20	50.2	292.6
22	51.5	280.8
22.8	51.6	275.9
24	51.4	268.6
26	49.9	256.9
28	47.3	246.2
30	43.7	236.7
32	39.3	228.4
34	34.3	221.3
36	28.9	215.0
38	23.2	209.4
40	17.3	204.3
42	11.3	199.6
44	5.2	195.1
45.7	0	191.6

Note: Ground track should be drawn directly on map using data presented. It may then be traced on clear plastic. The plastic overlay is then repositioned for each orbit.

Manned Soviet space stations of the Salyut series are generally launched into orbits with an inclination of about 51.5° and a period close to 90 minutes. The radio amateur satellite ISKRA 2 was released from Salyut 7 in May 1982. Although ISKRA has ceased operating (satellites in this low orbit reenter the atmosphere and burn up after a few months), it's likely that future Soviet amateur spacecraft may be placed in similar orbits. Data on the ISKRA 2 ground track has therefore been included for reference.

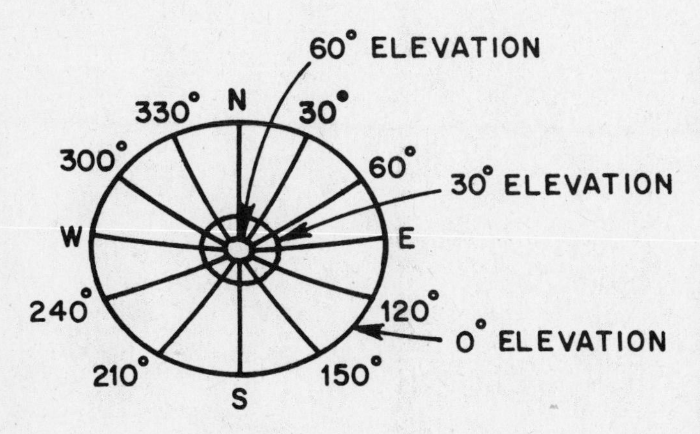

Spiderweb Data
Satellite: ISKRA 2
Mean altitude: 345 km
Ground Station: 30° N

	Latitude (°N)/Longitude (°W)		
Azimuth	*0° elevation 2051 km*	*30° elevation 529 km*	*60° elevation 187 km*
0° (north)	48.4/90.0	34.8/90.0	31.7/90.0
30°	45.4/77.0	34.1/87.1	31.5/89.0
60°	37.7/69.7	32.3/85.1	30.8/88.3
90° (east)	28.3/68.9	29.9/84.5	30.0/88.1
120°	19.7/73.1	27.5/85.4	29.1/88.3
150°	13.7/80.6	25.9/87.4	28.5/89.0
180° (south)	11.6/90.0	25.2/90.0	28.3/90.0
210°	13.7/99.4	25.9/92.6	28.5/91.0
240°	19.7/106.9	27.5/94.6	29.1/91.7
270° (west)	28.3/111.1	29.9/95.5	30.0/91.9
300°	37.7/110.3	32.3/94.9	30.8/91.7
330°	45.4/103.0	34.1/92.9	31.5/91.0

Note: Spiderweb should be drawn directly on the map using data presented. It may then be traced on clear plastic. The plastic overlay is then repositioned at the user's location.

ISKRA 2

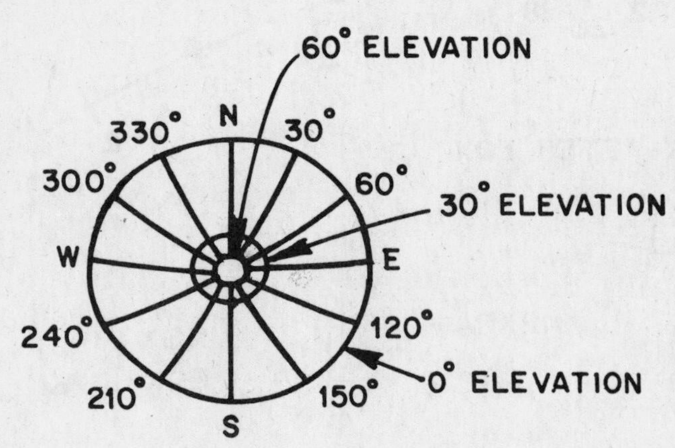

ISKRA-2

h = 345 km

46° N

Spiderweb Data

Satellite: ISKRA-2

Mean altitude: 345 km

Ground Station: 46° N

	Latitude (°N)/Longitude (°W)		
	0° elevation	*30° elevation*	*60° elevation*
Azimuth	*2051 k m*	*529 km*	*187 km*
0° (north)	64.4/90.0	50.8/90.0	47.7/90.0
30°	60.8/71.1	50.1/86.3	47.4/88.8
60°	52.4/63.3	48.2/83.8	46.8/87.9
90° (east)	43.0/64.4	45.8/83.2	46.0/87.6
120°	34.9/70.5	43.5/84.3	45.1/87.9
150°	29.5/79.5	41.8/86.8	44.5/88.8
180° (south)	27.6/90.0	41.2/90.0	44.3/90.0
210°	29.5/100.5	41.8/93.2	44.5/91.2
240°	34.9/109.5	43.5/95.7	45.1/92.1
270° (west)	43.0/115.6	45.8/96.8	46.0/92.4
300°	52.4/116.7	48.2/96.2	46.8/92.1
330°	60.8/108.9	50.1/93.7	47.4/9.12

Note: Spiderweb should be drawn directly on the map using data presented. It may then be traced on clear plastic. The plastic overlay is then repositioned at the user's location.

Space Shuttle (STS-9)

Ground Track Data

Satellite: STS-9 (Shuttle 9)
Mean altitude: 250 km   Period: 89.4 min   Inclination: 57°

Time after ascending node (minutes)	*Subsatellite point Lat. (°N)*	*Long. (°W)*	*Time after ascending node (minutes)*	*Subsatellite point Lat. (°N)*	*Long. (°W)*
0	0.0	0.0	24	56.4	263.9
2	6.7	356.1	26	54.2	250.8
4	13.5	352.1	28	50.7	239.4
6	20.1	347.8	30	46.1	229.9
8	26.6	343.1	32	40.8	222.1
10	32.8	337.7	34	35.0	215.5
12	38.8	331.5	36	28.8	209.9
14	44.3	324.2	38	22.4	205.0
16	49.2	315.3	40	15.8	200.6
18	53.1	304.6	42	9.1	196.5
20	55.8	292.0	44	2.4	192.5
22	57.0	278.1	44.7	0.0	191.2
22.4	57.0	275.6			

Note: Ground track should be drawn directly on map using data presented. It may then be traced on clear plastic. The plastic overlay is then repositioned for each orbit.

40 min
30 min

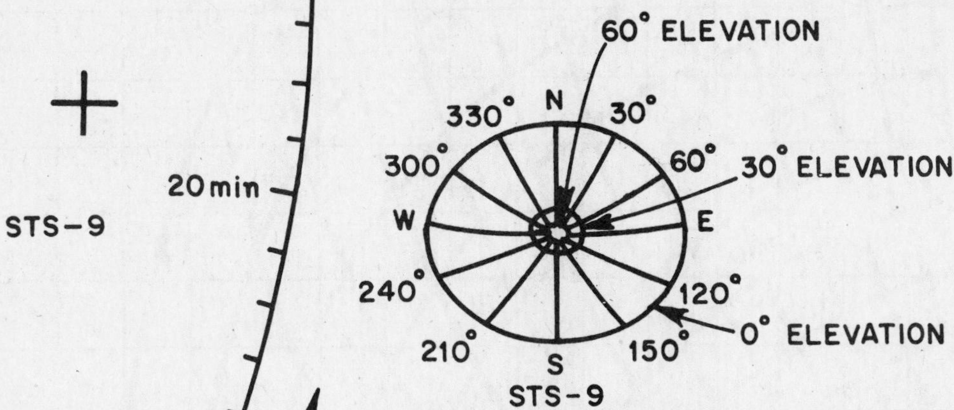

h = 250 km = 135 nm
center: 30° N, 90° W

10 min

60° ELEVATION
N
330°
30°
300°
60°
30° ELEVATION
W
E
240°
120°
0° ELEVATION
210°
150°
S
STS-9

h = 250 km = 135 nm
CENTER: 46° N, 90° W

Spiderweb Data

Satellite: STS-9 (9th Shuttle)
Mean altitude: 250 km
Ground Station: 30° N, 90° W

	Latitude (°N)/Longitude (°W)		
	0° elevation	*30° elevation*	*60° elevation*
Azimuth	*1,756 km*	*396 km*	*138 km*
0° (north)	45.8/90.0	33.6/90.0	31.2/90.0
30°	43.3/79.2	33.1/87.9	31.1/89.3
60°	36.8/72.9	31.7/86.4	30.6/88.8
90° (east)	28.8/71.9	29.9/85.9	30.0/88.6
120°	21.3/75.3	28.2/86.5	29.4/88.8
150°	16.1/81.9	26.9/88.0	28.9/89.3
180° (south)	14.2/90.0	26.4/90.0	28.8/90.0
210°	16.1/98.1	26.9/92.0	28.9/90.7
240°	21.3/104.7	28.2/93.5	29.4/91.2
270° (west)	28.8/108.1	29.9/94.1	30.0/91.4
300°	36.8/107.1	31.7/93.6	30.6/91.2
330°	43.3/100.8	33.1/92.1	31.1/90.7

Note: Spiderweb should be drawn directly on the map using data presented. It may then be traced on clear plastic. The plastic overlay is then repositioned at the user's location.

Spiderweb Data

Satellite: STS-9 (9th Shuttle)
Mean altitude: 250 km
Ground Station: 46° N, 90° W

	Latitude (°N)/Longitude (°W)		
	0° elevation	*30° elevation*	*60° elevation*
Azimuth	*1756 km*	*396 km*	*138 km*
0° (north)	61.8/90.0	49.6/90.0	47.2/90.0
30°	58.9/74.7	49.1/87.3	47.1/89.1
60°	51.9/67.6	47.7/85.4	46.6/88.4
90° (east)	43.8/67.8	45.9/84.9	46.0/88.2
120°	36.7/72.9	44.1/85.7	45.4/88.5
150°	31.9/80.8	42.9/87.6	44.9/89.1
180° (south)	30.2/90.0	42.4/90.0	44.8/90.0
210°	31.9/99.2	42.9/92.4	44.9/90.9
240°	36.7/107.1	44.1/94.3	45.4/91.5
270° (west)	43.8/112.2	45.9/95.1	46.0/91.8
300°	51.9/112.4	47.7/94.6	46.6/91.6
330°	58.9/105.3	49.1/92.7	47.1/90.9

Note: Spiderweb should be drawn directly on the map using data presented. It may then be traced on clear plastic. The plastic overlay is then repositioned at the user's location.

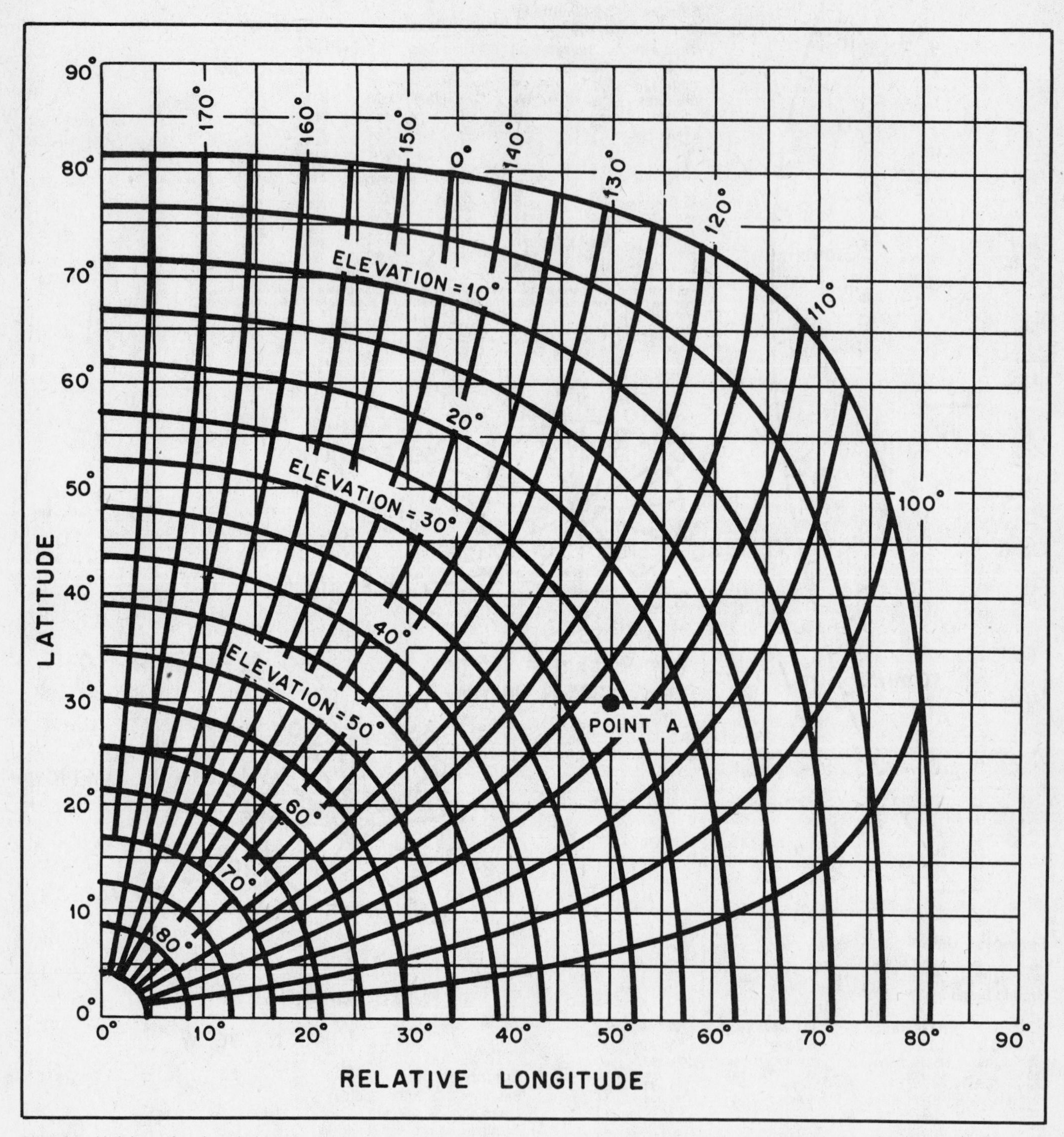

Chart for obtaining azimuth and elevation directions from a ground station to a geostationary satellite. See pp. 5-8 and 5-9 for instructions.

Appendix C Conversion Factors and Constants

CONVERSION FACTORS

The following values have been established by international agreement and are exact as shown. There is *no* roundoff or truncation error.

1 foot = 0.3048 meters
1 statute mile = 1609.344 meters
1 nautical mile = 1852 meters

Some additional conversion factors:

Length 1.000° of arc at surface of earth
= 111.2 km
= 69.10 statute miles
= 60.00 nautical miles

Mass 1.000 kg = 6.852×10^{-2} slugs

Force 1.000 N = 0.2248 pounds
1.000 kg (force) = 2.205 pounds
(at surface of earth)

SELECTED CONVERSION PROCEDURES (to four significant digits unless indicated otherwise)

1) To convert from statute miles to kilometers, multiply by 1.609
2) To convert from kilometers to statute miles, multiply by 0.6214
3) To convert from inches to meters, multiply by 0.0254 (exact)
4) To convert from meters to inches, multiply by 39.37

CONSTANTS

Mass of earth
$M = 5.98 \times 10^{24}$ kg $= 4.10 \times 10^{23}$ slugs

Mean Earth-Sun distance
1 AU $= 1.49 \times 10^{11}$ m $= 92.6 \times 10^{6}$ statute miles

Mean equatorial radius of earth
$R_{eq} = 6.378 \times 10^{6}$ m = 3963 statute miles

Mean radius of earth
$R = 6.371 \times 10^{6}$ m = 3959 statute miles

Sidereal day = 1436.07 minutes

Solar constant
$P_o = 1.38$ kW/m^2

Solar day = 1440 minutes (exact)

Speed of light in vacuum
$c = 2.9979 \times 10^{8}$ m/s

Stefan-Boltzmann constant

$$\sigma = 5.67 \times 10^{-8} \frac{\text{joules}}{\text{K}^4\text{m}^2\text{s}}$$

Gravitational Constant

$$G = 6.67 \times 10^{-11} \frac{\text{m}^3}{\text{kg-s}^2}$$

$$= 3.44 \times 10^{-8} \frac{\text{ft}^3}{\text{slug-s}^2}$$

$$GM = 3.986 \times 10^{14} \frac{\text{m}^3}{\text{s}^2} = 1.408 \times 10^{16} \frac{\text{ft}^3}{\text{s}^2}$$

Abbreviations
K = Kelvins
kg = kilogram
m = meter
N = Newton
s = second

Appendix D

FCC Rules governing the Amateur Satellite Service

Subpart H — Amateur-Satellite Service

General

§97.401 Purposes.

The Amateur-Satellite Service is a radiocommunication service using stations on earth satellites for the same purposes as those of the Amateur Radio Service.

§97.403 Definitions.

(a) *Space operation*. Space-to-earth, and space-to-space, amateur radio communication from a station which is beyond, is intended to go beyond, or has been beyond the major portion of the earth's atmosphere.

(b) *Earth operation*. Earth-to-space-to-earth amateur radiocommunication by means of radio signals automatically retransmitted by stations in space operation.

(c) *Telecommand operation*. Earth-to-space amateur radio communication to initiate, modify, or terminate functions of a station in space operation.

(d) *Telemetry*. Space-to-earth transmissions, by a station in space operation, of results of measurements made in the station, including those relating to the function of the station.

§97.405 Applicability of rules.

The rules contained in this subpart apply to radio stations in the Amateur-Satellite Service. All cases not specifically covered by the provisions of this Subpart shall be governed by the provisions of the rules governing amateur radio stations and operators (Subpart A through E of this part).

§97.407 Eligibility for space operation.

Amateur radio stations licensed to Amateur Extra Class operators are eligible for space operation (see §97.403(a)). The station licensee may permit any amateur radio operator to be the control operator, subject to the privileges of the control operator's class of license (see §97.7).

§97.409 Eligibility for earth operation.

Any amateur radio station is eligible for earth operation (see §97.403(b)), subject to the privileges of the control operator's class of license (see §97.7).

§97.411 Eligibility for telecommand operation.

Any amateur radio station designated by the licensee of a station in space operation is eligible to conduct telecommand operation with the station in space operation, subject to the privileges of the control operator's class of license (see §97.7).

§97.413 Space operations requirements.

An amateur radio station may be in space operation where:

(a) The station has not been ordered by the Commission to cease radio transmissions.

(b) The station is capable of effecting a cessation of radio transmissions by commands transmitted by station(s) in telecommand operation whenever such cessation is ordered by the Commission.

(c) There are, in place, sufficient amateur radio stations licensed by the Commission capable of telecommand operation to effect cessation of space operation, whenever such is ordered by the Commission.

(d) The notifications required by §97.423 are on file with the Commission.

Technical Requirements

§97.415 Frequencies available.

The following frequency bands are available for space operation, earth operation, and telecommand operation:

Frequency Bands

kHz
7000-7100 14000-14250

MHz
21.00-21.45 28.00-29.70
144-146 435-438[1]

GHz
24-24.05

[1]Stations operating in the Amateur-Satellite Service shall not cause harmful interference to other stations between 436 and 438 MHz (See International Radio Regulations, RR MOD 3644/320A).

Special Provisions

§97.417 Space operation.

(a) Stations in space operation are exempt from the station identification requirements of §97.84 on each frequency band when in use.

(b) Stations in space operation may automatically retransmit the radio signals of other stations in earth operation, and space operation.

(c) Stations in space operation are exempt from the control operator requirements of §97.79 and from the provisions of §97.88 pertaining to the operation of a station by remote control.

§97.419 Telemetry.

(a) Telemetry transmission by stations in space operation may consist of specially coded messages intended to facilitate communications.

(b) Telemetry transmissions by stations in space operation are permissible one-way communications.

§97.421 Telecommand operation.

(a) Stations in telecommand operation may transmit special codes intended to obscure the meaning of command messages to the station in space operation.

(b) Stations in telecommand operation are exempt from the station identification requirements of §97.84.

§97.423 Notification required.

(a) The licensee of every station in space operation shall give written notifications to the Private Radio Bureau, Federal Communications Commission, Washington, DC 20554.

(b) *Pre-space operation notification.* (1) Three notifications are required prior to initiating space operation. They are:

First notification. Required no less than twenty-seven months prior to initiating space operation.

Second notification. Required no less than fifteen months prior to initiating space operation.

Third notification. Required no less than three months prior to initiating space operation.

(2) The pre-space operation notification shall consist of:

Space operation date. A statement of the expected date space operations will be initiated, and a prediction of the duration of the operation.

Identity of satellite. The name by which the satellite will be known.

Service area. A description of the geographic area on the Earth's surface which is capable of being served by the station is space operation. Specify for both the transmitting and receiving antennas of this station.

Orbital Parameters. A description of the anticipated orbital parameters as follows:

Nongeostationary satellite

(1) Angle of inclination
(2) Period
(3) Apogee (kilometers)
(4) Perigee (kilometers)
(5) Number of satellites having the same orbital characteristics

Geostationary satellite.

(1) Normal geographical longitude.
(2) Longitudinal tolerance.
(3) Inclination tolerance.
(4) Geographical longitudes marking the extremities of the orbital arc over which the satellite is visible at minimum angle of elevation at 10° at points within the associated service area.
(5) Geographical longitudes marking the extremeties of the orbital arc within which the satellite must be located to provide communications to the specified service area.
(6) Reason when the orbital arc of (5) is less than that of (4).

Technical Parameters. A description of the proposed technical parameters for:

(1) The station in space operations; and

(2) A station in earth operation suitable for use with the station in space operation; and

(3) A station in telecommand operation suitable for use with the station in space operation.

The description shall include:

(1) Carrier frequencies if known; otherwise give frequency range where carrier frequencies will be located.

(2) Necessary bandwidth.

(3) Class of emission.

(4) Total Peak Power.

(5) Maximum power density (watts/Hz).

(6) Antenna radiation pattern.[1]

(7) Antenna gain (main beam).[1]

(8) Antenna pointing accuracy (geostationary satellites only).[1]

(9) Receiving system noise temperature.[2]

(10) Lowest equivalent satellite link noise temperature.[3]

(c) *In-space operation notification.* Notification is required after space operation has been initiated. The notification shall update the information contained in the pre-space operation notification. In-space operation notification is required no later than seven days following initiation of space operation.

(d) *Post-space operation notification.* Notification of termination of space operation is required no later than three months after termination is complete. If the termination is ordered by the Commission, notification is required no later than twenty-four hours after termination is complete.

****

Article 41 — Amateur Stations

****

Sec. 6. Space stations in the Amateur-Satellite Service operating in bands shared with other services shall be fitted with appropriate devices for controlling emissions in the event that harmful interference is reported in accordance with the procedure laid down in Article 15. Administrations authorizing such space stations shall inform the International Frequency Registration Board (I.F.R.B.) and shall insure that sufficient earth command stations are established before launch to guarantee that any harmful interference that might be reported can be terminated by the authorizing Administration.

[1]These antenna characteristics shall be provided for both transmitting and receiving antennas.

[2]For a station in space operation.

[3]The total noise temperature at the input of a typical amateur radio station receiver shall include the antenna noise (generated by external sources (ground, sky, etc.) peripheral to the receiving antenna and noise re-radiated by the satellite), plus noise generated internally to the receiver. The additional receiver noise is above thermal noise kT_oB.
Referred to the antenna input terminals, the total system noise temperature is given by
$T_s = T_a + (L-1)T_o + LT_r$
where: T_a: antenna noise temperature
L: line losses between antenna output terminals and receiver input terminals
T_o: ambient temperature, usually given as 290° K
T_r: receiver noise temperature. This is also given as $(NF-1)T_o$, where NF is receiver noise figure.

Glossary

Notes

1) All terms are defined as they apply to space satellites.

2) Where a term is followed only by a synonym in parentheses, see the synonym for definition.

---

access range (maximum access distance)

acquisition circle: On a map or globe, a circle drawn about a particular ground station and keyed to a specific satellite. When the subsatellite point is inside the circle the satellite is in range.

acquisition distance (maximum access distance)

altitude: The distance between a satellite and the point on the earth directly below it. Same as *height*.

AMSAT: Registered trademark of Radio Amateur Satellite Corporation.

anomalistic period: The elapsed time between two successive perigees of a satellite.

AOS (*a*cquisition *o*f *s*ignal): The time at which a particular ground station begins to receive radio signals from a satellite. For calculations, AOS is assumed to occur at an elevation angle of 0°.

apogee: The point on the orbit where satellite-geocenter distance is a maximum.

argument of perigee: The polar angle locating the perigee point of a satellite in the orbital plane; drawn between the ascending node, geocenter and perigee; and measured from ascending node in direction of satellite motion.

ARRL (*A*merican *R*adio *R*elay *L*eague): Binational (U.S. and Canada) membership organization of radio amateurs.

ascending node: Point on satellite orbit (or ground track) where subsatellite point crosses the equator from southern hemisphere into northern hemisphere.

ascending pass: With respect to a particular ground station, a satellite pass during which the spacecraft is headed in a northerly direction while it is in range.

AU (*A*stronomical *U*nit): Mean sun-earth distance = 1.49×10^{11} m

autotransponder: A computer-like device aboard a spacecraft designed to receive and respond to uplink signals directed to it. Radio-5 and Radio-7 contain autotransponders. Also called Robots.

azimuth: Angle in the local horizontal plane measured clockwise with respect to North.

BOL (*B*eginning *o*f *L*ife): Usually used in reference to a satellite parameter that changes over time such as solar-cell efficiency.

boresight: The direction of maximum gain of a spacecraft antenna. Also refers to point on earth where maximum signal level from aforementioned antenna occurs.

classical orbital elements: A set of orbital elements usually including ascending node longitude and epoch (time and date), nodal period, inclination, eccentricity, argument of perigee. Because these elements are earth-referenced and based on geometric properties, they're especially useful for intuitively picturing an orbit.

Codestore: An onboard digital memory system that can be loaded with data by ground stations for later rebroadcast in Morse or other codes.

coverage circle: Region of earth which is eventually accessible for communications to a particular ground station via a specific satellite.

DBS (*D*irect *B*roadcast *S*atellite): Commercial satellite designed to transmit TV programming directly to the home.

delay time: Either:

transponder delay time: The elapsed time between the instant a signal enters a transponder and the instant it leaves it, or

path delay time: The elapsed time between transmitting an uplink signal to a satellite-borne transponder and receiving the downlink.

descending node: Point on satellite orbit (or ground track) where subsatellite point crosses the equator from northern hemisphere to southern hemisphere.

descending pass: With respect to a particular ground station, a satellite pass during which the spacecraft is headed in a southerly direction while it is in range.

downlink: A radio link originating at a spacecraft and terminating at one or more ground stations.

eccentricity: A parameter used to describe the shape of the ellipse constituting a satellite orbit.

eirp: *e*ffective *i*sotropic *r*adiated *p*ower

elevation: Angle above the local horizontal plane.

elevation circle: On a map or globe, the set of all points about a ground station where the elevation angle to a specified satellite is a particular value.

EME: Abbreviation for Earth-Moon-Earth. Usually refers to a communication mode that involves bouncing signals off the moon.

epoch time: A reference time at which parameters describing satellite motion that vary are specified.

equatorial plane: An imaginary plane, extending throughout space, which contains the earth's equator.

EQX (ascending node)

ESA: *E*uropean *S*pace *A*gency. A consortium of European governmental groups pooling resources for space exploration and development.

footprint: A set of signal-level contours, drawn on map or globe, for a high-gain satellite antenna. Usually applied to geostationary satellites.

geocenter: center of the earth.

geostationary satellite: A satellite that appears to hang motionless over a fixed point on the equator.

ground station: A radio station, on or near the surface of the earth, designed to receive signals from, or transmit signals to, a spacecraft.

ground track: Path traced out by subsatellite point over the course of one complete orbit.

inclination: The angle between the orbital plane of a satellite and the equatorial plane of the earth.

increment: Change in longitude of ascending node between two successive passes of specified satellite. Measured in degrees West per orbit (°W/orbit).

IPS: *I*nterpreter for *P*rocess *S*tructures. A high level, FORTH-like, computer language used on AMSAT satellites. Developed by Dr. K. Meinzer, DJ4ZC.

Keplerian orbital elements: A set of orbital elements usually including mean anomaly, RAAN, inclination, eccentricity, argument of perigee, and mean motion, all specified at a particular epoch (time/date). The Keplerian set of orbital elements is particularly useful for calculations involving orbital motion.

line of nodes: The line of intersection of a satellite's orbital plane and the earth's equatorial plane.

LNA: *L*ow *n*oise *a*mplifier. LNA is a commercial term for a device that radio amateurs generally refer to as a low noise rf preamp.

longitudinal increment (increment)

LOS (*l*oss *o*f *s*ignal): The time at which a particular ground station loses radio signals from a satellite. For calculations, LOS is assumed to occur at an elevation angle of 0°.

maximum access distance: The maximum distance, measured along the surface of the earth, between a ground station and the subsatellite point at which the satellite enters one's range circle. (Corresponds to a 0° elevation angle).

mean anomaly: An angle that increases uniformly with time, used to indicate where satellite is along its orbit. Usually specified at epoch time when Keplerian orbital elements are used.

mean motion: A constant included in the set of parameters referred to as the Keplerian orbital elements. The number of complete orbits a satellite makes in one day.

NASA: *N*ational *A*eronautics and *S*pace *A*dministration, the U.S. space agency.

nodal period: The elapsed time between two successive ascending nodes of a satellite.

node: Point where satellite ground track crosses the equator.

orbital elements: A set of independent parameters that completely describes an orbit. Six are needed for an elliptical orbit, four for a circular orbit. Two sets are in common use by radio amateurs: classical elements and Keplerian elements.

orbital plane: An imaginary plane, extending throughout space, that contains the satellite orbit.

OSCAR: *O*rbital *S*atellite *C*arrying *A*mateur *R*adio.

OSCARLOCATOR: A tracking device designed to be used with a satellite in a circular orbit.

PCA (*p*oint of *c*losest *a*pproach): Point on ground track, during specific orbit, where satellite passes closest to a particular ground station.

perigee: The point on the orbit where satellite-geocenter distance is a minimum.

period: The amount of time it takes for a satellite to make one complete revolution about the earth. See anomalistic period and nodal period.

RAAN (*R*ight *a*scension at *a*scending *n*ode): The angular distance, measured eastward along the celestial equator, between the vernal equinox and the hour circle of the ascending node of the spacecraft. One of the Keplerian orbital elements.

range circle: Circle of specific radius on the surface of earth centered about particular ground station.

reference orbit: The orbit following the first ascending node of a given UTC day.

s.a.s.e.: *S*elf *a*ddressed *s*tamped *e*nvelope.

Satellabe: A tracking device for circular orbits. Similar to OSCARLOCATOR but with added features.

Satellipse: A tracking device designed to be used with a satellite in an elliptical orbit.

satellite pass: Segment of orbit during which satellite "passes" nearby and in range of particular ground station.

s/c: Abbreviation for spacecraft.

sidereal day: The amount of time it takes the earth to rotate exactly 360° about its axis with respect to the "fixed" stars. The sidereal day contains 1436.07 minutes. (See solar day)

slant range: Distance between satellite and a particular ground station. Usually varies with time.

solar constant: Incident energy 1 AU from the sun falling on a surface of unit area oriented perpendicular to direction of radiation.

$P_o = 1.38 \text{ kW/m}^2$

solar day: The solar day, by definition, contains exactly 24 hours (1440 minutes). During the solar day the earth rotates slightly more than 360° about its axis with respect to "fixed" stars. (See sidereal day)

spiderweb: Set of azimuth curves radiating out from a particular location and concentric elevation or range circles about the location, all drawn on a map or globe.

SSP (*s*ub*s*atellite *p*oint)

subsatellite path (ground track)

subsatellite point: Point on surface of earth directly below satellite.

telemetry: Radio signals, originating at a satellite, that convey information on the performance or status of onboard subsystems. Also refers to the information itself.

TLM: Short for telemetry.

transponder: A device that receives radio signals in one segment of the spectrum, amplifies them, translates (shifts) their frequency to another segment of the spectrum and retransmits them.

true anomaly: The polar angle that locates a satellite in the orbital plane; drawn between the perigee, geocenter and current satellite position, and measured from perigee in direction of satellite motion.

TVRO: *TV r*eceive *o*nly. A TVRO terminal is a ground station set up only to receive downlink TV signals from 4-GHz or 12-GHz commercial satellites.

uplink: A radio link originating at a ground station and directed to a spacecraft.

window: Overlap region between acquisition circles of two ground stations referenced to a specific satellite. Communication between the two stations is possible when subsatellite point is within the window.

ϕ3 TRACKER: A tracking device designed to be used with a satellite in an elliptical orbit.

Amateur Radio — Your Ticket to Space Communications

"CQ CQ CQ... Calling all Amateur Radio stations... this is W5LFL aboard the Space Shuttle *Columbia*..."

Thousands of people around the world tuned in to hear Astronaut Owen Garriott operate his ham radio station from earth-orbit in December of 1983 — but only licensed Amateur Radio operators were permitted to transmit directly to him from their home stations. Who were these hams? Among them were King Hussein of Jordan, JY1; Senator Barry Goldwater, K7UGA; NBC Science Editor Roy Neal, K6DUE; and thousands of lesser known but equally eager people in cities, towns and villages around the world.

Similarly, space buffs and students on all continents have for years monitored telemetry beacons and conversations on Amateur Radio satellites orbiting the earth, day after day, month after month. But only licensed Amateur Radio operators are permitted to communicate directly with the orbiting computer "ROBOTs" aboard the Soviet Amateur Radio satellites RS-5 and RS-7 or leisurely chat with other hams across the world via the high-orbiting AMSAT-OSCAR 10 satellite.

Ham radio by its very nature is space communication. Not only do hams bounce signals off our largest orbiting satellite, the moon, but they also bounce signals off meteor trails, aurora and, most commonly, the ionosphere. Amateur Radio is science, a hobby, a sport, a public service and exciting *fun*.

How can *you* get involved? Simple! To be allowed to transmit through Amateur Radio satellites you need an FCC license called the "Technician Class," the second rung on a five-rung ladder of license classes, each requiring a little more knowledge and skill, and each giving greater privileges. To earn the Technician Class license you'll need to learn a few relevant FCC rules and regulations, fundamental radio theory and the Morse code at the slow "recognition" speed of only five words per minute. More than 400,000 hams in the U.S. have done it (ages 8 to 80, students, homemakers, businesspeople and engineers — both electrical and sanitation) and you can do it, too.

The American Radio Relay League, the binational, not-for-profit membership organization of radio amateurs in the United States and Canada, will help you get started. The ARRL will put you in touch with one or more than 6000 volunteer instructors and 2000 radio clubs across the country (at no obligation and at no cost to you). In addition, ARRL provides basic radio articles and news of what's happening in the Amateur Radio world in the monthly journal *QST* (available through membership in ARRL). Those with access to a shortwave receiver can hear ARRL's free-access, on-the-air bulletins and Morse code practice, transmitted around the world from the Maxim Memorial Station, W1AW, at ARRL Headquarters in Newington, Connecticut. The ARRL also provides a multitude of other services to its members and represents the U.S. radio amateur both nationally and internationally.

It's up to you. The world of active space communication from your home isn't science fiction or the dream of some imaginative futurist — it is a reality here and now. For the name of a volunteer instructor near you, or for further information on obtaining an Amateur Radio license, write to ARRL Instructor, 225 Main St., Newington, CT 06111 USA.

INDEX